Lecture Notes in Civil Engineering

Volume 158

Lecture Notes in Civil Engineering (LNCE) publishes the latest developments in Civil Engineering—quickly, informally and in top quality. Though original research reported in proceedings and post-proceedings represents the core of LNCE, edited volumes of exceptionally high quality and interest may also be considered for publication. Volumes published in LNCE embrace all aspects and subfields of, as well as new challenges in, Civil Engineering. Topics in the series include:

- Construction and Structural Mechanics
- Building Materials
- Concrete, Steel and Timber Structures
- Geotechnical Engineering
- Earthquake Engineering
- Coastal Engineering
- Ocean and Offshore Engineering; Ships and Floating Structures
- Hydraulics, Hydrology and Water Resources Engineering
- Environmental Engineering and Sustainability
- Structural Health and Monitoring
- Surveying and Geographical Information Systems
- Indoor Environments
- Transportation and Traffic
- Risk Analysis
- Safety and Security

To submit a proposal or request further information, please contact the appropriate Springer Editor:

- Pierpaolo Riva at pierpaolo.riva@springer.com (Europe and Americas);
- Swati Meherishi at swati.meherishi@springer.com (Asia - except China, and Australia, New Zealand);
- Wayne Hu at wayne.hu@springer.com (China).

All books in the series now indexed by Scopus and EI Compendex database!

More information about this series at http://www.springer.com/series/15087

Łukasz Piątek · Soon Heng Lim ·
Chien Ming Wang · Rutger de Graaf-van Dinther
Editors

WCFS2020

Proceedings of the Second World Conference on Floating Solutions, Rotterdam

Springer

Editors
Łukasz Piątek
Faculty of Architecture
Warsaw University of Technology
Warsaw, Poland

Chien Ming Wang
School of Civil Engineering
The University of Queensland
Brisbane, QLD, Australia

Soon Heng Lim
Floating Solutions LLP
Singapore, Singapore

Rutger de Graaf-van Dinther
Rotterdam University of Applied Sciences
Rotterdam, The Netherlands

ISSN 2366-2557 ISSN 2366-2565 (electronic)
Lecture Notes in Civil Engineering
ISBN 978-981-16-2258-8 ISBN 978-981-16-2256-4 (eBook)
https://doi.org/10.1007/978-981-16-2256-4

This Springer imprint is published by the registered company Springer Nature Singapore Pte Ltd.
The registered company address is: 152 Beach Road, #21-01/04 Gateway East, Singapore 189721, Singapore

Committees

Review Committee

Dr. Ir. Rutger de Graaf-van Dinther, Research Centre Sustainable Port Cities, Rotterdam University of Applied Sciences; Co-founder of Blue21
PE. Soon Heng Lim, President SFSS, Society of Floating Solutions (Singapore)
Ir. Maarten Flikkema, Space@Sea Coordinator and Consultant at FBS
David Kirkwood, M.Arch., Architect PII ARB/RIBA, Blue21
Prof. Mr. Maarten Claringbould, Emeritus Professor of Maritime Law at Leiden University and Of Counsel with Van Traa Advocaten
Dr. Ir. Dries Hegger, Assistant Professor at Utrecht University
Fen-Yu (Vicky) Lin, M.Sc., Researcher and Project Coordinator at Blue21
Dr. Ir. Floris Boogaard, Professor at Hanze University of Applied Sciences Groningen and CEO at INDYMO
Dr. Ir. Jan van Kessel, Senior Engineer Floating City Development at Blue21 and Chairman of Offshore Technology at the Royal Institute for Engineers in the Netherlands (KIVI)
Robert Newman, M.Arch., Co-founder at Floatopolis and Architect at Foster + Partners
Dr inż. arch. Łukasz Piątek, Faculty of Architecture, Warsaw University of Technology
Prof. Chien Ming Wang, School of Civil Engineering, The University of Queensland
Ir. Karina Czapiewska, Cofounder of DeltaSync and Blue21
Dr. Alona Armstrong, Senior Lecturer in Energy & Environmental Sciences and Deputy Director of Energy Lancaster at Lancaster University
Barbara Dal Bo Zanon, Arch., Architect and Researcher at Blue21
Ir. Olaf Waals, Manager Offshore at Maritime, Research Institute Netherlands
Jeroen van Dijk, Watermanagement, Hogeschool Rotterdam
Rui Lima, M.Sc., Researcher and Engineer at INDYMO and Researcher at MARE—Marine and Environmental Sciences Centre
Ir. Pieter Ham, Cofounder of Finch Floating Homes and Ph.D. candidate at Faculty of Civil Engineering and Geosciences at Technical University of Delft

Organizing Committee

Committee Chair

Fen-Yu (Vicky) Lin, Researcher and Project Coordinator at Blue21

Co-chairs

Maarten Flikkema, Space@Sea Coordinator and Consultant at FBS
Łukasz Piątek, Faculty of Architecture, Warsaw University of Technology
Rutger de Graaf-van Dinther, Research Centre Sustainable Port Cities, Rotterdam University of Applied Sciences; Co-founder of Blue21
Koen Olthuis, Founder of Waterstudio.NL

Strategic Advisors

Chien Ming Wang, Professor at the School of Civil Engineering, The University of Queensland
Karina Czapiewska, Cofounder of DeltaSync and Blue21
Soon Heng Lim, President SFSS—Society of Floating Solutions Singapore

Communications, IT, Paper Management and Marketing Coordinator

David Kirkwood, Architect PII ARB/RIBA at Blue21

Publicity and Public Relations

Ankie Stam, Architect at Waterstudio.NL

Digital Coordinator

Nathalie Mezza-Garcia, Founder and CEO of Seaphia

Paper Management Assistant

Guus Gorsse, Mechanical Engineering, TU Delft

General Administration

Rowenna Reus-Das, Research Assistant at Marine Research, Institute Netherlands —MARIN

IT and Paper Management Assistant

Jeroen van Dijk, Watermanagement, Hogeschool Rotterdam

Marketing and Venue Logistics Coordinator

Anna Vendemia, Architect at Waterstudio.NL

Marketing and Design Assistant

Neha Gupta, Building Technology, TU Delft

Sponsors

Paving the Waves: The Second World Conference of Floating Solutions could not have been achieved without the support and expertise of our sponsors. We wish to thank Resilient Rotterdam, GICON® and ICE Marine Design for their contributions. It has been of great value to have such innovators supporting the initiatives of the conference.

Foreword

In the face of climate change, overpopulation and increasing demand of resources, many cities are seeking for more space for living, working, energy, food production and recreation. The most common and traditional way to create land has been land reclamation. However, such methods have raised more and more questions about its environmental impact, cost-efficiency and climate adaptability on sea level rise. An alternative to creating land is to build floating structures that could be of use for various types of activities. Floating development can adapt to sea level rise, can be built fast and towed to wherever needed. It is less disruptive to the marine ecosystems and can provide new habitats for marine life to flourish.

By creating space on water, or expanding coastal cities on water, floating development can add value to sustainable urban development by making cities more resilient. Through creating adaptive "land", excessive nutrients that will otherwise be wasted could be turned into energy and a circular urban metabolism could be fostered. This could contribute to developing Sustainable Cities and Communities (the SDG #11). With proper design, engineering and monitoring, our oceans and seas could be used more responsibly and sustainably which contributes to Life Below Water (the SDG #14).

"Paving the Waves" symbolizes the main debate in floating urban development at present. On one hand, it refers to "paving the waves for the future". It is a synonym for taking the obstacles away for the wider global application of floating development. This includes the need to deal with regulations and the encouragement of standardization to bring floating solutions from experiments and small-scale local and national solutions to mainstream urban components to increase the livability of waterfront cities around the globe. The theme also reflects on the open question of how cities will bring urban planning beyond the waterfront as the question of whether we can take advantage of the space on water is still an open discussion. Are we expanding the urban fabric of our cities to cover the water or will new urban configurations and forms originate? Some environmentalists see floating developments as a potential threat to marine ecosystems. For them "Paving the Waves" might symbolize the threat that marine systems will be urbanized. These two meanings of "Paving the Waves" are central to the debate in the Second

World Conference on Floating Solutions (WCFS2020). In addition to the more technical topics, the conference also covers issues related to affordability, social impact and ecological impact. With thought leaders and entrepreneurs in floating urban development from all over the world, WCFS2020 aims to explore how floating solutions could contribute to climate adaptation, the energy transition and social justice. It is essential to bring different perspectives on this topic together and take progressive steps towards achieving a sustainable future on the water. We wish to thank all of those who have participated to making this vision a reality.

Rutger de Graaf-van Dinther
Co-founder of Blue21 and Blue
Revolution Foundation
Rotterdam, The Netherlands

Koen Olthuis
Founder of Waterstudio.NL
On behalf of the PtW Organizing Committee
Rijswijk, The Netherlands

Preface

Summing up a passing year is rarely as easy and unambiguous as in December 2020. The COVID-19 pandemic has swept the world, unimaginably changing our lives and, despite optimistic forecasts for the future, our everyday life is still far from normal. The Second World Conference on Floating Solutions 2020 could not remain isolated from this global catastrophe. Instead of being held in Rotterdam, a symbolic place for building on water, the international academic, business, and activist community had to meet symbolically in a virtual space to discuss the latest developments in floating solutions.

The disruptive time of the pandemic is reflective and put the conclusions of WCFS2020 in a new context. This year nature once again profoundly proves that the progress of civilization has not made our existence safe and secure once and for all. Our Anthropocene world can still be a dangerous place, and we should prepare to face it. Doesn't it correspond with the conference message sent by city planners, architects and designers about the flood-resilient floating settlements and cities for a climate-change era?

After the first shock of the virus outbreak, it was science and technology that came to our aid. Internet connectivity enabled us to not only sustain a minimum level of economic activity but also to protect our valuable social ties, weakened by social distancing. Arising from scientific method and invention is a vaccine that gives us hope for a return to normality. In these uncertain times, this conclusion should be encouraging for all scientists and researchers, including the ones working every day on hydrodynamics, stability, eco-engineering, aquaculture or energy for floating solutions. We do believe that these efforts will continue towards successful realization of many new floating projects and contribute to climate-resilient growth and ecological progress.

Finally, in the time of lockdowns and limited mobility, we realized how flexible an individual human being can be. We quickly adapted ourselves to the new circumstances and regulations, which seemed impossible to implement and follow. This must be inspiring for the conference participants working on policies, regulations and guidelines, who aim at shaping a habitable floating built environment—an environment to which humans will have to adapt. We shall see if the pandemic experience will make societies and leaders more open to the innovative forms of living and working on the water.

The structure of this book corresponds to the narrative of the three groups of professionals active during WCFS2020: designers, researchers and policy-makers. In the consecutive three sections, they present the state of our imagination, knowledge and organizational capabilities in the field of floating solutions. This way of presenting the work aims to emphasize the highly interdisciplinary nature of the conference. Hopefully, it will appeal to the reader who will find this book something more than a collection of papers presented at a virtual venue.

Warsaw, Poland — Łukasz Piątek
Rotterdam, The Netherlands — Rutger de Graaf-van Dinther
Singapore — Soon Heng Lim
Brisbane, Australia — Chien Ming Wang

Background

WCFS was established in 2019 and was initiated by the Society of Floating Solutions (Singapore), a group of multi-disciplinary professionals with the vision of creating space for Singapore on water as an eco-friendlier alternative to land reclamation. The objective of WCFS2019 was to bring international experts and leaders together to disseminate recent research and developments in floating structures on both inland water bodies and offshore, that are for energy harvesting, aquaculture and farming, leisure activities, infrastructure, industrial plants, real estates and cities. A strong emphasis of the conference was on eco-sustainability of living, playing and working offshore. Going on from the many findings of last year's conference, PTW2020 aimed to bring these two worlds together;

Creating conversations between the urban and the offshore and exploring how floating solutions could create a common ground for the two in the face of 21st century urban challenges.

The Netherlands was chosen to host the Second World Conference on Floating Solutions. Despite the eventual digital form of the conference, it was an honour to be selected as host nation. The Netherlands is a country which holds a long and fascinating history with water. It is home to the world's largest floating community and has wide expertise and knowledge in maritime engineering, water management and innovations in sustainable urban development.

Acknowledgements

"Paving the Waves" wishes to acknowledge the great work which has been made to achieve this year's virtual conference. The co-organizers of the conference include Blue21, the Society of Floating Solutions (Singapore), Space@Sea (EU Horizon2020-funded research project), the Blue Revolution Foundation and Waterstudio.NL.

With special thanks also to our sponsors, Resilient Rotterdam, GICON® and ICE Marine Design, and our supporting organizations, Maritime Research Institute Netherlands (MARIN), The Netherlands Water Partnership, The World Ocean Council, Seaphia, and The American Society of Engineers (ASCE).

RESILIENT ROTTERDAM

GICON®

ICE MARINE DESIGN

Seaphia

MARIN

NWP
NETHERLANDS WATER PARTNERSHIP

ASCE
AMERICAN SOCIETY OF CIVIL ENGINEERS

WORLD OCEAN COUNCIL
The International Business Alliance for Corporate Ocean Responsibility

Contents

Design

Mega Floating City "Green Float": Concept and Technology Innovations

Masaki Takeuchi and Ikuo Yoshida

Abstract The world is currently facing the sea level rise due to climate change. And we are facing many sorts of problems of energy, water, food, garbage, air pollution and so on. Fortunately, however, 70% of the earth's surface is covered with the ocean. The ocean has a tremendous capacity and a big influence on the global environment and the sustainability of human beings. We Shimizu are heading new challenges for the better future utilizing the potential of the ocean. "Green Float" is a new model for eco friendly Mega Floating City. The Green Float concept embodies two areas of innovation. The one is "Green innovation" and the other is "Float innovation". By integrating "Green Innovation" and "Float Innovation", a super high value-added Mega Floating City can be realized. In order to perform a basic verification of the technology, several hydraulic float model experiments and numerical computations about wind waves and tsunamis were conducted. We got a public third party's certification concerning about "structural safety of floating building". In addition, we continue to research and development the "City Grid System (city Cell, Module, Unit) of mega float city" based on Green Float.

Keywords Mega floating city · Green float · Zero carbon · Zero emissions

1 Outline of Mega Floating City "Green Float"

The man-made island is 3000 m in diameter and actually floats on the ocean. There is a 500-m-wide lagoon around the periphery, so the net land area has a diameter of 2000 m. The skyscraper city is 1000 m tall, and the city in the sky begins 700 m above sea level. The middle levels comprise a plant factory that makes the island city self-sufficient in food. The ground level supports such facilities as a natural farm, a

M. Takeuchi (✉) · I. Yoshida
Shimizu Corporation, Tokyo, Japan
e-mail: michaeltakeuchi2020@gmail.com

I. Yoshida
e-mail: ikuoyoshida@shimz.co.jp

Ł. Piątek et al. (eds.), *WCFS2020*, Lecture Notes in Civil Engineering 158,
https://doi.org/10.1007/978-981-16-2256-4_1

playground, and a beach resort. The peripheral area of the city in the sky is a residential zone that is inhabited by 30,000–50,000 people. The central area is a business zone and commercial zone. The business zone is intended mainly for the research, development, and branch offices of companies engaged in biotechnology, renewable

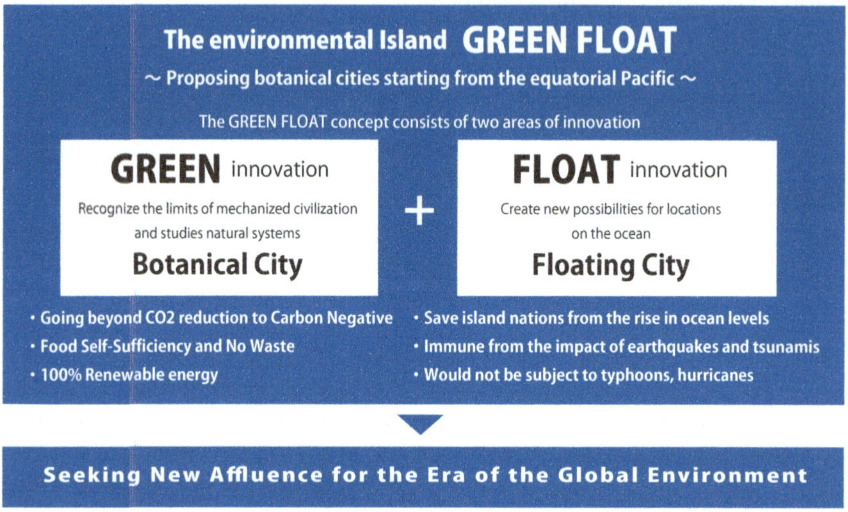

Fig. 1 Integration of GREEN & FLOAT innovations

Fig. 2 Mega floating city GREEN FLOAT

energy, and marine enterprises. The commercial zone is intended for shopping, restaurants, sky green parks and pools. The lagoon waterside on ground level will afford residents the opportunity to enjoy the lagoon and green forests (Figs. 1 and 2).

2 Project Summary

Green Float is a new model for environmentally friendly Mega Floating City, under development by Shimizu Corporation, Japan. The concept of "Green Float" embodies two areas of innovation. The one is "Green innovation", achieving self-efficient zero carbon and zero emission cities by utilizing the potential of the ocean, and the other is

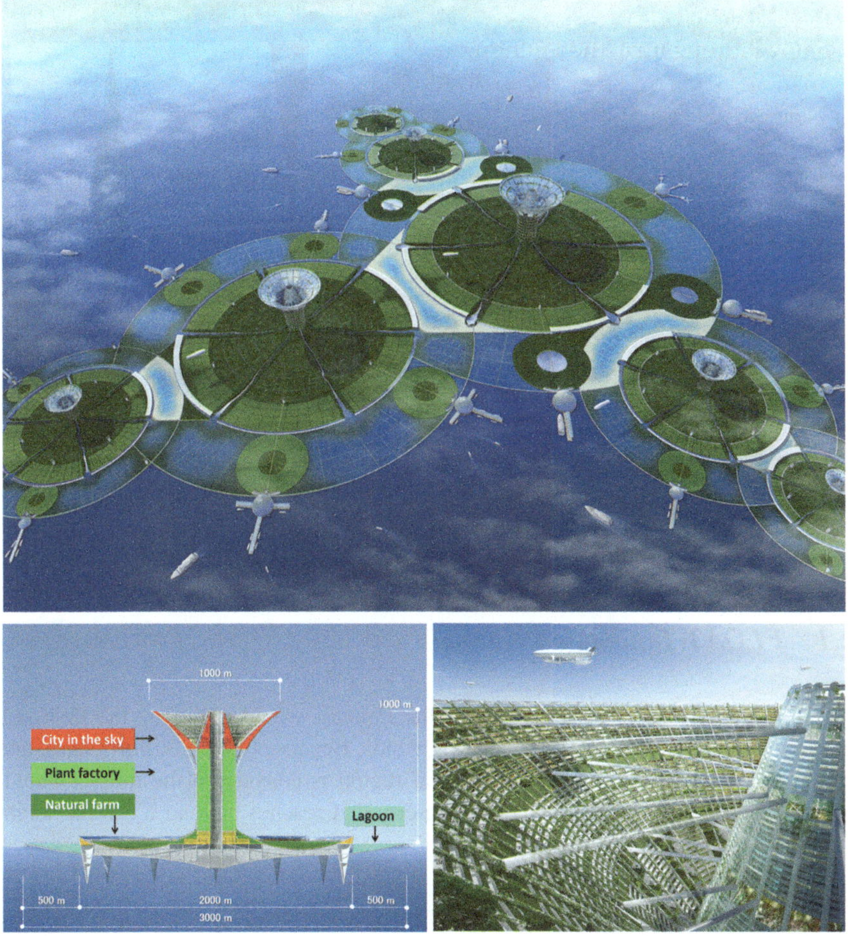

Fig. 3 (Top) GREEN FLOAT; (bottom left) Section; (bottom right) City in the Sky

Fig. 4 (Left) View from a residence; (right) Lagoon resort beach

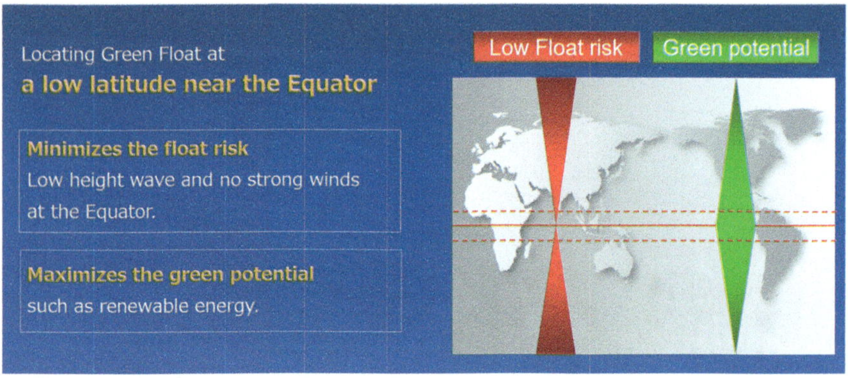

Fig. 5 Float risk & Green potential

"Float innovation", unaffected by sea-level rise and ensure high flexibility for city growth. By integrating "Green Innovation" and "Float Innovation", a super high value-added Mega Floating city can be realized (Figs. 3 and 4).

2.1 FLOAT Innovation

Locating the city at a low latitude near the Equator would minimize the float risk posed mainly by wave heights and strong winds. Because typhoons, hurricanes, and cyclones rarely occur there, the equatorial region is considered the most advantageous for a super-high-rise mega floating city that reaches beyond a certain height (Fig. 5).

2.1.1 New Location

For an existing large harbor city that faces difficulties in securing vacant land, it would be possible to extend the city functions offshore far beyond the limits of landfill. For an island country that is short of land area, new areas for development could be secured. Compared with landfill, floating development would be possible that is environmentally and ecologically friendly.

2.1.2 Unaffected by Rising Sea Levels Due to Climate Change

There are many island nations in the world facing the rising sea level crisis due to climate change. A mega floating city is one way to save such countries.

2.1.3 Avoiding the Influence of Earthquakes and Tsunamis

When a tsunami reaches a shallow shore, the wave height increases. However, a megacity floating offshore is completely immune from the impact of earthquakes and tsunamis.

2.1.4 Transferable and Flexible City

Because of durability issues, many cities and buildings on land cease to be functional within their lifetimes due to changes in the times. By contrast, a mega floating city can be moved more offshore locations or to different regions, cities, countries decades later. By relocating and continuing to use the lifetime of matter, the life cycle cost can be reduced.

2.2 GREEN Innovation

Zero carbon of CO_2, zero emission of resources and waste, and self-sufficiency of water and food are essential conditions for the future cities. These missions will be technically realized by utilizing the feature of Float City.

2.2.1 Pleasant and Comfortable Temperature

In the Pacific islands just below the equator, surface temperatures are 32 °C year-round. With the sky city of GREEN FLOAT rising 700–1000 m above the sea level, the temperature there would be a refreshing 26 °C all year round.

Because the temperature drops by 0.6 °C for every 100 m of height, it will be 6 °C lower at 1000 m.

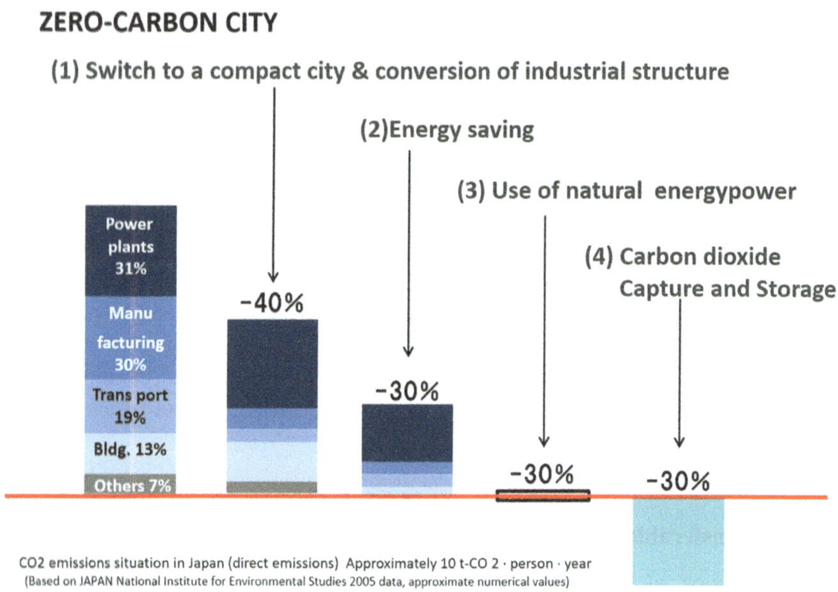

Fig. 6 CO_2 reduction of GREEN FLOAT

Fig. 7 Space Solar Power & GREEN FLOAT

2.2.2 Zero Carbon and Renewable Energy

The space solar power system is under development by JAXA in Japan. The system continues to send large amounts of electricity from space for 24 h. It is sent by microwave or laser wave. Green FLOAT, which is located on the sea, is suitable as a base site for the safety power reception.

In addition, because the sea surface is at a much different temperature than the seabed, the efficiency of power generation based on the Ocean Thermal Energy Conversion also improves. Taking the CO_2 emissions data for Japan as a benchmark, Figure shows how Green Float could reduce such emissions (Figs. 6 and 7).

2.2.3 Fresh Water

The annual precipitation is high in the equatorial Pacific, but precipitation levels differ considerably between the rainy and dry seasons. By storing water in the high-rise mega floating city and controlling water use throughout the city, it would be possible to supply as almost same as the amount of water required for living.

2.2.4 Food Self-sufficiency

By harnessing the power of the sun as much as possible, providing a vertical vegetable factory would contribute positively to food self-sufficiency, food mileage, and food traceability (Fig. 8).

2.2.5 Zero Emissions and Recycling

Human waste in the form of water, garbage, and CO_2 can be nutritious for plants. By harnessing the height of the sky city to redistribute human waste to lower levels,

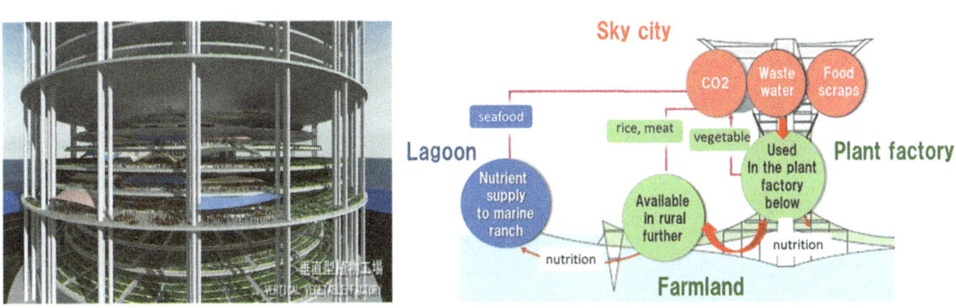

Fig. 8 (Left) Vertical vegetable factory; (right) resource recycling

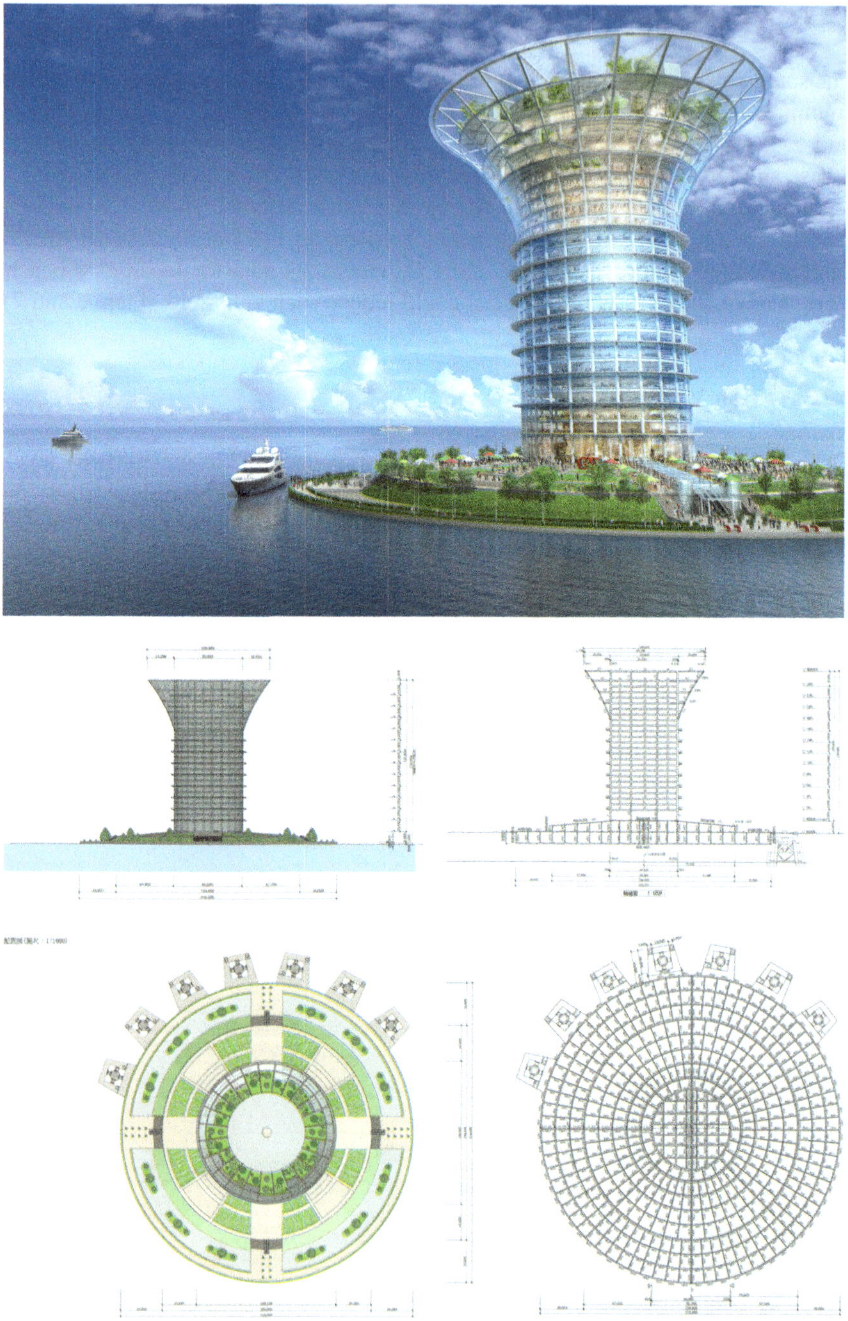

Fig. 9 Green Float II

plants and other living things would be supplied with nutrition. This raises the possibility of complete resource recycling from waste to food.

3 Validation of Float Technology

Green Float II (200 m in diameter; Fig. 13), which is roughly one tenth the scale of Green Float (2000 m in diameter) is taken as the latest technology-embodied model. Assuming a bay location, which is the extension of urban area, authors have conducted experiments and analyses of the response to tsunami and strong winds (Fig. 9).

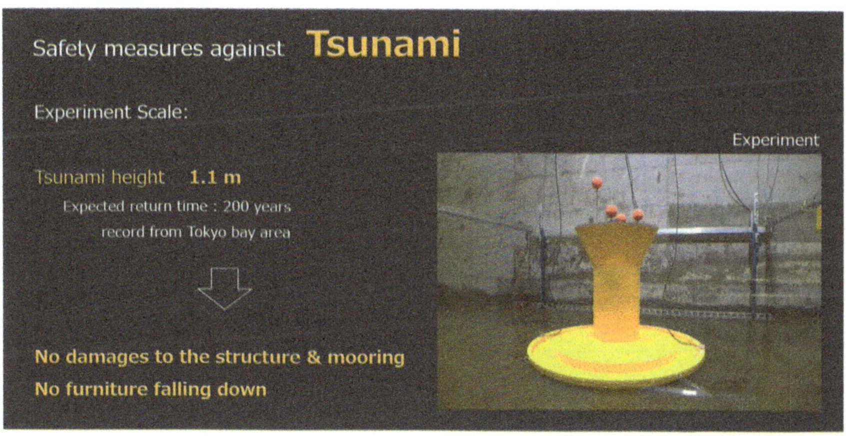

WCFS 2020

EXPERIMENTAL CONDITIONS

Made on 1/250 Scale. Whole structure is assumed to be RIGID.

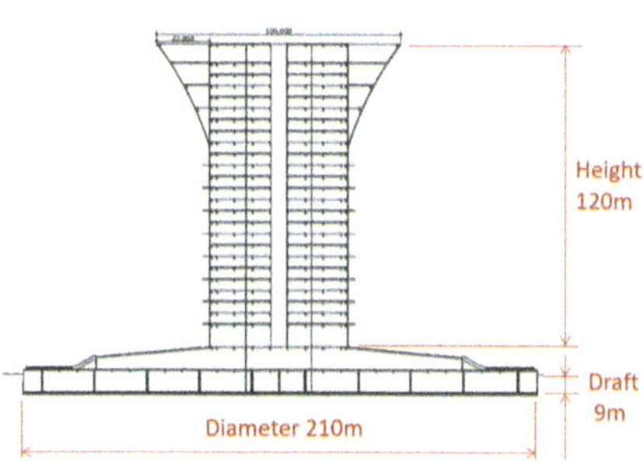

Items	Full size	1/250model
Diameter of floating body	210m	840mm
Height of floating body	10m	40mm
Draft	9m	36mm
Displacement	320,000t	19.98kg
Height of gravity center	19.41m	77.7mm
Radius of gyration	60.0m	240mm
Height of building	120m	480mm

Fig. 10 Tsunamis experiments of GREEN FLOAT II

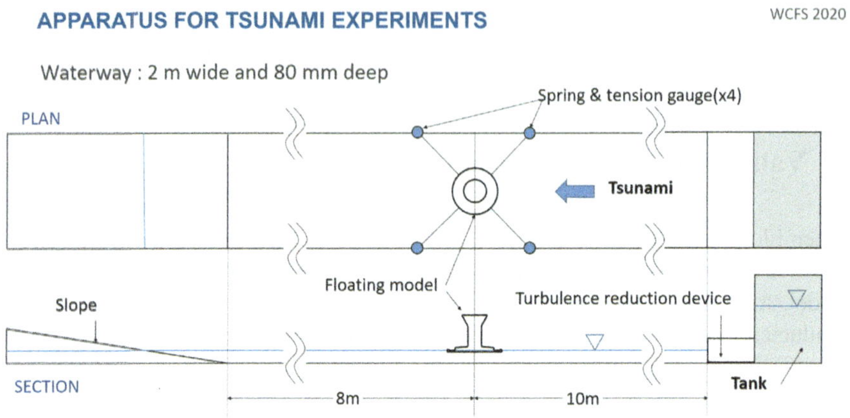

Fig. 11 Apparatus for tsunami experiments

Table 1 Tsunami experiment results

RESULTS: MEASURED MOORING FORCES

WCFS 2020

Cases (Wave type)	Sight	Max. height	Rise time	Maximum majored mooring force
Observed/ estimated wave	Port of Shimizu	3.6m	15min	67,172 kN
	Port of Sendai	7.7m	15min	67,445 kN
	Onagawa	13.0m	15min	67,728 kN
Bore	Tokyo bay	1.1m	–	19,078 kN
	Port of Shimizu	3.6m	–	65,995 kN
	Port of Sendai	7.7m	–	180,524 kN
	Onagawa	13.0m	–	302,024 kN

Design Condition of Fender's Load **27,300kN**

3.1 Earthquakes and Tsunamis

Authors assume the meteorological and oceanic conditions of a large Japanese city. Buildings and structures with floating isolation are extremely safe because they experience very little seismic force in the event of an earthquake. In addition, authors found experimentally that the tsunami does not exceed the breakwater of 1 m even with the tsunami of 10 m or more, which far exceeds the tsunami assumption in the assumed location. This is because the water level gradually increases due to the tsunami inside the bay. Of course, the structure itself is extremely safe (Figs. 10, 11 and Table 1).

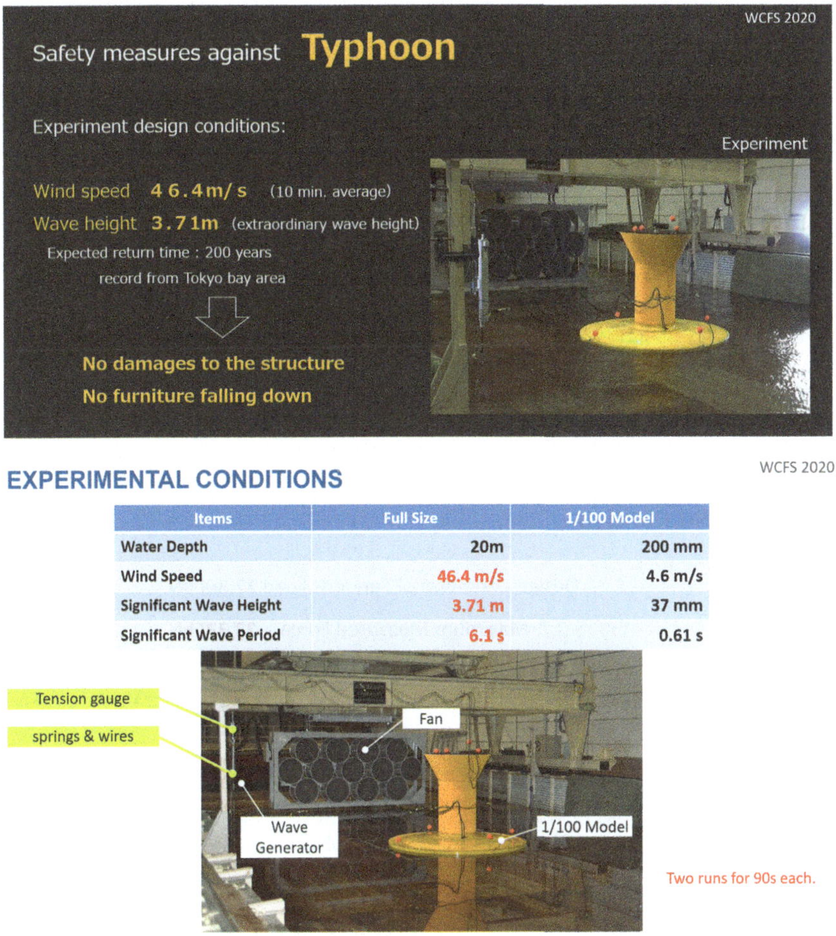

Fig. 12 Typhoon experiments of GREEN FLOAT II

3.2 Typhoon (Waves and Winds)

Again, authors assume the meteorological and oceanic conditions of a large Japanese city. Even assuming the direct hit of a typhoon bigger than the Ise Bay typhoon, structural safety was secured sufficiently for the floating structure (lower part), the skyscraper (upper part), and the mooring facilities. The acceleration also caused no furniture to fall over (Figs. 12 and 13).

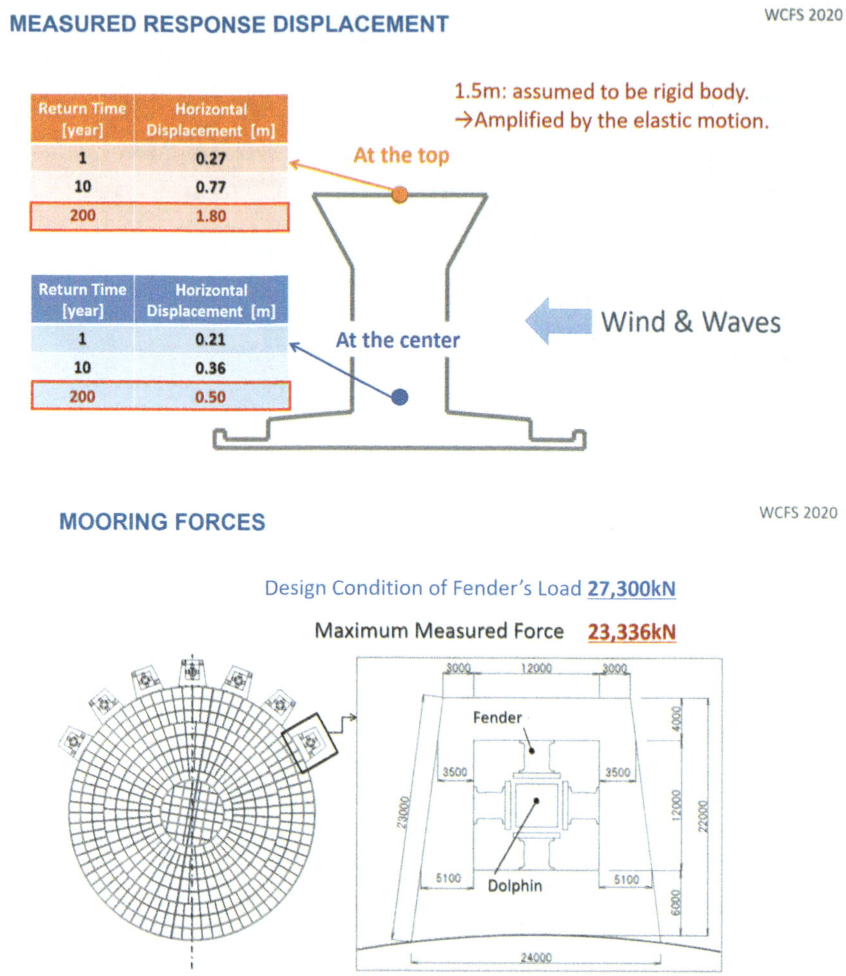

Fig. 13 Typhoon experiments of GREEN FLOAT II: (above) measured response displacement; (below) mooring forces

3.3 *Certification from Official Third Party*

Green Float II has received AIP from Nippon Kaijikyokai as certification by a third-party organization that is concerned with the structural safety of floating structures (Fig. 14).

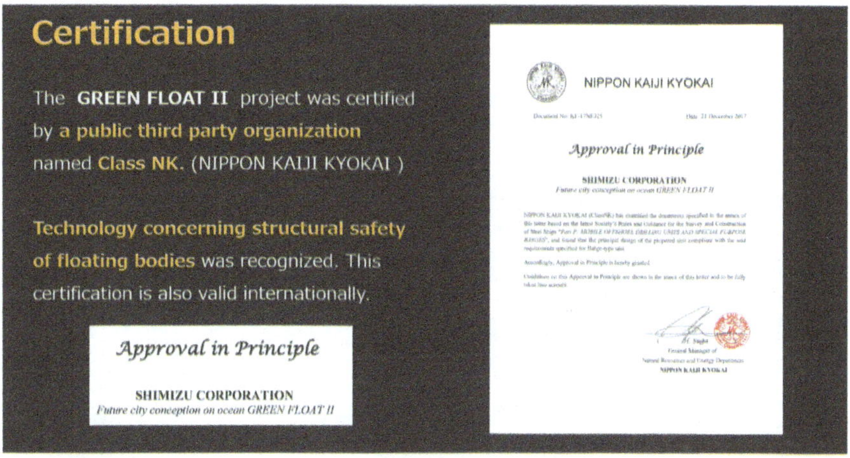

Fig. 14 AIP certification of GREEN FLOAT II

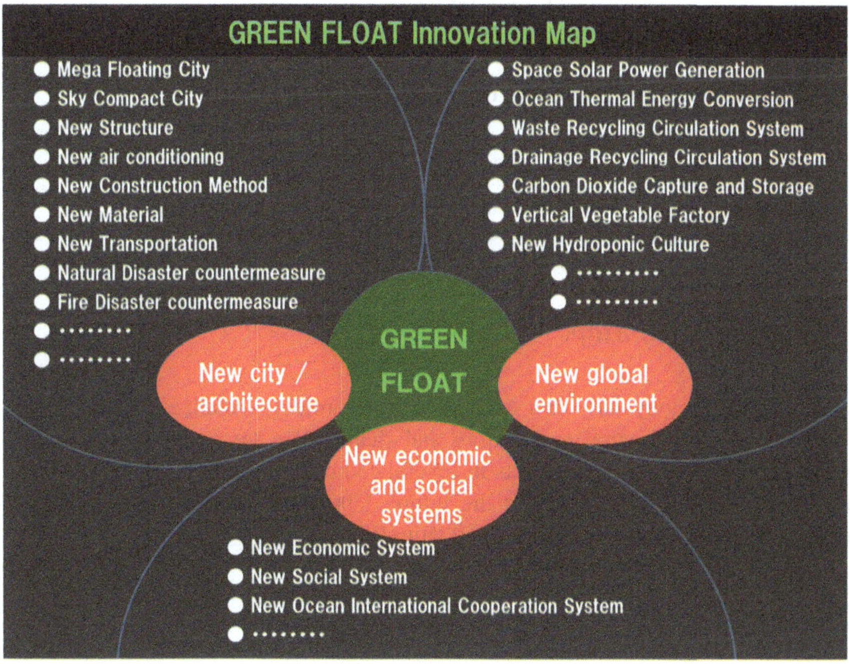

Fig. 15 R&D Road Map of GREEN FLOAT

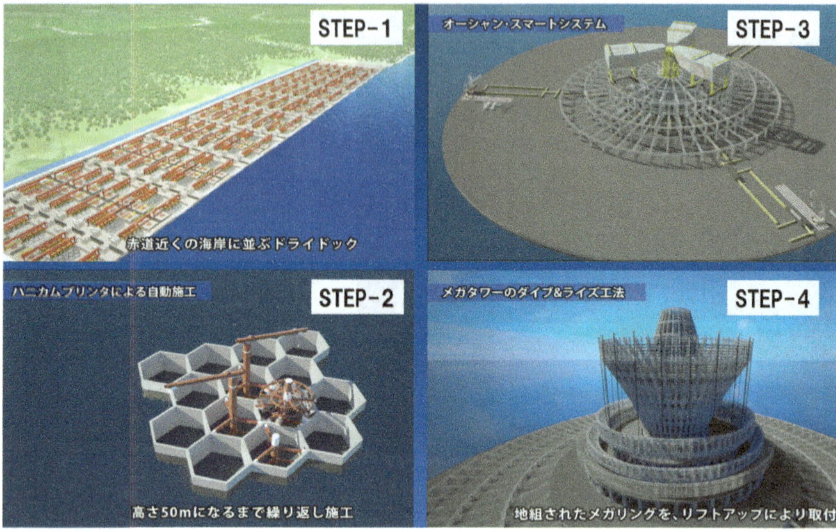

Fig. 16 Construction STEP of GREEN FLOAT

4 Research and Develop of Technology

The GREEN FLOAT project can be the most valuable project by integrating technological innovations in many technical fields. Therefore, I believe that promotion based on the "Comprehensive Technology Innovation Map" will be important.

4.1 R&D Road Map

After the technology has been established, the project is realized using different variations of architectural applications and shapes. We are working with government agencies and private projects in several countries at the schematic design and technical demonstration level. In the future, I believe that new cities and architectural solutions can be realized by floating (Fig. 15).

4.2 Construction Method

A mega floating city requires advances in aspects such as unitization and robot construction more so than a city on land. Although the future construction plans of Green Float remain at the idea stage at present, authors will try to improve the technology through future model projects (Fig. 16).

5 Conclusion and Next Challenge

In order to realize it as a concrete city, it is desirable that it is modeled as a basic "floating city grid system". The cells, modules, and units that can be established as a real city and rational as floating technology are shown below. This will be refined and enhanced by applying it to concrete plans in the future.

5.1 Conclusion

Through technical verification with the one-tenth-scale prototype, the structural safety of the mega floating city has been demonstrated for Japanese weather conditions and the assumed bay location. In the future, we plan to create model projects and pursue technical verification and technological development based on a specific location and specific sea and planning conditions.

5.2 City Grid System of GREEN FLOAT

After establishing and verifying the basic technology, GREEN FLOAT has begun the challenge as the next stage. As a floating city of a certain scale, what is a system that seems reasonable from the viewpoints of urban planning, structural dynamics, construction planning, and maintenance management? This is a study of the City Grid System of GREEN FLOAT as an advanced form of GREEN FLOAT. We believe that it will be constructed by integrating the two systems of Float System and Green System, inheriting the idea of GREEN FLOAT. Focusing on durability, it is premised on a concrete floating structure.

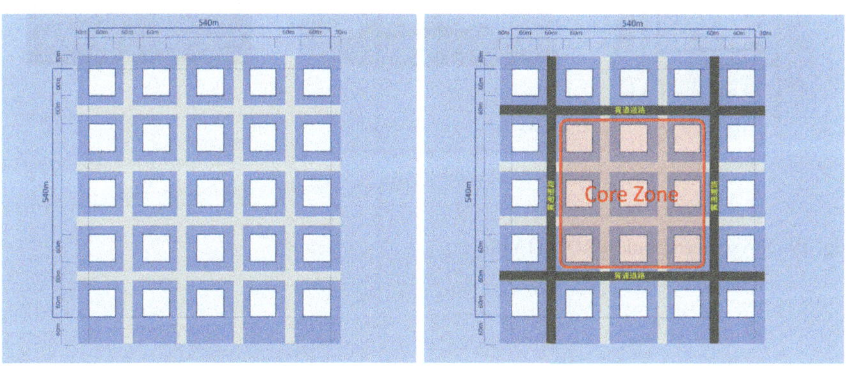

Fig. 17 (left) City Unit Standard GRID; (right) City Grid Standard ZONING

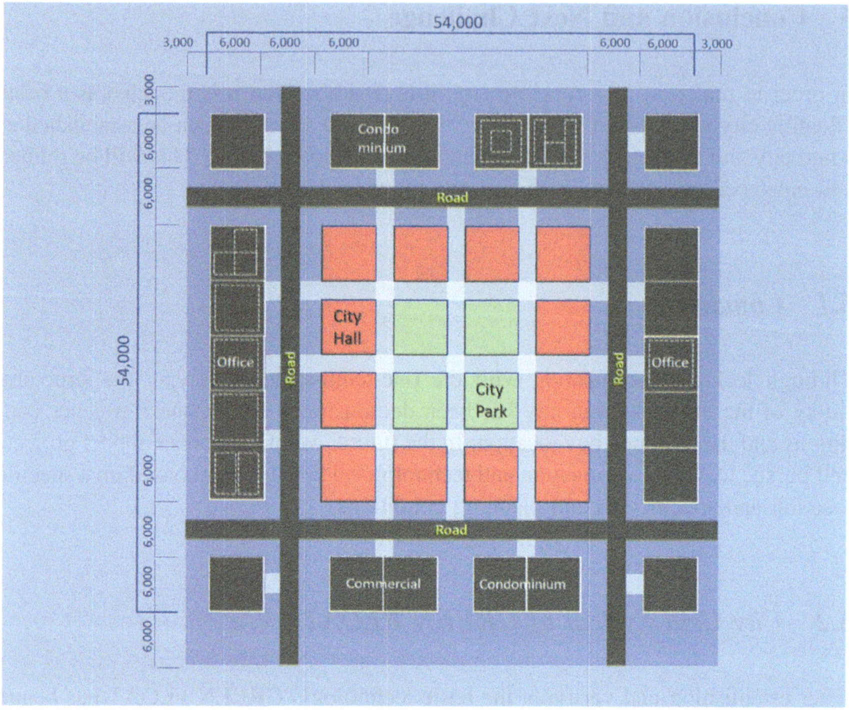

Fig. 18 City Unit Standard PROTOTYPE

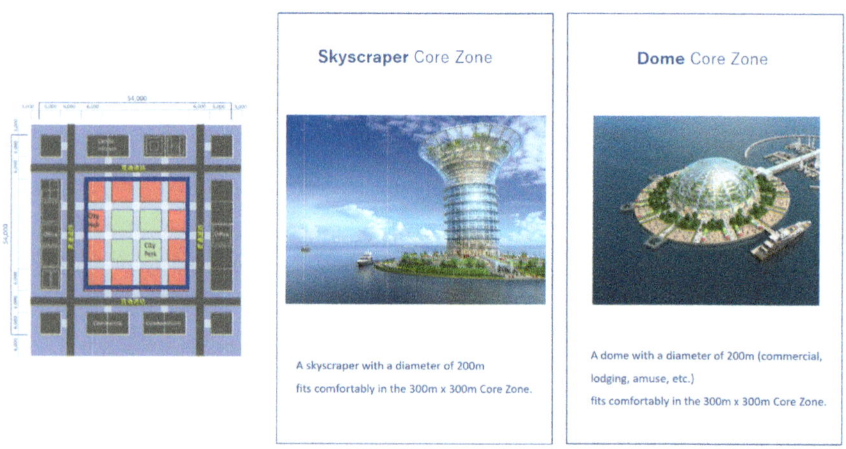

Fig. 19 Skyscraper/Dome PROTOTYPE

Fig. 20 Buildings PROTOTYPE

5.2.1 Float System of City Grid

As a system of Floating Structure, Cell, Module, Unit and City are defined as follows.

City Cell

Minimum unit that can be easily manufactured on docks around the world. Cell size is 60 m × 60 m. The size of 60 m × 60 m is a little small for a land-based city grid. However, for example, in Portland, USA, city planning is functional and attractive with a 60 m × 60 m City Grid (Figs. 17, 18, 19 and 20).

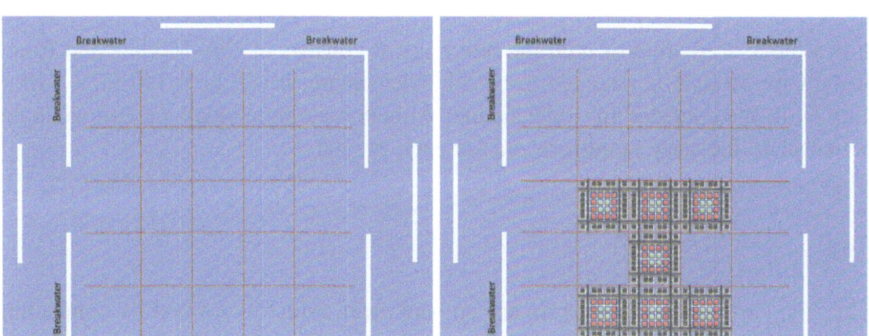

Fig. 21 (Left) Float City Grid; (right) Float City 1st Stage

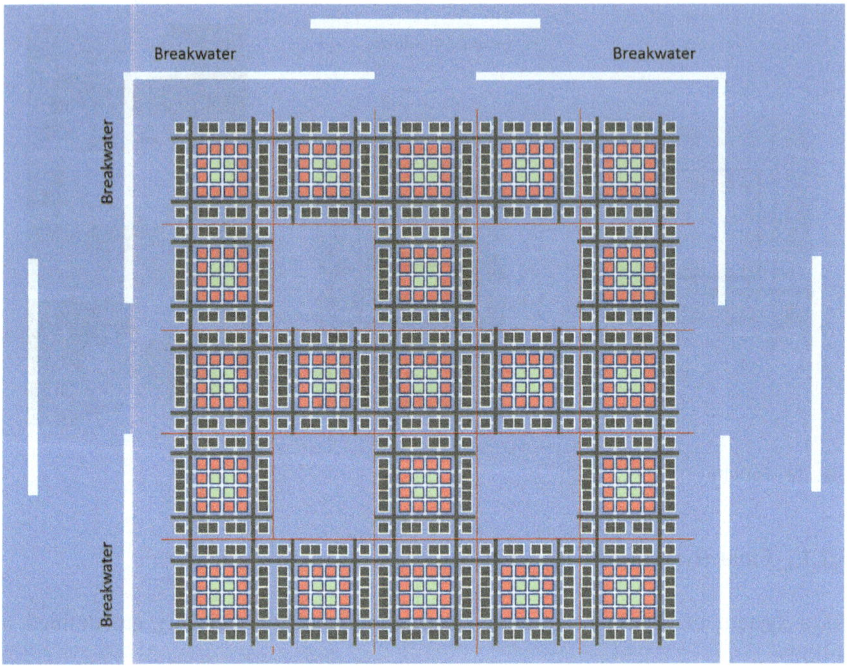

Fig. 22 Float city final stage

City Module

Several floating City Cells are connected for the Building Site.

- Floating structure connected in the dock. ex 60 m × 18 m, 60 m × 240 m (1)
- Floating structure connected by offshore connection. ex 300 m × 300 m (2)

City Unit

It is a block consisting of multiple buildings. It consists of City Module and Road. If the City Module is assumed to be 5 blocks square, it will be a 540 m × 540 m City Unit. We consider this to be an approximate scale for a block. Of course, in the actual plan, the scale is customized for each project.

City

The area surrounded by the primary embankment, which is formed by connecting multiple City Units, is assumed to be the Floating City. The embankment will be constructed in advance on this scale. City Units will grow and be connected in conjunction with the development of the city. Unlike landfill, small-scale land growth is possible according to demand (Figs. 21 and 22).

5.2.2 Green System of City Grid

As a Green System, we are considering whether it is appropriate to implement the following environmental items corresponding to Module, Unit and City of Floating System.

Zero Carbon

Renewable energy supply of appropriate scale linked with the scale of Float System proliferation and development. (Solar power generation, wind power generation, tidal power generation for self-sufficiency and Smart Grid)

Zero Emission

Creating an appropriate scale of water supply and sewerage and resource recycling system in conjunction with the scale of proliferation and development of the Float System.

Flexible Infrastructure

Floating structures of water and sewage, electricity, gas infrastructure system are easy to renew and repair by taking advantage of floating structure.

Future Activities

In parallel with technical studies, we will continue to take on the challenge of creating concrete projects by reflecting marketability and business profitability around the world.

Pond Urbanism: Floating Urban Districts on Shallow Coastal Groundwater

Kristina Hill and Greg Henderson

Abstract Shallow coastal groundwater is rising as sea levels rise, but most coastal cities do not have an extensive network of pumps to handle this, and subsidence risks can increase with additional pumping. Rising groundwater can remobilize capped soil contaminants, bring pollutants into coastal waters, cause foundations to heave, block sewer pipes, and increase soil liquefaction risks. Conventional development is likely to fail, requiring floodable development strategies. Our research analyzed the risk of rising groundwater in the San Francisco Bay Area, and identified opportunities for floodable development within existing policy restrictions. We combined this spatial and policy analysis with a study of a prototypical design for floating urban blocks in artificial ponds. Our goal was to identify a strategy for conserving protected coastal habitat while creating sustainable urban districts. Our proposed design uses shared decking, supported by pontoons, to provide flexibility for placing pre-fabricated housing units on a stable platform in an artificial, excavated pond. A series of artificial ponds could produce multiple benefits as housing, wildlife habitat, and space for flood management. The region's largest city, San Jose, has agreed to permit a version of this design, which could be used to adapt many similar urban districts.

Keywords Groundwater · Sea level rise · Flood adaptation · Coastal ecosystems · Urban planning

K. Hill (✉)
University of California, 301 Wurster Hall, Berkeley, CA, USA94720
e-mail: kzhill@berkeley.edu

G. Henderson
ArxPax Labs Inc., Los Gatos, CA, USA
e-mail: greg@arxpax.com

Ł. Piątek et al. (eds.), *WCFS2020*, Lecture Notes in Civil Engineering 158,
https://doi.org/10.1007/978-981-16-2256-4_2

1 Shallow Coastal Groundwater Rising as Sea Levels Rise

Relative sea levels are rising, and the rate of this change is accelerating [1]. Governments have required consideration of this phenomenon over the last two decades, with periodic updates to the rate and magnitude of sea level rise that public agencies should anticipate [2]. For example, the State of California released new guidance to public agencies in 2018, recommending that all agencies consider the possibility of an extreme scenario of 3.1 m when planning long-term investments in infrastructure (Fig. 1) [3].

Groundwater is also expected to rise within a kilometer or more of the shore (Fig. 2) [4, 5]. Vulnerability studies have been used to project the need for adaptation to sea level rise in coastal areas since the late 1980s [6, 7]. Very few vulnerability studies have incorporated the need to consider rising groundwater as either a source of underground flooding of infrastructure, or emergent flooding at the surface [8]. Most coastal cities do not have maps that represent the depth to shallow, unconfined groundwater. This may be because it has not been recognized as either a resource or a threat, or because it is not managed by public agencies, or simply because no data are available. In addition to the potential for groundwater to rise as sea levels rise, high rates of groundwater extraction have caused historical land subsidence in many coastal cities, increasing the risk of surface flooding by a combination of seawater, rainwater and groundwater [4]. Maps of vulnerability to coastal flooding from sea level rise typically do not incorporate information about land subsidence or rising groundwater, even when they are produced by data-rich

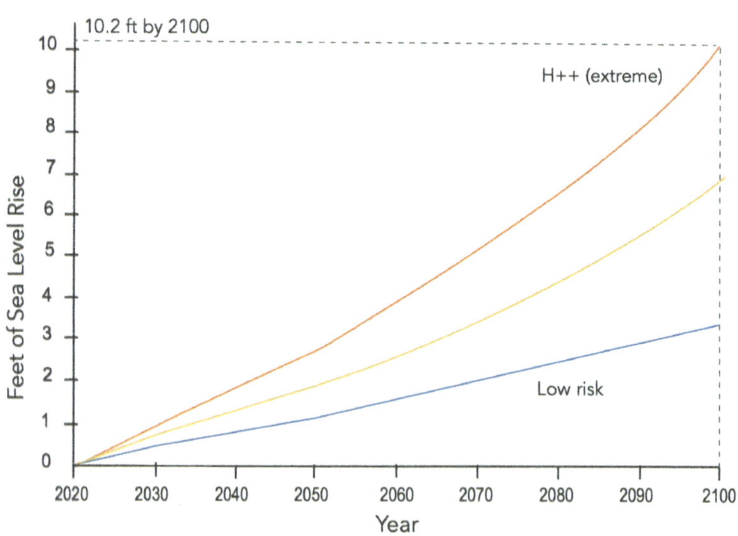

Fig. 1 Sea level rise projections provided by the State of California in guidance to public agencies, 2018, illustrated here using a graph instead of the original tables (credit: K. Hill)

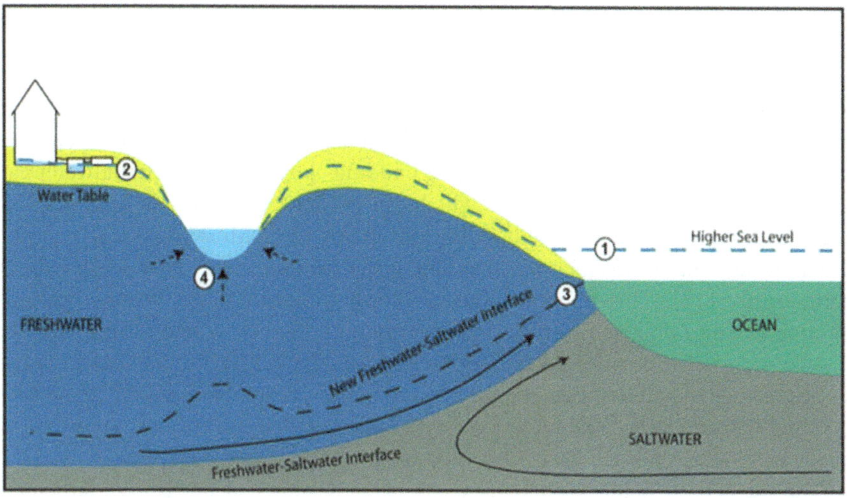

Fig. 2 The United States Geological Survey uses this cross-section to show that rising sea levels will produce rising groundwater levels in coastal areas

government agencies or non-profit research entities.[1] In 2012, the US Geological Survey published one of the first groundwater modeling studies of a coastal city, New Haven, which demonstrated that the impacts of 1 m of sea level rise could raise groundwater levels as far as 5 km from the current shoreline. Discharge to rivers and streams would also increase, creating flooding impacts [9].

1.1 *Impacts of Rising Groundwater*

Coastal cities on relatively-flat nearshore topography are often at significant risk of impacts from rising groundwater. Outside of the Netherlands, most coastal cities do not have extensive networks of pumps that could be used to adapt to rising groundwater. Since subsidence risks typically increase with groundwater extraction, pumping may be a maladaptive strategy for many coastal cities [10]. In the absence of extensive mechanical drainage networks that can be enlarged to remove more groundwater, cities will face considerable risks from rising groundwater.

Rising groundwater can infiltrate underground stormwater pipes and sanitary sewer pipes, as well as gas lines [11]. It can cause vertical displacement of the foundations of buildings and roadway infrastructure [12]. Rising groundwater has also been shown to cause roadway surfaces to require more frequent replacement

[1]For example, sea level rise maps produced for the San Francisco Bay Area do not yet incorporate either land subsidence or rising groundwater. Neither does the City of New York or Miami-Dade County. All three regions are highly vulnerable to coastal flooding.

[13]. In seismically-active regions, rising groundwater can increase the risk of soil liquefaction [14]. In addition, rising groundwater can cause significant health impacts. It can remobilize legacy soil contaminants that have been capped from above but not lined from below [15]. This remobilization can carry pollutants into coastal waters, as well as allow the vapor components of soil pollutants to penetrate sewer lines and enter living spaces inside buildings [16]. Shallow groundwater can also increase humidity in buildings, which can encourage the growth of molds that cause respiratory problems for building occupants [17]. Most importantly, rising groundwater can cause emergent flooding at the land surface, either seasonally or permanently [18]. This phenomenon can occur as groundwater moves upwards through existing drainage infrastructure, or through the soil column. Since surface flooding may also be affected by rainstorms, wind-driven waves and ocean storm surges, it can be difficult to identify rising groundwater as a driver of surface flooding without well data and process-based models that quantify the contribution of rising groundwater to creek flows and local surface flooding.

The impacts of rising groundwater can occur whether or not coastal structures have been raised to block high tides and waves, since the saltwater/freshwater interface in groundwater exists at depths that are typically below the depth of seawall foundations or the compacted soils under levees. Since freshwater inputs to groundwater come from the landward side of any coastal structures, the volume these flows add to rising groundwater would be very difficult to reduce without extensive pump networks. Traditional coastal protection structures such as levees and seawalls must be designed to allow stormwater and the water of rivers to discharge to saltwater bodies. Some use tide gates to allow water to flow through, and others use pumps to push water through or over the coastal levee or wall [19]. Both rising groundwater and more intense rainstorms must be discharged to avoid flooding on the inland side of coastal levees. The capacity of tide gates and pump discharge systems can be overwhelmed if they must convey an increasing volume of groundwater [8].

1.2 Floodable Development: Living with High Water

The alternative to conventional development that is vulnerable to temporary and permanent flooding is to consider floodable development strategies. Floodable development strategies adapt buildings and vehicle or pedestrian circulation networks to emergent water that is either temporary or permanent [19]. Temporary flooding that exceeds the capacity of urban storm drains can be accommodated with or without mechanical pumps by using detention ponds, parking structures, and plazas; raingardens and parks; underground tunnels, and other hybrids of conventional and green infrastructure [20]. Temporary floodwalls can be installed, either automatically or using human labor, as is done when river flooding is anticipated in Cologne, Germany [21]. Electrical equipment in underground tunnels can be removed before extreme flood events, as occurs in New York City before major storm surge events. Terraced or "tiered" development with waterproof underground parking has been used to successfully adapt to temporary flooding in the Hafencity district of Hamburg in Germany [22].

Rapid increases in the extent of surface and groundwater flooding are difficult to accommodate with conventional urban district designs. Innovation in architecture and infrastructure that responds to new permanent flooding is occurring rapidly. In Japan and the Netherlands, underground drainage tunnels and surface canals have been used successfully for decades and centuries, in combination with mechanical pumps [23, 24]. Newer approaches have been developed that rely less on mechanical pumping. Examples include floating urban residential districts, such as the Steigereiland neighborhood of Amsterdam's Waterbuurt, in which three-story homes float on the existing open water body of the IJsselmeer [25]. In addition, new artificial ponds have also been excavated to expose the groundwater surface and convert it to a residential amenity, while using the excavated sediment to build adjacent areas. This method was used to construct the Nesselande housing development in Rotterdam [26].

2 San Francisco Bay Area Case Study

A recent study of current depths to groundwater estimated that 72 km^2 of land adjacent to the San Francisco Bay is vulnerable to seawater flooding, with only 1 m of sea level rise (Fig. 3) [8]. The same study showed that an additional 105 km^2 of land is likely to be vulnerable to the impacts of rising groundwater, using the same scenario for sea level rise. The estimated cost of raising all of the existing coastal structures around the San Francisco Bay edge to protect from a sea level rise of 1 m to be is between 37 and 77 billion USD$ [27]. This same study showed that the decision to provide 0.5 m of additional height on levees and seawalls to prevent wave overtopping in extreme events would impose an additional cost of 50–75 billion USD$. These estimates provided two useful insights: first, that the region is quite vulnerable to rising groundwater, and second, that adding height to protective structures to prevent flooding during extreme storm surge events could double the cost of raising those same structures to cope with typical Bay water levels, with only 1 m of sea level rise. Adapting to the potential for flooding by wave overtopping and high groundwater using floodable development would be significantly less expensive than building coastal systems to prevent all flooding (Fig. 4).

Like all US states, the State of California requires reviews of environmental impacts for projects that would alter topography by adding levees or ponds to adapt to flooding (mandated by the California Environmental Quality Act, CEQA). These reviews can require expensive documentation and introduce time-consuming delays. The State of California has provided an alternative regulatory pathway for adaptation projects that respond to hazards driven by soil movement driven by seismic activity. Property owners can vote to establish a Geologic Hazard Abatement District (GHAD). Once established, the actions taken by an elected

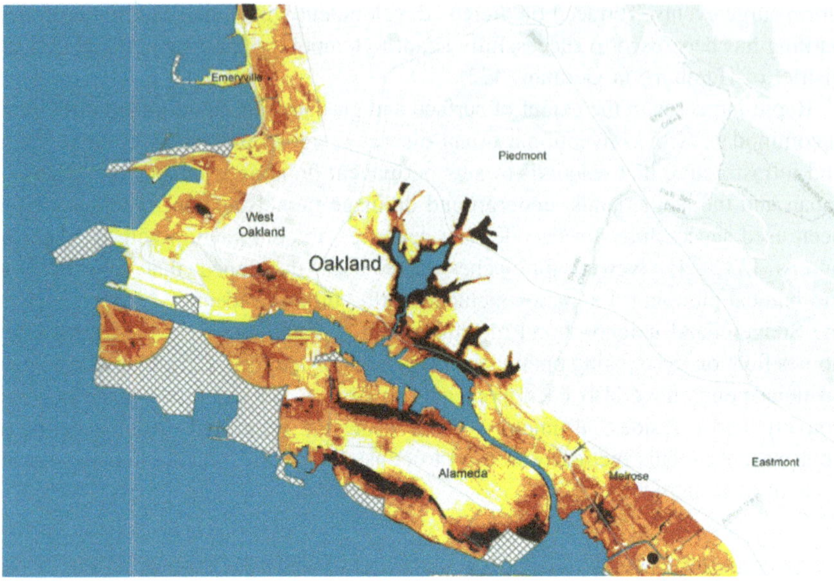

Fig. 3 Depth to groundwater under the cities of Alameda and Oakland, California, estimated by interpolation among well data points [8]. Black and red areas are likely to have groundwater within 1 m of the surface during wet conditions in 2020. Orange areas are likely to have groundwater within 2 m of the surface, and yellow are greater than 2 m in depth

board established for the GHAD are exempt from state-level environmental reviews. This means that land can be re-graded, canals and ponds can be installed with tide gates, and floating districts can be constructed without the costs and delays associated with typical environmental review processes. All of those actions would still require permits from Federal regulatory agencies, but at least one layer of jurisdictional regulatory reviews could be bypassed by organizing as a GHAD. GHADs also have access to shared insurance pools, and can establish taxes on their members' land value in perpetuity. Most other property taxation schemes require a new vote every 20 years to be renewed, which introduces uncertainty into long-range planning. To date, most GHADs have been established in hilly areas prone to landslides, but at least two coastal examples exist in areas where sand erosion or soil liquefaction create development hazards.[2]

The formation of a special governance district such as a GHAD allows local property owners to organize urban districts that provide greater flexibility to modify their landscape for flood adaptation. This is critical in the American system of land use laws, where authority is vested primarily in the property owner and the local jurisdiction, rather than in regional planning agencies. Adaptation projects at the urban district scale may achieve greater economic efficiency by combining lower

[2]Property owners in Malibu Beach and Pajaro Dunes established GHADs in coastal areas.

Fig. 4 Existing seawall along the coast of the City of Alameda, California, on San Francisco Bay (photocredit: K. Hill)

coastal protective structures with floodable urban districts on the landward side. Even with new coastal protection, large areas may be vulnerable to emergent flooding from groundwater. This emergent flooding is likely to begin as a seasonal phenomenon, becoming permanent as sea levels continue to rise [5].

Government agencies in the San Francisco Bay Area have reconstructed thousands of hectares of coastal wetlands (16,000 ha, with a goal of reaching 40,000 ha) [28]. Reconstruction of coastal wetlands is intended to support the characteristic biodiversity of the region, including wading birds, crabs and other species endangered by the loss of bayland ecosystems as coastal areas were filled for urban and industrial development. This regional-scale conservation effort also provides protection from wave energy for coastal roads and developed areas. In 2016, a regional ballot measure was used to establish a tax to create funding for continued wetland restoration by a new government agency called the San Francisco Bay Restoration Authority (SFBRA). The SFBRA will construct tidal marshes, along with ecotones including submerged aquatic vegetation and shellfish beds. The blueprint guiding reconstruction of coastal wetlands is known as the Baylands Ecosystem Habitat Goals Update (BEHGU) [28]. This scientific review also notes that managed ponds have habitat value for diving ducks, and can contribute to improved Bay water quality through denitrification.

Studies of sea level rise impacts on tidal marshes have indicated that these ecosystems are vulnerable to collapse [29]. The BEHGU study identified several areas where tidal wetlands could transition inland in the San Francisco Bay Area, but local cities have designated the largest of these transition zones for housing development. As sea levels continue to rise, managed ponds may provide an armature for constructing new coastal wetlands. A band of ponds can be moved inland, by excavating new sites, and using the excavated soil to construct tidal wetlands within the former pond zone. This would allow complex urban land uses to co-exist with baylands ecosystems as sea level and groundwater continue to rise.

In addition to ecological value, coastal lagoons and freshwater ponds in the Bay Area have provided significant recreational benefits, attracting residential development investments. For example, the City of Oakland built its downtown around a 56-ha saltwater lagoon (Fig. 5) that provides boating opportunities as well as bird habitat. This lagoon is tidal, although tides are muted by tide gates at a pumping station that was constructed to discharge floodwaters in an extreme rain event. Similarly, Shoreline Lake in the city of Mountain View provides community event spaces and boating in a 20-ha artificial pond. Saltwater is pumped from San Francisco Bay and discharged back to the Bay via a creek. The City of Berkeley maintains a saltwater lagoon known as the Aquatic Park, which receives stormwater

Fig. 5 The downtown area of Oakland, California, was built around a saltwater lagoon known as Lake Merritt, which was designated as the first urban bird sanctuary in the United States in the 19th century (photocredit: K. Hill)

discharge from the City's urbanized watershed and has tidal openings to the Bay. All of these existing ponds rely on pumping facilities and tide gates or other structures to mute the natural tidal range and maintain a minimum water depth suitable for recreation and habitat value. Ponds are a well-established element of urban development around the Bay edge, although they occur infrequently.

3 Proposed Design: Pond Urbanism

Based on our analyses of conditions in the San Francisco Bay Area over the last decade and longer, we separately developed elements of a prototypical design for introducing floating urban blocks in artificial ponds. Our primary goal is to identify a strategy for adaptation to flooding, driven by sea level rise, that creates multiple benefits, including protection of human health and conservation of protected coastal habitat. We were both concerned that conventional housing is likely to fail over the next several decades as a result of surface flooding and groundwater impacts. One of us (Kristina Hill) developed a proposal for linking canal systems to networks of artificial ponds as a floodable development strategy for the San Francisco Bay Area, shown in Fig. 6 [30]. She developed this idea before and during regional design competition in 2018.[3] Previously, her co-author, Greg Henderson, developed and patented a three-part foundation system.[4] This system, called the SAFE Foundation System (Self-Adjusting Floating Environment), was proposed to support an urban, infill, mixed-use residential project in a flood plain.[5] We have combined these proposals here in an effort to identify the maximum potential of a floodable district with multiple benefits.

Our proposed design uses artificially-excavated ponds as an environment for new mixed-use urban districts. Housing and commercial structures would be supported by the SAFE Foundation System, to provide design flexibility for where and how to place modular, pre-fabricated units for housing and small commercial shops, and extend networked infrastructure (sanitary sewers, drinking water, electric power, etc.) to stable platforms inside a controlled water body. These modular structures would be placed on shared decking to provide surface area for public spaces as well as for light-weight buildings.

Initially developed for earthquakes, the SAFE Foundation System's three parts consist of a containment vessel which holds a buffer medium, which in turn supports a construction platform. The construction platform is made of prefabricated SAFE modules, 3 ms by 6 m by 3 m in depth. These modules are then assembled

[3]Hill was a member of the ABC Team, which included a several professional design firms. They proposed a floating district called the Estuary Commons for a district of the city of Oakland in California. (website).

[4]US Patents: 8777519, 9103118, 9398878, 9790702, 10081960, 10711478 Additional Patents Pending.

[5]http://arxpax.com/.

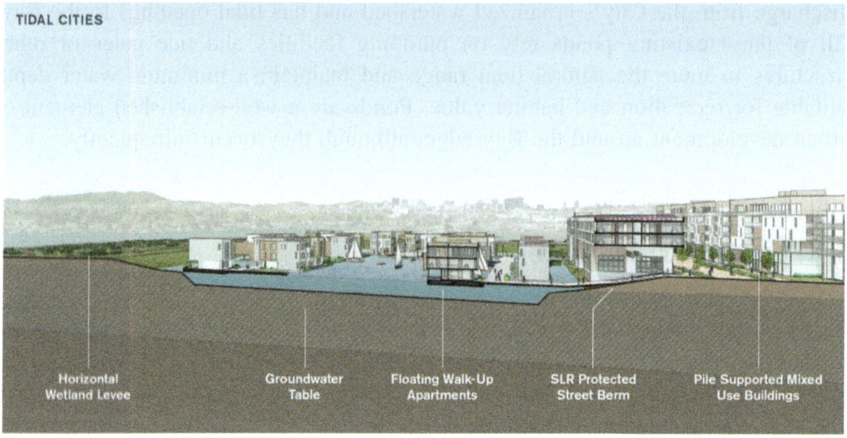

Fig. 6 This is an early conceptual sketch of one type of floating urban district that was proposed for Oakland, California by K. Hill and the All Bay Collective, in the Resilient by Design Bay Area Challenge in 2018 (credit: All Bay Collective). This proposal did not incorporate the SAFE Foundation system designed by Greg Henderson and ArxPax, which was developed on a separate schedule

into a broad, stable platform. Because floating objects rise and fall with the water level, SAFE provides a natural method of flood protection. The limiting factor on what can be built on the SAFE Foundation is the mass of the displaced water. Here it is helpful to imagine a large cruise ship. The ship is in essence a mixed-use building designed to move on the sea. The ship may house 5000 people and be 75 m tall but only draft 9 m below the water level. When optimized for stability[6] rather than movement, SAFE can be much shallower and still support vast loads.[7] These building and infrastructure dead loads are balanced with ballast. Relatively light live loads including cars, people, furniture, etc. have near zero effect but can be balanced dynamically if necessary. The buffer medium, in this case water, separates the containment vessel from the construction platform thereby decoupling the supported structure from the earth. This decoupling provides protection from seismic forces much like traditional base isolation but with greater cost effectiveness. New research on wave damping suggests that seismic seiche waves in ponds can be reduced by "cloaking" or using baffles, making it possible to limit the impact of seismic waves on floating structures [31].

The containment vessels themselves consist of precast concrete, drain rock and filtration systems. They are designed to be permeable and use ground water for

[6]Stability of a floating object, or its moment of inertia in the water plane increases by a factor of eight when the width of the object is doubled.

[7]The largest floating object in the world, the SR 520 floating bridge in Seattle, Washington, USA weighs approximately 700,000 metric tons providing more than 80,000 m^2 of buildable area.

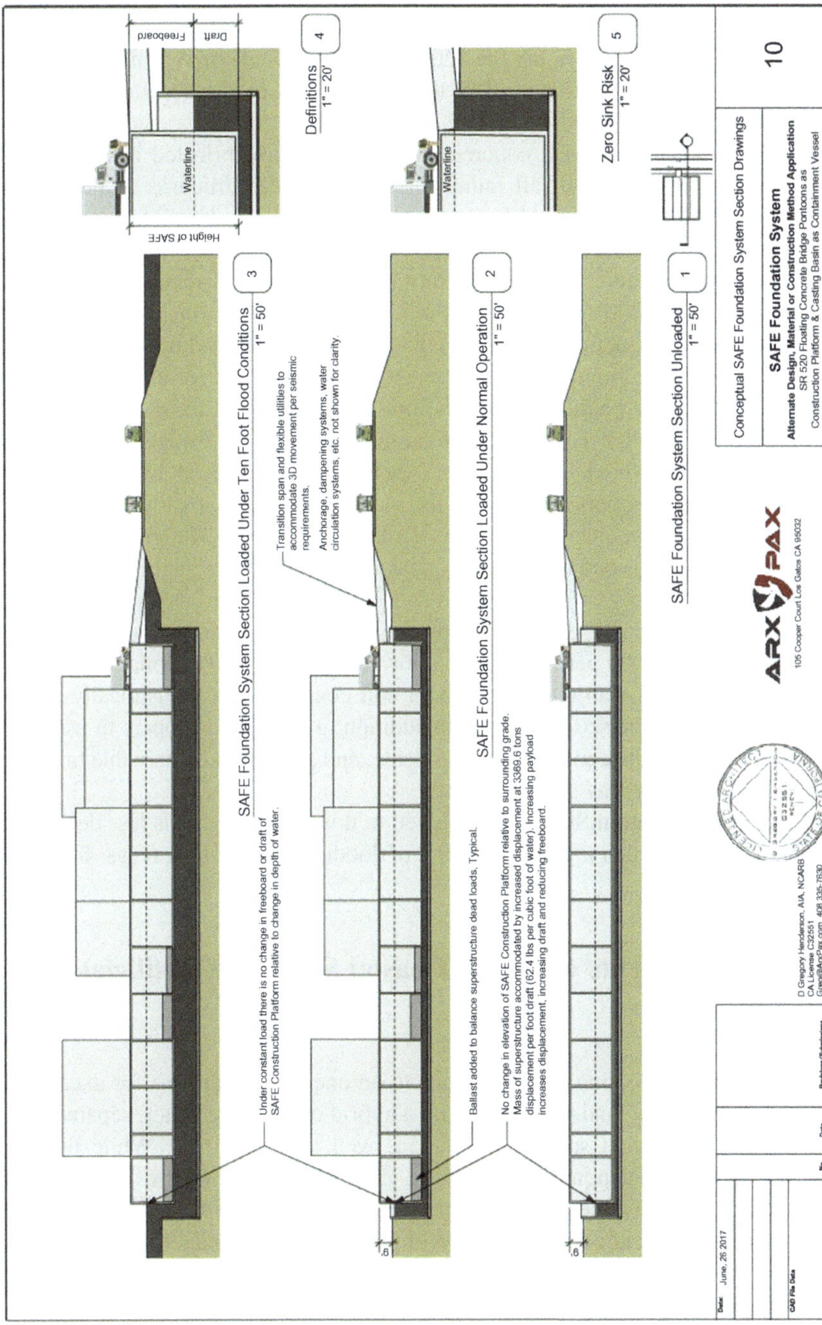

Fig. 7 Cross-section of floating district design proposed by ArxPax for San Jose, California

flotation. In the case of this floodable urban district proposal, a system of artificial ponds would act as the containment vessels (Fig. 7).

To understand the technical aspects of the SAFE Foundation System one can refer to the systems proven in the SR 520 Bridge in the State of Washington, USA. SAFE optimizes the solutions used in the construction of the 2350 m bridge for scalability, repeatability, and cost.

Like the bridge, the SAFE concrete pontoons are prefabricated and post tensioned together into a broad raft rather than a thin linear structure. Scaled down from 11,000 to 35 tons the SAFE modules are designed to be prefabricated and transported short distances to the building site and assembled using typical construction equipment rather than floated into place over long distances.

In the bridge, 70,000 tons of ballast was prepositioned during construction to maintain a level construction platform. Ballast is added or moved once construction is complete to best accommodate final design loads. Designed for a magnitude 9.0 earthquake, the utility connections on the bridge are designed to be flexible at the two transition spans that connect the floating portion of the bridge with the stationary bridgeheads fixed to the shore. Anchorage is achieved using opposing spring lines, technology borrowed from the marine industry that allows vertical but not horizontal position changes for ships docked along piers. The ground water is circulated using pumps and a filtration system to maintain water quality and prevent stagnation.

For urban areas at risk of flooding, the economics are very compelling. SAFE has a cost goal of \$1000 per m^2. This is a lower cost than conventional deep piers, grade beams and mat slab required in poor soil conditions in a seismically-active zone like the San Francisco Bay Area. In addition, it allows developers to use land that would otherwise be considered unbuildable, and provides a sustainable tax base for a municipality even as sea levels rise.

The SAFE Foundation System is designed to float in water that is less deep than the structure itself, so that it presents no risk of flooding from sinking (Figs. 8 and 9).

4 Multiple Benefits of a New Coastal System of Artificial Ponds

Our proposal is for floating urban districts to be one of several uses for excavated ponds that form a new landscape type with a hybrid of land and water, separated by wide earthen levees that support marshes as well as local roads. While floating structures and artificial ponds are not new, the use of an extensive landscape of managed ponds to produce multiple benefits in an urban coastal region is a novel proposal in the US. A honeycomb pattern of ponds that are managed with tide gates can provide zones for safe housing, benefits to water quality, habitat, recreation, and opportunities for flood management. Our intention is for the ponds to be constructed with a greenbelt of vegetated parks on the landward side, which can filter

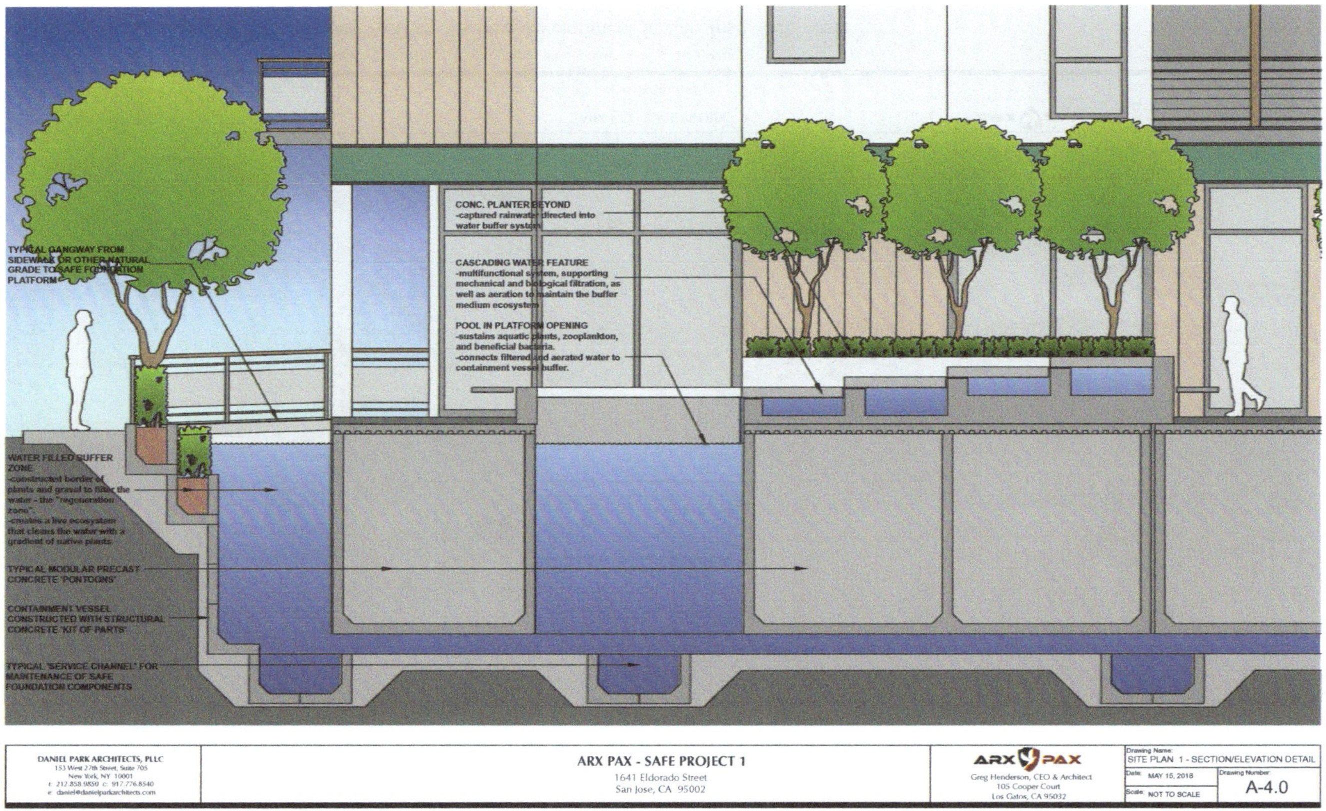

Fig. 8 Detail of cross-section showing pontoons supporting the construction platform inside the containment vessel in the ArxPax design for San Jose, California

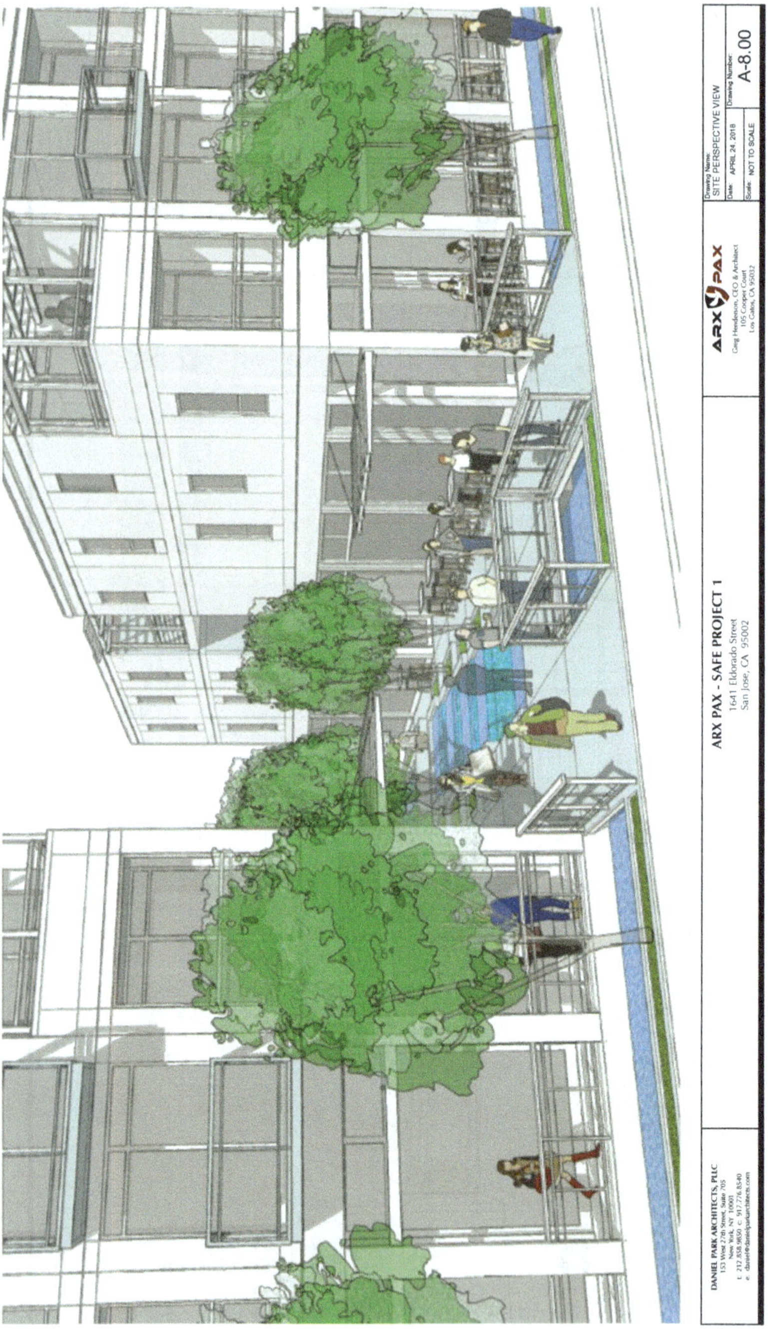

Fig. 9 Perspective drawing of the proposed ArxPax floating development in San Jose, California

stormwater runoff and remove some contaminants before they reach the ponds. Tributaries can pass through these pond landscapes, and overflow into the ponds during flood events when the discharge volumes exceed the capacity of openings and pumps in coastal protection structures. Ponds can be designed to receive up to a meter of floodwater from tributaries or from overtopping of coastal structures in an extreme event.

Floating structures could provide protection from damage associated with earthquakes. The network of levees that separate the ponds would need to be designed to perform during soil liquefaction events, or to fail in ways that minimize risks to life and property. A system of small canals or underground drainage pipes could be used on the landward side of the ponds to convey groundwater to the pond system, protecting existing housing from rising groundwater for several additional decades (Fig. 10).

Major transportation routes along the coast of the Bay are likely to be raised on berms or causeways as sea levels rise, which will create new barriers to stormwater and groundwater discharge from the inland side of these roadways. Pond systems

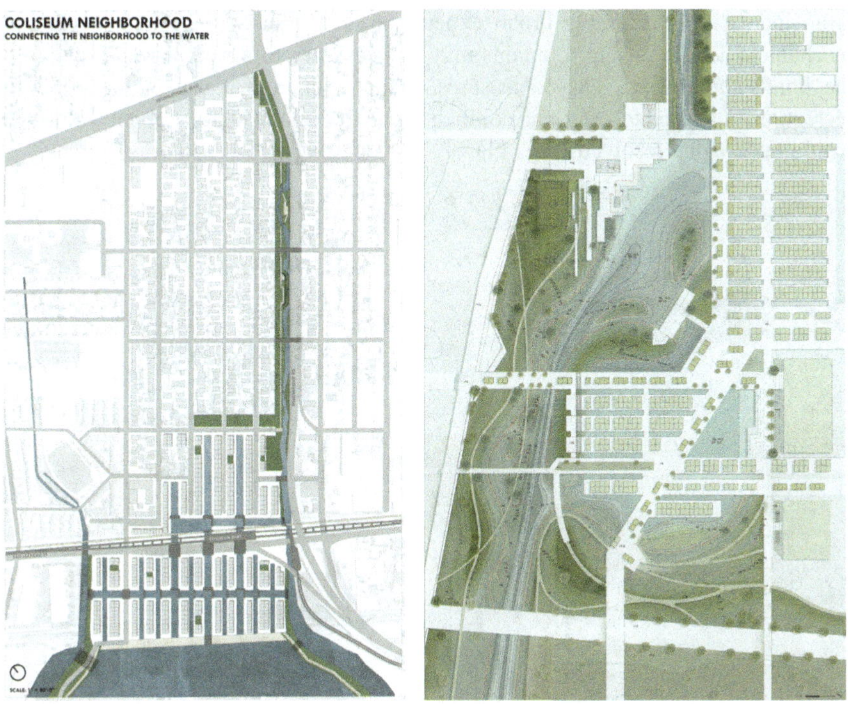

Fig. 10 These plan sketches illustrate the concept of a tidal city. The plan on the left proposes a series of floating platforms on the edge of an artificial pond, with canals to drain high groundwater from the land behind that would allow people to stay in existing homes as long as possible (Juanita Ballesteros, Sarah Abrams, and Greta Aalborg). The plan on the right proposes a floating urban district at the mouth of a tributary (Yuling Chen and Daria Kiefer)

can be used to manage the freshwater flooding that would result from these constraints on discharge. Alternatively, roadways raised on causeways will allow saltwater flooding to pass below and extend into existing urban districts. A series of saltwater ponds can be used to manage these new tidal flows and create multiple benefits. Transportation routes may also be placed underground in tunnels. In that case, pond systems can be used to create a zone of managed floodwaters that flow above and across these underground structures, while allowing room for ventilation structures.

Finally, managed ponds can provide much-needed recreation opportunities for people living and working in urban districts. Kayaking and stand-up paddle boarding are popular activities in ponds where wave energy is low. Boardwalks and piers can provide areas for fishing, bird watching and water access for people with walking disabilities. Communal events such as weddings and other family celebrations are often held in small public event centers with a water view, in addition to commercial businesses such as restaurants and cafes. Paths that provide walking loops around small waterbodies are popular for lunchtime exercise on weekdays, as well as weekend walks with family and friends. Particularly during the recent pandemic, the ability to walk or bike outdoors in a setting with high aesthetic value is an increasingly desirable urban experience. Tidal flushing using tide gates or mechanical aeration using fountains may be necessary to maintain water quality in the ponds themselves. These measures, along with human use of the pond for recreation, may create seasonal conflicts with management for habitat value that would need to be resolved (Fig. 11).

Fig. 11 Perspective sketch of a Tidal City proposed for Oakland, California (Illustration credit: All Bay Collective)

5 Conclusions

Multiple benefits can be provided by a single excavated pond that supports floating residential units. These benefits can be generated at a regional scale using a landscape of independent or inter-connected ponds. If only a half or a third of the ponds in an adapted coastal landscape are used for housing, that would still represent a significant social and economic benefit over the total loss of conventional housing units in a permanently-flooded environment. As sea levels continue to rise, an original band of ponds could be filled to reconstruct tidal wetlands, using sediment from the excavation of new ponds on the inland side of the original pond systems (Fig. 12). Adapting urban districts using floodable development strategies, such as ponds and canals, also allows coastal protective structures to be designed more economically, since they could be sized for typical conditions instead of being sized to prevent overtopping in exceptional events.

The San Francisco Bay Area's largest city, San Jose, has expressed interest in permitting a version of this design for floating structures on excavated ponds. That

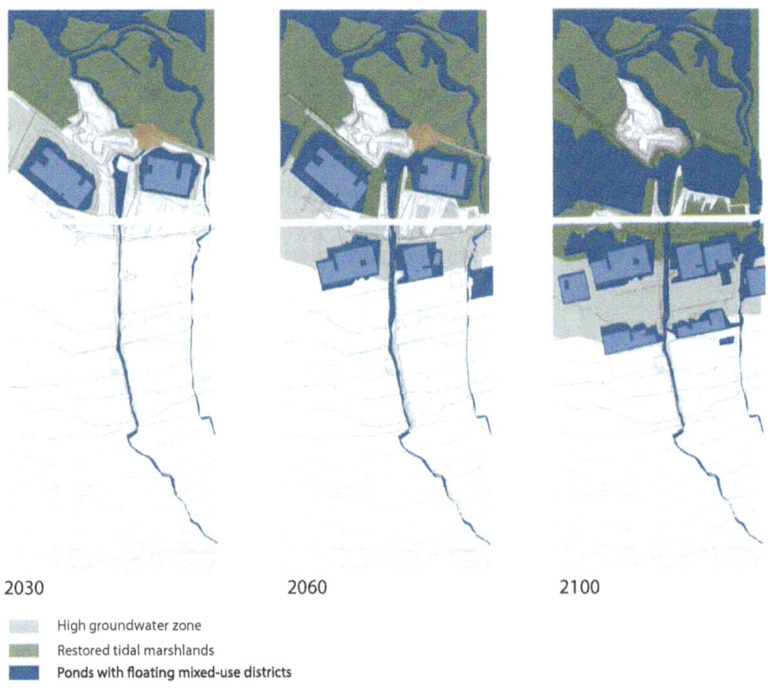

Fig. 12 As sea level and groundwater continue to rise, a new band of ponds can be excavated on the inland side. The original ponds can be filled using the excavated material to establish a new edge of reconstructed tidal marshes, eelgrass and shellfish (Illustration: Tomas McKay and Kristina Hill)

administrative flexibility may allow the first protoype of this design to be built. Once it is better understood, a system of ponds and floating structures could be used as a unique form of floodable development in many similar urban districts with relatively-flat nearshore topography. In addition, the regional agency tasked with developing a coordinated understanding of sea level rise vulnerability (the San Francisco Bay Commission on Development and Conservation, known as BCDC), has committed to developing vulnerability maps that reflect rising groundwater. These new maps will allow many cities around the region to identify new groundwater flooding risks that affect more land than seawater-driven flooding alone, and could trigger a reconsideration of the role that floodable development should play in adaptation strategies, alongside coastal structures such as levees and floodwalls.

Conserving the biodiversity of the region's aquatic and tidal ecosystems, the safety and number of housing units, and human health will require an approach that emphasizes multiple benefits, rather than flood mitigation alone. In addition, the public and private cost of coastal adaptation must be optimized. Multiple benefits can be achieved by using landscape-based adaptation strategies that bring blue and green infrastructure systems together. Our conclusion is that, although they are unconventional, floodable development that uses a landscape-scale system of ponds and canals could allow many coastal cities to develop districts that are adapted to more intense rainfall, higher seawater and rising groundwater. Extreme storm surge events that are particular to geographic regions may create limits on this strategy, such as typhoons and hurricanes. Coasts with unconsolidated bluffs or bedrock hills may not have much space to introduce pond systems. On the other hand, regions with high-value biodiversity resources, seismically-active coastal regions, and relatively-flat urban coastal regions that have experienced significant land subsidence may find this adaptation approach particularly useful.

References

1. Intergovernmental Panel on Climate Change (2013) Climate change 2013: the physical science basis. Contribution of Working Group I to the fifth assessment report of the intergovernmental panel on climate change. Cambridge University Press, Cambridge, United Kingdom and New York, NY, USA
2. Griggs G, Árvai J, Cayan D, DeConto R, Fox J, Fricker HA, Kopp RE, Tebaldi C, Whiteman EA (California Ocean Protection Council Science Advisory Team Working Group) (2017) Rising seas in California—an update on sea level rise science. California Ocean Sciences Trust
3. California Natural Resources Agency and California Ocean Protection Council (2018) State of California sea-level rise guidance: 2018 Update. https://opc.ca.gov/webmaster/ftp/pdf/agenda_items/20180314/Item3_Exhibit-A_OPC_SLR_Guidance-rd3.pdf. Accessed 2 July 2020
4. Nicholls RJ, Cazenave A (2010) Sea-level rise and its impact on coastal zones. Science 328:1517–1520

5. Rotzoll K, Fletcher CH (2012) Assessment of groundwater inundation as a consequence of sea-level rise. Nat Clim Change 3:477–481. https://doi.org/10.1038/nclimate1725
6. Leatherman SP (1990) Modelling shore response to sea-level rise on sedimentary coasts. Prog Phys Geogr Earth Environ 14:447–464. https://doi.org/10.1177/030913339001400402
7. Leatherman SP (1987) Beach and shoreface response to sea-level rise: Ocean City, Maryland, U.S.A. Prog Oceanogr 18:139–149
8. Plane E, Hill K, May C (2019) A rapid assessment method to identify potential groundwater flooding hotspots as sea levels rise in coastal cities. Water 11:2228. https://doi.org/10.3390/w11112228
9. Bjerklie DM, Mullaney JR, Stone JR, Skinner BJ, Ramlow MA (2012) Preliminary investigation of the effects of sea-level rise on groundwater levels in New Haven, Connecticut. U.S. Geological Survey, Reston, VA, USA, p 46
10. Nicholls RJ (1995) Coastal megacities and climate change. GeoJournal 37:369–379. https://doi.org/10.1007/BF00814018
11. Wittenberg H, Aksoy H (2010) Groundwater intrusion into leaky sewer systems. Water Sci Technol 62:92–98. https://doi.org/10.2166/wst.2010.287
12. Herle I, Herbstová V, Kupka M, Kolymbas D (2010) Geotechnical problems of cultural heritage due to floods. J Perform Constr Facil 24:446–451. https://doi.org/10.1061/(ASCE)CF.1943-5509.0000058
13. Knott J, Daniel JS, Jacobs J (2017) Assessing the effects of rising groundwater from sea-level rise on coastal road pavement. Roads & Bridges News. https://www.roadsbridges.com/eyeing-rise. Accessed 2 July 2020
14. Yasuhara K, Murakami S, Mimura N et al (2007) Influence of global warming on coastal infrastructural instability. Sustain Sci 2:13–25. https://doi.org/10.1007/s11625-006-0015-4
15. O'Connor D, Zheng X, Hou D et al (2019) Phytoremediation: climate change resilience and sustainability assessment at a coastal brownfield redevelopment. Environ Int 130:104945
16. Cernansky R (2016) It came from beneath: detecting and mitigating vapor intrusion. Environ Health Perspect 124. https://doi.org/10.1289/ehp.124-A141
17. Wu H, Wong J (2020) Current challenges for shaping the sustainable and mold-free hygienic indoor environment in humid regions. Lett Appl Microbiol 70:396–406. https://doi.org/10.1111/lam.13291
18. Hoover D, Odigie K, Swarzenski P, Barnard PL (2017) Sea-level rise and coastal groundwater inundation and shoaling at select sites in California, USA. J Hydrol Reg Stud 11:234–249
19. Hill K (2015) Coastal infrastructure: a typology for the next century of adaptation to sea-level rise. Front Ecol Environ 13:468–476. https://doi.org/10.1890/150088
20. Van Herk S (2014) Delivering integrated flood risk management: governance for collaboration, learning and adaptation, Delft University of Technology, Doctoral dissertation
21. Engel H (1997) The flood events of 1993/1994 and 1995 in the Rhine River basin. In: Destructive Water: Proceedings of IAHS Conference, Anaheim, CA, 1996 239:12
22. Serre D, Barroca B, Balsells M, Becue V (2016) Contributing to urban resilience to floods with neighbourhood design: the case of Am Sandtorkai/Dalmannkai in Hamburg. J Flood Risk Manage. https://doi.org/10.1111/jfr3.12253
23. Graaf R, Hooimeijer F (2008) Urban water in Japan. Taylor and Francis, London
24. Reh W, Steenbergen C, Aten D (2007) Sea of land, 1st English edn. Wormer, Noord-Holland
25. Projectbureau IJburg, Ontwikkelingscombinatie Waterbuurt West (2012) Floating Amsterdam: the development of IJburg's Waterbuurt. Municipality of Amsterdam. http://www.monteflore.com/wp-content/uploads/2017/03/Floating-Amsterdam-LR.pdf. Accessed 2 July 2020
26. Gemeente Rotterdam (2008) Ontwerp Bestemmingsplan: Nesselande, 2 Partiele Herziening
27. Hirschfeld D, Hill K (2017) Choosing a future shoreline for the San Francisco Bay: strategic coastal adaptation insights from cost estimation. J Mar Sci Eng 5(3):42. https://doi.org/10.3390/jmse5030042

28. Goals Project (2015) The Baylands and climate change: what we can do. Baylands ecosystem habitat goals science update 2015 prepared by the San Francisco Bay Area Wetlands ecosystem goals project. California State Coastal Conservancy, Oakland, CA. https://baylandsgoals.org/wp-content/uploads/2019/06/Baylands-Complete-Report.pdf. Accessed 2 July 2020
29. Blankespoor B, Dasgupta S, Laplante B (2014) Sea-level rise and coastal wetlands. Ambio 43:996–1005. https://doi.org/10.1007/s13280-014-0500-4
30. Hill K (2018) Amatures for Coastal Resilience. In: Mossop E (ed) Sustainable coastal design and planning. CRC Press, London, pp 415–434
31. Zou S, Xu Y, Zatianina R et al (2019) Broadband waveguide cloak for water waves. Phys Rev Lett 123: https://doi.org/10.1103/PhysRevLett.123.074501

Floating Modular Housing to Address Demand and Affordability

Jagmeet Khangura and Jason Haney

Abstract Increasing population, urbanization, and desire for coastal living creates an annual demand for housing units in coastal cities around the world that is not currently being met. The United States alone needs approximately 5 million new apartments by 2030, and the problems of housing availability and affordability are greatest in the coastal cities. These are also some of the most expensive cities in which to build housing due to limited land and constraints on materials and labor. The rate that housing must be produced in coastal cities is challenged with current construction and land use practices. Globally, a few companies are producing floating housing; none appear to be focusing on permanent, large-scale floating housing. Black & Veatch proposes developing floating housing communities constructed using innovative modular approaches to provide affordable and climate resilient housing. Housing construction costs can be significantly reduced compared to traditional, land based, "stick built" construction methods. Housing built in low-cost locations on floating platforms can be moved to urban areas and moored at shore or in open water to provide thousands of new housing units at affordable costs. Although similar solutions have been successfully deployed by others, no one appears to be addressing the affordable housing need at the scale proposed here.

Keywords Modular · Housing · Coastal · Affordable · Resilient

J. Khangura (✉)
Black & Veatch Corporation, 2999 Oak Road, Suite 490, Walnut Creek, CA 94597, USA
e-mail: Khangurajk@bv.com

J. Haney
Black & Veatch Corporation, 11401 Lamar Avenue, Overland Park, KS 66211, USA
e-mail: Haneyjd@bv.com

Ł. Piątek et al. (eds.), *WCFS2020*, Lecture Notes in Civil Engineering 158,
https://doi.org/10.1007/978-981-16-2256-4_3

1 Introduction

The demand for affordable housing continues to rise, caused by a multitude of social, economic, and environment factors. The current approaches to building adequate housing is not sufficient and there is a critical need for innovative approaches to lower costs and expedite the design and construction process of housing resulting in more people benefitting from housing that is affordable, healthy and climate-resilient. Large-scale, floating modular housing communities located in protected bodies of water, such as bays and harbors, would be a breakthrough in affordability, climate resilience and healthy living. Though a few companies are producing floating housing, no one appears to be building at a large enough scale to address the affordable housing need. Floating housing communities constructed using innovative modular approaches can provide affordable and climate resilient housing. Housing construction costs can be significantly reduced by building in low-cost locations on floating platforms that can be moved to urban areas and moored at shore or in open water to provide new affordable housing units. Although there are challenges and barriers to developing large-scale modular floating housing communities, the need for housing is great enough to warrant addressing them as a viable solution.

While the challenges of providing affordable housing are known, what is less known is how climate change, flooding, and sea level rise will exacerbate housing affordability. In addition to increasing availability of affordable housing in the near term, floating housing will also ensure availability of housing in the future because of its inherent resilience against rising seas due to climate change, seasonal flooding, earthquakes, and other natural disasters.

The future will require new approaches to building homes, communities, and cities. Traditional sea level rise mitigation solutions such as sea walls require major investments. A report released by Center of Climate Integrity concluded that seawall cost to protect coastal communities in the United States would be more than $400 billion by 2040 [1]. While affluent cities will be able to pay for this infrastructure, disadvantaged communities will likely struggle with funding. For hundreds of small coastal and tidal communities, the costs will far outstrip their ability to pay, making retreat and abandonment the only viable option unless enormous amounts of financing emerge in a very short time. In 178 small communities, the cost of building basic coastal defenses is more than $100,000 per person [1].

A recent independent study by First Street Foundation that takes into account rainfall and flooding along smaller waterbodies in addition to sea-level rise identified higher flood risk than previously determined by the United States Federal Emergency Management Agency (FEMA). A total of 14.6 million properties across the United States were found to be at substantial risk [2].

2 Housing Demand

Increasing population, urbanization, and coastal living creates an annual demand for housing units in coastal cities that is not currently being met. In the United States alone, the national shortage of affordable housing is staggering. The United States needs 4.6 million new apartments by 2030; no state or major metropolitan area has an adequate supply of housing for its poorest renters [3]. The National Low-Income Housing Coalition reports that the coastal states of Washington, Oregon, California, Texas, and Florida have 30 or fewer affordable and available rental homes per 100 renter households at or below 30 percent Average Median Income (AMI)[1] [4].

The problems of housing availability and affordability are greatest in coastal cities. A majority of cities with the highest cost of living are coastal cities [5]. Furthermore, 23 out of 25 cities in United States that are most densely populated and fastest growing cities are also coastal [6]. The problem is exacerbated by increasing urbanization and the desire to live near the coast, where the danger of rising sea and flooding is increasing because of climate change. Three coastal regions in California alone (Southern California-Los Angeles Region, San Francisco Bay Area and San Diego County) need to construct over 90,000 units annually until 2030 to meet the very low and low income housing requirements determined by the state of California)[2] [7]. Globally, it is projected that land currently home to 300 million people will experience chronic flooding by 2050, and land currently home to 200 million people will be permanently under the high tide line by 2100. Within the United States, Zillow estimated in 2017 that 1.9 million homes would be flooded with a 6 foot rise of sea level [8]. An internal analysis conducted by Black & Veatch, using updated United States National Oceanic and Atmospheric Administration (NOAA) data estimates 2.2 million homes will be lost with a 5 feet sea level rise predicted to occur in the 21st Century [9]. Many of these homes are in the most socially vulnerable communities. Of those homes affected, about half are in the 40–100th percentile of the United States Centers for Disease Control and Prevention's (CDC) Social Vulnerability Index (SVI).[3]

[1]Average Median Income is a metric calculated by the Unites States Department of Housing and Urban Development (HUD) to determine the income eligibility requirements of federal housing programs. The value calculation is based on the estimated area median family income.

[2]Calculated based on Regional Housing Needs Allocations for Very Low and Low Income. Total housing required for the cycle divided by number of years in cycle.

[3]Social vulnerability refers to the resilience of communities when confronted by external stresses on human health, stresses such as natural or human-caused disasters or disease outbreaks. Reducing social vulnerability can decrease both human suffering and economic loss.

3 Floating Housing

Floating modular housing communities can be a breakthrough in resilient, sustainable and affordable living. The building blocks of the communities would include modular housing systems already being produced and proven in land-based development by companies such as HBG.[4] The modular steel structures most applicable for deployment are purpose-built, factory manufactured, structural steel modular buildings utilizing high-grade, heavy gage structural steel and certified products. These structural-steel modular building systems are precisely engineered, can meet American Society for Testing and Materials (ASTM's) steel standards and can employ aspects such as a steel moment frame. These modules are often built to the size of a standard shipping container to maximize the use of methods already designed for transporting shipping containers. Additionally, these modules have the ascetic appearance of a shipping container while negating the structural modifications necessary when using shipping containers as building materials.

These systems are a radical departure from traditional and antiquated construction methods because 90% of the project takes place in a controlled factory environment before the steel modules are delivered to the project site. Highly resistant to fire, mold, water damage, termites, and the standard wear and tear of traditional construction, the modules are built to last. Once delivered to the site, the remaining ten percent of project work takes place, craning and placing living modules into their final location and hooking up all utility connections. Floating housing could extend the use of this innovation by deploying 10–1000 individual living units on a floating base ("vessel").

3.1 Resilience

Communities around the world face a variety of threats ranging from sea-level rise, seasonal flooding, thunderstorms, hurricanes, earthquakes, wildfires, excessive heat and droughts. The severity and frequency of these threats will continue to change in the future thus future housing solutions need to be resilient to meet the current challenges but also agile to meet the future needs of the communities. Though not suitable for every natural threat, like hurricanes, floating housing is inherently resilient to many of the regional threats faced by communities around the world. In addition to sea-level rise, floating housing is resilient to earthquakes, wildfires, extreme temperatures and droughts.

Floating housing is inherently adaptable to diurnal, seasonal and long-term sea level rise. It is impacted little by seismic activity since mooring and anchoring systems can be designed to limit movement. It is resilient to fire threats in regions

[4]HBG is a California based firm that designs, manufactures, and builds modular affordable housing.

with wildfires since the water body serves as natural fuel break, an absence of fuel for the fire [10]. It is resilient to extreme temperatures because the high heat capacity of water helps regulate temperature, limiting the increase and decrease in the local ambient temperature. Floating housing will benefit from this ability and reduce the impact of changes in the regional climate resulting in extreme heat. Onboard water treatment would provide resiliency to regional droughts.

3.2 *Sustainability*

Reducing the environmental impact of housing can be a central characteristic of floating housing; it can be incorporated into the full life cycle of the solution. One example, if determined to be practical, is using repurposed existing marine vessels and other materials such as structural steel to minimize the environmental impact of new building materials.

Traditional housing construction requires transportation of materials and equipment to each site using land-based transportation such as trucking. Transportation via trucking produces higher pollution than marine transportation; therefore, eliminating the need for ground-based transportation would significantly reduce the pollution associated with housing [11].

Renewable energy and a sustainable infrastructure system can be integrated into the design. In addition to including traditional design components such as solar panels, solar water heaters, rain water harvesting, and energy efficient appliances, floating housing can also incorporate more advanced components such as heat pumps using sea water, solar shading, on-board battery system, and microgrid[5] controls to optimize energy usage. The microgrid would optimize the loads, generation, and energy storage to maximize renewable energy usage while minimizing cost of energy [12].

Efficient utilities such as on-board waste water treatment and fresh water production similar to those used on advanced cruise ships can eliminate the need to extend underground municipal infrastructure requiring significant capital investment to adapt to sea level rise [13].[6][7]

---

[5]A microgrid is a group of interconnected loads and distributed energy resources within clearly defined electrical boundaries that acts as a single controllable entity with respect to the grid. A microgrid can connect and disconnect from the grid to enable it to operate in both grid-connected or island-mode.

[6]Advanced waste water treatment process used by cruise ships provide a level of treatment similar to municipal treatment plants [15]. Smaller systems of 200,000 gallons per day (equivalent to 730 units respectably using per capita approach) can cost up to 35% more depending on the nitrogen removal than a small central wastewater treatment of 1.5 million gallons per day [16, 17].

[7]Steam desalination or reverse osmosis is used to produce potable water, in addition to cruise ships, both technologies are also used to provide water for communities with limited water supplies. Historically the cost for reverse osmosis has been high compared with traditional processes but they are becoming more prevalent as cities have no choice as their water supply shifts to

3.3 Cost Reduction and Affordability

McKinsey identified ways to reduce the cost of affordable housing by 20–50%. Increasing land availability was identified as the most effective way to reduce cost as it is the largest real-estate expense. Use of efficient procurements, value engineering (standardizing design) and industrial approaches, such as assembling buildings from prefabricated components manufactured off-site would result in construction cost reduction by 30% while also shortening the construction schedule by 40% [14].

Total cost to develop housing typically includes land cost, permits and city fees, professional services, legal fees, construction financing fees, construction labor and materials. Depending on the jurisdiction, market rate housing developments may be required to pay additional fees to support community subsidies for affordable housing. Furthermore, in the United States affordable housing developments receiving public funding in forms of grants and tax breaks, while having reduced permitting and city fees, may have higher legal and professional services fees in order to manage the complex process. Total cost for affordable housing in California is over $500 per square foot in several markets, reaching as high as $1170 in San Francisco, California [20].

The cost of land is highly variable and contributes a significant portion of the final cost, being as high as 21% in San Francisco in 2019 [21]. This is likely to increase further as the value of land used for single-family housing in the United States increased almost four times faster than inflation between 2012 and 2017, with most coastal regions in California seeing increases of 50% and above [22].

Floating housing would provide a large source of new affordable housing because the cost basis is disconnected from several of the local factors that increase housing cost such as land, labor, and materials. Floating housing eliminates land use in high cost urban areas where the cost of land is prohibitively expensive. A benefit of developing floating housing is that it can be fabricated and constructed off site, using lower cost labor than the labor at the urban location where the housing is needed. Globally sourced materials and construction labor in low-cost locations allows labor shortages in high cost markets to be addressed. Housing can be built in the lowest cost locations using repeatable modular designs and building at large scale. Modular and repeatable construction enables cheaper construction and repeatability of new locations. Disconnecting the construction of housing from

---

sources with increased saline content (brackish aquifers, seawater) and other pollutants that require more advanced treatment to meet the required quality standards [18, 19]. Sandia National Laboratories and the United States Department of Interior, Bureau of Reclamation identified cost reduction and decentralization of the potable water supply as components of the Desalination and Water Purification Technology Roadmap [19]. The increased use desalination in lieu of the traditional method and the move towards decentralized supply will mean that the cost to produce potable water for floating housing will not be higher than land-based housing.

a specific location also allows for parallel development of multiple projects, thus reducing the risk and impact of development delay.

Floating housing vessels would be built at large-scale, cost efficient, and high-quality global shipyards thus allowing the simultaneous construction of multiple vessels, which could be deployed to separate locations, or to one community consisting of multiple vessels. Shipyards produce somewhat similar vessels (e.g., cruise ships, accommodation barges), but none target the affordable housing market. Common designs unaffected by local site constraints, leveraging a global supply chain and logistics network, and maximizing module construction techniques will lead to manufacturing economies of scale and drive down costs. The cost-effective global shipping industry can then deliver the vessels to coastal locations worldwide.

By eliminating land use, maximizing modular and repeatable construction, and rapidly and flexibly deploying housing to multiple coastal markets, development risk will ultimately be minimized, and housing delivery efficiency increased. Significant savings are realized by reducing need for land and using efficient value engineering and industrial approaches. The costs will decrease further with increased production, allowing the solution to become cheaper than traditional construction in additional coastal locations worldwide. Our internal estimates and research with marine construction companies indicate that floating housing could be built at a cost significantly lower than traditional housing in most high cost markets, ranging from $200 to $320 per square foot.

4 Design and Construction

Black & Veatch specializes in innovative infrastructure development and has the unique capability to leverage experience in various markets to integrate all aspects of development, permitting, design, and construction and has pioneered large-scale industrial floating infrastructure solutions. Experience and lessons learned from previous large-scale floating infrastructure projects provide a general path forward to address design and construction challenges of floating housing.

Several unique considerations must be addressed when planning for the development of large-scale floating housing when compared to planning for land-based infrastructure:

- Minimizing the footprint of ancillary systems because of the lack of real estate.
- Minimizing the weight.
- Minimizing interconnections between building modules and the floating system to avoid complexity.
- Ensuring equipment is accessible for maintenance and egress.

These requirements will drive the planning decisions around the key design criteria such as:

- Technology selection.
- Equipment selection.
- Hull configuration.
- Marinization of design.
- Selection of shipyards and topside fabrication.
- Safety (fire and explosion).
- Regulatory impacts (local government).
- Classification society requirements.
- Mooring and associated marine infrastructure.
- Power and utilities.
- Dredging.

Requirements for space efficient design drives reductions in the system components as compared to land-based systems. Modular design and construction techniques are ideal for floating applications because these techniques ensure the fewest pieces of equipment and the smallest footprint. Additionally, less equipment results in a better spaced layout for easy safety escape routes and maintenance.

Marinization of the land-based technology and systems for a floating environment is necessary. Floating vessels are subjected to six degrees of motion: pitch, roll, yaw and heave, sway and surge. With motion comes acceleration and acceleration-induced additional forces. All systems on a floating structure must be capable of withstanding the motion-induced forces and perform safely. Marinization requires studies to examine system operability, mechanical design, safety, materials, and maintenance.

Hull and topside structures must seamlessly interface. Floating housing is a combination of the building and shipping industries; these industries are different. The topside structures are primarily a land-based building that interfaces with the hull, which is a shipping system. Generally, shipyards design the hull system but lack the capabilities to design the topside structures, if not standard marine topside structures. For a successful floating housing project, significant physical and design interfaces between the topside structures and the hull need to be managed carefully by dedicated interface teams.

The vessel's marine classification requirements must be determined early, including whether the unit will be built to classification society requirements and/or classified. Early identification is essential to the development of the basic design and approach to be taken during construction. Although classification requires a more complex and rigorous approach during design, construction, and final deployment, floating housing projects may benefit from oversight by classification societies during construction. Classification societies provide an independent review of standards, hazard analysis, and assurances that the finished project will conform to recognized marine industry practices established by each classification agency.

If the structure is to be classed as a vessel under the current standards and regulatory approach, it must be designed by a naval architect and built to relevant marine classification society rules and regulations. Initial considerations included how to involve the local regulatory agencies to gain approval for the facility since it will be classified as a vessel and not a building. Classing the floating structure as a vessel will result in the requirement that the vessel's license be renewed periodically, which requires the vessel to be inspected. For standard vessels, this involves towing to a dry dock for inspection. Since this floating structure cannot be moved easily, special arrangements must be made and planned for accordantly in the operation of the facility.

Close communications and coordination must be maintained with all the primary and secondary governmental agencies having regulatory and jurisdictional responsibility over vessels or facilities sited on the water in an urban jurisdiction. An example that is common on such projects is the intersection between the local fire department and the federal maritime agency. For example, the federal agency may develop rights of way with the local fire department to give the fire department authority to provide emergency response. The local fire department may also accept modifications to the design to comply with the rigorous fire safety code. For one project supported by Black & Veatch in the United States, the result was an approved fire suppression system that will use external water for fire suppression instead of the charged pipe required in a standard land-based building

Though further analysis is necessary to quantify the specific design criteria and maintenance requirements of the vessel, considering the local environmental conditions where the units are placed, there are examples of recent projects that show that proper design and maintenance can extend the life of vessels for up to 50 years. The floating base of the housing vessel can be either of steel or concrete construction. Steel barges are commonly used for the transportation and storage of goods, and marine construction and maintenance activities. Concrete barges are in limited usage and becoming popular for floating houses, offices, and kiosks and are moored in sheltered locations on a permanent basis. Table 1 provides a simple indicative comparison between large flat decked concrete and steel barges having the same deck area and loading capacity per square foot.

A recently completed floating Fire Station in San Francisco is a representative example of how a floating structure can be designed and maintained to extend useful life up to 50 years. The project included the following design guidance and maintenance activities for a steel barge [24]:

- Include sacrificial steel entire perimeter in contact with water. Additional sacrificial steel in the splash zone. Thickness variation on inside.
- Special long-lasting coating 100% solid polyurethane by Zebron. Used on icebreakers. Could possibly last 40–50 years. Coating applied in ship yard under controlled environment with rigorous quality control program.
- Cathodic protection system to slow the steel corrosion rate.
- Divers to check the steel barge condition of its surface coating and check for corrosion every 5 years.
- Replace anodes for the cathodic protection system and clean the surface coating every 5 years to prolong the life of the barge.

Table 1 Comparison concrete barges versus Steel barges

Attribute	Concrete barge [23]	Steel barge
Weight	Heavy compared with steel	Light compared with concrete
Draft	Larger draft due to weight. Hence, increased water depth may be required (compared with steel barges)	Shallower draft when compared with concrete barges
Movability	Large towing power required for large barges, and hence limitations apply on the towing of large concrete barges	Fully loaded large barges can be towed in the sea using tugs
Durability	Well-constructed barges are more durable than steel barges	Less durable than concrete. Durability is subject to the level of periodic maintenance
Maintenance	Very low maintenance. However, difficult to repair if needed	Periodic maintenance required. The extent of onsite repairs and maintenance activities can be limited
Constructability	Limitations regarding the construction of large barges due to the capacity of available dry/floating docks. Labor intensive construction. Quality control can be difficult	Already an established industry, hence construction is much easier than concrete barges. Can be built to very high standards
Movements (roll, yaw pitching, sway, heave & surge)	Less movement for waves & wind due to bulkiness	Expect more movement than concrete barges
Mooring	Large forces likely on the mooring restraints due to bulkiness. Hence, larger guide/mooring piles than for steel barges may be needed. There may need to replace moving parts of the moorings (rollers and guides) on a regular basis	Mooring loads are more favorable than those of concrete barges due to lightweight. Easier to modify the mooring points if needed
Berthing	Large forces may apply on the berthing structure compared with a steel barge	Relative low forces when comparing with the concrete barges
Other	Limited experience and limited suppliers. Lower initial construction costs [23]	Already established industry

5 Challenges and Barriers

Although technically feasible, notable challenges and barriers will need to be addressed for floating housing. These include regulation and permitting, desirability, community acceptance, and financing. Regulation and permitting will be a challenge in some markets as floating housing is unique to most regions and not covered under existing rules. New regulations and permitting processes will need to be created in conjunction with the local authorities in each market. The use of floating housing is not culturally accepted in some parts of the world. In the United States, for example, floating housing does not exist except for the limited use of individual houseboats in specific locations. Financing may also be a challenge as it is an esoteric asset class (real estate + marine), perceived future values are unclear, and depreciation is different than traditional real estate.

Technical challenges of the proposed self-sustainability of floating housing were assessed as less of a barrier due to the ability to use existing commercial solutions. Although the cost of implementing commercial solutions such as those used by modern cruise ships for power generation, water desalination, and wastewater treatment require future study, addressing the other barriers were prioritized higher.

To specifically address and overcome the challenges and barriers to large scale floating housing, a pilot project consisting of a small-scale housing development is proposed. The pilot project will provide the opportunity to test design assumptions, develop the necessary technologies and create templates for regulatory approvals. The performance and monitoring plan, developed in conjunction with regulatory agencies, investors, host city and other stakeholders, will improve the understanding of the remaining technology risks, benefits and impacts. Initial research also highlights the need for prospective residents to see and experience the housing solution being developed and a pilot project provides an avenue for this experience.

Potential locations for the pilot deployment include high density coastal regions with limited available land and high housing costs. Initial deployment in protected waters such as bays, harbors, deltas, etc. provides a location with lower risks. Therefore, the U.S. regions of highest priority are primarily located in urban areas such as Portland, Oregon; Seattle, Washington; San Francisco, Los Angeles, or San Diego, California; New York, New York; Miami, Florida; Boston, Massachusetts; or Washington D.C. Globally, this includes cities like Hong Kong and Singapore. Other target areas include any city with access to major waterways, located on a major river or lake, and regions exposed to rising seas and/or flooding.

6 Design Concepts

Black & Veatch developed multiple design concepts to visualize ideas for high density modular floating housing. The design concepts offer tangible material for communicating the feasibility and aesthetics of floating housing to potential residents and community planners to overcome the expected obstacles of desirability and community acceptance. Initial informal, qualitive research with community planners and architects indicates a method for greater community acceptance involves designing floating housing that fits visually into the existing land-based architecture and is a familiar structure instead of a novel or unique floating structure. With this premise, these initial concepts mimic land-based high-density housing to an extent necessary to ease acceptance. Considering that the goal of the pilot project is to study the major challenges and barriers, a simplified initial design was proposed. This includes assumptions that the pilot project will be located near public transportation, negating the need for vehicle parking, and include shore connections for utilities.

The design concepts were developed by multiple designers using a standardized set of design criteria. The primary objective was to create a housing solution that set a new standard of affordability while also being aesthetically pleasing in many locations globally. Designers were instructed to consider spatial integration (the design must aesthetically fit into the surrounding environment and offer an inviting place for people to live and not simply be the maximum use of allotted space to house as many people as possible), functionality, innovation and originality, sustainability, general aesthetics, and constructability (affordability). Each design was created using some base guidelines for high density housing covering a range of apartment sizes common in the United States. The height of the structures was limited to comply with height restrictions that exist in some coastal regions. Typical building code requirements for multifamily housing in California, United States were used to ensure each design would be in line with typical land-based designs. The design criteria included:

- Residential housing facility of 15–30 units.
- Max height of facility limited to 30 feet (assumed up to three stories are allowable).
- Individual housing units between 400 and 700 square feet.
- A floating platform (assumed 150–300-foot length and 50-foot width).
- A floating platform permanently moored to the shore or jetty with connections to land-based utilities (power, water, sewer, etc.).
- Assumed location in a harbor relatively protected from extreme weather events thus no consideration necessary for extreme ocean weather, storm surges, typhoons, etc. However, consideration of safety by design techniques that could improve the resiliency of the structure to mitigate future climate or weather risks should be considered.
- The target market is moderate income customers, most likely singles or couples.
- Include standard apartment amenities and services comparable to complexes of similar size on land.
- Primary access to the facility will be from shore via a gangway.
- Vehicular parking and access not included on the floating structure. Assumed location is accessed via public transit.
- Use of modular construction techniques to increase constructability.

- Use of California, United States building codes for multifamily manufactured and factory-built housing programs.

6.1 Design Concept 1

Design Concept 1, illustrated in Figs. 1, 2 and 3, seeks to provide sustainable efficient accommodation for the future. A decommissioned barge is utilized to provide a mixture of 1-bed (20 units, 400 square feet each) and 2-bed (8 units, 700 square feet each) apartments, 28 units in total. The presented apartment mix is indicative only and can be set as needed—the modularity of the design allows for total flexibility. Each housing unit has an external space in the form of a terrace. The main structure is created from modular units which can be prefabricated thus reducing time, cost and waste.

Design Concept 1 also provides a central common house with co-working and social areas. The common house is a central facility for the use of all inhabitants providing communal functions such as multi-purpose space as well as co-working office space or even a coffee shop and gym. The building is in the middle of the development facing a public plaza.

Energy efficiency, renewable generation and battery storage are key to the energy strategy while options for rainwater harvesting, desalination and waste water treatment provide the potential for self-sufficiency. Renewable energy and sustainable infrastructure systems are an integral part of the concept. It is estimated that the development would generate surplus energy from the solar roof that could be fed into the grid or stored in hydrogen cells for use by the wider community e.g. for public transport. Starting from a single barge, Design Concept 1 can be extended

Fig. 1 Design concept 1

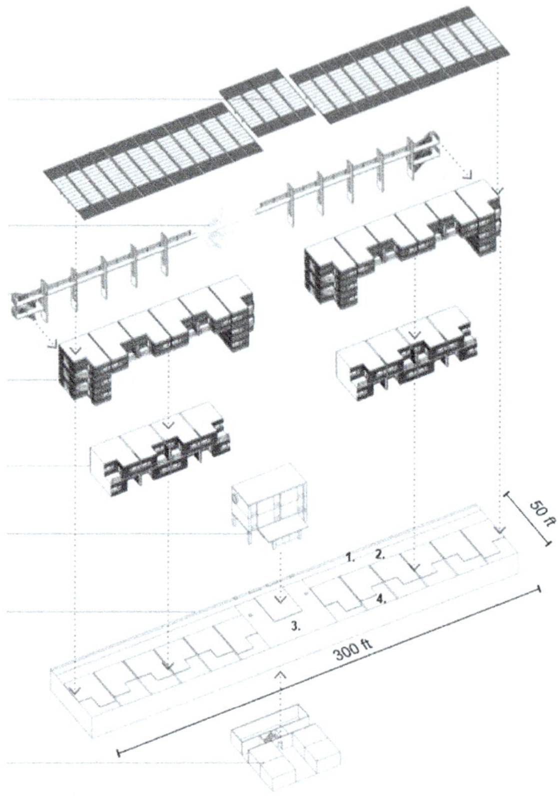

Fig. 2 Design concept 1 schematic layout

Fig. 3 Design concept 1 perspective view

into a community of vessels, incorporating options for plug and play commercial units. The sustainable strategy components could include:

- Renewable energy: PV panels and solar water heating tubes
- Rain water harvesting

- Cooling strategy from the sea/heat pump
- Solar shading: overhanging 'solar roof' also acts as louvres providing shadow for the housing units below as well as the deep reveals on facade
- High spec insulation
- Energy efficient appliances and smart energy management within housing units

There is also an opportunity to develop a self-sufficient community scenario where the services running below the housing units instead of feeding directly into the grid were linked to facilities in a central plantroom. Such solution provides additional efficiency and lack of need for connection to city-wide infrastructure systems.

6.2 *Design Concept 2*

Design Concept 2, illustrated in Figs. 4, 5 and 6, is a harborside floating development containing 30 dwelling units (1-bed, 750 square feet) on each of three standard size commercial barges (10 units per barge). The construction is all prefabricated by offsite manufacturers and transported into places via traditional shipping practices for final outfitting at its final anchorage. Costs are minimized by

Fig. 4 Design concept 2

Fig. 5 Design concept 2 schematic layout

Fig. 6 Design concept 2 perspective view

the reuse of materials and the reused barges in lieu of land development. The structure is constructed of individual modular units, or repurposed shipping containers, can have overcladding as shown or be painted to fit the local environment.

The enclave is intended to be a community onto itself with amenities such as common use spaces like the roof decks shown or additional spaces for fitness facilities, communal gardens, common use pavilions and parking for shared vehicles or rideshare services. There are also spaces provided adjacent to, or on a separate floating structure for public functions like a farmers market or coffee shops open to the general population.

As shown, the community also utilizes sustainable features with onboard power generating PV panels as cladding that also have single batteries integrated for storing power are dispersed to provide off-the-grid power to each room. The barge hull can also house larger batteries as part of a microgrid that connect to the grid or make the community entirely self-sufficient. Rainwater is captured and utilized for building systems.

This community is flexible and resilient, it can be added to or its location/orientation modified easily. It has applications as temporary housing that can be transported or added to over time.

6.3 Design Concept 3

Design Concept 3, illustrated in Figs. 7, 8 and 9, is initiated on the concept of traditional Chinese Sanheyuan, a type of dwelling structure with individual units on 3 sides and a yard in the middle. This kind of dwelling unit creates an introverted space allowing residents the ability to enjoy a peaceful time at home. To integrate this concept into the floating house design, modular units, such as containers, are arranged to form into the shape of a Sanheyuan, i.e. modules enclose a yard on 3 sides. A total of 3 Sanheyuans (Group A/B/C) are designed. Each group includes 6 or 9 standard living units stacked in 3 stories. A total of 24 units (1-bed, 639 square feet) are provided in this design. The individual units can be painted different colors or clad in different finish materials to make the entire community colorful and vibrant. Backup power will be provided by solar panels placed on the rooftop to provide backup power. To make some variations in each group, the yard is either open to sea or open to land and the yard design can be further developed based on local style preferences (e.g. Chinese, Japanese, western) to make each group unique. Within each group, connecting corridors, outdoor stairs and yards provide a communal space for the neighborhood to communicate and enjoy the beautiful sea view and sunshine together.

The three groups are connected by a covered corridor and a central public space. The public space includes a small 3 story building which can house a variety of community spaces such as a pub, coffee shop, laundry, grocery store, gym, book bar, etc. On each level, an outdoor terrace is designed to provide a space for residents to enjoy the fresh air and sunshine.

6.4 Design Concept 4

Design Concept 4, illustrated in Figs. 10, 11 and 12, uses modular living units installed within a separate free-standing structural framework on the barge. The design incorporates common exterior circulation paths for occupants, saving construction costs of numerous stairs and maintenance on excessive interior spaces. The modular living units are sized to accommodate the use of shipping containers if preferred and the design includes 2 different unit floor plans (1-bed) of 520 square feet and 640 square feet each. The larger unit can accommodate a washer/dryer, larger kitchen, desk area, additional storage closets, walk-in bedroom closet, and

Fig. 7 Design concept 3

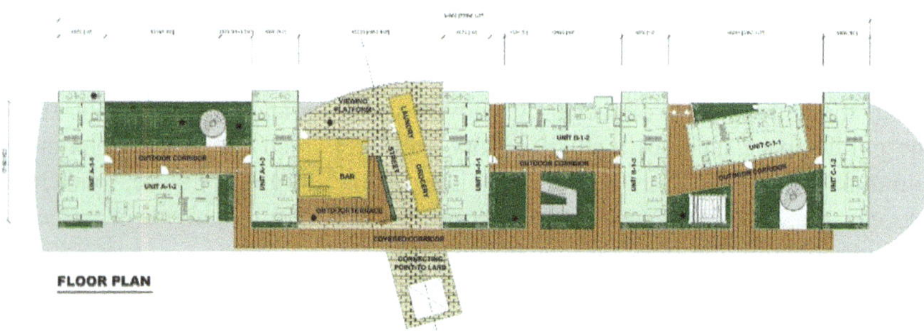

Fig. 8 Design concept 3 schematic layout

larger patio. The structural steel framework is completely independent of the condo units. The modular living units can be fabricated off site while the structural framing is installed on the barge. The modular living units are designed to be transferable to future barges for reuse in the future. The layout of the living units and structural framing can easily be scaled (larger or smaller).

One key element of this concept is the use of a pre-engineered metal rainscreen and roof spanning over the living units. This structure is intended to provide some

Fig. 9 Design concept 3 perspective views

Fig. 10 Design concept 4

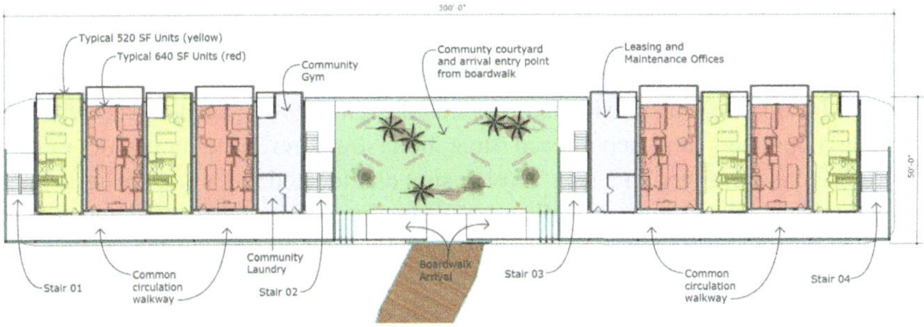

Fig. 11 Design concept 4 schematic layout

Fig. 12 Design concept 4 perspective view

weather protection for occupants on the shared outdoor circulation walkways and stairs. The additional screening structure will also provide another layer of weather protection for the living units, prolonging their life expectancy. The design of the rainscreen and the color of the modular living units can easily be changed to reflect local architectural design aesthetics anywhere in the world.

The living units are grouped in two housing clusters with a community courtyard in between. Within each cluster, a unit can be used for community functions such as a gym, laundry, or leasing/maintenance office.

6.5 Design Concept 5

Design Concept 5, illustrated in Figs. 13, 14 and 15, is an eclectic configuration of geometry using modular units atop the floating structure. To maximize a sustainable design solution, the illustrated version of this design uses salvaged and refurbished shipping containers and repurposed floatation structures. The units are unique stateroom plans arranged in a split-level design and located in habitat clusters. Each unit includes insulated interior walls to preserve the exterior corrugated appearance with minimal cosmetics. This design focuses on scalability and uses module stacking design relationships to allow for different layouts and provide minimal post-engineering considerations. The structure includes multi-level decks with a commons area to foster social integrations and the upper pedestrian promenade deck includes green spaces, community gardens, areas for social gatherings, and occupant exercise opportunities. The central circulation core in-between the habitat clusters also provides for a commons area. Renewable energy sources in the facility include artistic and sculptural styled vertical wind turbines, and photovoltaic

Fig. 13 Design concept 5

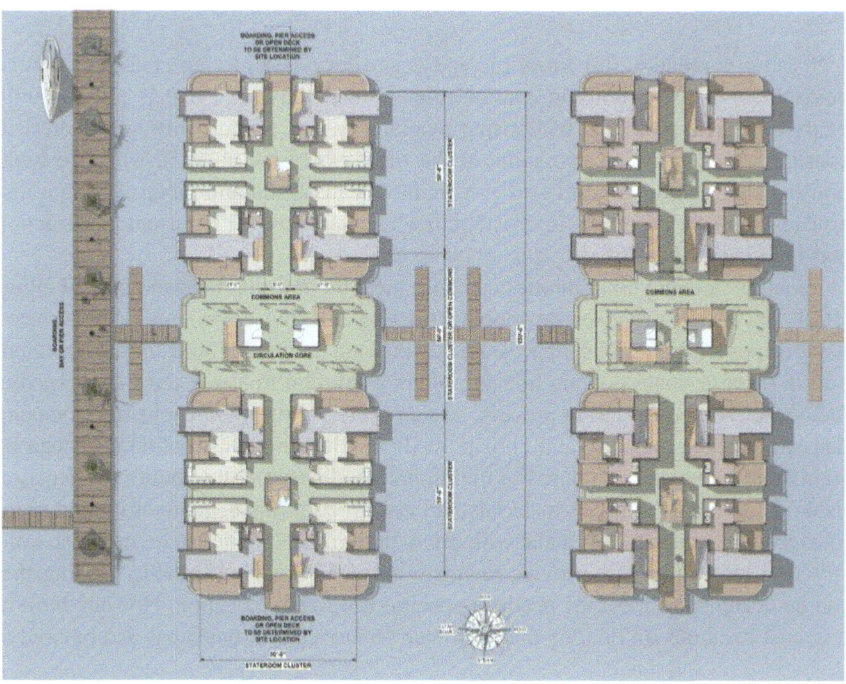

Fig. 14 Design concept 5 schematic layout

Fig. 15 Design concept 5 perspective view

assemblies integrated passively to maximize building systems efficiencies. Daylighting is optimized with glass openings, shading deck extensions and Solar Tubes.

7 Conclusion

Increasing population, urbanization, and desire for coastal living creates an annual demand for housing units in coastal cities around the world that is not currently being met. The problems of housing availability and affordability are greatest in coastal cities which are also some of the most expensive cities in which to build housing due to limited land and constraints on materials and labor. The rate that housing must be produced in coastal cities is challenged with current construction and land use practices.

Floating housing communities can help meet the housing needs in coastal cities. Black & Veatch proposes developing floating housing communities constructed using innovative modular approaches to provide affordable and climate resilient housing. Large-scale floating housing offers a solution to lower costs and expedite the design and construction process, so that more people can benefit from housing that is affordable, healthy and climate-resilient. Housing construction costs can be significantly reduced compared to traditional, land based, "stick built" construction methods. Globally sourced materials and construction labor, modularity and manufacturing economics of scale will drive down costs. Innovative modular construction techniques can provide resilient, sustainable and affordable housing that can be designed to meet the needs of specific geographic location. Housing built in low-cost locations on floating platforms can be moved to urban areas and moored at shore or in open water to provide thousands of new housing units at affordable costs. Floating housing is inherently resilient against rising seas due to climate change, seasonal flooding, earthquakes, and other natural disasters. Renewable energy and a sustainable infrastructure system can be integrated into the design with an on-board battery system, and microgrid controls to optimize energy usage.

The experience gained from developing other floating infrastructure projects can be applied to a proposed pilot floating housing project which offers an opportunity to test assumptions and overcome challenges and barriers to large scale floating housing developments.

References

1. LeRoy S, Wiles R et al (2019) High tide tax: the price to protect coastal communities from rising seas. Cent Clim Integrity Resilient Anal. Available via https://www.climatecosts2040.org/files/ClimateCosts2040_Report.pdf. Accessed 01 July 2020
2. First Street Foundation (2020) The first national flood risk assessment: defining America's growing risk. First Street Foundation. Available via https://assets.firststreet.org/uploads/2020/06/first_street_foundation__first_national_flood_risk_assessment.pdf. Accessed 01 July 2020
3. Walter C (2017) Apartment demand: national multifamily housing council vision for 2030. In: Paper presented at the National Multifamily Housing Council OPTECH conference & exposition, Las Vegas, NV, 25–27 Oct 2017
4. Aurand A et al (2019) The gap: a shortage of affordable homes. National Low Income Housing Coalition. Available via https://reports.nlihc.org/sites/default/files/gap/Gap-Report_2019.pdf. Accessed 01 July 2020
5. Titus JG et al (2009) Coastal sensitivity to sea-level rise: a focus on the mid-atlantic region. U. S. climate change science program synthesis and assessment product 4.1. U.S. Climate Change Science Program. Available via https://www.globalchange.gov/sites/globalchange/files/sap4-1-final-report-all.pdf. Accessed 01 July 2020
6. Council for Community and Economic Research (C2ER) (2020) The council for community and economic research. Available via http://coli.org/quarter-1-2019-cost-of-living-index-released-3/. Accessed 01 July 2020
7. California Department of Housing and Community Development (2020) Regional housing needs allocation and housing elements. Available via https://www.hcd.ca.gov/community-development/housing-element/index.shtml. Accessed 20 Aug 2020
8. Bretz L (2017) Climate change and homes: who would lose the most to a rising tide? Zillow. Available via https://www.zillow.com/research/climate-change-underwater-homes-2-16928/. Accessed 01 July 2020
9. Kulp SA, Strauss BH (2019) New elevation data triple estimates of global vulnerability to sea-level rise and coastal flooding. Nat Commun 10:4844. Available via https://www.nature.com/articles/s41467-019-12808-z#citeas. Accessed 01 Sept 2020
10. Green LR (1977) Fuelbreaks and other fuel modification for wildland fire control. U.S. Department of Agriculture, Forest Service, Agricultural Handbook No. 499
11. Hecht J (1997) The environmental effects of freight. Organisation for Economic Co-operation and Development. Available via http://www.oecd.org/environment/envtrade/2386636.pdf. Accessed 01 July 2020
12. Garcia VY, Dufo-Lopez R, Bernal-Agustin J (2019) Energy management in microgrids with renewable energy sources: a literature review. Dig J Mol Med. https://doi.org/10.3390/app9183854

13. Azevedo de Almeida B, Mostafavi A (2016) Resilience of infrastructure systems to sea-level rise in coastal areas: impacts, adaption measures, and implementation challenges. Dig J Mol Med. https://doi.org/10.3390/su8111115
14. Woetzel J, Ram S, Mischke J, Garemo N, Sankhe S (2014) A blueprint for addressing the global affordable housing challenge. McKinsey Global Institute, MicKinsey & Company. Available via https://www.mckinsey.com/~/media/McKinsey/Featured%20Insights/Urbanization/Tackling%20the%20worlds%20affordable%20housing%20challenge/MGI_Affordable_housing_Full%20Report_October%202014.pdf. Accessed 10 Sept 2020
15. King County Department of Natural Resources and Parks Wastewater Treatment Division (2007) Cruise Ship Wastewater Management Report. Available via https://your.kingcounty.gov/dnrp/library/wastewater/csi/0708-CruiseShipWMreport.pdf. Accessed 09 Sept 2020
16. Benefield LA (2002) Rule development committee issue research report draft: residential flow rates. Washington State Department of Health wastewater management program. Available via https://www.doh.wa.gov/Portals/1/Documents/Pubs/337-103.pdf. Accessed 10 Sept 2020
17. Barnstable County Wastewater Cost Task Force (2010) Comparison of costs for wastewater management systems applicable to cape cod: guidance to cape cod towns undertaking comprehensive wastewater management planning. Available via https://sp.barnstablecounty.org/ccc/public/Documents/208%20Draft/Chapter%203%20Appendices/Appendix%203B.pdf. Accessed 09 Sept 2020
18. Luthy RG, Wolfand JM, Bradshaw JL (2020) Urban water revolution: sustainable water futures for California Cities. J Environ Eng 146(7):04020065
19. U.S. Bureau of Reclamation Sandia National Laboratories (2003) Desalination and water purification technology roadmap: a report of the executive committee. Desalination & Water Purification Research & Development Program Report #95. Available via https://www.usbr.gov/research/dwpr/reportpdfs/report095.pdf. Accessed 05 Sept 2020
20. Reid C (2020) The costs of affordable housing production: insights for California's 9% low-income housing tax credit program. Terner Center for Housing Innovation. Available via http://ternercenter.berkeley.edu/uploads/LIHTC_Construction_Costs_2020.pdf. Accessed 08 Sept 2020
21. Li R (2019) SF is one of the most expensive places in the world to build housing. Here's why. San Francisco Chronical. Available via https://www.sfchronicle.com/bayarea/article/SF-is-one-of-the-most-expensive-places-in-the-14888205.php. Accessed 07 Sept 2020
22. Joint Center for Housing Studies of Harvard University (2019) The State of the Nation's Housing 2019. Available via https://www.jchs.harvard.edu/sites/default/files/Harvard_JCHS_State_of_the_Nations_Housing_2019.pdf. Accessed 06 Sept 2020
23. Burnside OH, Pomerening DJ (1984) Survey of experience using reinforced concrete in floating marine structures. Ship Structure Committee report SSC-321. Available via http://www.shipstructure.org/pdf/321.pdf. Accessed 07 Sept 2020
24. San Francisco Public Works (2016) San Francisco fire department: earthquake safety and emergency response (ESER) 2014 Fire Station 35-Boathouse Presentation to Port Commission, 12 July 2016. Available via https://www.sfearthquakesafety.org/uploads/1/9/4/3/19432507/2016-0712_presentation_for_port_final__edit_.pdf. Accessed 05 Sept 2020

The Design, Construction and Evaluation of a Pilot Project of a Bahay Kubo Inspired Floating Home

Pieter H. Ham

Abstract In the overpopulated deltas of the Philippines people live in are-as that see floods regularly. The floods are being caused by a com-bination of tides, heavy rainfall and land subsidence. The demand for safe and affordable housing is immense, yet available dry land is scarce. By implementing floating homes in vacant former rice fields, demanded new building space is becoming available. To come to a sustainable design that fits in the Pampanga Delta, traditional building designs as the Bahay Kubo have been analysed. Many aspects of this design correspond with modern sustainable development goals. By means of parametric building simulations, key aspects of the Bahay Kubo have been used to provide the home with good performances in indoor climate and structural behavior. Now the first pilot building has been built, the home is being tested for validating the parametric models and to evaluate the building design. The first round of test results has led to proper insights in indoor climate, user friendliness, and affordability. Initial design improvements have been made and will be used in upcoming developments such as the construction of a floating neighborhood and the construction of floating classrooms.

Keywords Floating architecture · Flood events · Vernacular · Pilot building · Affordability

1 The Pampanga Delta

The situation in coastal areas of the Pampanga Delta in the northern part of the Manila Bay Area in the Philippines resembles many urbanized areas of Southeast Asia. Analysing the housing situation, and generating and evaluating implementable designs of housing types can thus be translated to other regions that are dealing with similar problems [1].

P. H. Ham (✉)
Delft University of Technology, Stevinweg 1, 2628 CN Delft, The Netherlands
e-mail: p.h.ham@tudelft.nl

Ł . Piątek et al. (eds.), *WCFS2020*, Lecture Notes in Civil Engineering 158,
https://doi.org/10.1007/978-981-16-2256-4_4

The housing situation in the provinces Pampanga and Bulacan is highly affected by floods which are mainly the result of incoming seawater by tidal movement and by overflow of rivers after heavy rainfall. A worsening trend in these floods is detected. Worsening trends are caused by a combination of channel filling by sediments from floods and lahars, deforestation, rapid urbanisation, channel encroachment and sea level rise. Moreover, nowadays the main impact on these worsening trends is ground subsidence [2, 3]. Groundwater overuse causes land around northern Manila Bay to subside with a subsidence rate of around 5 cm per year. This land sinks relatively fast since extraction rates are higher than recharge rates. Rapid urbanisation has a big impact since it causes an increasing demand for groundwater. Land subsidence occurs all over large urbanized parts in Southeast Asia (Fig. 1) [4].

Where growing rice on rice fields was once one of the biggest sources of income, it is nowadays no longer possible due to salinization and high water levels. For this reason, over the last decades many rice fields have been transformed into fish ponds. Due to the increase in flood events, fish ponds overflow regularly. As a consequence, many of the fish ponds have been abandoned, and thereby large flooded areas currently don't have any spatial function anymore [1] (Fig. 2).

1.1 Housing Situation in the Pampanga Delta

Another phenomenon that affects the housing situation in Bulacan and Pampanga is rapid urbanization. In the following 30 years, more than 60% of the world's population is expected be found in cities and urban areas. This fast and often unplanned development results in serious problems due to scarcity of land, congestion and poverty. The current deficits in housing are in the low cost sectors and the housing

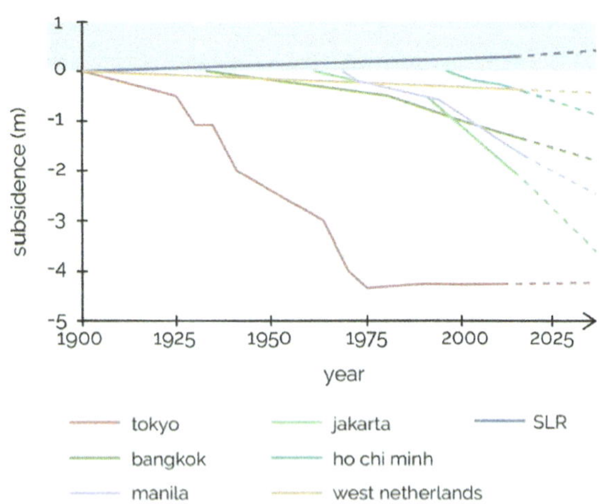

Fig. 1 Ground subsidence in Southeast Asian cities [4] (adjusted by author)

Fig. 2 Flooded streets and ground subsidence in the Pampanga delta

backlog has been estimated to will have reached over 10 million units in the Philippines in 2030 [5]. This rapid urbanization not only occurs in Metro Manilla. Also smaller settlements in the Manila Bay Area, like the previous described coastal areas, face this problem.

Figure 3 shows the municipality of Hagonoy in the province of Bulacan. Settlements are developed along the riverside. The river was of high importance for daily life of people, since it was used for transportation, fishing industry and sewage system. People settled in family owned compounds along those rivers. Over multiple generations this family-owned land has been divided over multiple family members, who built their houses on these plots. Due to this rapid urbanization, these areas are now fully occupied with buildings. Expansions to areas further away from the city are not possible (anymore), since these areas are permanently exposed to flooding [1].

A negative side effect is that squatters settle themselves in self-built structures above the river. These informal settlements obstruct the waterways, which results in an increase of the flood risk. Furthermore these provisionally built structures are often highly vulnerable for typhoons. Due to an increase in intensity of typhoons in the near future these structures will become even more insecure to inhabit [6].

Fig. 3 Hagonoy (Bulacan). Settlements along the riverside

This combination of scarcity of buildable land, a relatively large housing backlog, an increase in flood events and an increase in typhoon intensities, creates an enormous demand for safe, sustainable and affordable housing.

2 Analyses of Vernacular Architecture

An answer to developing these safe, sustainable and affordable houses that are suitable for the Pampanga Delta area can be found by analysing vernacular building types. Vernacular buildings are ancient buildings which are not designed or built by formally-schooled architects and engineers. They rely on designs, skills, craftsmanship and traditions of local builders [1].

2.1 The Bahay Kubo

The Bahay Kubo (which literally means cube house) is a Philippine building type of which its architectural principles were created in the pre-colonial era. This ancient building type provides a practical template for designing sustainable, climate-conscious, energy-efficient houses and buildings. As most vernacular architecture in the Southeast Asian regions, the Bahay Kubo is built from low-cost, readily and locally available materials—in this case, mostly bamboo and nipa leaves. The Bahay Kubo is specifically designed for a tropical climate and this is reflected in multiple design principles like a tall pitched roof, large roof overhangs, wooden stilts and large openings in the facades [1].

Nowadays the Bahay Kubo has disappeared from the streets and a more foreign Western-style architecture model has been adopted. These homes mainly consist of a combination of concrete and corrugated steel plates. In contrast to the Bahay Kubo, these contemporary buildings need air conditioning to create a comfortable indoor climate, which entails relatively high costs for homeowners. Because of these relatively high energy costs and growing environmental awareness, many of today's designers and engineers re-examine the design principles of the Bahay Kubo and its approach to the contemporary design challenges in a simple way.

2.2 Parametric Analyses

Main vernacular design-strategies of the Bahay Kubo are a hipped roof, large roof overhangs, (wooden) stilts, and large openings in the facades [7]. The common goals of these design strategies are to create a comfortable indoor environmental quality by enhancing natural ventilation and by protecting the building from direct sunlight. By means of parametric analyses of these four design principles, four main 'lessons learnt from the Bahay Kubo' have been created.

First, it is essential to create sufficient openings in the facades to increase indoor thermal comfort by an increase of air velocity. By creating large openings in the façade from the ground floor up to 0.85 times the façade-height, a significant increase in thermal comfort level can be gained [7].

Secondly, by creating large overhangs an increase in comfort level can be gained. Large overhangs or eaves influence thermal comfort by creating an increase in air velocity and a decrease in radiant temperature. Eave lengths of 0.6 times the floor height under an angle of 15° are giving a proper protection against direct sunlight and thus a significant increase in thermal comfort level [7].

Thirdly, placing the building on stilts influences the thermal comfort level in a positive way. By increasing in stilt height, the wind speed for natural ventilation increases with an exponential decay. An optimum of the height of the stilts mainly depends on costs and local flood levels [7].

Finally, a hipped shaped roof influences the comfort level positively. By increasing the angle of the roof, an increase in comfort level can be gained, since it allows heat to dissipate and it lowers the area subjected to solar radiation. As well by adding openings in the roof, buoyancy driven ventilation can be enhanced. Furthermore, a hipped roof improves the water drainage. An angle close to 30° is advised to decrease wind pressures during strong winds [7] (Fig. 4).

3 The Design and Construction of the Pilot Building

These lessons from analyzing the Baha Kubo need to be translated to current situations in order to come to a buildable design. In this paragraph these translations and the design of the pilot building are described.

3.1 Pilot Building Design

First, building on stilts has been a proper solution for enhancing natural ventilation and for protection against floods. However, due to increasing land subsidence, and thus increased and more severe flood events, stilts become less advantageous. Furthermore, available land in the coastal areas of Bulacan and Pampanga to build the demanded new homes is nowadays scarce. A solution for this lack of available space may come from the previously described vacant former fish ponds and rice fields. By implementing floating structures or floating homes in these areas, there will be new space available. Another side effect is that people do not have to move to places far away from family or work, since these vacant fish ponds and former rice fields are located close to currently built settlements. Furthermore, these

Fig. 4 Vernacular design principles of the Bahay Kubo

floating houses can provide a dry and safe place for living without being vulnerable for daily floods [1].

By constructing locally prefabricated modules of locally available renewable materials and recyclable materials a floating foundation can be built. By prefabrication, the production can take place in controlled and dry climate conditions, without the impact of floods. This has a positive influence on the quality of the building and can reduce building costs [8]. A number of parameters are important for the design of the floating foundation. Due to the lack of heavy machinery and a construction site that is relatively hard to access, it is important that the floating foundation consists of light weight building modules that can be carried and transported by hand. By simply connecting these modules on the water, a stiff and stable floating platform can be created. The intended material must be locally available and, of course, as sustainable and affordable as possible. By making use local, renewable and recyclable products, both the CO_2 emission and the costs can be kept as low as possible. The floating foundation consists of nine timber modules filled with recycled plastic barrels. By predicting the depth, and tilting behaviour, it is possible to create a structure in a way that the timber elements in every load case remain above the water level. Upon this floating foundation, prefabricated panels can be installed and can be connected to each other. Like the lessons learnt from the Bahay Kubo these panels can be opened to create a naturally ventilated indoor climate [1] (Fig. 5).

On top of the wall panels, a prefabricated hipped roof can be placed. The shape of the roof can be copied directly from the Bahay Kubo. However, large eaves will form a weak part of the structure during strong winds. Typhoon resiliency becomes more important, since intensities of typhoons are increasing. Openings in the roof enhance natural ventilation and natural light. In case of typhoons they can be closed. Eaves that protect the building from direct sunlight and from monsoon rains can be folded inwards. Since the building is totally made of locally produced and locally available prefabricated building parts, it can easily be repaired with new elements in case of damages by typhoons.

A first pilot home of this design has been built in Macabebe (Pampanga) on a vacant former rice field (Figs. 6 and 7).

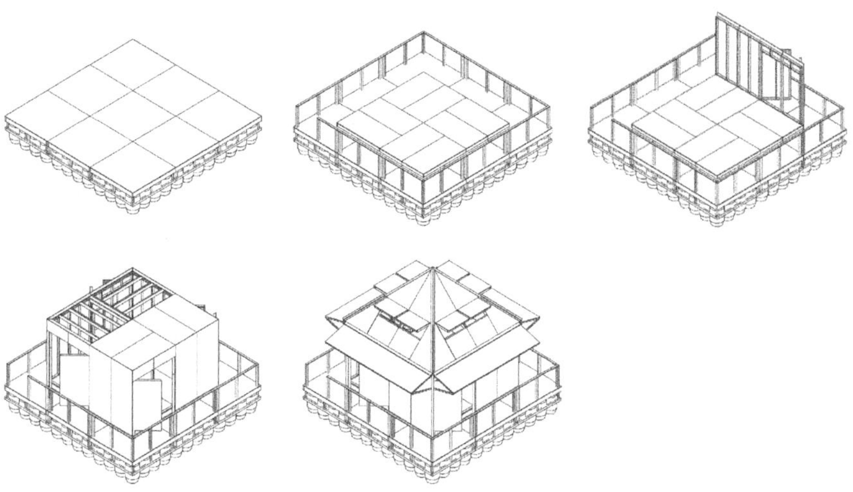

Fig. 5 Design of the locally prefabricated floating pilot home

Fig. 6 Floating foundation modules

Fig. 7 The pilot home on a former rice field in Macabebe (Pampanga)

4 Evaluation of the Pilot Building

Within this pilot project, the floating home is evaluated in multiple ways. The technical performance is evaluated by monitoring the indoor environmental quality. The user experience has been evaluated by multiple families who have lived in the pilot house.

In this paper, key lessons that follow from the evaluation of the pilot building are elaborated and potential design improvements are generated.

4.1 Evaluation of Indoor Environmental Quality

The main parameters that determine the indoor environmental quality are thermal quality (comfort), air quality, sound quality, lighting quality, ergonomics and cleanliness. As the house is located in a tropical climate and the main goal is to provide a comfortable indoor quality without using artificial air-conditioning, thermal comfort is given priority in this study.

To evaluate the thermal comfort level, it has been measured how the indoor temperature relates to a certain thermal comfort range. For this comfort range, the ASHRAE adaptive comfort model for the Manila region is used. The temperature is considered as acceptable when it is between a lower comfort limit of 22.3 °C and an upper comfort limit of 30.5 °C. By also taking natural ventilation into account, comfort limits increase with a number that depends on the wind speed [9].

To monitor the thermal comfort level, the following parameters are measured:

- Dry-bulb temperature measured with HOBO U12-012 data logger and IoT Monitoring Devices
- Operative temperature measured with HOBO U12-012 data logger and IoT Monitoring Devices
- Surface temperature measured with Thermochron I-button
- Relative humidity measured with HOBO U12-012 data logger and IoT Monitoring Devices Air velocity measured with Extech SDL350 hotwire anemometer.

The measured results show that the comfort performance of the building is satisfactory for most of the time, except for the attic floor. During the hottest days the attic floor exceeds the upper comfort level with relatively high numbers. From the in situ measurements it can also be seen that the air velocity is relatively low in the attic compared to the other spaces in the house. Figure 8 shows the dry bulb temperature in multiple zones in the home. As well the upper comfort limit for each zone is shown [9].

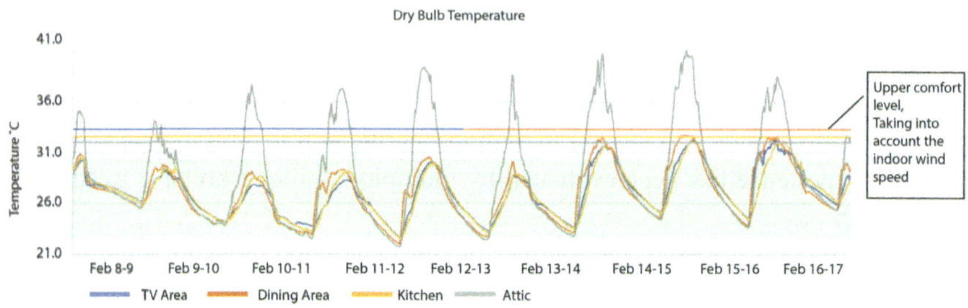

Fig. 8 Dry bulb temperature in each zone of the pilot home [9]

4.2 User Experience

Besides the indoor environmental quality, the user experience is an important aspect for evaluating the floating pilot home. A first test round in which a family of 4 people has inhabited the floating home for a time period of 5 months has been completed. Through semi structured interviews, with open ended questions main topics of the benefits and main topics of improvements of this floating home have emerged.

The first residents of the floating home stated that they prefer living in this home over living in their current house. The main reason for this statement is that unlike their own home situation, this home stays dry during flood events. The interviews also substantiate the comfortable indoor temperature, since the residents describe the indoor climate as cool and comfortable with a pleasant airflow through the building. Other benefits of the floating pilot home are mainly in terms of ambiance, view and material use.

In addition to these advantages, there are also a number of points for improvements. The first residents have been interviewed about how they experienced possible motions of the floating home. Residents indicate that motions can be felt with two different causes. The first of these is wind induced movements. Strong tropical storms cause roll, pitch and sway motions. Sway motions have been limited by increasing the stiffness of the bamboo mooring system, by adding bamboo diagonals.

Secondly, movements of people on the floating foundation cause roll and pitch motions. The resident indicates that both people induced and wind induced movements were noticeable, but that they never made the resident feel unsafe. For further research, it is advised to monitor the motions of the floating home.

Other points for improvements are the design of the sleeping zone in the attic and in the design of the openings.

4.3 *Lessons Learnt for Further Developments*

From the results of monitoring the thermal comfort of the pilot house it can be stated that the comfort performance of the attic space needs to be improved. To achieve this, a number of design improvements have been proposed. Climate models show that with these measures a desired temperature can be achieved in every room in the home.

Solar heat gain can be reduced by lowering the attic to the height of the ground floor door frame and hence the attic is protected by the foldable eaves.

The air velocity in the attic can be increased by adding extra openings at the bottom level of the attic zone.

Heat dissipation can be improved by providing openings at the top of the roof.

Air velocity can be better controlled by providing operable louvers incorporating principles of so-called Dutch door and collapsible doors [9].

For future projects, it is advisable to build configurations of multiple interlinked floating homes. The floating platform hereby widens, which increases the stability of the floating object. The house will be less sensitive to wind and movements of people. Since the foundation consists of a modular system, it is relatively simple to expand the floating foundation without changing the design.

It is advised to include these lessons in the development of upcoming floating projects such as envisioned floating neighborhoods and floating schools (Figs. 9, 10 and 11).

Fig. 9 Proposed design improvements [9]

Fig. 10 Floating classroom design

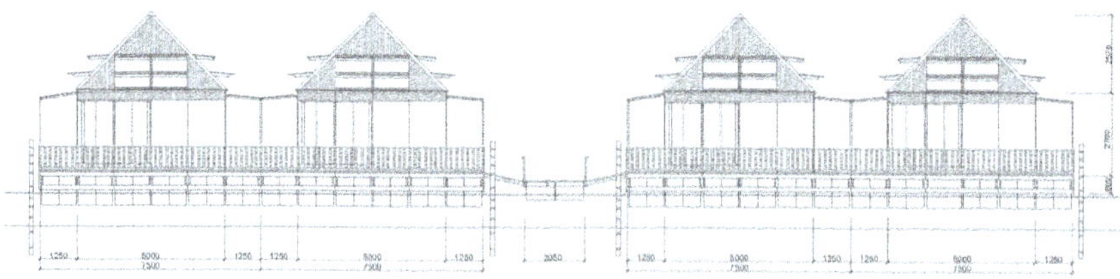

Fig. 11 Configuration of connected floating homes

5 Conclusions

Analysing vernacular design principles offers a rich repertoire of architectural and engineering knowledge in the field of design, innovations, and low-tech techniques. It can be an useful tool to use these old-age low-tech techniques for contemporary design challenges. By translating these vernacular principles to design, energy efficient buildings can be created, that are perfectly suited for their location. By testing this design with a pilot project, these vernacular design strategies are evaluated. In the pilot home a comfortable indoor climate has been created, however there is room for improvement in the attic floor. For following projects it is advisable to place more openings in the facades and to create openings in a way that heat can dissipate from the rooftop.

Connecting several floating homes together makes the floating platform more stable and less susceptible to wind and movements of people.

By incorporating the lessons from the pilot project in upcoming floating developments, a better contribution can be made to improving living conditions for residents of flood prone areas such as the Pampanga delta.

References

1. Ham P (2018) Vernacular architecture inspires flood-proof housing. TU Delft Deltalinks
2. Nippon Koei Co., Inc. (2001) Monitoring and planning of flood control works on the Pasac Delta (including. Unpublished report submitted to Pinatubo Hazard Urgent Mitigation Project, DPWH, Government of the Philippines, Quezon City
3. Rodolfo KS, Siringan F (2006) Global sea-level rise is recognised, but flooding from anthropogenic land subsidence is ignored around Northern Manila Bay, Philippines. Disasters:118–139
4. Deltares—Taskforce Subsidence (2015) Sinking cities, an itegrated approach towards solutions. Retrieved March 2016, from Deltares: www.deltares.nl/app/uploads/2015/09/Sinking-cities.pdf
5. Villegas BM (2017) inquirer.net. Retrieved 9 Feb 2018, from http://opinion.inquirer.net/108054/meeting-backlog-housing
6. Mei W, Xie S (2016) Intensification of landfalling typhoons over northwest Pacific since the late 1970s. Nat Geosci 9:753–757
7. Spittka R (2019) Passive design guidelines for Philippine housing based on the Bahay Kubo. TU Delft, Delft
8. Badir Y, Kadir MR (2002) Industrialized building systems construction in Malaysia. J Architect Eng:19–23
9. Benoy R (2020) Floating house, The Philippines, recommendations for improvement of the building design by evaluating the indoor environmental quality of the pilot floating house project. Delft University of Technology, Delft

Floating Architecture and Conversion of Offshore Structures: A Chronicle of Knud E. Hansen Designs

Carmelo Cascino and Francesca Arini

Abstract Alteration of the physical environment, and the scarcity of land for urban growth in proximity of the coast as well as rising sea levels are endangering the prosperity of entire communities around the world. These represent some of the major challenges our society will be facing in the near future. However, water can also provide alternative scenarios for human habitability. In this regard, floating structures appear as a realistic solution to adapt to the ongoing transformations. Although projects on water, ranging from a single building to entire neighborhoods, have increased significantly in the last decade, thanks to technological advancement, visions of water civilizations have been around since approximately 1950, when research in this field was undertaken for military and economic purposes. Therefore, interest in buoyant structures has appeared from time to time. It was during the 80s that the Danish ship design firm Knud E. Hansen was involved by a visionary entrepreneur in a project of conversion of an offshore platform into a luxury resort. This experience emphasized the multidisciplinary nature of floating structures in a hybrid of naval engineering and civil architecture.

Keywords Buoyant structures · Naval design · Sustainability · Conversion · Hospitality

C. Cascino
University of Genoa, Genoa, Italy

C. Cascino (✉) · F. Arini
Knud E. Hansen, 67 Akti Miaouli Str, 11537 Pireas, Greece
e-mail: cca@knudehansen.com

F. Arini
e-mail: fra@knudehansen.com

Ł. Piątek et al. (eds.), *WCFS2020*, Lecture Notes in Civil Engineering 158,
https://doi.org/10.1007/978-981-16-2256-4_5

1 Introduction

Knud E. Hansen is a Danish ship design company located in Elsinore, in the North East of Zealand island. Founded in 1937, Knud E. Hansen has been providing design and consultancy services for nearly a century. More than 750 vessels have been built based on their projects, without counting the large number of conversions, hull lines development and feasibility studies. Knud E. Hansen has contributed to the history of Naval Architecture with significant innovations and pioneering concepts.

Floating Architecture is a rapidly growing industry. As a discipline, it requires the contribution of different expertise, from Urban Design to Civil and Naval Architecture, operating at different scales: the city and the building. The latter condenses all the typical functions of land-based constructions with the physical qualities of ships such as buoyancy, stability, resistance to marine elements etc.

Furthermore, within the spectrum of floating buildings, the work of Knud E. Hansen includes a focus on the area of hospitality on water. In the 80s, the Danish company developed several projects for hotel vessels, which even today are still relevant and innovative, deserving to be spread and celebrated. The present study intends also to highlight the importance of creating an anthology of Floating Architecture; an organized and exhaustive collection of the most relevant works, innovations and authors.

Knud E. Hansen activity in this intriguing field has touched, both directly and indirectly, sensitive subjects such as the question of the environmental impact and the decommissioning procedures for obsolete offshore platforms. The strategy proposed concerns the 'extension of their career' by assigning a new function in order to enhance structural qualities and materials and reduce the need of new-builds. The concept was applied through conversion of the existing, and the contextual integration of the new function. Knud E. Hansen adopted this principle to design its Hotel Vessel (KEH 87075), described within this work.

However, history shows a number of token attempts in this direction, that provide the background scene. For instance, the unusual attempts to establish 'micro-nations', in truth never recognized, employing offshore structures, such as the Principality of Sealand (Fig. 1a, b) in the North Sea declaring independence from the UK in 1967 [1, 2], and the Republic of Rose Island (Fig. 1c, d, e) in the Adriatic Sea created in 1968 [1, 3]. Similarly, in 1975, on the occasion of the Okinawa Ocean Expo, the Aquapolis was launched (Fig. 1f, g), a 10,000 m^2 floating city designed by the Japanese architect Kiyonori Kikutake, standing on a crossover between an oil rig and an aircraft carrier [4–6]. Operating within the 'Metabolism Architecture' movement, the architect intended to explore the idea of metropolis in the ocean, toward realization.

Converting disused drilling rigs still represents a hot topic to investigate and experiment within new forms of architecture. Interesting visuals produced by XTU Architects, describe a post-oil era, where rigs are transformed into habitable platforms with vegetation, waterfalls, renewable energy sources etc. (Fig. 1h).

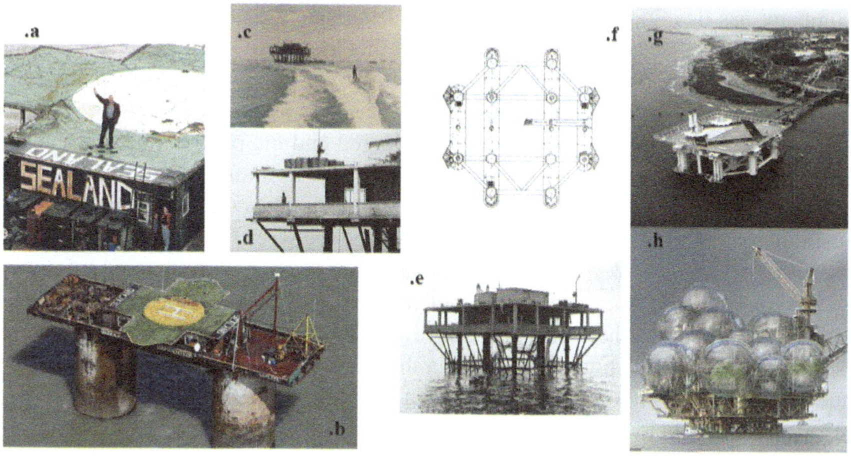

Fig. 1 Collection of offshore structures: **a** principality of Sealand, world's smallest nation (Reproduced from dailymail.co.uk et al. 2017); **b** Sealand (Reproduced from BBC et al. 2015); **c** Isola delle Rose (Reproduced from ioerimini.com 2016); **d**, **e** platform built by Giorgio Rosa (Reproduced from Domusweb 2020); **f** Aquapolis Lower Hull Plan (Kiyonori Kikutake Japan 1975); **g** Aquapolis, Okinawa Ocean Expo 1975 (Reproduced from worldfairphotos.com); **h** post-oil era, reuse of drilling rigs (XTU Architects)

2 Buoyant Accommodations from Past to Present, Differentiation, and Main Features

The study analyzes a brief selection of floating accommodation for commercial purpose that has appeared in the market in the last forty years. The aim is to comprehend the evolution that has occurred in this field and keep track of the most recent advances and designs. It also represents the opportunity to position Knud E. Hansen's work to an ideal timeline and find elements of comparison with other projects.

Surprisingly, a large variety of constructions and technical solutions have emerged. Proposals for floating hotels; accommodation for offshore workers and student villages; the type of foundations (barge, pontoons, concrete base, etc.); the environment of employment (onshore, offshore, North Sea etc.); the implementation (new build, conversion); materials and dimensions. The sustainability of these various projects is a further element of differentiation.

2.1 *Four Seasons Great Barrier Reef Resort, the Pioneer of Hospitality on Water*

As mentioned previously, the present work aims to bring to light Knud E. Hansen designs and expertise in the area of Floating Architecture. During the 80s, the company has developed several proposals of buoyant structures, and among them luxury resorts. Perhaps it is not a coincidence that the world's first floating hotel, the 'Four Seasons Great Barrier Reef Resort', in Townsville, was launched in 1988. This is a decade that has seen a growing interest for floating constructions. A wide spectrum of business opportunities was envisaged and the potential of water as the bases of coastal cities in the future. Knud E. Hansen has a long record of innovative and revolutionary projects. In this regard, three hotel vessels coming from the Elsinore archives of Knud E. Hansen produced between 1986 and 1988 rate a mention. At a brief look, they shared similarities in terms of space organization and architectural forms. They were designed in the same period and according to the main criteria in use. The fact that the historical framework strongly influences the boundaries of design, should also explain the analogies between the works of KEH and Doug Tarca, who was instrumental in the development of the Australia reef resort described below.

Tarca was a visionary man and the town developer of Townsville, in the northeastern coast of Australia. He believed permanent accommodation on the reef would be a revolutionary attempt to attract tourism on the area, and a way to enhance a unique environment inscribed in the World Heritage List from 1981. Tarca's idea was to create a sustainable floating building in a sensitive area, that fulfilled all necessary requirements for the preservation of the reef marine park. The building was placed inside a circular reef protected from damaging ocean waves (Fig. 2a).

Overall, the construction was composed of seven storeys above the main deck. Foundations consisted of a barge type hull, 90 m long, with an oil rig-style anchoring system, and without a propulsion system. The five-star hotel offered about 200 guest rooms conceived like cruise ship cabins, and many attractions such as a 100 seat theater, conference center, library, restaurants and bars, sauna, gym and floating tennis court (Fig. 2b), 50-seat underwater observatory, nightclub, and disco. Guests could also enjoy diving, fishing, and other underwater exploration direct from the hotel. The floatel was accessible by both helicopter and taxi boat service (Fig. 2c, d) [7].

Below the main deck, was the "machinery space" of the hotel, with engineering systems such as desalination plants, air-conditioning systems, diesel generators, etc. organized along a corridor running on the center line, from side to side of the hull. Furthermore, the resort introduced expedients to increase its environmental qualities and fulfill all requirements to operate in a very sensitive ecosystem. For instance, the hull was treated with nontoxic paint; sewage and other liquids were processed on board before being discharged; trash was incinerated and taken to the mainland. The mooring system allowed the platform to move on a circular path reducing possible damaging effects of shading the surrounding area under the hotel. In addition, the structure was equipped with its own marine lab, monitoring the overall impact on the ecosystem [8, 9].

The reef resort revealed the advantages of buoyant structures, in certain areas, in terms of landscape preservation rather than land-based constructions. It overcame the concept of permanency of buildings in favor of moveable accommodation on

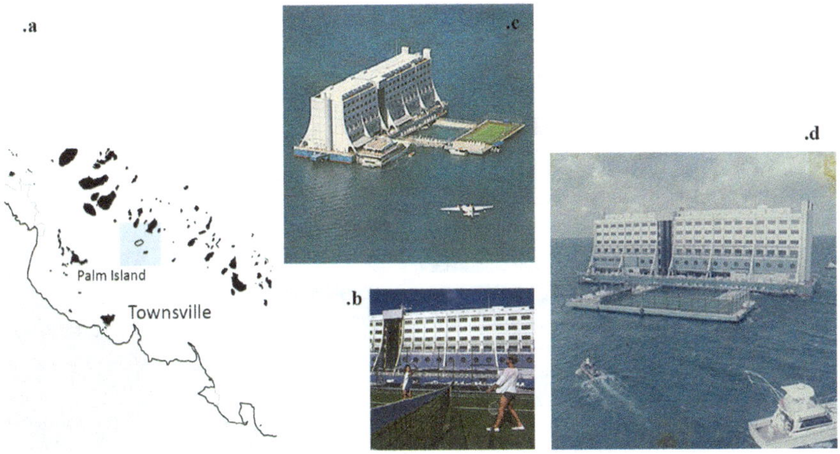

Fig. 2 The four seasons great barrier reef resort: **a** John Brewer Reef on map; **b** resort facilities (Reproduced from ABC Science 2018); **c** John Brewer Floating hotel 1988 (Reproduced from Design.tel); **d** view of the resort in operation (Reproduced from Traveller.com 2018)

water; in other words, 'adaptation' to the environment rather than transformation, which is a basic principle of Floating Architecture.

The 'Four Seasons Great Barrier Reef Resort' of Townsville was sold couple of years after opening due to financial issues, and towed away to another country.

2.2 *Further Attempts*

In 1988, the Australia reef resort introduced a hybrid architecture, deeply related to maritime design, and promoted the concept of 'paving the waves.' It acted as an incubator of innovations of marine design and technologies, environmental standards and policy, that was essential to build up the necessary know-how to implement further constructions.

Floating accommodations offer a different client experience and an alternative to traditional hotels in coastal cities. The mooring location can be decisive for customer appreciation, and consequently the business success. For instance, the NDSM Wharf in Amsterdam is a revitalised industrial area offering cultural events and modern life (Fig. 3a). Here, along the pier, adjacent to the old shipyard facilities, Amstel Botel (2015) is docked (Fig. 3b), providing a unique view over the IJ river to the historical center from the singular suites laying inside the letters on the roof (Fig. 3c). The Botel, with its silhouette emulating ocean liners, became a characteristic element of the city skyline and an attraction to all those cruising across the IJ [10].

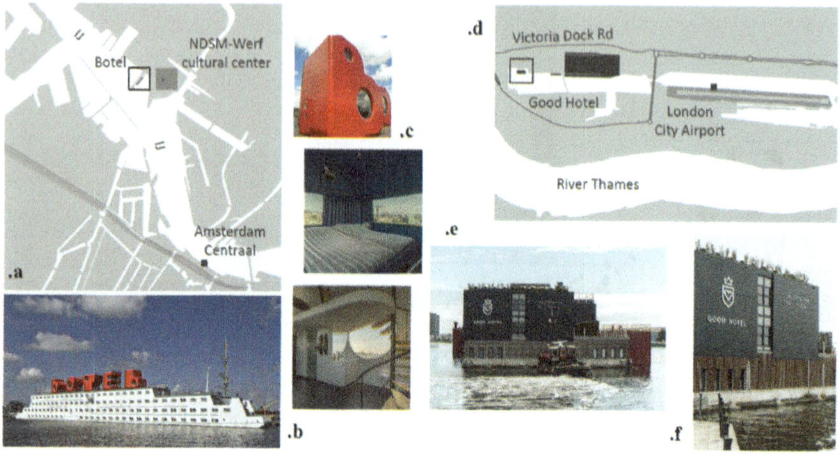

Fig. 3 Floating hotels: **a** Botel on map; **b** Botel view in IJ river (Reproduced from ndsm.nl); **c** Botel suites inside the letters (botel.nl images); **d** Good Hotel site map; **e** Good Hotel London being towed to site (Good Hotel images); **f** Good Hotel in Royal Victoria Dock (Reproduced from Good Hotel images)

Similarly, England hosts the Good Hotel London, which was launched in 2017 and located at the historical Royal Victoria Dock on the river Thames (Fig. 3d). The structure has an interesting history. A floating prison in Amsterdam before being transformed into a boutique hotel and carried away on a submersible ship carrier across the North Sea to the current location, it is managed by a non-profit organization within a program of social rehabilitation for the unemployed and will be temporarily in UK (Fig. 3e, f). Overall the floatel has a capacity of 144 rooms, with most of them overlooking the river, and the common areas are configurated in an open plan living room of a contemporary style [11, 12].

The above two cases revealed the advantages of floating buildings specifically in social and economic growth in certain areas. They are convertible structures, respectively a river cruiser and a concrete-box-foundation floating prison before refurbishment. Compared to on-land constructions, they adapt to different environments with minimal impact, and they can always be relocated. However, the Amstel Botel and Good Hotel are suitable in calm water rather than the open sea, which would require elements of a much higher standard of safety and durability.

Nowadays, the interest around floating hotels is steadily increasing, evidenced by the numerous conceptual projects appearing on the market.

2.3 *Floating Accommodations and Aquatic Communities*

Hospitality represents one of the several sectors where buoyant structures are employed. For instance, the fleets of accommodation support vessels for the oil and gas industry operating worldwide, or the containers of the student housing compound in Copenhagen harbor. These examples are remarkable not only for the architectural solutions they offer, but also for the social models they introduce. They are examples of 'machine for living in', using a Le Corbusier aphorism to describe cruise vessels; complex organisms of people and activities functioning on water. In other words, forms of aquatic communities [13].

Accommodations for offshore workers are commonly configured according to barge or semi-submersible schemes. The first operates in relatively shallow and calm waters, while the second is designed to resist the harshest environmental conditions in the world.

The semi-submersible type consists of an operating deck located above the sea level, away from the disturbing action of waves. It is connected through vertical columns to pontoons floating below the ocean surface and the wave actions. Platforms use thrusters or anchors to stay in place and ensure stability and comfort to people onboard [14]. Among companies operating in this field, Floatel International Group have a fleet of five semi-submersible hotels that can be employed from moderate to harsh environment. The Triumph launched in 2016 is a small island, 180 m long and 80 m wide, to accommodate 500 people (Fig. 4a, b). It offers generous recreational areas, a cinema and café, gymnasium, workshops, storage and a helideck [15]. Moreover, barges offer a similar service and amenities, albeit often in more sheltered environs. They are autonomous living units ensuring

Fig. 4 Floating accommodation and aquatic communities: **a**, **b** the semi-submersible Floatel Triumph in operation (Floatel International images 2018); **c** Urban Rigger exterior details (Urban Rigger images); **d** Urban Rigger interior arrangement; **e** Urban Rigger in Refshalevej 155

life onboard for several days, flexible structures which are also employable for various purposes such as hospital or temporary shelter in case of natural disasters [16].

For example, the high demand of low-cost student housing in Copenhagen is the driving force behind the Urban Rigger, a floating apartment block designed by the Danish firm BIG. The unusual configuration consists of six containers fixed on a concrete floating base, forming a two-level construction (Fig. 4c). It hosts twelve fully furnished residences of 23–30 m^2 (Fig. 4d) with an additional kitchen, storages and laundry in the basement. The unit includes communal spaces such as the roof terrace and the courtyard which give access to the kayak dock and the bathing platform [17].

Overall it is a compact, fully equipped, multifamily floating building. It can readily be replicated to a desired number in order to create a community on water. For instance, six residential units are available at Refshalevej 155 (Fig. 4e).

Urban Rigger contributes to a larger project of revitalization of a former industrial area, taking advantage of water as the building ground for an innovative and sustainable residential complex.

The range of floating solutions in the market hold the promise of larger communities living afloat.

3 KEH 87075

Founded in Denmark in 1937, Knud E. Hansen has left its mark on thousands of vessels, becoming a world recognized leader in its field [18, 19]. Along the way, KEH has been challenged with special projects; unusual structures such as floating buildings, aiming to investigate new opportunities of human life on water.

In the 80s, the company was commissioned for a number of studies related to hotels standing on a floating base. It is not clear what are the motivations that drove the comparatively high number of requests during that decade, but the interest toward aquatic settlements has always been there, arising occasionally to slowly disappear once again. However, customers recognize KEH's quality of being able to look ahead, the character and expertise to undertake such designs and turn futuristic visions into feasible projects.

During the renovation of the company archive in Lundegaarden, the Elsinore-based headquarters in Denmark, documentation of three works came to light (Fig. 5). Despite fragmented information, a reasonable number of drawings related to project KEH 87075 were found, digitalized and secured.

The design of the Hotel Vessel is dated 1987, and was ordered by Sibbern International, a trading company with Norwegian origins. There are not significant details available about the negotiation: what we know is that the private enterprise requested Knud E. Hansen technical assistance to develop the project. The idea, which never left paper, was an attempt to launch an alternative business in

Fig. 5 Sketch of an hotel vessel from the company archive, year 1998

hospitality. The hotel was intended for the Caribbean market, offering five star comfort based on Sheraton standards.

KEH carried the design up to the 'contract level', and all departments were involved, from Stability and Machinery to Structure and Industrial Design. The collection of technical drawings related to the KEH 87075 floatel design widely illustrates its general features to the reader.

What was Sibbern Intl. vision and challenge to KEH? their intention was to convert a second-hand Aker-H3 mobile semi-submersible offshore drilling rig into the groundwork for a floating building. 'It was the first time dealing with such a request', commented Ole Olsen, a senior naval architect who played a role in the project. 'There was no design manual yet; such a project was not common, and no specific procedures or standards were available. In general, any proposals of this kind were analyzed case by case', he explained further. Certainly, the design process was characterized by an interdisciplinary approach, where Naval Design and Civil Architecture were equally involved, bringing together their respective know-how.

To acquire a second-hand offshore structure was relatively cheap, but what about the entire cost of the operation? Why not to start from a new one? Why not a vessel? Perhaps it was the size of the investment that persuaded Sibbern Intl. to later end the project. Unfortunately, the answers to these questions have been lost to time. The company had the intuition to make hospitality business out of an oil rig and wanted to understand its feasibility.

Similarly, the conversion of cruise ships as floating resorts supports the Sibbern idea. Compared to a fixed, land-based hotel of similar capacity, the regulatory and operational standards to be met are quite different; stationary installations follow the regulations in force in the city or area in which they are built. Furthermore, no propulsion is required, and in most cases the supply of electricity, water etc.

is connected to the municipal systems. Therefore, we can assume that the operating costs are far lower, representing an additional factor of differentiation. Nevertheless, floating buildings have to fulfil the marine requirements to remain on water, explaining why, very often, floatels are old ships laid up and not able to sail by themselves.

Nowadays, regenerating offshore platforms could be seen as the attempt to encourage a 'circular economy' in this sector, since decommissioning costs are extremely high. The principle of extending the lifecycle of products would have a positive impact on the market, both for the finances of sellers and buyers. It would also contribute to increase the quality of the natural environment, reducing waste and the demand of raw materials to build new structures.

However, coming back to 1987, a compromise of budget and technical issues drove Sibbern Intl. to explore and evaluate design solutions.

Knud E. Hansen's proposal was a stationary hotel, anchored in shallow water, and protected by the port facilities. It could withstand moderate weather conditions, and the foundation of the building was a former offshore semi-submersible drilling platform. Its conversion represented both the core and the hardest part of the job. It emerged from drawings that technical solutions radically changed in just a few months, from May to July. The issue was to integrate the existing structure, composed of pontoons, columns and top deck, into the new building.

Designers decided to preserve the two ballasted pontoons (Fig. 6b), increasing their volume to ensure buoyancy and stability. Later, in a second stage, due to the need of extra space and while keeping their function, the pontoons became an integral part of the barge, forming the new foundations (Fig. 6c).

The capacity generated inside the pontoons, was separated into a number of technical compartments to accommodate machinery and services, essential for the

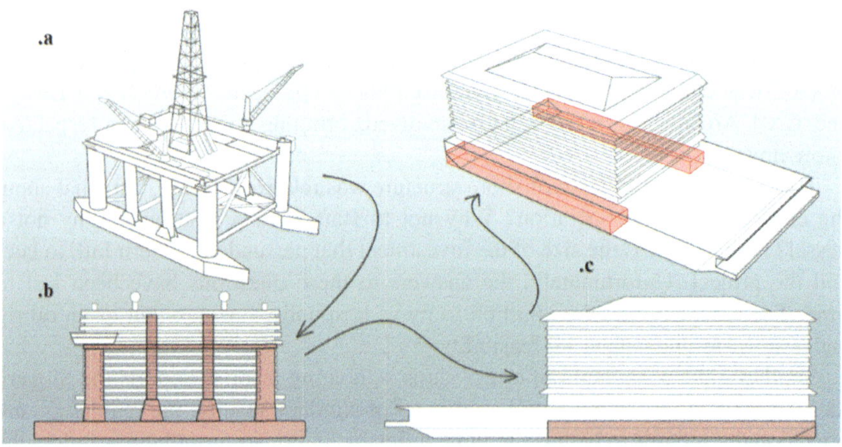

Fig. 6 KEH 87075, evolution from semi-submersible offshore platform to Hotel Vessel: **a** Aker-H3; **b** KEH 87075 Hotel Vessel first proposal; **c** KEH 87075 Hotel Vessel final proposal

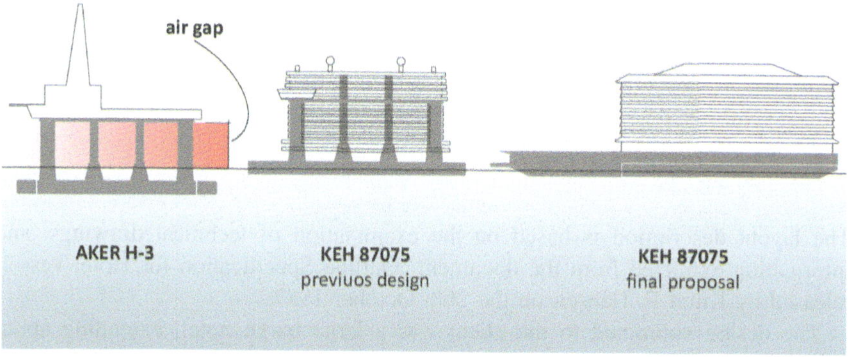

Fig. 7 KEH 87075, evolution from semi-submersible offshore platform to Hotel Vessel

hotel to function and be self-sufficient. It appeared to be a better solution than having a separate 'utility barge'.

Initially the vertical columns kept their structural purpose. Similar to land based constructions, they sustained the existing operating deck (meanwhile transformed into a sort of Upper Garden) and the additional storeys. The building appeared to unfold around them (Fig. 6b). Afterwards, the structural scheme changed, transitioning to a barge-like version. The exiting operating deck and vertical pillars were cut away and a brand new superstructure was laid above the pontoons, free from the old architectural framework (Fig. 6c). The conversion from offshore drilling platform to inshore floatel caused a significant reduction of draught and 'air gap' (the distance between water surface and platform). In calm water, the columns lose their essential function to keep the operating deck suspended and protected over the waves. Instead, the latest configuration resulted in a twelve story building, covering the maximum height of approx. 50 m above the construction line (Fig. 7).

Moreover, the overall weight was estimated to be 21,600 tons, similar to a light-weight cruise vessel construction. This was comprised of the 7,500 tons existing structure, with the extra steel, accommodation, machinery, and outfit contributing about 14,100 tons.

3.1 Aker H-3

The offshore rig Aker H-3 was developed in 1960s by the Norwegian enterprise Akers Mekaniske Verksted. It was the beginning of oil and gas discoveries in the North Sea. Akers gained expertise in the conversion and preparation of drilling platform to serve in harsh conditions, before they developed their own design, which remains one of the most successful in the Oil and Gas history.

It belongs to the second generation of semi-submersible platforms and was composed of two pontoons with four columns on each and the operating deck.

Some of them have been converted to accommodation platforms or floating production units exhibiting a high degree of flexibility and adaptation.

3.2 The Floatel Layout

The layout description is based on the examination of technical drawings, and information extracted from the document 'Outline Specification for Hotel vessel' released by Knud E. Hansen on the 26th October 1987.

The design submitted to the client was a large barge hotel, extending about 148 m along the base line, and 63 m wide. The drilling platform evolved into the new construction; some components disappeared while others remained, but were not evident. Visually, it was composed of two volumes, the barge and the superstructure. The profile was minimal, simply characterized by row windows interrupted by horizontal elements of superstructure. In the vertical plane, it was divided into twelve stories: the first three, counting from the bottom, were the largest, approximately 8000 m^2 each, contained technical, public and entertainment areas and the entrance hall. Above, there were eight floors of cabins and, at the top, an open deck, and a huge glass roof (Fig. 8).

While functioning, the hotel vessel was connected to the mainland road system by means of permanent constructed roadway. Two access ways to the floatel were provided: one two-lane access for taxis and guest cars with pedestrian entry at level 3, and one single-lane access for the service purposes at level 2.

The main entrance and the reception area were arranged at Level 3 in the fore part, while the aft was occupied by the sundeck and the pool. This level was partially open for about half of the platform, while the enclosed space sprawled out on a surface of around 3700 m^2. The internal layout was organized with the reception area dealing with the various hotel services, such as administration and accountancy, lounge, front desk and luggage office, rentals and showcases. A large lobby/atrium led to the restaurants and coffee shop, whit a breath-taking view of the

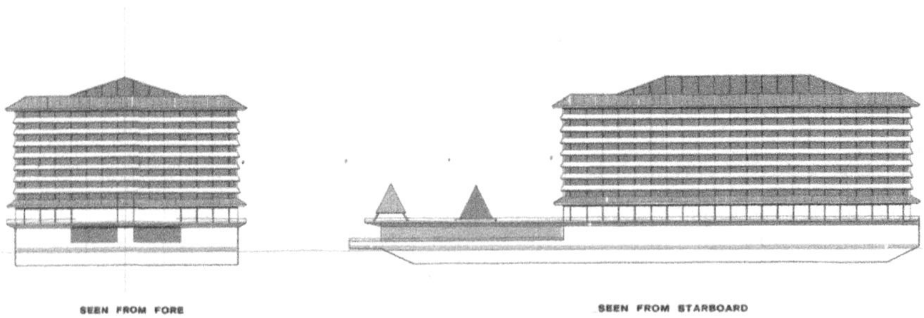

Fig. 8 KEH 87075, fore and starboard views from KEH archive

sea; the English Pub and recreational areas. The lobby was characterized by a large opening deck surrounded with decorative railing, the scenic lift and grand staircase heading to Level 2, enhancing the connection with the exhibition area and ballroom. The outdoor area could seat 200 guests in deck chairs and a few more in way of the pool bar. A shading structure made of wood and thatched roof was also planned, and planters and potted palm trees were arranged under skylights and between sunshades (Fig. 9).

The existing towing and the new mooring arrangement were positioned on Level 2. Openings of the barge were furnished with watertight doors, whilst indoors bulkheads with fire doors separated the different public and service spaces. The lower lobby is connected to the upper level via an opening and as mentioned previously, the staircase, lights and scenic lifts formed one of the main attractions of the atrium. It also contained meeting rooms and an exhibition area on the Portside, a theatre for 90 persons and a discotheque in Caribbean décor forward. Towards the aft was a ballroom for about 750 persons and the Marina deck area covered with wood and equipped with bollards etc. for the mooring of small yachts. The wall against the deck was decorated with art and maritime objects creating a harbour atmosphere.

The organization of the service area at Level 2 Starboard side, was crucial for hotel management. The personnel entrance and supply zone were arranged forward, then two distinctive flows, 'dirty and clean', leading respectively to duties and utility rooms. The first by way of the galley gave access to the blocks of service stairs and lifts; the second headed to the control office, sanitary installations, changing rooms and employee cafeteria and lounge (Fig. 10).

Level 1 was the smallest, approximately 3400 m^2. It was accessible through the four blocks of service stair and lift arranged within the old pontoons. Along the external edges were water ballast tanks, a desalination plant, sewage buffer tanks, fresh-water buffer tanks etc. In the middle, across the corridor, were planned workshops, laundry, wine and liquor room, storages, and other utilities (Fig. 11).

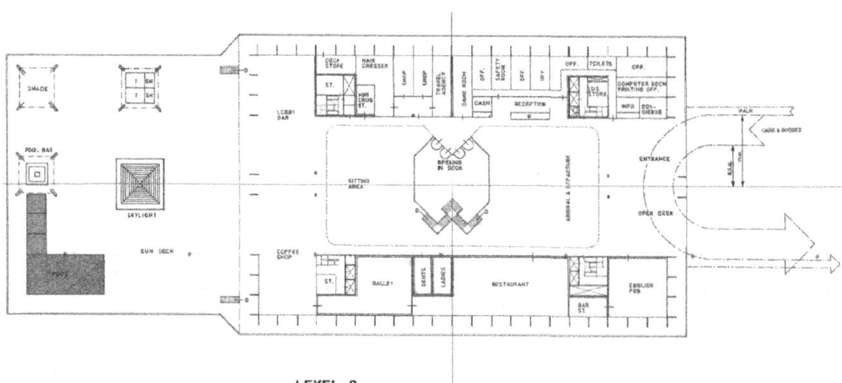

Fig. 9 KEH 87075, level 3 entrance deck. Drawings from KEH archive

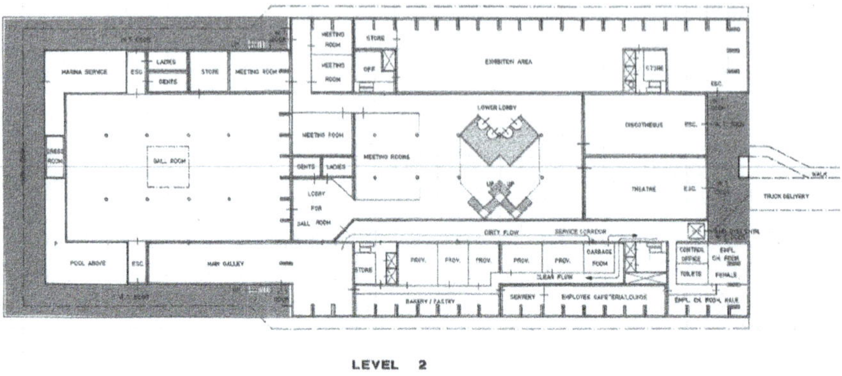

Fig. 10 KEH 87075, level 2 ballroom and exhibition area. Drawings from KEH archive

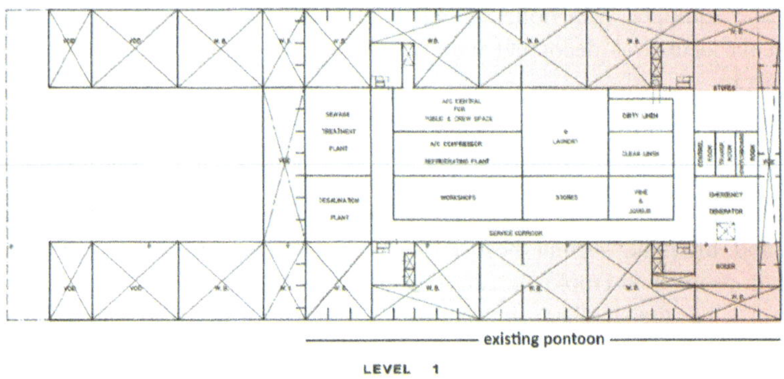

Fig. 11 KEH 87075, level 1 machinery deck. Drawings from KEH archive

The typical guest cabin level consisted of a ‘ring’ of suites around an atrium, covering an area of about 3700 m^2. Each floor was provided with four service rooms for personnel to perform their duties: these included storage of clean linen, storage of soiled linen, pantry, ice-vending machine, trash chute and sanitary room. They were interconnected with the other levels through the aforementioned service block of stairs a lifts. Guests reached their suites taking the glass-enclosed elevators in the atrium. There were eight levels of rooms, with an average of 52 per floor: 44 standard two-room suites, 4 standard single suites and 4 standard two-room corner suites. Each suite was accessible through a large door from a hallway or a balcony to the living room furnished with sofa bed, table with chairs and dresser. The corridor with ample-sized closets and kitchen table and refrigerator, led to a very large bedroom, dressing room and the bathroom. Outside the bedroom was a balcony with table and chairs, accessible through a sliding door (Fig. 12).

The hotel vessel was partially self-sufficient, and designed to be environmentally sensitive. It was equipped with a solar hot water heating system, desalination plant,

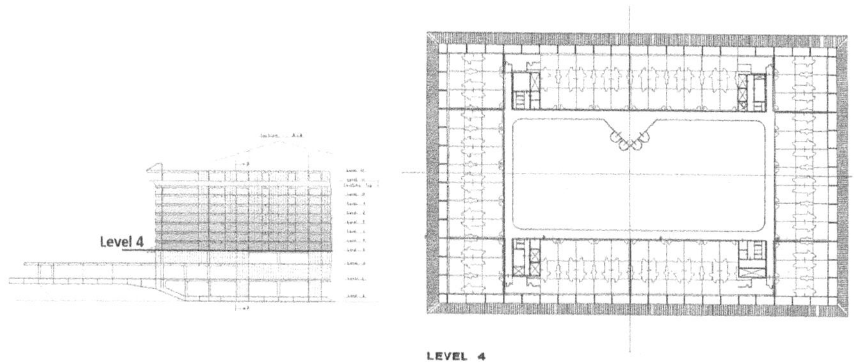

Fig. 12 KEH 87075, level 4 typical guest cabin deck. Drawings from KEH archive

Table 1 Comparison between the Australia reef resort and Knud E. Hansen hotel vessel

		Four season great barrier reef resort, Townsville, Australia	Sibbern international floatel KEH 87075
Year		1988	1987
Hull type/foundations		Barge	Semi-submersible rig converted into a barge
L.o.a.	m	Approx. 89	147.4
Length superstructure	m	Approx. 89	Approx. 96
Breadth	m	Approx. 26	63
H. max		N/A	47.8
Draught	m	N/A	Approx. 4.5
Weight		N/A	21.600*
Nr. deck/storey	t	7	12
Nr. cabin/room		approx. 200	416
Regulatory requirements		N/A	Det Norske Veritas +1A1 R5 Barge GND Hotel Barge

a. Weight of KEH floatel 21,600 tons is composed of 7,500 tons of the existing drilling platform and 14,100 tons of the additional parts after conversion

and sewage treatment plant. There was also a trash treatment system composed of a vacuum machine at all levels and chutes to the garbage room, where trash was compacted, bagged and removed to a disposal site.

The principal particulars of the Sibbern International Floatel are compared to the Four Seasons Great Barrier Reef Resort in the Table 1.

4 Conclusions

Knud E. Hansen's experience acquired during the 1987 conversion of the offshore platform Aker H-3 into a Hotel Vessel, provided not only a design solution that fulfilled all client requests and regulations, but also a way to generate benefit from the reuse of existing structures which have reached the end of their original life. For the previous owner the decommissioning costs were severely reduced through the sale. For the buyer, making a new construction based on the existing platform offered not only financial benefits, but also significant benefits to environmental impact, by reducing the need for raw materials and resulting wastage from breaking down the old platform. The oil rig life-cycle was altered through assigning a second function to increase its overall longevity.

Today we could talk about strategies of Life-cycle Design pertaining to efficiency and sustainability. Certainly, the issue of what to do with disused drilling facilities, beyond scrapping, is a design field to investigate. There are 184 offshore rigs in the North Sea, and 26 in the Mediterranean area [20] that later on might be reemployed to a wide range of uses, such as hotel, student accommodation etc.; from single to multiple units, offshore or protected inside a lagoon. Furthermore, despite being thirty years old, the Knud E. Hansen Hotel Vessel still can be considered an avantgarde project, and an expression of the fundamentals of Floating Architecture, i.e., 'adaptation' to the natural environment, 'preservation' of landscape from destructive actions, 'mitigation' of the overall impact of human activities to the Earth. Goals that the company achieved by designing a self-sufficient floating building which used alternative sources of energy, and systems for the waste management and treatment on board. Not least, the hotel could be towed to different locations whenever it was required. The 'absence of permanent foundations' is a peculiar feature of buoyant structures, allowing flexibility to adapt to various external factors, from economic contraction to the turnover of tourism seasons [21].

The work of 'search and screening' carried out in the first part of the research has been essential in order to identify the various commercial floating buildings and classify them according to architectural criteria. The collection also provides the reference background to contextualize this among other projects and find elements of comparison and details useful in describing its architectural profile.

The foundation is laid to engage in new challenges, share experiences and build knowledge. Collaboration will surely lead to more feasible designs, enabling the implementation of habitable structures on water in the near future.

Acknowledgements We would like to thank Christine Harris, Finn Wollesen Petersen, Morten Skrydstrup, Ole Olsen and Tristan Andrewartha for their assistance.

References

1. Olthuis K, Keuning D (2010) Sea of opportunities. Float! building on water to combat urban congestion and climate change. Frame Publishers, Amsterdam NL, pp 233–235
2. Principality of Sealand History. http://www.sealandgov.org. Accessed 5 May 2020
3. Comoglio G (2020) Isola delle Rose: an Esperanto republic. https://www.domusweb.it/en/from-the-archive/gallery/2020/01/29/isola-delle-rose-an-esperanto-republic-on-domus.html. Accessed 5 May 2020
4. Olthuis K, Keuning D (2010) Momentum. Float! building on water to combat urban congestion and climate change. Frame Publishers, Amsterdam NL, pp 273–2765
5. Nyilas A (2018) Beyond utopia: Japanese metabolism architecture and the birth of mythopia. Routledge, New York USA
6. Papadakis N (1977) The international legal regime of artificial islands. Sijthoff International Publishing Company, Leyden NL
7. Smith C (2018) The bizarre story of Australia's floating hotel and its 14,000 km round journey to North Korea. https://www.abc.net.au/news/science/2018-06-14/the-bizarre-story-of-australias-floating-hotel/9849482?fbclid=IwAR2z2bNpQX1-dev_c0kzcIwciCsoEJlxKdKgb8K-vMKaUdEP87rxZngJzIs. Accessed 15 May 2020
8. Smith C (2018) Seasteading: could artificial floating cities be a lifeline for low-lying Pacific nations?. https://www.abc.net.au/radionational/programs/scienceshow/seasteading-artificial-floating-nations/9824972. Accessed 22 May 2020
9. Bastin J, Mc Ginnity P (1988) Impact assessment case study the floating hotel at John Brewer Reef. Paper presented at VII annual meeting of the international association for impact assessment, Griffith University, Queensland, 5–9 July 1988
10. Botel Amsterdam. https://www.botel.nl/. Accessed 25 May 2020
11. Good Hotel. https://goodhotel.co/our-story. Accessed 25 May 2020
12. Shurvell J (2017) Former Dutch floating prison becomes a boutique London hotel in 2017. https://www.forbes.com/sites/joanneshurvell/2017/06/30/former-dutch-floating-prison-becomes-a-boutique-london-hotel-in-2017/. Accessed 25 May 2020
13. Le Corbusier (1986) Eyes which do not see. Toward a new architecture. Dover Publication, USA, pp 85–105
14. Chandrasekaran S (2017) Introduction to offshore platform. Dynamic analysis and design of offshore structures. Springer Nature, Singapore, pp 2–40
15. Floatel Triumph. http://www.floatel.se/fleet/floatel-triumph. Accessed 30 May 2020
16. Speight JG (2015) Drilling barges and drillships. Handbook of offshore oil and gas operations. Elsevier, USA, pp 71–106
17. Urban Rigger. https://www.urbanrigger.com/category/story/. Accessed 10 June 2020
18. Bruce P (2017) Knud E. Hansen 80 years. Ferry Publication, Isle of Man
19. Peter B (2012) Knud E Hansen A/S: 75 years of ship design. Lily Publications, Isle of Man
20. Statista (2020) Number of offshore rigs worldwide as of January 2018 by region. https://www.statista.com/statistics/279100/number-of-offshore-rigs-worldwide-by-region/. Accessed 15 June 2020
21. Cherubini RA, Botta A, Djuric I (2016) Floating squares. A mediterranean model. Franco Angeli, Milano, Italy

Design of Havfarm 1

Emiel Mobron, Torgeir Torgersen, Suji Zhu, John Riis, and Morten Bye

Abstract Very Large Floating Structures are an important part of urban floating development and offshore food production. To increase the production of salmon in Norway the offshore fish farm Havfarm 1 is built. The Vessel is 385 m long and has capacity for 10,000 ton salmon. To accommodate the fish nets the structure has an open bottom, which limits the torsional stiffness of the structure. This increases the natural periods to be of the same order of magnitude as typical wave periods in moderate sea states, introducing the need to capture the dynamics with coinciding natural periods and excitation periods. A new method is developed allowing for accurate estimation of fatigue utilization around the natural periods by including the effects of hydroelasticity. Furthermore, the use of well-known structural details from the oil and gas industry is incorporated in the steel-design. Together, this led to a robust steel-design that was verified and approved by DNV-GL and built for operation in Norway in 2020. The design of Havfarm 1 shows that structural dynamics around the natural period can be captured and practically used in the structural analysis. The building of Havfarm 1 provides a benchmark and opportunity for the development of other VLFS.

Keywords Aquaculture · Very-Large-Floating-Structures · Steel · Fatigue · Hydroelasticity

E. Mobron (✉) · S. Zhu · J. Riis · M. Bye
7Waves AS, Slettaveien 21, 1515 Son, Norway
e-mail: emiel.mobron@7waves.no

S. Zhu
e-mail: suji.zhu@7waves.no

J. Riis
e-mail: john.riis@7waves.no

M. Bye
e-mail: morten@7waves.no

T. Torgersen
NSK Ship Design, Sjøgata 5, 9405 Harstad, Norway
e-mail: torgeir.torgersen@nsk.as

Ł. Piątek et al. (eds.), *WCFS2020*, Lecture Notes in Civil Engineering 158,
https://doi.org/10.1007/978-981-16-2256-4_6

1 Introduction

Floating structures are found in various forms and sizes. While medium sized boathouses and offshore platforms are a common sight, Very Large Floating Structures (VLFS) in offshore environments rarely leave the drawing board and large challenges come with realizing projects of this scale. Apart from the size, VLFS distinguish themselves from common ships and platforms by their low stiffness (soft) structure. This gives rise to large natural periods that can be excited in moderate sea states and hence make the structure more susceptible to structural dynamics. When a structure is excited in its natural period, resonance occurs which is characterized by excessive vibrations. This phenomenon is well known from for example the Tacoma bridge, shown in Fig. 1.

For VLFS, dynamic loading is induced by wave forces acting on the structure as it floats in the water. Waves are characterized by an amplitude and period, with short-wave periods roughly ranging from 2 to 20 s. Since the structural natural periods of VLFS can be of the same magnitude as typical waves in moderate sea states, the structure can be excited, giving rise to large dynamic loads. This may affect the fatigue lifetime considerably.

Attention should therefore be paid to an accurate estimation of the dynamic response. This is important for the structural design because overestimation would lead to very large plate thicknesses. This can cause problems for manufacturing and obviously leads to a high steel weight. The consequently high building cost can be a major obstacle in going from concept to building stage as the project simply would get too expensive.

Fig. 1 Tacoma bridge excessive vibrations. *Source* https://www.thestructuralengineer.info/education/bridge-management/bridges/tacoma-narrow-bridge-collapse

Furthermore, the assessment of structural dynamics is part of the global strength documentation, which for offshore structures is to be approved or verified by a classification society such as Lloyds, Bureau Veritas or DNV-GL. Class approval is a necessary part (prerequisite) of the documentation required to obtain insurance and financing (loan) in the maritime world. In fact, one of the main obstacles to realize urban floating developments on a global scale is to be able to document a satisfactory level of safety, as part of making these projects bankable.

2 Background

Havfarm 1 is a VLFS that is designed and built in light of the Norwegian government's aim to have a five-fold increase of the production of salmon in Norway by 2050.

Traditional fish farming is located in the fjords (see Fig. 2), where environmental conditions are benign. However, since space in the fjords is limited and aquaculture in sheltered water affects the local environment, solutions must be found to move production further out at sea. This poses challenges for the structure that support the fish-net as traditional constructions are not able to handle the large loads from current and waves. New designs have to be developed that are able to withstand the environmental loads at sea, while being able to accommodate large quantities of salmon.

Fig. 2 Traditional fish farming in Norwegian fjord. *Source* https://businessportal-norwegen.com/2019/02/07/fische-in-bewegung/

Havfarm 1 is a 385 m long steel construction that is built to fulfil these requirements. 7Waves has been contracted by NSK Ship Design to perform the structural design and engineering of Havfarm 1.

One of the main challenges in the structural design and analysis of Havfarm 1 is the fatigue assessment. The vessel has an open bottom structure to accommodate the fish nets, as illustrated in Fig. 3, which decreases the global torsional stiffness. This, together with the sheer size and weight of the structure gives rise to a natural period around 3 s for torsion around the longitudinal axis, as illustrated in Fig. 4. Moderate sea states are characterized by similar wave periods, and hence the dynamic response at these periods is of importance in fatigue assessment.

To accurately estimate the fatigue utilization the hydro elastic response should be included in the hydrodynamic analysis. This comes with major computational challenges, as pointed out in Ng and Jiang [1] and no solid engineering practice exists for structure like Havfarm.

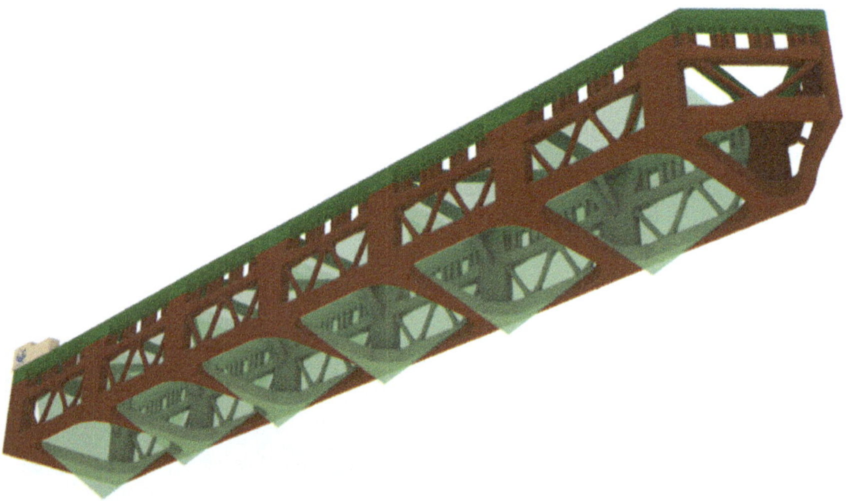

Fig. 3 Havfarm 1 open bottom structure with nets

Fig. 4 Torsional eigenmode of Havfarm 1

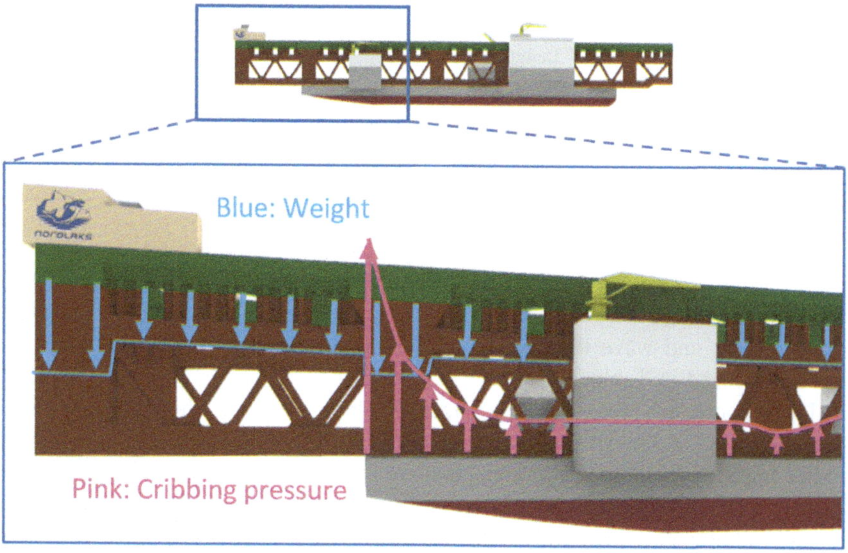

Fig. 5 Transport weight and high cribbing pressure due to overhang

Furthermore, the regulatory framework is not well suited for documenting the global strength in a straightforward manner. However, the DNV-GL offshore standards allow for direct calculation using finite-element methods (FEM). This basically opens for taking any shape and documenting the safety level for an offshore location. The offshore standards therefore create large opportunities for development of innovative offshore structures like Havfarm 1.

Another challenge in the structural design and analysis of Havfarm 1 is the transport from the building location to the operating site. The vessel is built in China but will operate in the northern part of Norway. Towing the vessel is not an option due to the large wave heights that can be encountered on this long voyage. Transport on a heavy-lift vessel (HTV) is therefore necessary, making Havfarm 1 the largest cargo ever transported by vessel. For this, the 275 m long Boka Vanguard from Boskalis is used, the largest HTV that exists today. Since Havfarm 1 is 110 m longer than the HTV, the structure will have a large overhang at the bow and aft ship, posing challenges for the structure itself, as well as the wooden beams that support it on the deck. This is illustrated in Fig. 5, showing the weight distribution of Havfarm 1 and the cribbing pressure on the deck of the HTV.

3 General Design

Havfarm 1 is a steel structure with six net areas, designed to accommodate a maximum of 10,000 ton salmon. The net areas have a square shape and are placed in a longitudinal arrangement. The vessel is permanently moored, and a turret is

Table 1 Havfarm 1 main dimensions [2]

Main dimension	
Length over all (m)	385.0
Beam (m)	59.50
Depth (m)	37.75
Net volume (m^3)	69,000 per net cage, 414,000 total
Net draught (m)	56.00

placed at the (triangular) bow section, connecting Havfarm 1 to an array of drag anchors. This way, the vessel is free to change heading with respect to dominant wind, current and wave directions (weathervaning). The main dimensions of the vessel are listed in Table 1.

To support the six nets, the steel construction uses the principle of a semi-submersible, known from oil rigs. In this configuration, most of the buoyancy comes from the (ballasted) pontoons, that are located at the bottom of the vessel, as shown in Fig. 6. The columns connect the pontoons to the deck structure and provide additional buoyancy/ballast capacity. A corrugated bulkhead is placed between the upper- and mid-pontoon, damping the waves and providing protection for the fish inside Havfarm. Longitudinal trusses are placed between the mid-pontoon and lower pontoon.

The connections between the different elements are similar to well-known structural details from the oil and gas industry. This is done in order to keep the design, which is complex in itself, as simple as possible and to ease fabrication. Furthermore, it facilitates the structural analysis and approval process, as reference can be made to existing documentation (Fig. 7).

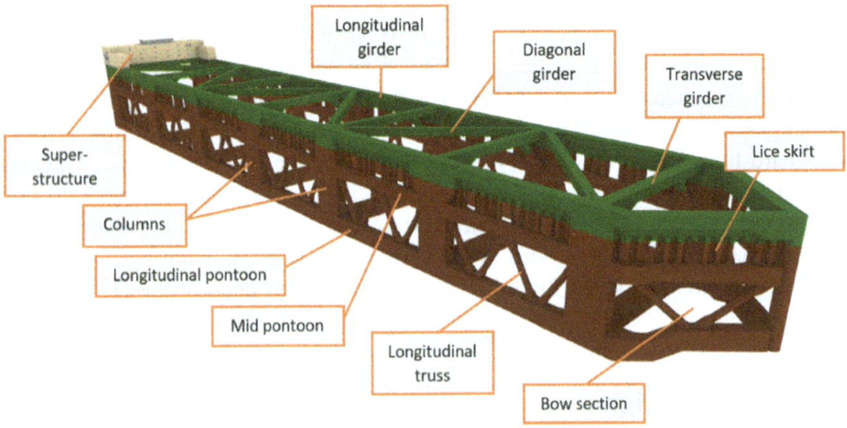

Fig. 6 Havfarm 1 overview of structural elements

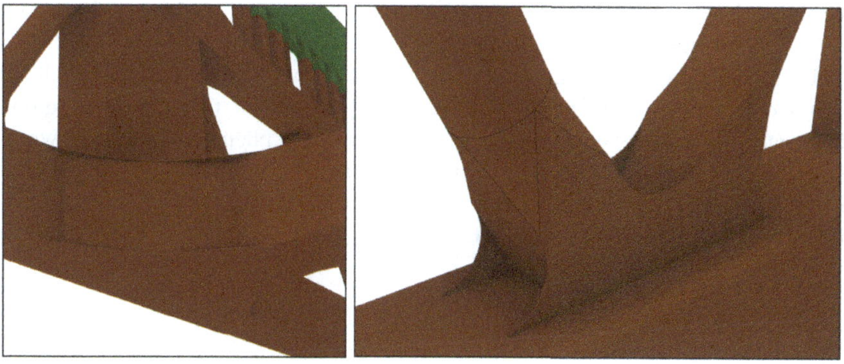

Fig. 7 Typical structural details. Left: connection transverse brace and lower pontoon; Right: connection longitudinal trusses to lower pontoon

4 Fatigue Assessment

In this chapter more details are given about the fatigue assessment used for Havfarm 1. First, an overview of existing methods is given, after which the newly developed method using a multi-body approach is described.

4.1 Existing Methods for Fatigue Assessment

To determine the fatigue utilization close to the natural period, different methods have been developed for various types of offshore structures.

It would therefore be natural to use the approach for oil rigs and apply it to Havfarm 1. The geometry of Havfarm is similar to an oil rig with pontoons, columns, braces and trusses. Also, the structural details and connections to the different members are similar. However, this approach treats the vessel as a rigid-body and it cannot capture the response around the structural natural periods (bending and torsion). This is acceptable for oil rigs, which have structural natural periods that are much smaller than typical wave periods and the fatigue contribution is negligible. However, the natural period in torsion of Havfarm is in the same order of magnitude as typical wave periods in the moderate sea states and fatigue contribution around the natural periods cannot be neglected.

A well-known approach that is used for jacket structures is the use of Dynamic Amplification Factor (DAF), which relates the dynamic response to the static response [3]. Jacket structures are fixed structures for which inertia loads are of minor influence compared to floating structures. It is therefore acceptable to assume the structure as a rigid-body and use DAFs to account for the dynamic effects. For Havfarm however, being a large and slender vessel with large ballasts volumes, the inertia loads are significant, the global stiffness is low and neglecting this would yield wrong results.

To include inertia loads, the global deformation of the structure should be included in the hydrodynamic analysis, i.e. the effect of the structure's global deformation on the wave pressure distribution. This phenomenon is known as hydroelasticity and has previously been studied for ULCS and VLFS. However, the method and results of these studies cannot be used for Havfarm due to its configuration as a column-stabilized unit, which behaves fundamentally different than a ship shaped vessel.

4.2 *Multi-body Approach*

A new method is developed that makes use of the principle of DAFs, but at the same time includes hydroelasticity. The principles of the method are described in this chapter.

4.2.1 Hydrodynamic Analysis

To include hydroelasticity in the hydrodynamic analysis two methods are used:

- Morison model.
- Panel model.

In the Morison approach, the excitation forces, added mass and damping effects can be applied to beam elements so that coupled hydrodynamic analysis can be performed. However, diffraction effects are known to be over-estimated for short waves [4].

With the panel method, the hydrodynamic loads are directly calculated by Wadam [5]. However, the global stiffness (hydroelasticity) cannot be represented exactly by this method.

To perform hydrodynamic analysis with the correct global stiffness, while at the same time not over-estimate the response for short waves, the solution is to use both approaches together in multi-body analysis. The multi-body analysis uses a combination of a panel model (Fig. 8) to determine the pressure distribution, added mass and damping, and a beam model (Fig. 9) to include inertia and drag contributions.

To account for different structural elements, the panel model is divided into a large number of separate bodies, as illustrated in Fig. 10. The pressure distribution of each body is integrated so that the excitation forces, added mass and damping are evaluated for each body. These load effects are acting at the body centre and transferred as point loads to the beam model used in the SIMO-Riflex time-domain simulation [6]. Now, the dynamic response can be accurately calculated for each body and for a given sea state and heading angle.

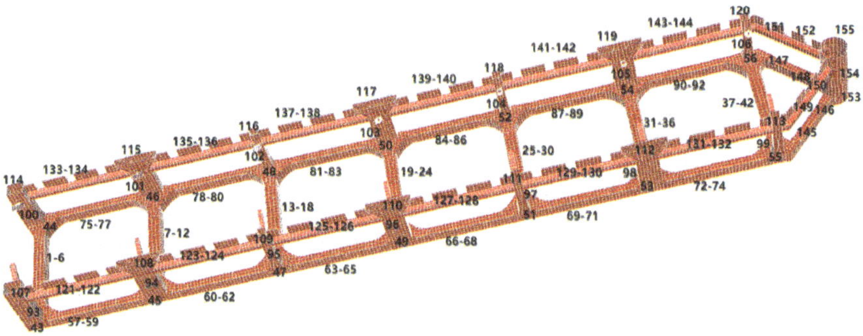

Fig. 8 Panel model

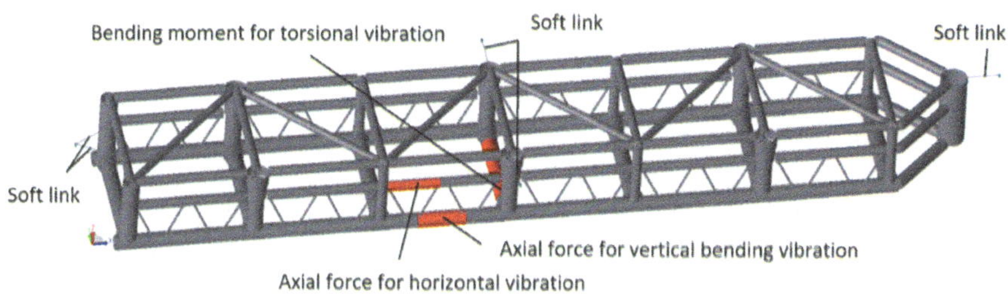

Fig. 9 Beam model

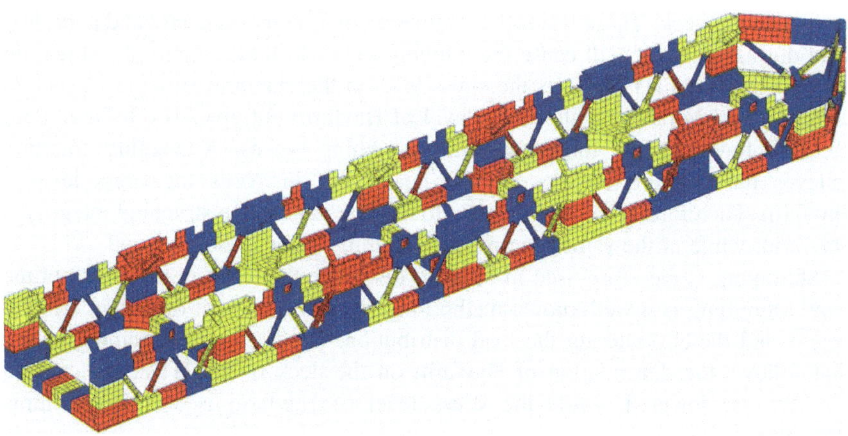

Fig. 10 Division of panel model into separate bodies

4.2.2 Dynamic Amplification Factor

The transfer functions for each body are first calculated for a flexible model, which represents the actual global stiffness of the structure. To determine the DAF, the rigid-body response should be determined in addition. This is done by using the same panel model, but an artificially stiff beam model, so as to replicate a rigid-body. DAFs are calculated as the ratio of the standard deviations of response between the flexible response and rigid-body response. Axial force and bending moment are calculated for a beam element in each structural member and the maximum DAF is used in fatigue assessment for this member, ensuring a conservative approach.

The dynamic behavior is different for every sea state, wave heading and structural element. Therefore, separate DAFs are determined for each structural member type (body), and for a large number of sea states and heading angles.

5 Transport

Havfarm 1 is transported by the HTV Boka Vanguard. This is a vessel that can submerge to a large draught, position the cargo above the deck and lift it as it pumps out ballast and emerges.

Since Havfarm is significantly longer than the HTV the large overhang at the bow and aftship create large bending moment and stress concentrations in the Havfarm structure. The vessel is supported by wooden cribbing beams that are placed on the deck. These beams have limited compressive capacity and a too high load during transport will cause the cribbing to crush. Furthermore, the still water deflections of the HTV affects the stress level in the Havfarm structure: a sagging condition is favorable for the stress level of Havfarm but gives rise to large compression forces in the fore and aftmost cribbing beams. A hogging condition relieves the pressure on the cribbing beams but increases the stress level of Havfarm. Therefore, a balance must be found that assures the structural integrity of Havfarm, while at the same time avoids crushing of the cribbing wood.

Shimming is therefore used in order to distribute the loads in the most optimal way. Shimming is a well-known method to increase the height of certain cribbing beams and thereby altering the load distribution. By creating a shimming pattern that follows the deformation of Havfarm on the deck of the HTV, the cribbing pressure are lowered, while the stress level of Havfarm is still within limits (Fig. 11).

Fig. 11 Havfarm on board of Boka Vanguard, departure from China

6 Results

The hydro elastic fatigue assessment made it possible to design Havfarm 1 for the offshore conditions that characterize the location in the northern part of Norway. The structure is designed for the local wind and current conditions and a significant wave height of 6.0 m. Large fatigue utilization was observed at connections between pontoons, braces, and columns, as well as around the lice skirt. Using the actual wave scatter for the location in combination with the DAF approach the fatigue life of 25 years could be documented.

The method using the DAF approach is verified internally and by DNV-GL and is found to give reliable yet conservative results.

Since the panel model exactly reflects the vessel geometry, the dynamic structural analysis in combination with the DAFs can be used directly for global fatigue assessment. Furthermore, the results can be used for local fatigue assessment using a sub-modelling procedure. This has made it possible to document sufficient fatigue life for Havfarm 1.

The design of Havfarm 1 is approved by DNV-GL and the vessel is built in China by Yantai CIMC Raffles Offshore Ltd (YCRO). The vessel is transported to the offshore location in Norway and operation is expected to start during the summer of 2020 (Figs. 12 and 13).

Fig. 12 Havfarm 1 launched at yard in China

Fig. 13 Havfarm 1 arrival in Norway after transport

7 Conclusion and Outlook

A method has been developed that makes it possible to accurately estimate the fatigue utilization of Havfarm 1, with special attention to the natural periods. DAFs have been calculated using a multi-body analysis comprising of a panel model in combination with a beam model. In addition, standard structural details from the oil and gas industry are used that relate to existing codes and standards. The multi-body fatigue assessment and use of well-known structural detail made it possible to verify the fatigue calculations and ultimately approve the complete vessel design.

The project is an example of crossover between technology originally developed for the oil and gas industry to new marine applications and can provide a benchmark for future projects.

When Havfarm 1 enters operation, the wave loads will be monitored on board the vessel with the aim of validating the results from the fatigue assessment. This is done by measuring the wave height with wave-radars, together with the structural response in real-time. This will give more insight into the level of conservativeness and provides a basis for calibration of the fatigue calculations for future use.

References

1. Ng CW, Jiang R (2019) Classification principles for very large floating structures. In: WCFS2019 Proceedings of the world conference on floating solutions, ISBN 978-981-13-8743-2
2. Nordlaks, Havfarm 1, https://www.nordlaks.no/havfarm/havfarm1. Accessed 11 June 2020
3. Horn J, Amdahl J, Haver S (2015) Dynamic amplification of drag dominated structures in irregular seas. In: Proceedings of OCEANS'15 MTS/IEEE Washington conference and exhibition. IEEE conference proceedings, ISBN 978-0-933957-43-5
4. Pereyra B, Wendt F, Robertson A, Jonkman J (2016) Assessment of first- and second-order wave-excitation load models for cylindrical substructures. In: Proceedings of international society of offshore and polar engineers, Rhodes, Greece
5. DNVGL (2017) Sesam user manual—Wadam, V 9.3-07, March, 2017
6. SINTEF Ocean (2017) Riflex 4.10.3 Theory manual, Sept 2017

Parametric Model for Generation and Analysis of Modular, Freeform Floating Island Networks, Constructed Using Flexibly Formed Buoycrete®

Diederik Veenendaal, Marco Bovio, Oscar Sainz Avila, and Guido Visch

Abstract A robust parametric model is presented that can generate different polygonal regular tilings or irregular networks of freeform, modular floating islands. The doubly curved geometry is possible by using Buoycrete, a neutrally buoyant, non-dissolvable concrete mix. This allows for floating bodies that are designed and optimized beyond what is traditionally possible with conventional construction. One output from the parametric model, a single module designed for North Sea conditions, is evaluated using diffraction and CFD analysis to inform the parametric model and to demonstrate the potential of our approach.

Keywords Parametric modelling · Flexible formwork · Buoycrete · CFD · Floating island

1 Introduction

The world's growing population, coupled with sea level rise due to climate change, has generated interest in land creation at sea. Modular floating islands can potentially create such space in an affordable and sustainable manner. Other driving

D. Veenendaal (✉)
Summum Engineering, Van Vollenhovenstraat 14, 3016 BH Rotterdam, The Netherlands
e-mail: diederik.veenendaal@summum.engineering

M. Bovio
LeadingEDGE Marine Engineering, Bernardus Gewinstraat 14C, 3031 SE Rotterdam, The Netherlands
e-mail: bovio@leadingedge.it

O. S. Avila · G. Visch
Boskalis Research and Development, Rosmolenweg 20, 3356 LK Papendrecht, The Netherlands
e-mail: oscar.sainz.avila@boskalis.com

G. Visch
e-mail: guido.visch@boskalis.com

Ł. Piątek et al. (eds.), *WCFS2020*, Lecture Notes in Civil Engineering 158,
https://doi.org/10.1007/978-981-16-2256-4_7

forces are the need for industrial space, logistical purposes, leisure or creating space in international waters with its own jurisdiction. The concept of large floating cities captured our imagination throughout history, and past decades have seen many proposals [1, 2], including, but not limited to those from: Paul Maymont (1926–2007); Koen Olthuis, Docklands/Waterstudio; ATDesignoffice; Blue21/DeltaSync; Vincent Callebaut; Oceanix; the Seasteading Institute; and the Venus Project, Jacque Fresco (1916–2017).

Their concepts consist of either single or multiple modular floating bodies, many circular or hexagonal in shape. An ongoing study, Space@Sea, concluded that while "a triangle shape [..] shows the highest flexibility", "the use of deck space for a square floater is more efficient." (Figure 1) [3]. Recent concepts feature more irregular networks of triangles, squares and pentagons (Fig. 2).

2 Parametric Model

A robust parametric CAD model is presented that can generate different patterns of multiple islands; both regular tilings and irregular tessellations. The model is set up in Grasshopper/Rhinoceros 3D.

Fig. 1 Triangular and square floating bodies: Space@Sea study by Blue21/DeltaSync with Waterstudio, ICE Marine Design Group, 2019

Fig. 2 Floating City Project with squares and pentagons, 2013 by the Seasteading Institute with Blue21/DeltaSync, and Blue Revolution with triangles, 2015 by Blue21/DeltaSync

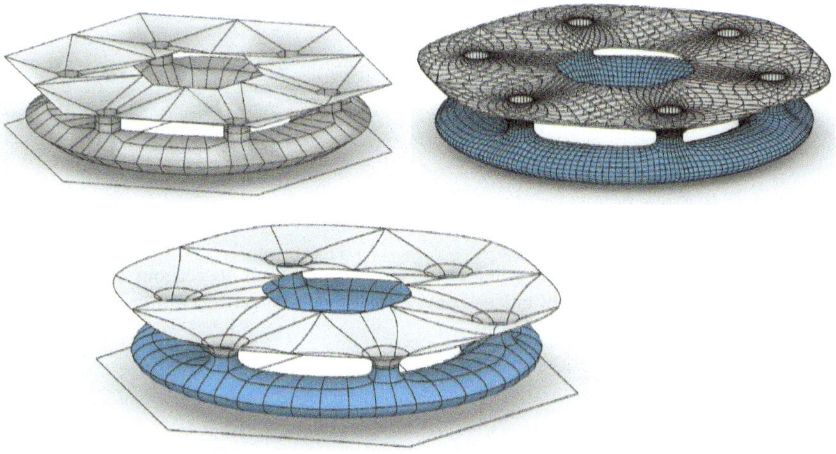

Fig. 3 Low-poly mesh, mesh from Catmull-Clark subdivision with internal deck structure, and reconstructed polysurface of the outer body

Each island (Fig. 3) has a polygonally shaped perimeter and deck area, so that it can be connected to its neighbors to form a network. An opening in the middle can be used for fish farms, optional quay walls for small ships, and to allow for more daylight entry to the seabed.

Each floater has a freeform toroidal body to form a sort of semi-submersible shaped island. Semi-submersibles are commonly used in the offshore industry and are characterized by lower wave loads and motions compared to full body floating objects because of their reduced form stability. The body is made using Buoycrete: a neutrally buoyant, non-dissolvable concrete mix.

Initially, a low-polygonal mesh is created, which is later refined through Catmull-Clark subdivision [4]. This allows for rapid generation and iteration of the model. The upper deck is determined by: the number of sides; the number of columns per side; the deck clearance; and, the inner and outer radius. The radius of the lower torus is governed by the column positions on the upper deck and the perimeter of the deck. The columns themselves can vary in radius, as can the cross section of the torus. The deck has ribs designed to follow lines of principal stresses [5]. The lines of the quadrilateral mesh are then used to construct surfaces, so the geometry is represented both as a mesh and as a polysurface (Fig. 3).

For multiple adjacent floating bodies, a tiling of regular convex polygons [6] or tessellation of irregular polygons can be used as the basis for their arrangement. The simplest regular tilings are triangular, square and hexagonal, and infinite combinations are possible by including dodecagons. The parametric model is set up for any tiling or tessellation to be used (Fig. 4).

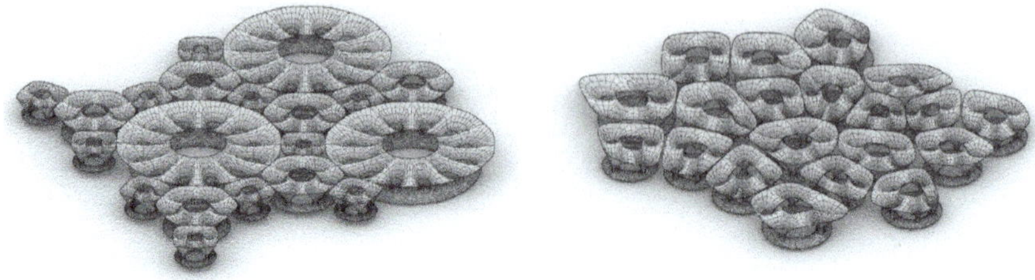

Fig. 4 An Archimedean tiling consisting of squares, hexagons and dodecagons, and an irregular Voronoi tessellation, consisting of unique convex polygons, both generated from the same model

3 Buoycrete

The complex, freeform geometry of each module is made possible by using Buoycrete, a neutrally buoyant, non-dissolvable concrete mix, with a patented work technology and method [7]. There are no comparable lightweight cement mixtures available on the market.

The high slurry pressures that arise when pouring regular concrete in a hard formwork are not apparent with Buoycrete underwater. Thus, inflatable and other flexible formworks made of inexpensive fabrics like geotextiles or plastic foils can be used to form predefined (complex) concrete shapes. When a complete formwork is inflated (filled with Buoycrete), the predefined shape becomes instantly apparent. Figure 5 shows the result of a first inflatable demonstration in the Boskalis laboratory, using a low-cost plastic formwork. After curing under water, the structure was removed from the basin and loads could be applied on top. To manufacture such a structure with regular concrete, above water, an expensive, milled or printed, rigid formwork would be needed to withstand the slurry pressures and avoid collapse of the formwork.

The mixture research is under constant development and with a tunable aggregate weight and percentage, the density and compressive strength can be tuned to a

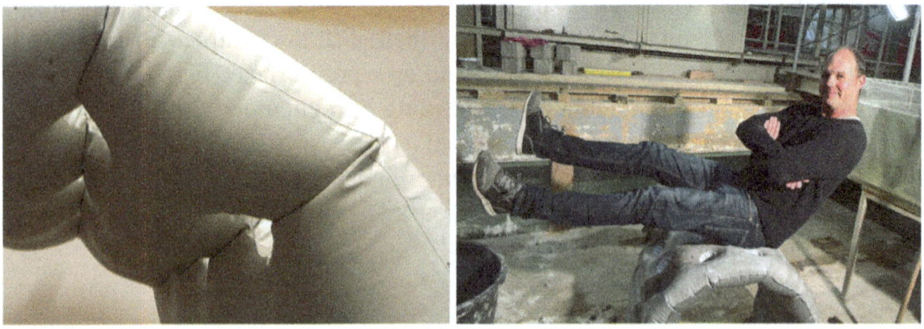

Fig. 5 Buoycrete demonstrator and 7-days UCS load test with 100 kg static load

Table 1 Buoycrete properties for density equal to that of water

Property		Value	Unit
Uniaxial compressive strength	UCS	35–40	MPa
Indirect tensile strength	ITS	>2	MPa
Young's modulus	E	5–6	GPa
Density	ρ	1000	kg/m^3

desired level. Table 1 shows properties for a density equal to that of water. A side-effect of our lightweight aggregate is the low modulus of elasticity.

4 Benchmark Study

One specific output for a floater (Fig. 3) is analyzed in further detail (structural, diffraction and CFD), to demonstrate the benefits of the parametric model and of freeform bodies compared to conventional ones. A location (Fig. 6, Table 2) and some design constraints (Table 3) are assumed for this benchmark.

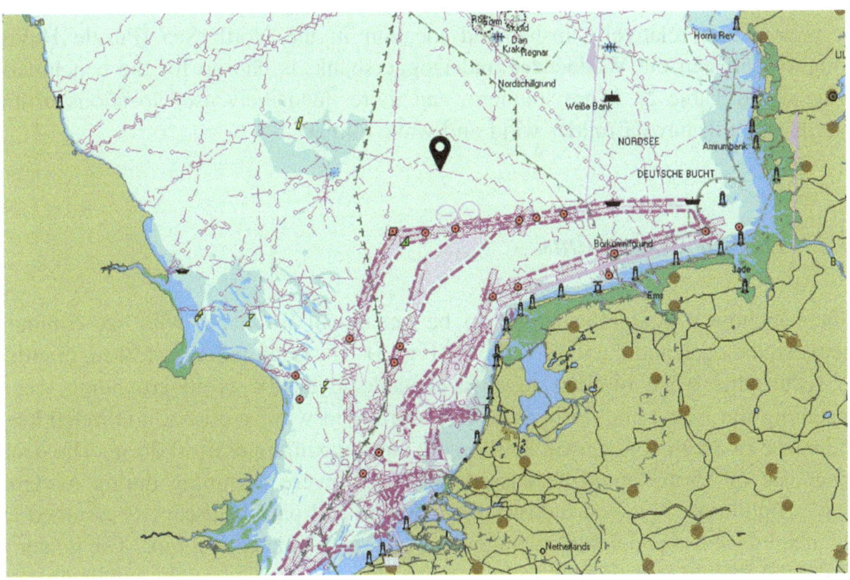

Fig. 6 Location for benchmark with coordinates 56° 50′ N, 4° 00′ E

Table 2 Metocean data for benchmark with coordinates 56° 50′ N, 4° 00′ E

Description			Quantity	Value	Unit
Water depth			WD	45	m
JONSWAP spectrum	Extreme wave	Wave height, peak period	H_s, T_p	9, 15	m, s
	Extreme swell	Wave height, peak period	H_s, T_p	1, 25	m, s
Surface current			Vc	1.5	knots

Table 3 Main features of the hexagonal benchmark floater

Description	Value	Unit
Total deck load	8.0	t/m^2
Displacement	502×10^3	t
Deck area	50×10^3	m^2
Inner, outer radius	50, 150	m
Deck clearance	15	m
Draft	30	m

4.1 Location

A remote and relatively unsheltered location in the North Sea (Puzzle Hole), roughly between the Waddenzee and Doggersbank, is chosen for the benchmark (Fig. 6 and Table 2). More southern, and more intensively used locations in the North Sea will have a milder wind and wave climate.

4.2 Design Constraints

The benchmark floater is chosen to be hexagonal in shape, with six columns between the deck and the toroidal body (Fig. 3). The outer radius is 150 m, in order to reduce the wave motions to an acceptable level. To avoid grounding when installing the island and during a severe storm and wave motions, a minimal keel clearance of 15 m is assumed. This results in a maximum draft of 30 m. The deck clearance of 15 m should be enough to avoid severe slamming during extreme wave conditions. A mixed industrial, residential and park landscape is assumed as the main deck use and a total average deck load is calculated at 8 t/m^2. The floater's structural weight is estimated to be 20% of the total displacement, comparable to other large floating objects, meaning the total load is 10 t/m^2. A preliminary structural analysis verifies that: this total load does not cause the Buoycrete's ITS (Table 1) to be exceeded; displacement stays within 1/125th of the deck width, in this case 80 cm; and, the self-weight stays below the 20%.

4.3 Design Criteria and Model Parameters

The described starting point of the floater design is established with some fast qualitative seakeeping and current load parameter checks. To assess each floater's design, several criteria are defined, and translated to parameters to be maximized or minimized. The parametric model automatically computes these parameters related to the geometry, hydrostatics and structural performance (Table 4). Their quantitative values can be evaluated and then tuned within the parametric model, the diffraction analysis and the CFD model.

The considerations behind each design criterion and their parameters are as follows:

- Seakeeping: Large mass moments of inertia result in lower acceleration levels and thus better seakeeping characteristics. A low area moment of inertia, or reduced stability, also reduces the acceleration levels and comfort on board of floating bodies.
- Load transfer (coupling forces): With a low stability, the wave induced pitching moment will be smaller. Thereby, a small pitch mass moment of inertia will reduce the needed coupling forces to counteract the rotational acceleration levels.

Table 4 Design criteria, corresponding parameters to be maximized/minimized, and specific values for the benchmark geometry

Criteria	Objective	Parameters	Quantity	Value	Unit
Seakeeping	Max.	Radius of gyration	k	70.8	m
Coupling forces	Max.				
Seakeeping	Min.	Area moment of inertia	I	6.62×10^6	m^4
Coupling forces	Min.				
Wave loads	Min.	Height of centre of buoyancy	z_{KB}	−19.5	m
Wave loads	Min.	Wetted surface area	S_w	72,633	m^2
Drift and current forces	Min.				
Drift and current forces	Min.	Transverse area	A_{px}	5600	m^2
Drift and current forces	Min.	Height of transverse centroid	z_C	−18.0	m
Global bending moment	Max.	Area moment of inertia transverse	I_{ss}	699×10^3	m^4
Global bending moment	Min.	Circumference of transverse	C_{ss}	573	m
Amount of buoycrete	Min.	Surface area of Buoycrete	S	173×10^3	m^2

- Wave loads: The wave motions and pressures reduce with greater distance from the waterplane, therefore a low centre of buoyancy generally results in lower wave loads. In our case, with a relatively low water depth, we expect the wetted surface area to have a negative influence on the wave loads because in high seas the wave velocities are far from zero around the keel line.
- Drift forces and current forces: Analogous to the wave particle velocities, the current velocity will also reduce at a greater distance from the waterplane. Therefore, the projected transverse area and distance from the waterline are of influence on the drift and current forces.
- Global bending moment: Bending moment stresses are calculated with the cross-sectional area moment of inertia of the structure. A large distance between the deck and torus, and a large diameter of the torus itself will increase the global stiffness and reduce the global material stresses.
- Amount of Buoycrete: Assuming an over dimensioned wall thickness of the structure's surface area because of impact force integrity, water intrusion and so on, a low surface area will result in less used Buoycrete material. Large surface curvatures, especially around the columns, will lead to smaller deformations and material fatigue during the intensive, cyclic wave loads.

5 Diffraction Analysis

For the hydromechanical analysis, a single module generated by the parametric model (Fig. 3), is studied in more detail as a first iteration. A lower centre of buoyancy, a small waterline area together with an optimized underwater volume results in reduced wave motion response and coupling forces between the island modules. As a comparison, a cylindrical floating island is modeled with an equal displacement, diameter, centre of gravity above waterline and radius of gyration. Figure 7 shows the AQWA models and Table 5 the hydrostatic properties of the freeform floating island (FFI) module and the cylindrical island (CI) floater.

The first order motions in Table 5 show the relative mild seakeeping behaviour of the freeform floating island in survival conditions. Shallow water effects result in the relative high surge response in comparison with heave. The low pitch response, in combination with a small waterline area, and reduced GM, will be beneficial for

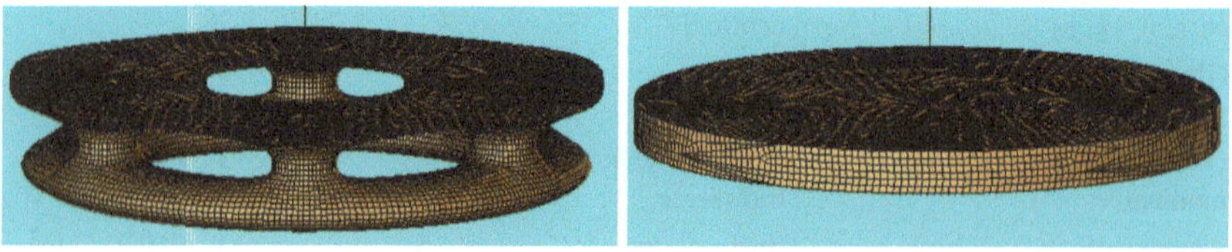

Fig. 7 Freeform floating island and cylindrical floater

Table 5 Hydrostatic properties of freeform floating island (FFI) and cylindrical island (CI)

Description	Values FFI	Values CI	Unit
Displacement	500	500	kton
Outer diameter	300	300	m
Vertical CoG wrt waterline	15	15	m
Mass radius of gyration	70	70	m
Vertical CoG wrt base	22	45	m
Metacentre height wrt CoG	795	1	m
Draft	7	30	m
Waterline area	70,686	3,555	m^2
Eigenperiod heave	50	22	s
Eigenperiod pitch	400+	16	s

the coupling forces between the different modules. Also in extreme swell conditions the motions' response does not show any extreme response. Even the surge response is relatively small. The cylindrical island on the other hand, clearly shows more heave and pitch response in both sea states. This is also clarified by the RAO's for pitch and heave for both floaters (Table 6, Figs. 8 and 9).

5.1 *Wave Elevation and Drift Forces*

There are significant amplitudes of the wave pattern during survival conditions (Fig. 9). Around the columns near the leeward side of the module, some peaks in wave elevation are found around 9.5 m. With a deck height of 15 m above waterline and a very mild heave response, the chances for severe deck slamming are minimal. Furthermore, the amount of wave obstruction is small (Fig. 10), which

Table 6 Significant amplitudes motion response surge, heave and pitch

	Significant amplitudes						Units
	Surge		Heave		Pitch		
$H_s = 9.0$ m, $T_p = 15.0$ s	*FFI*	*CI*	*FFI*	*CI*	*FFI*	*CI*	
Displacements	1.281	0.741	0.368	0.903	0.501	1.710	m
Velocities	0.599	0.266	0.184	0.358	–	–	m/s
Accelerations	0.296	0.106	0.098	0.158	–	–	m/s^2
$H_s = 1.0$ m, $T_p = 20.0$ s	*FFI*	*CI*	*FFI*	*CI*	*FFI*	*CI*	
Displacements	0.241	0.283	0.070	0.197	0.077	0.253	m
Velocities	0.074	0.083	0.022	0.061	–	–	m/s
Accelerations	0.027	0.025	0.009	0.020	–	–	m/s^2

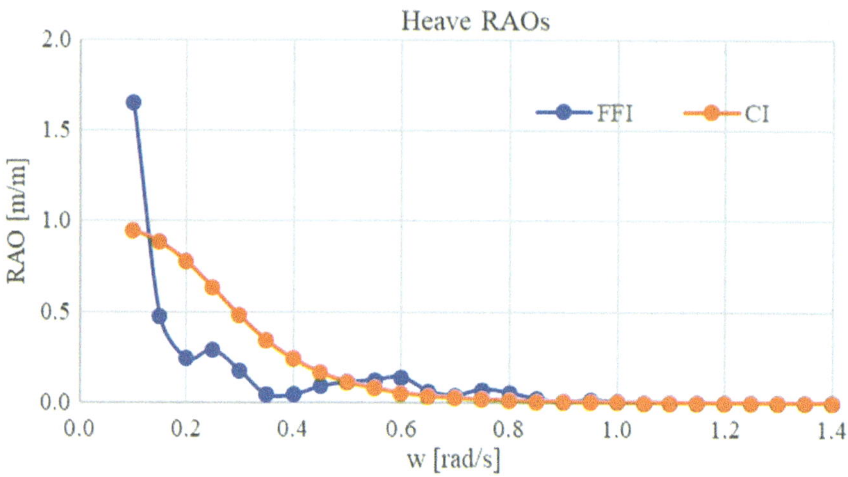

Fig. 8 RAO's heave, for freeform floating island (FFI) and cylindrical island (CI)

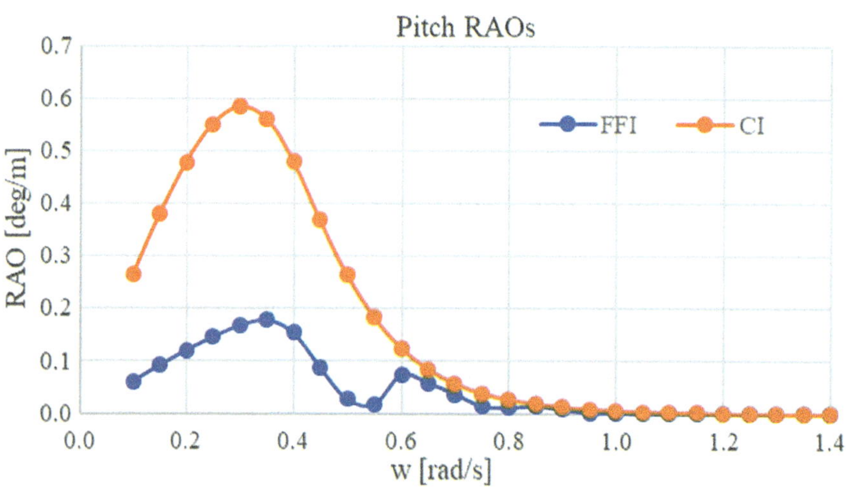

Fig. 9 RAO's pitch, for freeform floating island (FFI) and cylindrical island (CI)

means that wave drift forces are expected to be small, contrary to the wave pattern around the floating cylinder.

The wave drift forces show remarkable peaks and troughs between 0.3 and 0.9 radians per second. Hypothetically, the discontinuous transverse cross-section of the floating body in wave direction, with especially the large hole in the middle, could result in some standing wave phenomena and/or levelling out of wave forces at certain wavelengths. These mechanisms should lead to lower drift forces compared to solid box-shaped floating bodies, like the cylindrical floater (Fig. 11).

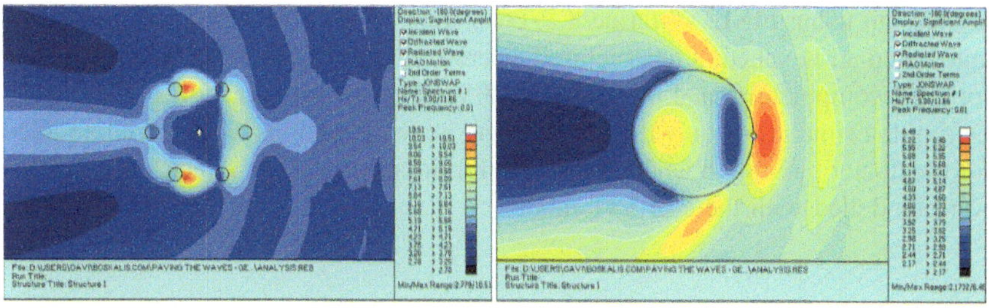

Fig. 10 Significant wave elevation

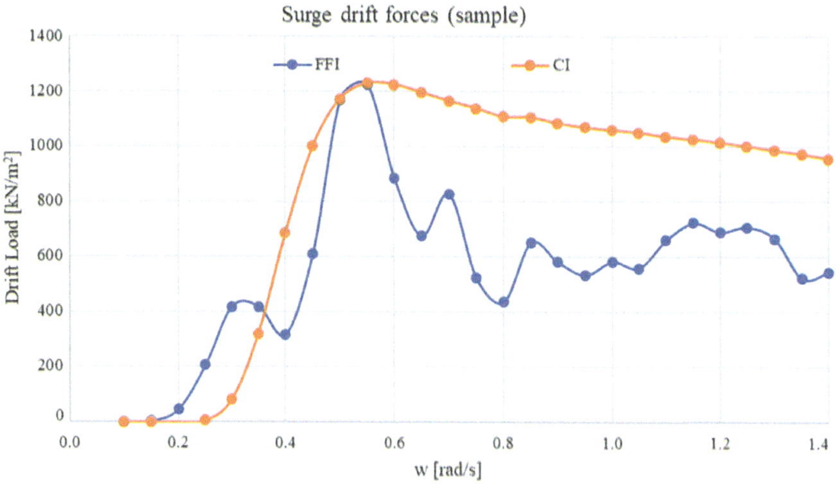

Fig. 11 Wave drift forces per unit wave height

In general, the relatively mild seakeeping behaviour, small waterline area, minimal wave obstruction and acceptable drift forces show the hydromechanical feasibility of the designed freeform floating island. Furthermore, the comparison with the cylindrical floater validates the presumed outcome of the optimization parameters as described in Sect. 4, nevertheless this is a first iteration of the complete analysis of the model. The expected relative low coupling forces could further be analysed in detail in a coupled multi-body diffraction calculation model.

6 CFD Analysis

The parametric model is coupled with a RANSE CFD solver in order to evaluate the hydrodynamic forces acting on the single floating element due to typical current streams occurring around the North Sea coastal areas. The CFD analysis was carried out using FINE™/Marine 8.1, consisting of: NUMECA HEXPRESS™ for automated all-hexahedral unstructured discretization of the flow domain; ISIS-CFD for flow computation; and, NUMECA CFView™ for visualization of the results. During the simulation, only the pitching and heaving motions were solved, to allow for the floater moving to its dynamic equilibrium position. The effects of turbulence were modelled with the k-ω SST model.

6.1 Computational Mesh

Two computational domains were used for this case study: one for a deep-water case and a second one for a shallow water environment, which should be more representative of a coastal installation of the floater. Particular attention has been dedicated to the latter scenario since the flow entrained in between the floater structure and the seabed was expected to generate strong interaction forces due to typical shallow water effects (i.e. blockage). This peculiar situation is generally resulting in an increment of those forces exerted by the flow on the structure—as well as on its hydrodynamic equilibrium—due to its local acceleration, hence on the required tension to be sustained by the mooring lines. To resolve all the relevant flow features, the domain volume was divided into small cells to generate the numerical mesh. Table 7 shows the relative dimensions of the computational domain when using the maximum length of the floater of 247 m as a reference value (L_{ref}), whereas Fig. 12 shows the resulting domains. An example of the computational mesh generated can be seen in Fig. 13.

6.2 Solver Setup

For a smoother convergence, the simulations were initialised with a zero flow speed and slowly accelerated to an imposed current speed of 1.5 knots (0.772 m/s) as this was deemed being representative of the tidal currents encountered in the installation area of interest. A “Volume of Fluid” (VoF) method was used to account for the

Table 7 Relative computational domain dimensions

Number of L_{ref} from origin				
Domain	Deep water		Shallow water	
Direction	Min.	Max.	Min.	Max.
x	−4	4	−4	4
y	−3	3	−3	3
z	−2	1	−2	1
No. of cells in mesh (millions)	2.6		11.7	

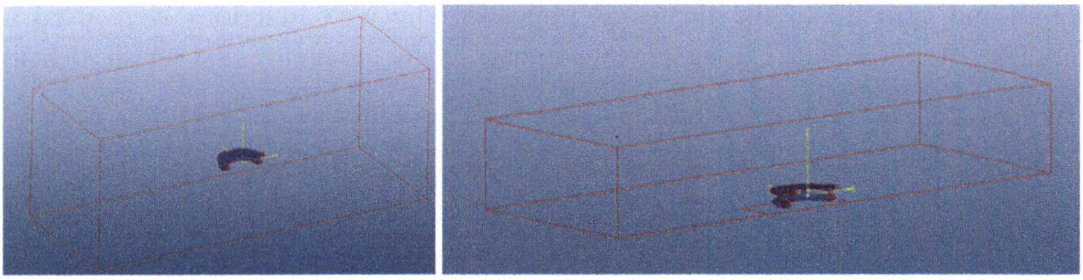

Fig. 12 Deep water (left) and shallow water (right) CFD domains

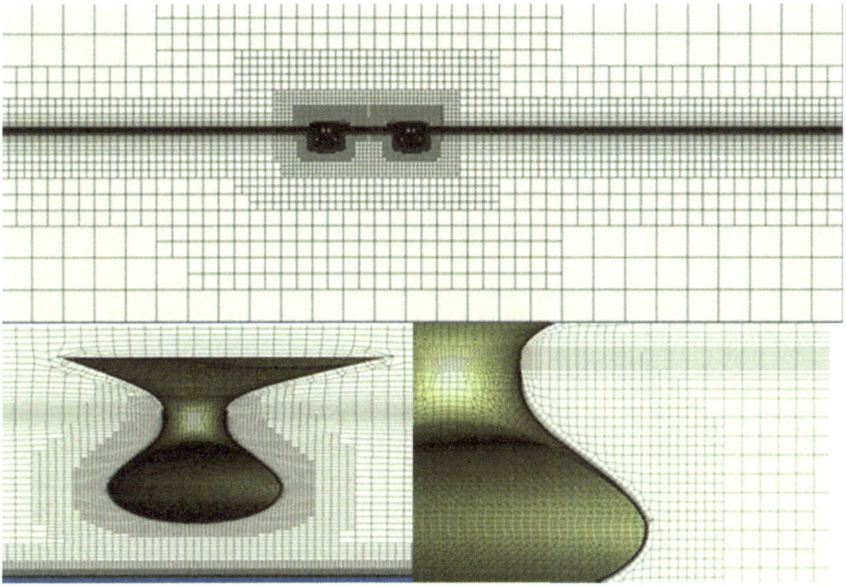

Fig. 13 Details of the computational meshes

free surface (i.e. both water and air flow are solved), for which the parameters are given in the Table 8 in accordance with the most recent ITTC recommended values.

The solver adopts the (Unsteady) Reynolds Averaged Navier Stokes (U)RANS equations to describe the flow motions and characteristics. These equations need a closure model for which the two-equations k-ω SST Menter turbulence model was used. The free stream turbulence quantities were initialized using the reference

Table 8 Fluid properties

Property	Temperature	Density	Dynamic viscosity
Unit	°C	kg/m^3	10^{-6} Pa·s
Water	19	1025.07	1103
Air	19	1.20	18.5

length and velocity of the floater as imposed. The wall functions were used to simulate the flow in regions close to solid walls, reducing the mesh density requirements in the boundary layer.

6.3 Results

From the results gathered for the floater in shallow waters and exposed to a tidal current of 1.5 kn (0.772 m/s), instabilities of the floater were detected. Strong blockage effects occur underneath (Fig. 14).

From ship design theory, it is generally accepted to consider potential shallow water effects relative to water depth h and depth Froude number $Frh = V\sqrt{g}h$, where $Frh < 1.0$ is subcritical and $Frh > 1.0$ is supercritical. Once operating near or approaching $Frh = 1$, corrections to the resulting hydrodynamic forces due to local flow effect will be required, usually based on water depth and current speed. Although in the case under examination the resulting depth Froude number is far from the critical region, however, this does not mean there are no shallow-water effects yet (Table 9).

Even for a small depth Froude number, the ratio h/T can be rather small, meaning the floating object has little under keel clearance. The seabed then restricts the flow under it and forces it to pass along the side. As a result, the flow will follow a more horizontal path, often characterised by larger curvature and, together with the mirroring effect of the bottom, this causes larger pressure gradients. Simultaneously, the increased flow speed past the floating object due to the proximity of the bottom means a lower pressure, causing an increased dynamic sinkage that typically must be restored by further volume immersion, as it has been shown

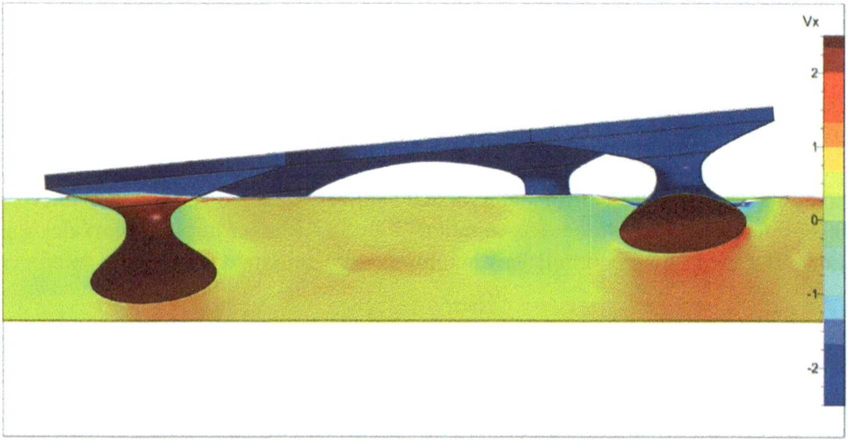

Fig. 14 Flow axial velocity (Vx) contour plot at the symmetry plane

Table 9 Shallow water case parameters

Variable	Symbol	Value	Unit
Current speed	Vc	0.722	m/s
Water depth	h	45.0	m
Floated draft	T	30.0	m
Depth Froude number	Frh	0.037	–
Under keel clearance	h/T	1.5	–

by the results. The computed dynamic trimming of the platform (5.5°) under external perturbations is clearly unfavourable, since in case of h/T values too close to the unity, there is a high risk of grounding events.

7 Conclusions

In this paper, a novel type of modular, freeform floating island has been proposed, made possible through parametric modelling and Buoycrete, a neutrally buoyant, non-dissolvable concrete mix. A parametric model allows for the rapid generation of variations of this concept, in the form of single modules, or tilings and tessellations of multiple islands. With Buoycrete such island networks become technically and financially feasible. Future research will give answers on technical aspects of the inflatable construction method for these large floating structures. Inflatable reinforcement integration, dimensional stability, inflatable injection methods are some aspects which are being researched at this moment.

One output from the parametric model, a first design iteration for a single module was evaluated using diffraction and CFD analysis. The hydromechanical analysis regarding seakeeping, drift forces and coupling forces, shows the feasibility of this initial freeform floating island design. However, more iterations of the complete analysis, including multi-body dynamics, could further improve the hydrodynamic characteristics and create more insights in the coupling forces. From the CFD analyses it was seen that the floater is too sensitive to external perturbations such as the hydrodynamic forces generated by the interaction with the seabed. This resulted in an unstable platform which is not yet adequate for the intended use.

Ongoing analyses—not published in this paper—are aimed at improving the conceptual design, such as increasing the inherent stability of the floater's initial configuration. This is done through a feedback loop from current results, further informing the parametric model.

Future studies in the next stages of development can focus in detail on the cost-effectiveness of the proposed structures, connections between modules, mooring systems, hydrodynamic performance, as well as structural and fatigue performance.

Fig. 15 Artist impressions of the proposed system of freeform floating islands (Matteo Covini)

Further iterations of the design will push the limit beyond what is traditionally possible with conventional construction, by using geometry to efficiently achieve characteristics desirable for large floating structures and cities (Fig. 15).

References

1. Mohie El-Din R (2015) Overview of futuristic floating cities and structures. Futuristic cities "aqua cities". In: International conference on IT, architecture and mechanical engineering (ICITAME2015) Dubai, UAE, May 2015
2. Watanabe E, Wang C, Utsunomya T, Moan T (2004) Very large floating structures: applications, analysis and design. Centre for Offshore Research and Engineering, National University of Singapore. CORE report no. 2004-02
3. Köpke K, Lindeboom R, Jak R (2018) Inventory of regulations D2.1, V2.2, Space@Sea report
4. Bhooshan S, El Sayed M (2011) Use of sub-division surfaces in architectural form-finding and procedural modelling. In: Proceedings of the 2011 symposium on simulation for architecture and urban design (SimAUD'11), San Diego, CA, USA, pp 60–67

5. Halpern A, Billington D, Adriaenssens S (2013) The ribbed floor slab systems of Pier Luigi Nervi. In: Obrębski J, Tarczewski R (eds) Proceedings of the international association for shell and spatial structures (IASS) symposium 2013, Wroclaw, Poland, Sept 2013
6. Grünbaum B, Shephard G (1987) Tilings and patterns. Freeman, San Francisco, CA, USA
7. Visch G (2019) Method and system for forming structures in fluid, especially under water. Patent NL2018969B1

The Evolution of Aquatecture: SeaManta, a Floating Coral Reef

Joerg Baumeister

Abstract With our oceans under increasing pressure, our coral reefs are amongst the first to suffer. Therefore, we were exploring to what extent we could merge our skills as aquatects with those of marine engineering and marine environment to design a floating artificial reef which provides ecological and economic stimulus to our reefs and oceans. The transdisciplinary system-based design approach revealed many opportunities for floating structures. Therefore, we suggest as a more general outcome the development of the concept "aquatecture" further. The specific outcome is a first of its kind floating structure with a total length of 60 m. Its pragmatically shaped aquatic design is driven by hydrodynamic and wind parameters, maximum water contact, and sufficient sunlight for the floating reef. Digital models were cross-compared with scaled physical models in wave tanks and a wind tunnel. The hybrid design technique was used to optimise the development process which lead to a Manta-like form, therefore the name SeaManta. The Sea Manta is anchored to allow to move as non-invasive structure in line with changing wind and wave conditions. Its major component is the integrated artificial reef which can be sustained through a process called electro-accumulation of minerals. A low voltage current is passed through the structure to help grow a limestone-like substance from the dissolved minerals in the water, supplemented by porous 3D printed ceramic. This will create a very strong reef structure with high levels of dissolved oxygen, letting the corals and fish flourish. Divers can enjoy this marine habitat below the water surface. Above water, hospitality and dive facilities, a swimming pool and interactive aquarium displays are inviting everybody who is thrilled by the synergy of marine wild-life's protection and fascination.

Keywords Aquatecture · Floating reef · Marine system design · Aquatic design · Dive attraction

J. Baumeister (✉)
SeaCities.org, Cities Research Institute, Griffith University, Gold Coast, QLD, Australia
e-mail: j.baumeister@griffith.edu.au

Ł. Piątek et al. (eds.), *WCFS2020*, Lecture Notes in Civil Engineering 158,
https://doi.org/10.1007/978-981-16-2256-4_8

1 Aquatecture

Especially under the current anthropogenic conditions, humankind should have the aim to adapt to the changing future. One crucial question thereby is about the relationship between humankind and the nature which has been discussed already by philosophers of ancient Greece: "Economists are concerned with the environmental constraints on human population growth; geographers and anthropologists seeking to understand global patterns of land use and culture, and ecologists and conservatists are concerned with human impacts on the environment" [1]. The task of future Architecture and Design will be to consider these and more aspects.

A systemic perspective helps to understand holistically the natural environment as dynamic interaction between biomass (like plants and animals), abiotic matter (like minerals), sun, wind, and water. The natural elements are interconnected to manmade built environment consisting out of system elements like structures (buildings), transport and infrastructure, people and community, and economy. This holistic perspective of natural and built system elements as part of one environmental system allows architects and designers to create environmental-friendly and -integrated buildings, cities, and landscapes [2].

When compared with land-based environment, the marine environment comprises very different conditions due to waves, buoyancy effects etc. It cannot be handled by architects who are educated to design land-based environments. And naval architects are focusing on the design of vessels or rigs. This is the reason why we are suggesting to bridge this gap with the introduction of the system- and marine-based term "aquatecture". It derives from merging the Latin expression aqua for water and architecture. Like architects, aquatects could operate on a smaller scale for floating structures and on a larger scale for floating cities or urban areas which will be flooded soon.

Since 1986, aquatecture has been used for the design of water-features in the landscape [3] and for buildings on floating platforms. Companies for the design of swimming-pools and bathrooms applied it too. A more holistic definition of aquatecture has been introduced by publishing "a system of floating aquaculture, manufacturing, transportation, and energy facilities for four water-borne structures that explore using the water environment for producing food, moving manufacturing onto the water; integrating rail, air, and sea-based shipping and transportation systems; and harnessing solar, wind, wave, and thermal energy sources for electricity" [4] for the New Tokyo Plan 2025.

The latter example can still serve as a role model to demonstrate how a specific aquatic site can be related to the design of a diverse built system which is still a challenging issue. Therefore, the following case-study will continue to test the question: How can aquatecture with a system-based perspective onto the marine environment create and apply new solutions for the anthropogenic challenges?

2 The Problem

What has been predicted by researchers already more than 50 years ago [5] is finally common sense: Burning of fossil fuels emits greenhouse gases which generate global warming and consequently sea level rise (SLR). Current anthropogenic greenhouse gas concentration has reached a concentration which resulted in earlier history in 15–25 m higher sea levels [6].

Fortunately, the melting of the north and south polar ice is a slower process than the increase of greenhouse gases and temperature. Unfortunately, these relevant physical ice sheet processes are still poorly understood to come up with proper predictions. The current (always upward developing) estimations are a SLR of up to 1.40 m or more by 2100 [7].

Out of pragmatic perspective, it can be argued that change of sea levels happened constantly in Earth's history due to the continuous change of climate (which makes the term "Climate Change" as a description of the current anthropogenic changes questionable). Greenhouse gas concentration changed always due to internal events such as volcanic eruptions or events coming from outside the earth like meteor strikes. The novelty is nowadays, that not nature-based processes but humankind is responsible for *the* increasing impact of greenhouse gas and rising sea level being responsible for endangered coastal cities, suffering refugees and climate wars, as well as increasing shortage of food and freshwater.

The rapidly changing climate impacts organisms too, including extinction and decreased population abundance, reduced genetic diversity, lower reproductive success and ability to disperse, and altered interspecies interactions [8]. Reef structures are thereby one of the most endangered living structures: Approximately 75% of the world's reefs are under threat due to human activity [9]. Reefs feed millions of people directly as source of food and income through tourism. The more the biodiversity of reefs declines, the more resources and the protection of the coast are lost [10].

Although dangers of reefs are nowadays more evident and discussed, SLR has to be seen as even bigger danger in the near future [11]. Many reefs already suffer from the gap between maximum growth rate of reefs and local rise of sea level. If the sea water level rises by more than 0.5 m in 2100, more than 75% of the reefs worldwide will perish [12]. By having the current estimations of up to 1.40 m SLR in mind, the probability seems to be very high that many reefs will decline in the next decades.

Experiments have shown that coral reefs could be taken out of their naturally dynamic equilibrium to be put into an alternative state [13]. In addition to the more general question about the future of aquaculture in Chap. 1 the following question has to be asked therefore: How can a floating coral reef be created to adapt automatically to SLR? Both questions are related because the floating reef investigation can serve as an example for aquaculture. And vice versa, aquacultural thinking can drive the floating reef design.

Both questions will be examined further in four steps: First there will be an investigation about possibilities how to grow reefs artificially (Sect. 3) and how to make reef structures floating (Sect. 4), before a systemic design development will lead to a proposal (Sect. 5) which will lead into a final discussion (Sect. 6).

3 Artificial Coral Reefs

Coral reefs are submerged structures that can be found almost anywhere in the deep sea. However, their most common distribution is in tropical latitudes where they are most visible because they are close to the surface of the sea [14].

Coral polyp organisms, living in symbiosis with algae, are extracting calcium carbonate crystals from the sea and creating a calcareous protective exoskeleton. The more the living colonies increase, the more the limestone deposit grows [15]. Due their mobility, the polyp larvae can spread and can attach at other submerged locations to found new colonies [16].

The dramatic depletion of coral reefs worldwide has driven several attempts to help to restore coral cover and reef ecosystem function. The perhaps most promising active restoration concept is the "gardening" concept which works as a two-step restoration process. The first step deploys mature coral larvae ex situ onto a support material. Large stocks of coral colonies in mid-water floating nurseries are thereby created from a coral nubbin. As soon as they reach suitable sizes, they are transplanted in a second step either onto the degraded reef or somewhere else [17].

The transplantation from the nursery to the reef can happen with different techniques. One technique starts with drilling holes into a substrate to attach corals horizontally as well as vertically which allows a maximum coverage. Other transplantation techniques use iron meshes (which are usually utilised in the construction industry) or ropes which are fixed to the sea floor [18]. The transplantation techniques have been tested at several coral reefs worldwide successfully [19]. This is the prove that new artificial reef systems can be introduced successfully.

The materiality of the artificial reef based on a limestone structure can follow the original slow composition process of natural limestone. But it can grow also faster with "Biorock" where electric reefs are grown by low voltage electrolysis of sea water. Biorock reefs have also a higher coral survival rate during natural bleaching events compared to a natural reef. Following the manufacturer, "Biorock reefs are growing limestone structures that get stronger with age, repair themselves, and are cheaper than concrete." It can protect floating steel structures also from corrosion [20].

Can these reef structures float? If ecological systems of healthy coral reefs (Fig. 1a) are endangered and floating nurseries for coral reefs' larvae are already applied (Fig. 1b) [21], there is no reason why the materials used by coral transplantation techniques (substrate, the iron meshes, rope) cannot be part of a floating structure (Fig. 1c). Principles of floating structures will be investigated in the following chapter to understand this better.

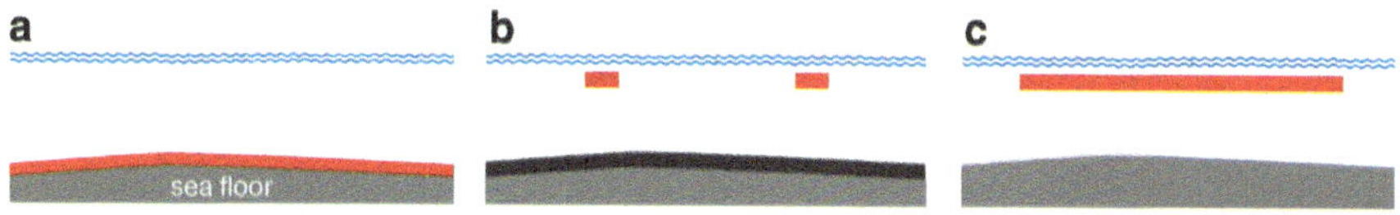

Fig. 1 **a** Reef with corals on bottom of sea floor (*left*). **b** Floating nurseries in use (*middle*) on top of dead coral reef. **c** Floating reef (*right*)

4 Floating Structures

The idea of a floating artificial reef is perhaps new. But floating structures are widespread starting from historic floating villages, floating bridges, floating ports and docks, offshore floating structures for oil exploration and extraction, floating entertainment facilities, to floating utility plants. Floating structures keep automatically the same distance to the water surface and have even more advantages: They can be moved so that the location of construction can differ from the location of deployment, they are earthquake resistant, and they can be less disruptive than other interventions if sunlight is permitted to penetrate [22].

The intense exposure to the complex maritime environment requires special considerations for floating structures:

- Horizontal forces of waves and wind demand proper mooring like a conventional connection with anchor chain and anchor to the seabed
- The structure's buoyancy force has to balance the static self-weight and payloads
- Dimensions depend on environmental forces, peak loading, and the consideration of possible accidental events like collisions with ships
- Either parts of the structure or the entire structure must be capable to move from the construction site on-land to be assembled on the sea
- The corrosive sea environment demands a good corrosion protection and an ongoing inspection [22, 23].

Preferred materials for floating structures are currently concrete or steel, the latter especially if light-weight and flexible constructions are desired. The shift of Additive Manufacturing (AM) from rapid prototyping to rapid manufacturing allows also the introduction of materials which can withstand the harsh marine environment often better. AM products out of ceramics could become for example attractive alternatives for floating structures [24, 25].

Floating structures are different to vessels: Vessels consist out of one or more hulls which are primarily shaped to reduce hydrodynamic drag when they move (Fig. 2a). In the opposite to that, floating structures are often anchored with the consequence that the shape is primarily driven by the purpose to carry payloads. The depth-to-width ratio determines thereby the classification either as pontoon (Fig. 2b), or as semi-submersible (Fig. 2c) floating structure. Hybrids of the different floating structures may unify advantages of types [22].

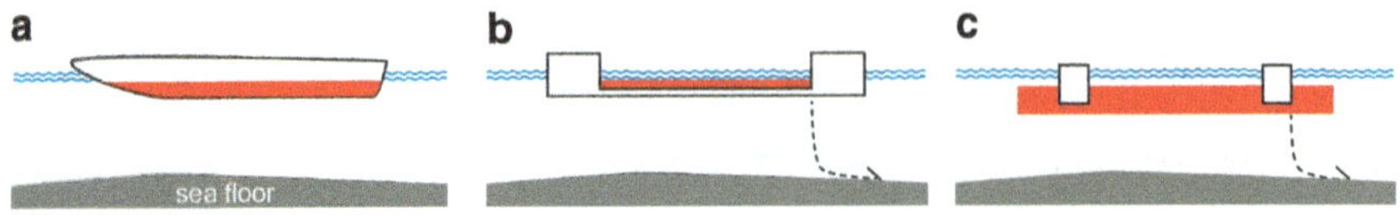

Fig. 2 **a** Vessel (*left*). **b** Pontoon (*middle*). **c** Semi-submerged (*right*)

In the discussed case of a floating reef, the volume needs to be semi-submerged to enable marine life. Being anchored in a (sometimes) rough environment, the shape of the semi-submerged structure should consider also horizontal forces of waves and wind. Therefore, the structure will be a hybrid of

- A semi-submerged structure with integrated buoyancy bodies to keep it sub-submerged and
- A hydrodynamically shaped vessel type.

These findings are giving us enough background to start the design process.

5 Design Development of SeaManta

Following the introductorily comments regarding aquatecture, the case-study of a floating coral reef will be developed as a holistic system where elements of the natural environment including waves, wind and the sun are relevant drivers of design [26]:

The aquatectural design starts with the development of the hybrid floating structure and will be shaped by the logic of three system elements of the natural environment which are water, wind, and sun. Following a request coming from a potential partner, the Arabic Gulf has been defined as area of operation. This enables a specific definition of maximum wave height, maximum wind speed and a location-based sun irradiation pattern.

Owing to the complexity of fluid dynamics, physical and digital models have been built and improved in various design steps. Hydrodynamically, the underwater "hull" has been developed to minimise the drag in accordance to the arrangement and dimension of the buoyancy bodies and the position of the chain (Fig. 3a). Aerodynamically, the structure above the water has been progressed in a way that minimises wind resistance and reduces vorticity to avoid sedimentation of desert sand (Fig. 3b).

The optimisation of sunlight irradiation for the underwater reef lead to two measures: A longitudinal cut enabled central illumination and the lateral expansion created underwater wings (Fig. 3c) which remind of the shape of a Manta, therefore the name SeaManta.

The main target of the SeaManta is the creation of a coral reef ecosystem, which makes Biomass to another important system element. The limestone should expose

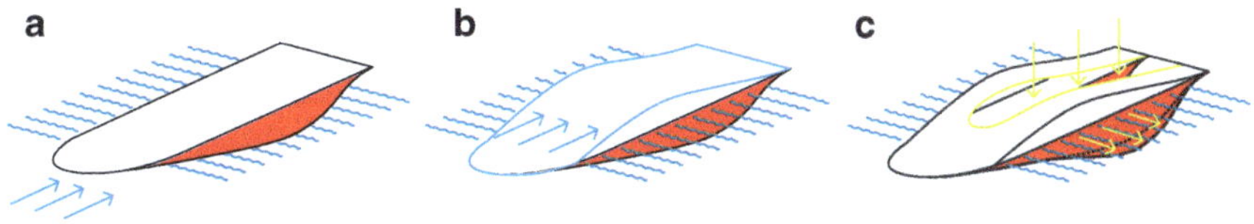

Fig. 3 Drivers of the design development: **a** Water current (*left*). **b** Wind flow (*middle*). **c** Sunlight (*right*)

the corals to the water and its related ecosystem as much as possible. Therefore, a load-bearing skeleton structure either out of metal and/or ceramics is suggested. The reef works as its infill out of limestone and reinforced by Biorock.

An "assisted colonisation" approach [27] enables the translocation of corals within their natural range. The procedure is well known due to successfully executed interventions such as coral nurseries and reef restoration [18]. More species like fish, lobsters, clams, seahorses, sponges, and sea turtles as part of the ecosystem will be provided with facilities to feed and breed and populate the reef. Therefore, the design should include protected breeding areas, aquariums for nurseries, elements to hide and integrated caves.

The creation of a diverse ecosystem requires a larger sized natural reef environment. At the other hand, the floating coral reef operates in a prototypical way which has certain risks, therefore the effort for construction and maintenance should be limited. The best compromise between a maximised ecosystem and a minimised structure will be in the middle. We suggest to start to test a floating reef structure with a length of 60 m and, due to the required hydrodynamic shape, 25 m width. This dimension of an average super-yacht allows also its construction in a dockyard which can be moved on site after completion (Fig. 4).

So far, the proposed floating structure has been only shaped by drivers of the natural environment. They, and the structural parameters, were also responsible for the proposed dimension of the SeaManta. However, its size creates also opportunities for functions for people. Therefore potential extra functions of the developed spaces will be explored by focusing on the benefits for community and economy.

Different communities will be attracted by the SeaManta. The one will be the scientific research community which will be important to improve the efficacy, and cost-efficiency of the floating reef and to manage risks. This requires permanent monitoring of the reef's ecosystem performance by researchers and constant adaptation of scientific models' predictions [28]. Therefore, the SeaManta will be fitted out with research facilities (1), an observation deck (2), a dive pool (3), coral and fish nursery (4), and research aquariums (5).

Another community will be the tourists. Coral reefs are the reason for 70 million tourist trips per year generating US$36 billion global economic value for tourism and recreation [29]. Therefore, the experience factor and uniqueness of the SeaManta can be considered as a potential income generator. Touristic functions for dive and non-dive tourism in all age groups should be included, therefore following

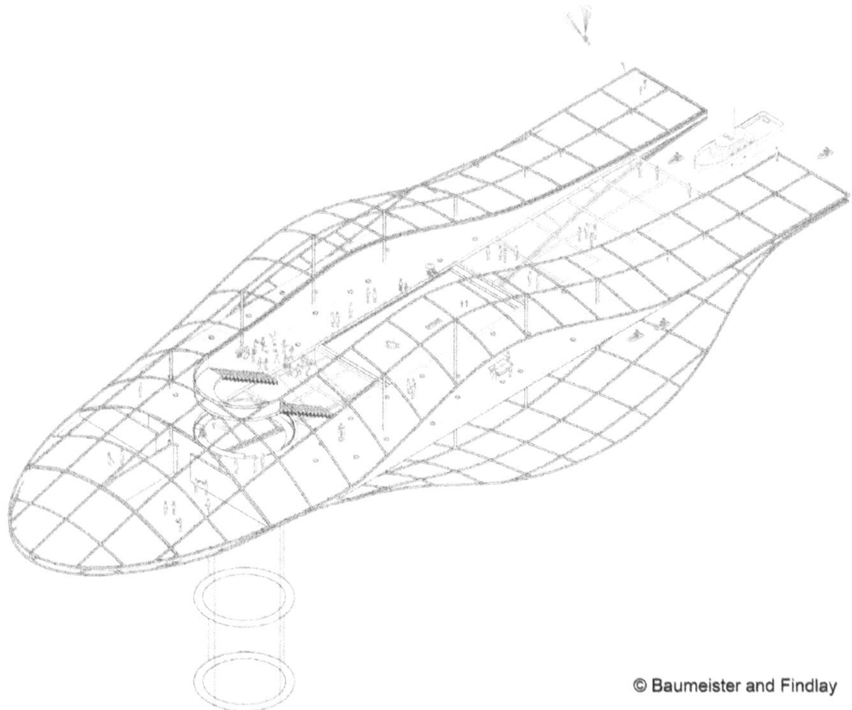

Fig. 4 Overall shape driven by parameters of natural environment

additional attractions are suggested: Arrival at the mooring (6), walking over a glass bottom (7), watching the underwater exhibition (8) along the paddling pool (9).

Tourists will require a third community group which will be business operators. Transport providers will make the SeaManta accessible and a dive school + shop operator (10) will be as essential as a food and beverage (11) and an event space (12). Either a research organisation or an extra operator will deal with the management of the SeaManta regarding safety, maintenance, and organisation of programs (Figs. 5, 6, and 7).

All communities together, the scientific, the tourists, and the businesses are creating additional opportunities: Business operators can run workshops with researchers to share research topics and experiences with tourists. At the same time tourists can be involved in feeding and monitoring activities. This will make the SeaManta to an ambassador promoting the co-existence of humankind with the natural marine environment in a sustainable way.

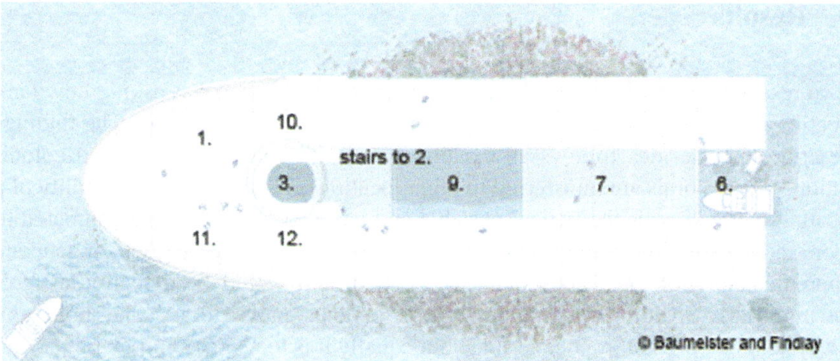

Fig. 5 Floor plan

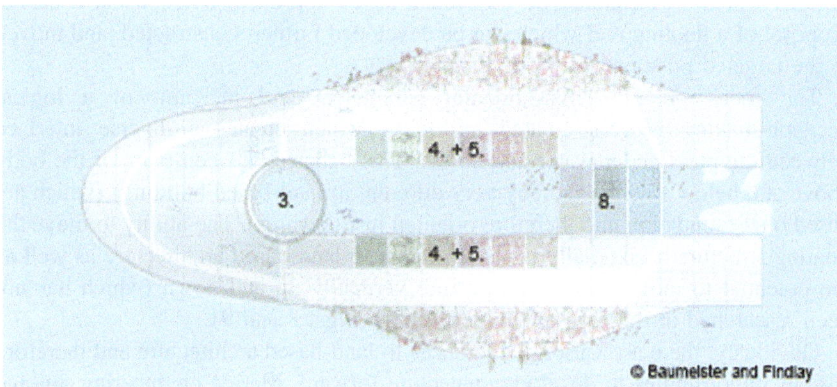

Fig. 6 Reef plan (underwater)

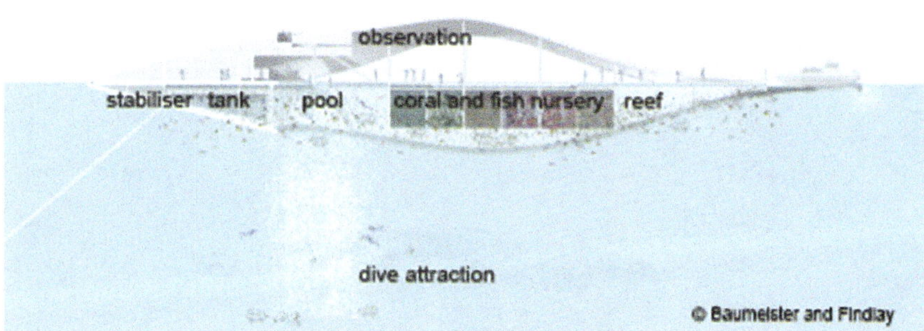

Fig. 7 Section view

6 Results

Two research questions were raised at the beginning. The one was asking how far a floating coral reef can be created which adapts automatically to SLR. The findings describe that the idea follows an existing successful concept of "assisted colonisation" where corals are transferred to other locations. Therefore, the possibility of a floating reef colony is theoretical possible and has also a higher value compared to more radical solutions like genetic modifications or geoengineering which generate unpredictable ecological consequences. A next step would be a reality-test following Mark Spalding's comment: "It is perhaps naive to try and hold any ecosystem in a pre-anthropocene state, but equally it is too early to proclaim the end of coral reefs" [29].

The second research question was about the potentials of future aquatectural developments. The described design experiment is based on logical step-by-step process which was enabled by its system-based approach. It results in a design proposal of a floating reef which can be developed further, constructed, and moved to the targeted position.

The focus on the environmental parameters and the aim of a logical decision-making process created an object which creates a diverse interface between the parts under water and above. This balance between parts of the body above and below water is thereby very different to land-based buildings (which are based on a foundation and therefore oriented just upwards). The ability to move the floating structure horizontally is also different to land-based architecture as well as the potential to move a floating structure vertically up and down (which has not been researched in the case of the SeaManta) (Figs. 8 and 9).

Obviously, there are various differences to land-based architecture and therefore various opportunities to develop aquatecture forward. Marine engineering, marine planning, and marine environment are existing fields of research and expertise, but

Fig. 8 Elevation above water surface

Fig. 9 Perspective below water surface

there is no corresponding area for water-based architecture. Therefore, this is to suggest an investigation how to evolve aquatecture further as a next step.

References

1. Chapin FS, Kofinas GP, Folke C, Carpenter SR, Olsson P, Abel N, Biggs R, Naylor RL, Pinkerton E, Stafford DM, Steffen W, Walker B, Young OR (2009) Resilience-based stewardship: strategies for navigating sustainable pathways in a changing world. In: Principles of ecosystem stewardship, pp 319–337
2. Baumeister J, Ottmann DA (2015) Urban ecolution—a pocket generator to explore future solutions for healthy and ecologically integrated cities. UWAP, Crawley. ISBN: 9 781742 589985
3. Wylson A (1986) Water environment: water features. Aquatecture:149–184
4. Aquatecture: Creating land in the water (1988) The Futurist 22(2):52. Washington
5. Meadows DH, Meadows DL, Randers J, Behrens WW (1972) The limits to growth: a report for the club of Rome's project on the predicament of mankind. Universe Books, New York
6. Jones N (2020) How the world passed a carbon threshold and why it matters. Yale E360. https://e360.yale.edu/features/how-the-world-passed-a-carbon-threshold-400ppm-and-why-it-matters. Accessed 8 Nov 2020
7. Pörtner, HO, Roberts DC, Masson-Delmotte V, Zhai P, Tignor M, Poloczanska E, Mintenbeck K, Nicolai M, Okem A, Petzold J, Rama B, Weyer N (2019) IPCC, 2019: summary for policymakers. In: IPCC Special Report on the Ocean and Cryosphere in a Changing Climate. In press. https://www.ipcc.ch/site/assets/uploads/sites/3/2019/11/SROCC_SOD_Ch04_FINAL.pdf. Accessed 8 Nov 2020
8. Segan DB, Murray KA, Watson JE (2016) A global assessment of current and future biodiversity vulnerability to habitat loss–climate change interactions. Global Ecol Conserv 5:12–21
9. Burke L, Reytar K, Spalding MD, Perry AL (2011) Reefs at risk revisited. Manage Environ Qual: Int J 22 (4). https://doi.org/10.1108/meq.2011.08322daa.003. Accessed 8 Nov 2020
10. Spalding M, Brown BE (2015) Warm-water coral reefs and climate change. Science 350 (6262):769–771
11. De'ath G, Fabricius KE, Sweatman H, Puotinen M (2012) The 27-year decline of coral cover on the Great Barrier Reef and its causes. Proc Natl Acad Sci 109(44):17995–17999

12. Perry CT et al (2018) Loss of coral reef growth capacity to track future increases in sea level. Nature 558(7710):396–400
13. Hughes TP (1994) Catastrophes, phase shifts, and large-scale degradation of a Caribbean coral reef. Science 265(5178):1547–1551
14. Smith SV (1999) Coral reef ecosystem. In: Alexander DE, Fairbridge RW (eds) Encyclopedia of environmental science, pp 1010
15. Barnes RS, Hughes RN (1999) An introduction to marine ecology. https://doi.org/10.1002/9781444313284. Accessed 8 Nov 2020
16. Shumway SW (1998) An introduction to the biology of marine Life. James L. Sumich. Q Rev Biol 73(2):224–225
17. Chan WY, Peplow LM, Van Oppen MJ (2019) Interspecific gamete compatibility and hybrid larval fitness in reef-building corals: implications for coral reef restoration. Sci Rep 9(1). https://doi.org/10.1038/s41598-019-41190-5. Accessed 8 Nov 2020
18. Rinkevich B (2014) Rebuilding coral reefs: does active reef restoration lead to sustainable reefs? Curr Opin Environ Sustain 7:28–36
19. Rinkevich B (2015) Climate change and active reef restoration—ways of constructing the "Reefs of tomorrow". J Mar Sci Eng 3(1):111–127
20. Hilbertz WH, Goreau TJ (1996) Method of Enhancing the Growth of Aquatic Organisms, and Structures Created Thereby. U.S. Patent No. 5,543,034
21. Levy G, Shaish L, Haim A, Rinkevich B (2010) Mid-water rope nursery—testing design and performance of a novel reef restoration instrument. Ecol Eng 36(4):560–569
22. Wang CM, Wang BT (eds) (2015) Large floating structures technological advances. Springer, Singapore. ISBN 978-981-287-136-7
23. Clauss G, Lehmann E, Östergaard C (1992) Hydromechanical analysis of offshore structures. In: Offshore structures, pp 145–332. https://doi.org/10.1007/978-1-4471-3193-9_3. Accessed 8 Nov 2020
24. Breddermann K, Drescher P, Polzin C, Seitz H, Paschen M (2016) Printed pressure housings for underwater applications. Ocean Eng 113:57–63
25. Park H, Kim E, Cho G, Jung Y, Zhang J (2019) Process development of fabricating ceramic core using 3D printing technique. Mater Chem Phys 231:382–387
26. Baumeister J, Findlay N (2020) SeaManta, a research exploration in aquatectural design. ISBN: 978-0-6483715-3-3. www.seacities.org. Accessed 8 Nov 2020
27. Coles SL, Riegl BM (2013) Thermal tolerances of reef corals in the Gulf: a review of the potential for increasing coral survival and adaptation to climate change through assisted translocation. Mar Pollut Bull 72(2):323–332
28. A Decision Framework for Interventions to Increase the Persistence and Resilience of Coral Reefs (2019) National Academies of Sciences, Engineering, and Medicine. The National Academies Press, Washington, DC. https://doi.org/10.17226/25424. Accessed 8 Nov 2020
29. Spalding M, Burke L, Wood SA, Ashpole J, Hutchison J, Zu Ermgassen P (2017) Mapping the global value and distribution of coral reef tourism. Mar Policy 82:104–113. https://doi.org/10.1016/j.marpol.2017.05.014. Accessed 8 Nov 2020

BLUECITY Lab: A Climate Adaptation Amphibious Lab

Gita Nandan, Zehra Kuz, and Tim Gilman-Ševčík

Abstract RETI Center, BlueCity LAB [BCL] creates a framework to transform the City's shoreline community with post-carbon solutions as the first off-grid, community-based, floating climate lab in the US. This climate responsive architectural innovation deploys design principles for amphibious structures supported by scientific climate adaptation research, with education and workforce training as its social benefit focus. BCL envisions an organic amphibious structural prototype; designed based on interactions with the surrounding environment and tidal flows. This hybrid between a vessel and a building provides a laboratory-like training atmosphere, creating net-positive loops related to energy, water, air, food, and aquamarine habitat. It cleans environmental systems through low-carbon biophilic concrete formwork, solar capture skins, aquatic kelp and mussel habitats, phytoremediation gardens, water-heat exchanger, anaerobic digestion, a desalination system, and more. BCL moors at the shoreline of a low-income community, extending the community's reach into New York Harbor for social, educational, vocational and environmental benefits, additionally serving as a model for others. The 13-acre waterfront of GBX Terminal in Red Hook Brooklyn is secured as the home.

Keywords Amphibious · Carbon-capture · Net-positive · Self-sustaining · Eco-restoration

---

G. Nandan (✉)
Thread Collective, 225 Troutman Street, Brooklyn, NY 11237, USA
e-mail: gita@threadcollective.com

G. Nandan · Z. Kuz
Pratt Institute, 61 St James Pl, Brooklyn, NY 11238, USA

Z. Kuz
Oasis Design Lab, 37 King Street, Norwalk, CT 06851, USA

T. Gilman-Ševčík
RETI Center, 153 Coffey St, Red Hook, Brooklyn, NY 11231, USA
e-mail: tim@reticenter.org

Ł. Piątek et al. (eds.), *WCFS2020*, Lecture Notes in Civil Engineering 158,
https://doi.org/10.1007/978-981-16-2256-4_9

1 What is RETI Center and the BlueCity Lab?

NYC must face the reality of climate change and develop new modes of equitable adaptation and resilient waterfronts—not in 100 years but now. It is predicted we will see, at minimum, six feet of sea-level rise by 2100. While political forces are negligent in building flood barriers to protect our marginalized communities, we, as a team of activists, planners, designers, engineers, and architects, must ask ourselves instead: How can we design and live with water now? What is a new paradigm for equitable urban coastal inhabitation for the next 50 years? The timing is critical.

The Resilience Education, Training and Innovation Center (RETI) was developed to weave together businesses, local government, workforce and community development organizations, and educational institutions, with the purpose of driving innovation and job creation to advance urban climate resilience and sustainability in New York City and beyond [1, 2]. In doing so, it addresses the socioeconomic inequity challenges that low-income coastal communities face, by providing education, job training and placement, and improved local infrastructure. Headquartered on a donated piece of privately-owned waterfront property in Southwest Brooklyn, RETI Center is establishing BlueCity Lab (BCL), a water-borne physical space and community support structure, where these entities work together to develop technologies, products, and services to help New York City become more sustainable and climate resilient [3]. The beneficial urban impacts are paired with robust programs to train the workforce needed to plan, build and operate these emerging industries. Human benefits within the built environment are synchronized with benefits to the natural environment and local ecosystem—promoting biodiversity, environmental advocacy, and an approach to our living environment in accord with positive anthropogenic impacts. RETI Center offers a substantial step forward in achieving a more resilient and equitable New York City in the face of rising sea levels, extreme weather events and climate change (Fig. 1).

While some floating developments prioritize the isolation of the facility, BCL cultivates a mutually beneficial relationship with the built and natural environments of the coastline it occupies. Primarily focused on low-income urban settings, our program will respond to local economic, environmental, and manufacturing conditions with positive inputs circulating to all engaged stakeholders. Underdeveloped resources will be leveraged to maximize positive impacts; such as to educate, train, and employ disenfranchised human populations; to restore and protect existing flora and fauna and support the return of greater native biodiversity; to harness natural environmental conditions for sustainable use and biome support; and to strengthen local supply resources for waste reduction and reuse, carbon capture and sequestering, waste-to-energy processes; and social justice benefits.

BCL's mission will be framed as a site for experimentation for the eco-conscious generation of students and young people to learn about and acquire the essential skills for navigating an equitable, sustainable, and greener economy and society. Drawing on foundational partnerships with public schools emphasizing

Fig. 1 Conceptual Rendering of BCL view from the Floating Gardens, credit: thread collective + Oasis design Lab

marginalized populations of color from early education through graduate studies, BCL will provide a demonstration model, classroom, workshop and public meeting venue for the community. A Green New Deal-inspired curriculum is already being developed in cooperation with the students and their schools to ensure empowerment and innovation [3] (Fig. 2).

BlueCity has benefited from a plethora of educational partnerships from across the Northeast region for current and future studies and programming. The BlueCity Lab has already figured out the long-term planning of our school partners with PS 676 Elementary School and South Brooklyn Community High School, two neighborhood schools incorporating RETI Center into their "Imagine Schools" education innovation program, sponsored by the New York City Department of Education and Lauren Powell Jobs' XQ Super-School program. The Governor's Island-based Harbor School expects to expand its network of waterfront public schools by establishing its first middle school for maritime education on site, and Brooklyn College has launched a place-based program culminating in a graduate-level residency fellowship at GBX through their Earth and Environmental Sciences department. Student design input has been factored into the concept, classroom and laboratory programming in development, as well as exhibition materials for public outreach and education. BCL will serve as a home for applied racial and environmental justice learning, in addition to a site for scientific research, experimentation and development (Fig. 3).

Fig. 2 BlueCity partnerships and stakeholders

Fig. 3 BlueCity conference and launch event, December 2018

1.1 New York

In the geological era of the Anthropocene, our planet is changing at a pace never observed before. Pollution, natural resources exploitation, habitat fragmentation, and climate change are only some of the threats our biosphere is facing. Living in an urban condition, it can be challenging to comprehend the scale and impact of such loss and the linkages to these critical ecological networks.

New York's (city and state) response to Sandy opened a new chapter for the City's coastline. Since the city's conception, industrial development and land-use forever changed the city's waterfront. The shoreline's original soft edges and estuaries gave way to reclaimed land and hard-lined bulkheads for waterways and transportation. Decades of neglect and abuse of natural resources and environments has become a threat and is responsible for the physical decline and socio-economic disparity in many of the city's underserved and endangered coastal communities and fragile biomes. New York City encounters serious challenges with its aging infrastructure in the face of increasing demand and climate change (Fig. 4).

BCL sees itself as a demonstration and a microcosm, a small representative site within much larger regional ecological networks. The experience and journey at BCL will foster deep connections to the fertile and complex ecological systems that support species survival on our earth.

1.1.1 New York City Site Selection Process

In collaboration with Anna Yie, a graduate student at Pratt Institute, Graduate Center for Planning and the Environment, an in-depth analysis was conducted for the site selection through mapping and superimposing critical factors; including wave action, water depth, public transportation/access, local educational facilities, waterfront usage, ownership, and edge conditions [4]. The resulting composite map highlights optimal locations within the NYC five boroughs. Among these sites, Red Hook, where our organization was originally founded, is the most appropriate to host BCL. In the future, BCL outposts will be distributed across the five boroughs, acting as drivers for coastal-edge-adaptation and transformation (Fig. 5).[1]

Red Hook South Brooklyn

The Red Hook neighborhood, part of Brooklyn Community District 6, was originally settled by Dutch colonists and registered as a Town of Brooklyn in the 1600 s. The development of Brooklyn and plans to create streets began in tandem with the shipping ports that sustained a busy and prosperous life. Paired maps below are drawn barely 100 years apart. The later map from 1897, after the construction of the Brooklyn Bridge, shows the complete transformation of the original low-lying marsh land into the current, highly urbanized environment (Fig. 6).[2]

---

[1]Capstone Presentation, Anna Yie, Pratt Institute Graduate Center for Planning and Environment, 2020, link.

[2]Link to historical maps Ratzer map https://redhookwaterfront.com/2016/01/red-hook-history-maps/, 1897 map, Wikimedia Commons https://commons.wikimedia.org/wiki/File:1897_Brooklyn_map.jpg.

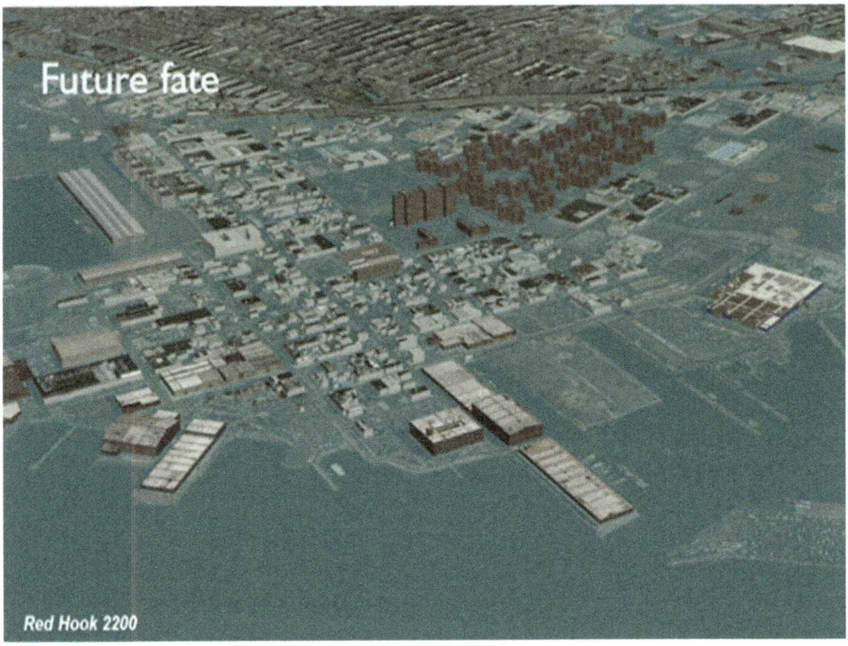

Fig. 4 Red Hook South Brooklyn, 2200 predicted sea-level rise with no mitigation measures taken approximately 12 ft

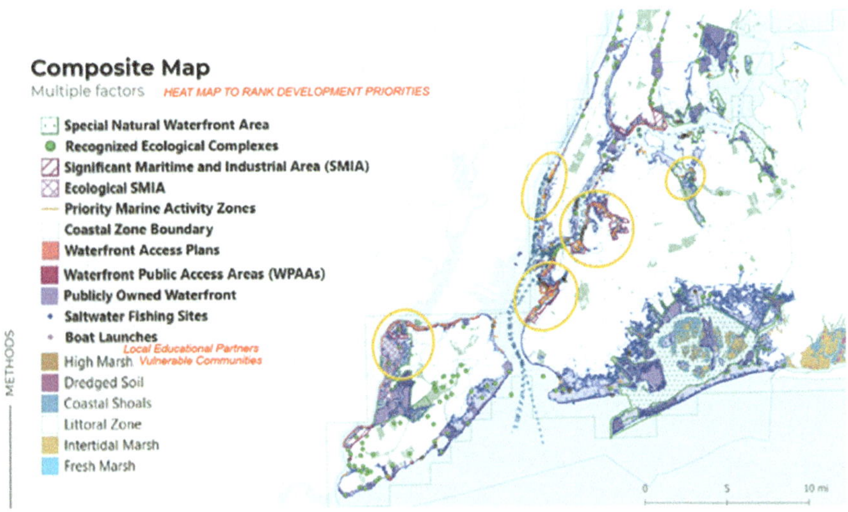

Fig. 5 Composite Map showing best options for BCL and floating developments within the five boroughs ([4] Reproduced from Anna Yie)

Fig. 6 Historical maps of Red Hook left (original marshlands from the 1600 s right) 1897 From Fluid Frontiers: stormwater management research in Red Hook Sewershed, Brooklyn: To the left is a map of Town of Brooklyn by Bernhard Ratzer 1766, to the right is a map of Brooklyn by Rand McNally & Co 1897

Red Hook Point became a shipping destination with the Atlantic Basin and the Erie Basin being developed during the middle of the nineteenth century. Shipping paved the path for the development of warehouses, catalyzing a boom in the industry. As a port, Red Hook's potential diminished when the Panama Canal was enlarged and the international shipping requirements changed. Unless there is a major commitment by the general consensus to revive and restore Red Hook as an international port, Red Hook will ride the current trajectory of gentrification.

Currently, the Red Hook neighborhood is home to about 11,000 people, with the vast majority residing in NYCHA public housing.[3] According to the most recent census report, in the surrounding areas of Red Hook and Carroll Gardens, more than 60% are White, nearly 20% Latino or Hispanic, 12% African American, 4.5% Native American and the remaining population are Asian, Pacific Islanders and other races. A full 57% of the Red Hook residents—approximately 6300—live in the New York City Housing Authority Red Hook Houses East and West. The racial makeup of NYCHA residents, by comparison, is 16.3% White, 51.8% Hispanic, 28.7% Black, 2.3% Asian, and 0.9% members of other ethnic minorities. Of those within Red Hook, 50% of youth are unemployed, and the average household

[3]NYCHA Fact Sheet, www1.nyc.gov/assets/nycha/downloads/pdf/NYCHA-Fact-Sheet_2020_Final.pdf.

income falls below $25,500.[4] This is a low-income federal and state subsidized housing complex that dominates the Red Hook landscape.[5]

During the catastrophic event of Superstorm Sandy in 2012, most of Red Hook was inundated by rising sea levels. All supply and distribution chains were disrupted; including electricity, gas, Wi-Fi, food, transportation, and health care services. Elderly people with disabilities were trapped in apartments on higher floors without elevator service. Sewers were backed up, allowing untreated wastewater mixed with floodwater to run on the streets. In total, 3100 Red Hook businesses employing 34,600 people were impacted by the storm. A number of retail businesses, both large and small, were also severely affected [5].

Low-lying topography, environmental pollution, and contamination resulting from the fill that hardened the original marshy wetland predominantly makes up the land conditions in today's Red Hook. Generally, the fill is composed of a variety of materials and contaminants; including building demolition debris, dredging spoils, and byproducts of industrial activities such as slag and foundry sand. Areas of historic fill are often contaminated with polycyclic aromatic hydrocarbons (PAHs), heavy metals such as lead, and petroleum products. Given the nature of filling activities, contamination is not contained within property lines and results in area-wide environmental impairment (Fig. 7).[6]

2 BlueCity Lab Integrated Design Process

Ms. Gita Nandan, R.A and principal of multidisciplinary design firm thread collective, and Zehra Kuz, founder of Oasis Design Lab, are leading the effort with RETI Center's Executive Director, Tim Gilman-Ševčík, PhD, for the conception and realization of BlueCity Lab, through an integrated community input-based design process. Over the past two years, community-based research has been conducted through workshops with local residents, non-profit organizations, and educational institutions to ensure a common vision for the design and programming of the facility. This has yielded a clear set of guidelines for annual operations and programming partnerships. BlueCity Lab is both a metaphorical vessel as an amphibious structure, but also as a community gathering space—a place to hold a core set of values and push an activist agenda for climate justice. Every discipline

---

[4]NYC Census Statistics, www.census.gov/quickfacts/redhooktowndutchesscountynewyork.

[5]Table PL-P3A NTA: Total Population by Mutually Exclusive Race and Hispanic Origin - New York City Neighborhood Tabulation Areas*, 2010, Population Division - New York City Department of City Planning, March 29, 2011. Accessed July 1, 2020.

https://www1.nyc.gov/site/nycha/about/reports.page *NYCHA Fact Sheet and Resident Data Summary,2019. New York City Performance Tracking and Analytics Department. Accessed July 1, 2020.*

[6]NYC Department of City Planning Brownfield Opportunity Area Study, 2014.

Fig. 7 The site for BCL, is located along the northwestern coastal edge of Brooklyn, NY, an area, which is once known as South Brooklyn. photo credit: William Ngo

will have a seat at the table; from architects and planners to educators, engineers, neighborhood leaders, biologists and industrialists.

Academic partnerships are a major focus of our process, giving the younger generation the opportunity to participate and direct research in tandem with our efforts. The RETI Center team is currently working with a variety of higher education institutions across the nation; including Pratt Institute, NYU, Brooklyn College, Princeton, and University of Pennsylvania. These all have contributed to the development of BCL and will continue to drive the design process and program once realized.

2.1 BCL Design Principles

The framework for the concept of BlueCity Lab is primarily based on how natural ecosystems sustain life and provide services. This can be divided into four areas, as described by the Millennium Ecosystem Assessment 2005: provisioning, such as the production of food and water; regulating, such as the control of climate and disease; supporting, such as nutrient cycles and crop pollination; and cultural, such as spiritual and recreational benefits.[7] These principles are in alignment with an

[7]Millennium Ecosystem Assessment 2005.

overlapping series of the UN's 17 Sustainable Development Goals (SDGs)[8] and support New York's OneNYC 2050 resolutions.

- Address relevant and predictive data available regarding SLR and climate change impacts from local, state, federal and international agencies such as the Intergovernmental Panel on Climate Change.[9]
- Design in concert with community input, community resident review through a series of workshops and design integration sessions.
- Establish a modular type of system(s) that allows for expansion and growth over time.
- Explore and include a variety of relationships to the water for both public experience and RETI programming: from above, protected water areas, open water access, and from below.
- Create a fluid yet responsive sectional design that will address natural system flows.
- Articulate architectural elements that allow for experimenting with building technologies and scientific means, such as an outdoor wet-lab, floating green walls, shade systems, and integrated, innovative power-generation devices.
- Expand the notion of off-the-grid, with all aspects architecturally integrated i.e. energy (solar, tidal, wind) water, waste, sewage: a net-positive, eco-positive addition to the anthropocenic landscape.
- Explore material options—provide an architecture that allows for education around waterborne and sustainable architectural materials.
- Emphasize the measurable impacts of climate change with on-site experiential demonstrations.
- Provide flexible gathering spaces that include the ability to host a variety of community events, and workshops.
- Create on-site habitats for biodiversity and marine life (bird nesting, fish habitat).
- Create net-positive ecological conditions related to waste, water, energy, food, and bio-habitat.
- Challenge current state and local policy, rules and regulations to scientifically prove that floating structures can be bio-beneficial entities [6, 7].
- Achieve third-party ratings such as LEED-Platinum, Living Building Challenge, Waterfront Alliance's Waterfront Edge Design Guidelines (WEDG).[10]

[8]UN Development Goals.

[9]IPCC, Climate Change 2007: Impacts, Adaptation and Vulnerability. Contribution of Working Group II to the fourth assessment report of the Intergovernmental Panel, 2007.

[10]Water Front Alliance WEDG Program.

2.1.1 Site and Environmental Analysis

Dorit Aviv, Director of the Thermal Architecture Lab at the University of Pennsylvania, and her graduate student, Mrinalini Verma, with Forrest Meggers, director of Princeton University's CHAOS Lab, together they are investigating present time local climate conditions and low-impact mechanical design options, along with predictive models of sea level rise. Research findings and assembled data will be fundamental in design development and will influence the architectural form. Our aim is for it to be not only responsive to the natural context, but to also partake in a dynamic feedback loop to restore and reinvigorate the local ecology. In addition, ongoing site research is investigating data and metrics over an extended period of time for wind and water salinity levels on the site [8] (Fig. 8).

2.2 Primary Lessons Learned

- Water temperatures within the Gowanus Bay remain consistent over long durations (between 42 DegF and 70 DegF at seabed) while nearby land-surface temperatures have greater fluctuation on a daily basis. This water body will allow for a heat sink/heat source, minimizing energy requirements to balance heating and cooling systems to control thermal comfort
- Wind directions, as observed at weather stations at LaGuardia and JFK airports, are from the south in the warmer summer months and the north-west in the cooler winter months. Hyper-localized observations are needed to further confirm these conditions, along with wind-speed, but natural ventilation will be a clear asset to create longer durations of comfort within the shoulder months without reliance on mechanical systems. This will influence the exact locations of openings and the methods through which the skin is operable
- Solar radiation is quite abundant due to lack of shade (a beneficial characteristic of coastal edge conditions), but a pitched south facing orientation remains to be the most productive for PVs
- Natural daylight, and the allowance for the greatest daylight autonomy will minimize power loads and maximize energy independence. Orientation and atmospheric conditions greatly impacts the level of daylight autonomy and will impact material choices, between lighter reflective options and darker absorptive ones (Table 1, Figs. 9, 10).

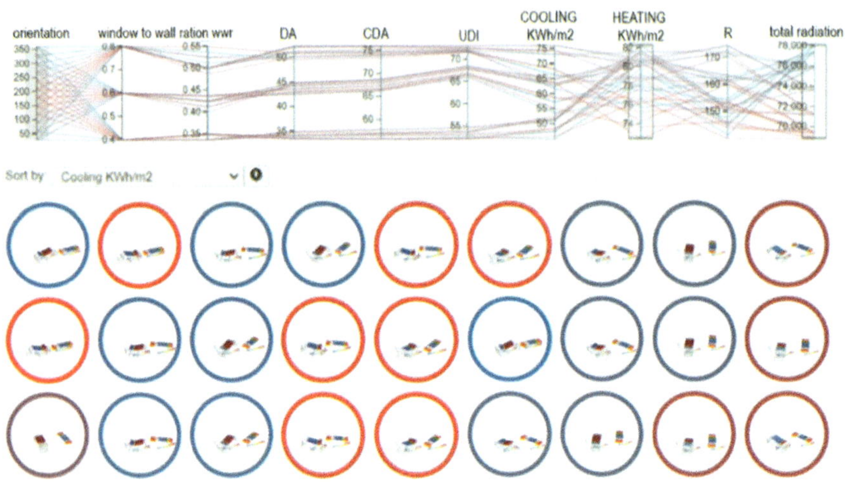

Fig. 8 Design Explorer: A study of different options based on orientation, heating/cooling loads, daylighting, floor plate shape and solar radiation on roof. Credit: Mrinalini Verma and Dorit Aviv, University of Pennsylvania ("Design Explorer | CORE Studio.". https://core.thorntontomase tti.com/design-explorer/)

Table 1 Cooling / Heating load with daylight autonomy figures for orientation study

Case	Cooling load (kWh)	Heating load (kWh)	Total energy (kWh)	Daylight autonomy (%)	Total solar radiation of panels (kWh)
WWR = 0.28 Orientation = 0 Window on shorter side, North–South Side	10,070	14,440	27,930	42%	78,095

2.3 BioDiversity Conditions

Marine Life and Water Quality: Gowanus Bay sits at the mouth of the Gowanus Canal and is a protected waterbody along the eastern edge of the 153-mile Hudson River Estuary; a man-made creation from the mid-nineteenth century transforming local tidal wetlands and freshwater streams. By the end of the nineteenth century, heavy industrial use had caused large amounts of pollutants to drain into the Gowanus Canal and flow out into the Bay, and therefore Hudson Estuary. The pollution draining from the industrial land usages has continued for over a century with no mitigation completed. By the 1990s, the canal was recognized as one of the most polluted bodies of water in the United States, currently listed as a Super Fund Site, with a cleanup finally underway. Owing to pollution with high ratios of fecal coliform, deadly quantities of pathogens, and a low concentration of oxygen, it is

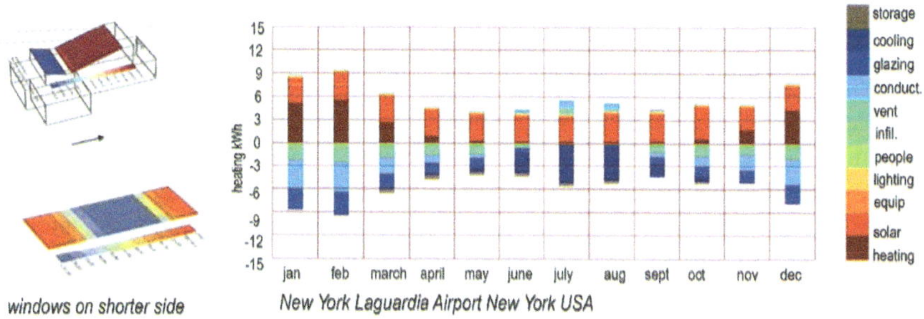

Fig. 9 Cooling/heating load with daylight autonomy calculations for orientation study, credit: Mrinalini Verma and Dorit Aviv, University of Pennsylvania (Studies performed in Ladybug/ Honeybee Roudsari, M. "Ladybug + Honeybee." Grasshopper. Np, nd Web 5 (2015), and Diva: Jakubiec, J. Alstan, and Christoph F. Reinhart. "DIVA 2.0: Integrating daylight and thermal simulations using Rhinoceros 3D, Daysim and EnergyPlus." In Proceedings of building simulation, vol. 20, no. 11, pp. 2202–2209. 2011)

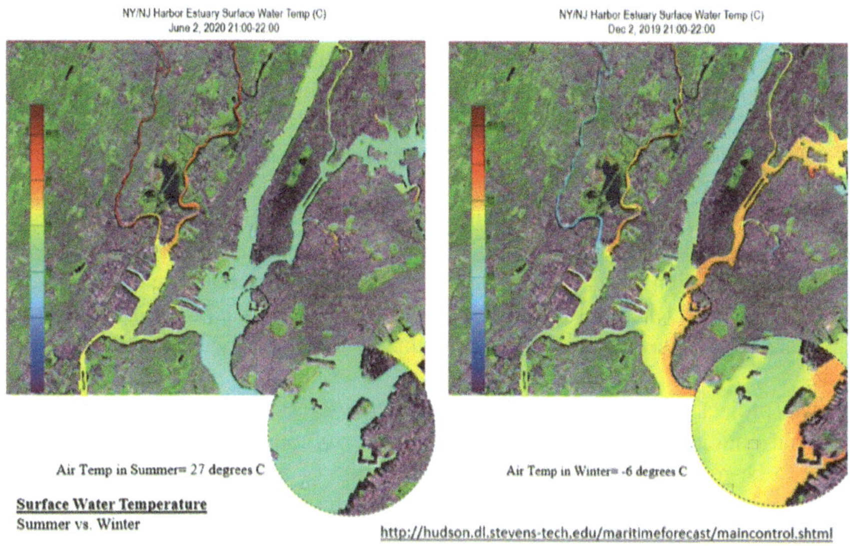

Fig. 10 Water surface temperature across the Hudson Bay Estuary (Maps retrieved from from "NYHOPS: Urban Ocean Observatory at Davidson Laboratory." https://hudson.dl.stevens-tech.edu/maritimeforecast/maincontrol.shtml) credit: Mrinalini Verma and Dorit Aviv, University of Pennsylvania

generally seen as incompatible with marine life. The Bay and its connection to the greater Hudson River allows for some dilution of pollutants. Despite or perhaps in spite of the great levels of pollution, there is a variety of marine life that visit, inhabit, and migrate through the river; such as trout, shad, bluefish, and mallard ducks. BCL's design addresses the need to improve water quality, cultivate salt marsh and soft edges, and provide habitat for promoting greater biodiversity. Stimulating positive, natural feedback loops to accelerate site restoration will be key to design decisions along the coastal edge.

Atlantic Flyway, Coastal Aviary/Waterfowl Conditions: Red Hook and the Gowanus Bay are directly within the Atlantic Flyway, with thousands of migrating birds stopping in backyards and our coastal edges as they seasonally migrate, both during the daylight and nighttime. Following the Audubon Society's mandate of preservation, monitoring, and education, BCL will not only support such migration patterns, but also ensure that architecturally, BCL is not an obstacle in the flight patterns. Considerations need to be taken with glass/envelope surfaces, green space provision, dark sky principles, and resting/nesting spaces for the migrating flocks. Additionally, improved water quality will stimulate aquatic food sources to nourish these flocks as they pass through.

3 BlueCity Lab Design

BCL is a composition of several materials and methods. Each building component is designed with a specific role and performance in mind, and yet each is integrated with the other. A post-and-beam stick frame structure sits atop a barge-like hull that is made of biophilic concrete, and acts as the buoyant foundation and thermal mass for heating and cooling.

The exposure of construction methodologies and connection details provides an important teaching tool, demonstrating how systems are erected and maintained. Echoing the design concept first conceived by Cedric Price and made famous by Renzo Piano and Richard Rogers' Centre Pompidou featuring building systems and structures on display, BCL will advance this concept in the current eco-conscious era.

The project will dive deep into Life Cycle Assessments (LCA) of multitude material options to inform the most low-impact solutions. All decisions will continue to be made through the lens of the creation of a net-positive local environment, cultivating increased biodiversity through net-zero carbon emissions as defined by the World Resources Institute: removing greenhouse gases from the atmosphere in a process known as carbon removal. For example, the concrete hulls can be manufactured by capturing and sequestering carbon dioxide, while making the material stronger and less expensive (Fig. 11).

3.1 The Hull

The hull supports the upper structure and is semi-submerged, floating within the tidal flow of the Gowanus Bay. 18'-8" wide × 65' long × 10' high, the bargelike structure will have a waterline at 5'-0". The thickness of the walls will allow for

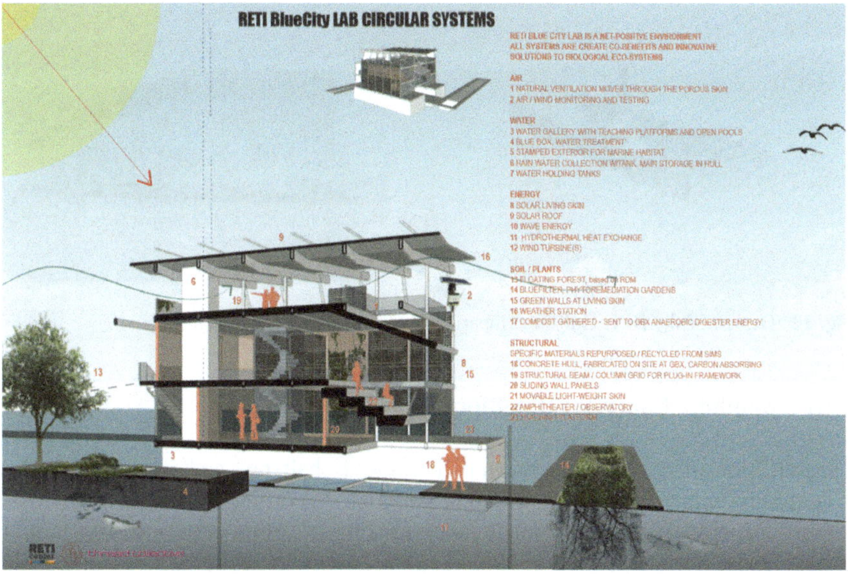

Fig. 11 Original conceptual programming/circular systems diagram for a larger iteration of BCL, thread collective + Oasis Design Lab

insulated cavities to store mechanical piping, and will act as a critical thermal sink, conducting heating and cooling. The shell of the hull will actively condition the interior space to be used as the laboratory for the BCL team and interns.

Based on age-old technology and principles of buoyancy, the hull is similar to the offshore concrete floating structures used by the navy and marine exploration for decades. In World War II, these floating military structures were fabricated on a large scale to form artificial harbors like Mulberry Harbour,[11] providing critical landing sites to facilitate the Allied Forces invasion in Normandy. Buoyancy, described by the law of physics as volume of water displaced by the volume of the object, keeps these superstructures afloat. The long life-span, corrosion resistance, and ready-made simple forms make the hull a durable base for marine environments (Fig. 12).[12]

[11]Remains of Mulberry Harbour caisson - Wikimedia Commons.

[12]Evans, J.; Walter, T.; Palmer, E. (2000). A Harbour Goes to War: The Story of Mulberry and the Men Who Made It Happen. South Machars Historical Society. ISBN 1–873,547-30–7. OCLC 59,573,968.

Fig. 12 Design of the BCL Hull in comparison to historic hulls of WWII

3.1.1 Biophilic Concrete: BlueBlock FLOATS

The selection of concrete for the hull became a challenge and motivated further research in order to achieve our net-carbon goal. According to the Catham House Report, cement is one of the global economy's most carbon-polluting industries. It is responsible for about 8% of global carbon dioxide (CO2) emissions in 2015.[13] One critical aspect for BCL was the creation/adaptation of a new type of concrete, which is developed based on a technology that can neutralize the carbon footprint and create a bio-beneficial ecological environment. Learning from leaders in this field such as Evelyn Tickle of Grow Oyster Reefs,[14] CarbonCure, ECOncrete, and partner GBX Terminal, we are creating a financially-accessible alternative to conventional concrete. Currently, the NY State Assembly is considering the Low Embodied Carbon Concrete Leadership Act,[15] a bill which requires the use of more ecologically sustainable concrete in State projects. BCL fabrication would embody the diverse applications and carbon reduction limits possible in the chemical composition and manufacturing processes [9].

In addition to its hull, BCL will be surrounded by biophilic concrete infrastructure, different forms of BlueBlocks, with varying uses; some are floating gardens, some are pathways for waterfront access and others for soft shore experiments. They will attract marine and marsh organisms which, in turn, engage in and accelerate breaking down the pollutants in the water.

3.1.2 The Stick Frame: Post-And-Beam Structure

The BCL is structurally composed of a post-and-beam system that sits above a cast biophilic concrete hull, which acts as a floating foundation. The system appears to

[13]Chatham House Report, Making Concrete Change, Innovations in Low-Carbon Cement and Concrete.

[14]Grow Oyster Reefs www.growoysterreefs.com/about.

[15]Low Embodied Carbon Concrete Leadership Act.

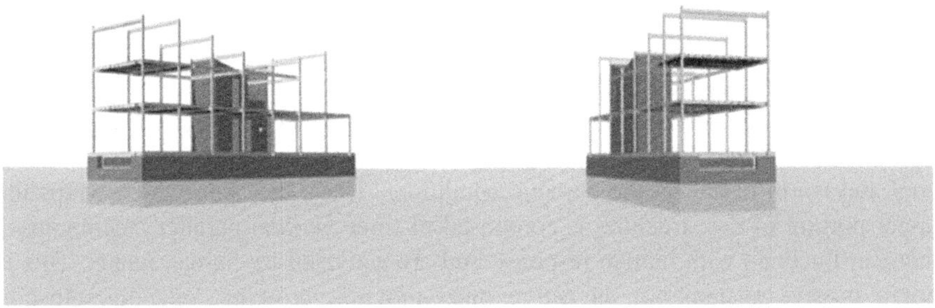

Fig. 13 Design in development BCL internal frame credit: thread collective + Oasis Design Lab

be simple, but is complex in the detailing and the methods of how it responds to local climate conditions. The goal of a net-carbon-positive material expression requires maximizing the efficiency of the material and the use of space while minimizing the energy load to sustain habitable space within, and sequestering carbon wherever possible.

Cross-Laminated Timber (CLT), a system that is rare in the urban context of New York City but has been proven to perform better than typical metal/steel/ concrete systems in terms of Global Warming Potential (GWP) CLT is a multi-layered structural wood product constructed by large panels made from solid wood and glued together in alternating directions of their fibers; the average GWP of CLT is approximately 30% better than steel/concrete, lowering our overall embodied energy.[16]

The stick frame structure allows the space between the Lab and the Multi-purpose/classroom to be entirely open to the elements and to the surrounding built and natural landscape. The short span between the column grid is to accommodate a variety of façade systems resulting from dialectical work between the engineers, the designers and the sponsors. This framework will create a structure for a skin that can be malleable and experimental in nature, changing over time as climate adaptation materials evolve, improve, and come to the marketplace (Fig. 13).

The tower is composed of the CLT framework and acts as the structural bracing for the stick frame structure that rests atop the hull. It also hosts the base-building core of the structure including the wet rooms, necessary storage, the vertical circulation and the internal guts of the building. Supply, waste plumbing lines, power lines as well as shafts and other necessary conduits for heating and cooling will be integrated into the CLT tower. The infrastructural elements are surface mounted, exposed, and color coded in order to display how the system works at a glance for the users and the visitors. All systems will be off-grid and self-contained on the BCL and its surrounding environment.

[16]www.sciencedirect.com/science/article/pii/S2352710219302542.

3.1.3 The Skin

The most critical architectural element of BCL is the spatial enclosure of the third level, the skin. Similar to the human body, these layers protect the vulnerable interior systems and yet actively engage with its environment, to keep the systems cool and warm based on the climatic conditions. Here, the skin, which wraps the upper portion of the structure, is composed of three vertical parallel "membranes" that can fluctuate with human response and are activated by human hands. This is not a passive system but an active one—moving, growing, opening, closing, breathing.

The outermost layer is outbound of the structural frame by three feet; a system of cabling is suspended between the 2nd floor catwalk and a light-weight beam system cantilevered from the roof structure. Along these cables, a shade structure system can be hung to provide shade and prevent heat-gain at various times of the day, keeping the solar radiation out. A planted green system will be experimented with to determine the viability of living walls in such marine environments. The catwalk also creates a zone for movement between the outer shade layer and the central protective layer, providing access for maintenance, measurement and experimentation.

This middle layer is the most robust and most complex, as it needs to behave as the insulating layer, allow for seasonal air movement, and maximize natural daylight to minimize power loads. Several options are being explored, with a combination of opaque to translucent, highly insulated walls. Customized ETFE[17] translucent panels with a solar PV capturing film surface to maximize power are being considered. An interior layer will consist of a day-light control device and winterizing insulation that can be moved into place when needed throughout the day.

Natural airflow is a critical part of the solution in order to lower heating loads and maximize comfort in the marine environment. By pushing the thermal definition of conditioned, occupiable spaces, BCL will explore the balance between temperature, utility, and energy usage to secure maximum efficiency (Figs. 14, 15).

4 Reaching Out

BlueCity Lab reaches out into the marine landscape beyond its architectural form through a series of BlueBlock Gardens. These are individual floating and semi-submersible biophilic concrete structures planted with salt-tolerant bio-beneficial ecosystems. These types of unique landscapes have been proven to provide a variety of multiple benefits; such as improving air and water quality,

[17]https://www.teflon.com/en/products/resins/etfe-resins Ethylene tetrafluoroethylene is a fluorine-based plastic. It was designed to have high corrosion resistance and strength over a wide temperature range. ETFE is a polymer and its source-based name is poly. It is also known under its brand name: Tefzel.

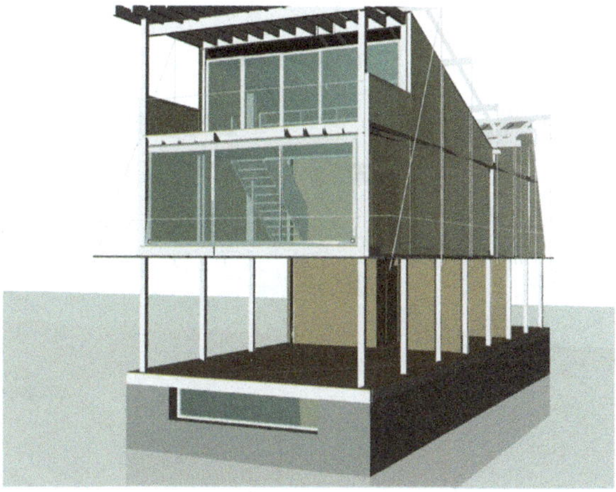

Fig. 14 Design in development, BCL north facade, credit: thread collective + Oasis Design Lab

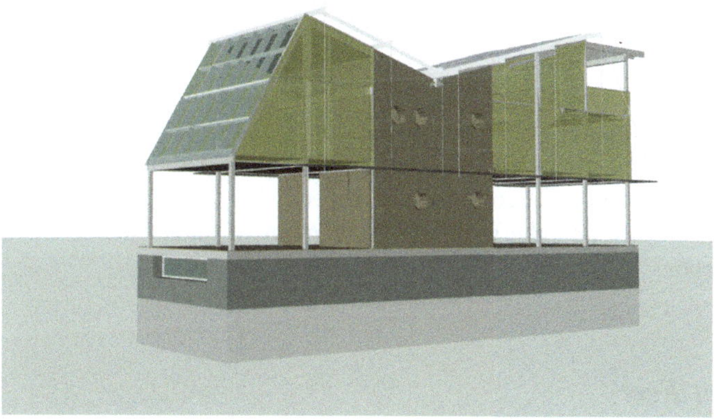

Fig. 15 Design in development, east facade, BCL credit: thread collective + Oasis Design Lab

increasing habitat for waterfowl, and supporting the flourishing of marine life. There are many precedents around the globe, dating as far back as 500 BCE when the Aztecs created 'Chinampa', a type of floating island for growing crops and capturing water that is now considered a world agricultural heritage.[18]

The design of the BlueBlocks Garden is multilayered, mimicking an integrated ecological system that supports life along a deep sectional column—with a bio-beneficial concrete anchor on the seafloor; undulating underwater edges indented to expand the marine habitat; and planted beds suspended in and above the water, providing habitat for mammals and insects. Every level of the floating garden

[18]Chinampa, An Urban Farming Model of the Aztecs.

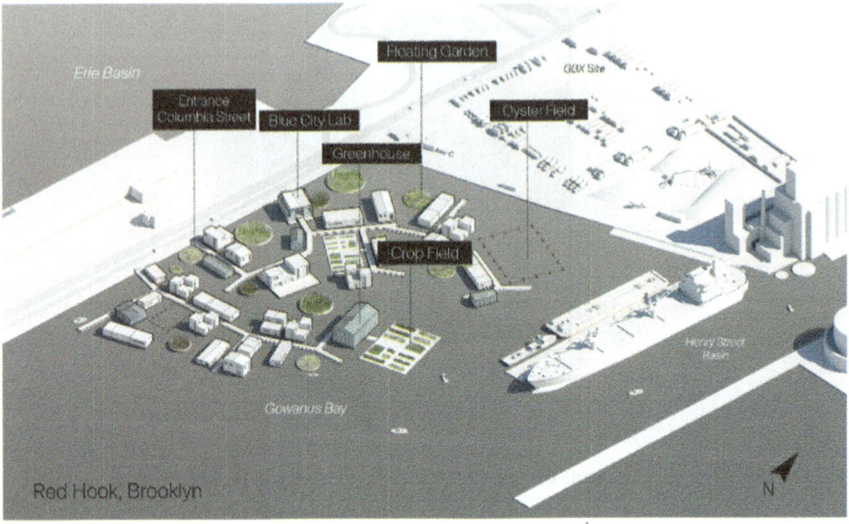

Fig. 16 BlueCity conceptual diagram, credit: Space&Matter

can be activated. Over time, the development of a dangling root system will increase water filtration of toxins and aeration by adding oxygen to the system and contributing to the clarity of the surrounding water. The levels are designed for habitation, as outdoor classrooms, and observation spaces, allowing people and wildlife to occupy an otherwise very rare moment, alone in the middle of our watery harbor.

4.1 Wet Feet

The floating structure requires stabilization. Typical anchoring systems, for conventional marine application on similar-scaled vessels would be spud pilings. The BCL will have "legs" that connect to the seafloor. These will support aquatic kelp and mussel habitats like vertical gardens, and will be made of hollow forms filled with waste glass provided by SIMS, the neighboring facility responsible for all of New York City's public waste recycling.[19] Circular economy principles such as the use of local resources and reducing waste through reuse and repurposing, will drive design and development decisions.

[19]Sims Recycling.

5 Concluding Remarks

As we plan the early phases of design and construction of BCL, we are also looking towards the future and how BCL can influence a 100-year plan for creating and supporting a new paradigm for New York City through waterborne structures. Our focus builds on a desire to model a new paradigm in equitable green manufacturing and employment opportunities [10]. In partnership with Matthijs Bouw of One Architecture and Urbanism, and Marthijn Pool of Space and Matter,[20] we are envisioning a larger urban scale model of BlueCity. This industrial and commercial district will be inclusive, and will expand the mission of BCL to a wider, more regional impact to create a community of innovative partners focused on mutual benefit and just climate adaptation (Fig. 16).

Acknowledgements Thank you to our founder Ron Shiffman, who has led this project from its inception, over a decade ago, through a reimagining with Pratt Institute's Resilience Adaptation Mitigation Program [RAMP]. Ron's drive to ensure equity is at the heart of building a just future with RETI inspires our work every day.

Thank you also to our partnering design team members that have supported the project as it develops: Persak Wurmfeld, Evelyn Tickle, Matthijs Bouw, and Marthijn Pool.

Thank you also to the many students, volunteers, and community partners who continue to support and advance our mission.

Finally, a very big thank you to John Quadrozzi of GBX who inspired our watery future and supports the vision and expansion of the project by opening his gateway for new life on New York Harbor.

References

1. Becker J, Karp R (2015) Resilience labor force analysis. South Brooklyn Industrial Development Center, Brooklyn NY
2. Mayer B (2014) Framework for the Southwest Brooklyn resilience education, training & innovation center: an integrated approach for an equitable and climate resilient city. Pratt Institute School of Architecture, Brooklyn NY
3. Pettifor A (2019) The case for the green new deal. Verso Books, London UK
4. Yie A (2020) Ahead of the tides: a case for increasing resilience in NYC. Pratt Institute School of Architecture thesis
5. South Brooklyn Industrial Development Corporation and WXY Architects (2015) Red Hook Brownfield opportunity area study. Final Report, NYC Department of City Planning, Brooklyn NY
6. Harkin B (2015) Floating structures: the ripple effects on aquatic habitat and hydrodynamics in the New York Harbor, NY/NJ Port Authority, New York NY
7. Harkin B (2014) Dredged materials and climate change pilot, NY/NJ Port Authority, New York NY

[20] One Architecture and Urbanism and Space&Matter.

8. Luo Y, Zhang L, Bozlar M, Liu Z, Guo H, Meggers F (2019) Active building envelope systems toward renewable and sustainable energy. Renew Sustain Energy Rev 104:470–491
9. Flower DJM, Sanjayan JG (2016) Handbook of low-carbon concrete. Butterworth + Heinemann, 1st edn
10. NYC Environmental Justice Alliance (2020) NYC climate justice agenda 2020: a critical decade for climate, health, and equity, New York NY, pp 48–64

Design and Engineering of an Energy Maintenance Hub Superstructure

Frank Adam, Peter Dierken, Moe Moe Aye, Falk Wittmann, Clemens Schmitt, and Alexandru Cobzaru

Abstract This paper aims to contribute the final design of a cost-effective and energy self-sufficient offshore wind operation and maintenance (O&M) platform so called Energyhu@Sea. The Energyhub@Sea concept is one of the four applications namely living, energy hub, farming, transport and logistic of the Space@Sea project funded by EU's Horizon 2020 research program. The energy hub concept integrates four different functions within the same space such as O&M services, accommodation facilities for service staffs, renewable energy extraction and smart storage system, and housing of spare parts. The primary energy for the entire hub is supplied from a medium sized wind turbine coupled with photovoltaic modules and the wave energy converter. A special focus is put on the development of a modular and standardized system, which can be easily adapted to different climates. Moreover, the design considers the needs of future users, and health, safety and environmental risk related issues. A final design for both the North Sea site and the Mediterranean Sea site based on a hypothetical offshore wind farm with a capacity of 100 units of 10 MW direct drive turbine is presented. The results indicate that Energyhub@Sea can be a promising solution for offshore wind parks to save time

F. Adam (✉) · M. M. Aye
Chair of Wind Energy Technology, University of Rostock, 18059 Rostock, Germany
e-mail: frank.adam@uni-rostock.de

M. M. Aye
e-mail: moe.aye@uni-rostock.de

P. Dierken · F. Wittmann · C. Schmitt
Energy Management and Offshore Wind Division, GICON GmbH, Leipzig and Rostock, Germany
e-mail: p.dierken@gicon.de

F. Wittmann
e-mail: F.Wittmann@gicon.de

C. Schmitt
e-mail: C.Schmitt@gicon.de

A. Cobzaru
ICEPRONAV Engineering SRL, 19A Portului Street, 800025 Galati, Romania
e-mail: alex.cobzaru@icepronav.ro

Ł. Piątek et al. (eds.), *WCFS2020*, Lecture Notes in Civil Engineering 158,
https://doi.org/10.1007/978-981-16-2256-4_10

and costs required for O&M activities. This study contributes useful inputs for industries towards the implementation of an offshore onsite-based O&M platform.

Keywords Energyhub@Sea · Offshore wind maintenance platform · Energy self-sufficient

1 Energy Maintenance Hub Concept

Offshore wind energy is undeniably receiving a growing attention due to its great potential contribution to meet the global sustainable energy goals. Unprecedented development in offshore technology has enabled the market transition continuously to farther offshore. Despite these impressive improvements, the levelized of cost of energy (LCOE) of offshore wind power in particular is still hampering the implementation of offshore wind farms in deeper waters. The looming challenges of logistics and transportation posed by deeper water sites are driving solution requirements for offshore operations and maintenance (O&M). In the future, an offshore on-site maintenance base will be an important counterpart to the land-based one to save transport costs and time required to the offshore wind parks and reduce downtimes. In this regard, a maintenance hub on a floating island aims to offer a technical solution for offshore wind O&M services targeting low LCOE.

The Energyhub@Sea concept is one of the four applications namely living, energy hub, farming, transport and logistic of floating islands considered in the Space@Sea project funded by EU's Horizon 2020 research program. The major goal is to develop a cost-effective and energy self-sufficient maintenance platform for offshore wind farms. The design concept encompasses a multitude of O&M services for offshore wind parks, accommodation facilities for crews and renewable energy extractions and housing of spare parts. The design process involved an initial site selection followed by defining the generic wind farm, anticipated offshore wind O&M services and required crews, successive design iterations of an accommodation platform for crews, analysing the energy demand for the entire hub, integrating renewable energy solutions like solar, wind and wave, developing the energy storage system, incorporating health, safety and environment (HSE) into the design, performing structural analyses and conducting a business case study. Studies related to the site selections, design iterations and the economic feasibility have been published in [1, 2]. Therefore, they will not be discussed here.

2 O&M Strategies for Offshore Wind Parks

Offshore wind turbines are complex mechanical, electrical and structural systems operating in a highly demanding environment. An offshore wind farm is typically designed for an average lifespan of 25 years. In accordance with relevant standards and guidelines such as DNVGL, BSH, IEC, etc., periodic technical inspections must be performed during its design life to detect failures in time and prevent unnecessary downtime of the wind farm. The following subsections explain maintenance services which will be undertaken by the energy hub for the nearby wind parks. These services generally include inspection, maintenance, repair and improvement.

2.1 Reference wind park

To order to define O&M service requirements, a hypothetical wind park consisting of 100 units of 10 MW rated capacity wind turbine was defined. A generic direct drive 10 MW wind turbine was defined by scaling up a Siemens 8 MW wind turbine (Siemens SWT-8.0–154) [3]. The wind park was envisioned to be located about 20–300 km far from the nearest harbor at the North Sea site and about 60 km for the Mediterranean Sea site.

2.2 O&M Services

Energyhub@Sea will provide O&M services both for the energy hub itself and nearby wind parks for their entire lifespan. Regarding O&M services for the energy hub itself, the basic maintenance activities necessary for solar panels, wind energy turbine, wave energy converter and energy storage system are primarily taken into account. Since they are medium-sized systems, the demand for maintenance and lifting actions are low and easy to handle. Therefore, no external equipment is necessary. On the other hand, regarding O&M services for nearby wind parks, handling and access demand are high, expensive and sometimes requires outsourcing technical expertise and equipment. The service team of O&M hub will undertake a range of service levels from performing regular maintenance tasks like cleaning, greasing, oiling up, underwater inspections up to major operations like replacement of components.

Generally, regular maintenance will be performed in the period between spring and early autumn, service and repair actions will take place in the remaining time only if the weather permits. Planned repair or retrofit measures will take place mainly during summer months from April to October. Depending on the size of components to be inspected and repaired, the lifting height and the water depth, the necessary type of cargo and equipment and repair time are different. Large components like rotor blade, nacelle and secondary steel components like tower segment are not stockpiled on the hub because special transport vessel and equipment will be acquired from onshore warehouse or logistic hub on demand.

2.3 Logistic of Spare Parts

Storage of spare parts plays a critical role for effectively supporting O&M services with continuous efficiencies when it is needed. There are many mechanical, electrical and structural components in each wind turbine and sufficient replacement parts and materials must be available at right times to minimize downtime. Accordingly, the energy hub aims to improve spare parts availability from service aspect and conduct a cost-effective inventory management from logistic aspects.

Essentially, some spare parts need to be repaired, while others have to be replaced when failures happen. In practice, the demand for all spare parts is not continuous since certain spare parts require only periodical inspection and exchange. For instance, gear oil and generator coolant are needed to be exchanged every 3 years while the renewal of rechargeable accumulator has inspection coverage over a 5-year period. On the other hand, some parts like filters and brakes require frequent inspection and maintenance. Nevertheless, it is not feasible and costly to have spare parts for each individual piece of components and equipment. Therefore, system components and failures rates of wind turbines and the regular maintenance work needed for them were analyzed so as to manage a systematic inventory. The storage capacity which can be stored on the hub were already summarized in the previous paper [1].

2.4 Service Team

All crews will accommodate on board. Daily working hours and day off for each personnel depend mainly on the region of operation, for instance, maximum continuous stay of 2 weeks is allowed for a crew who works more than 10 h per day according to German offshore working time regulations (Offshore-ArbZV) [4]. Therefore, a total of 12 h daily working hours which includes 2 h for crew transfer and waiting time for the service technicians with a maximum stay of 14 days was assumed in this study. Finally, it was assumed that a total of approximately 32 persons is needed in summer to provide preventive maintenance of 100 wind turbines and 14 persons in winter for necessary urgent maintenance at the North Sea site, if the weather conditions permits it. The amount of operating days increases in south European regions. The weather conditions in these regions allow nearly a continuous operation of the O&M hub throughout the year. Therefore, it was assumed that approximately 32 persons are enough to carry out the same activities in the Mediterranean Sea site both in winter and summer.

3 Final Design of Energyhub@Sea

The final design process adopted here was successively iterative as mentioned in our previous paper on the basic design of the energy hub [1]. It involved a variety of factors different from those customarily designed. The energy hub comprises multiple functions and the available areas are optimized to ensure an efficient co-use of space and deliver maximum performance in an integrated way. The high flexibility, modularity and structural integrity in the design allow downsizing or upsizing of accommodation module over time. For example, in the case of larger accommodation complexes, it can be stacked to form a multi storey building supplemented by coffee shops, gym etc.

In order to get the best possible efficiency, the quantity of space in terms of floor area was calculated as a starting point. Thereby the accommodation building was designed to meet a compromise between the architectural and engineering demands and ensure compliance with several standards and guidelines, for instance, American Bureau of Shipping (ABS) guide for building and classing accommodation.

The number of users, duration of stay and the minimum necessary space for the respective functions were properly considered. After making a benefit analysis together with work package 7 (WP7) of the Space@Sea project which focused on living, working and recreation from both offshore and urban perspectives, it was assumed an area of approximately 1242 m^2 will accommodate comfortably around 32 persons. Moreover, other necessary functions like a spacious patio surrounded by green area and other recreational activities were supplemented in the design to create green atmosphere from sociological and architectural aspects. Special attention was paid to integrate HSE in the design ensuring safe manning, emergency platform evacuation, safe escape routes and etc.

3.1 Energyhub@Sea at the North Sea Site

The O&M hub itself was divided into storage building, which lies directly on the floater and the accommodation and office building, which is supported by four columns on top of storage building. Figure 1 depicts an impression of how an Energyhub@Sea hub looks like in the North Sea site. All columns and the topside building are welded structures made of steel. All walls and doors were designed to be stable enough to withstand all environmental loads. The columns, house elevators and stairs and the two-story building accommodate at least 32 people, offices, kitchens, etc. Ships can be moored and unloaded at the quayside. Component exchange can also be done at the quayside.

A special structural feature of the North Sea version is the length of the supporting columns that position the building above the maximum wave height. Another adaptation to this location is the more efficient heating. On the other side,

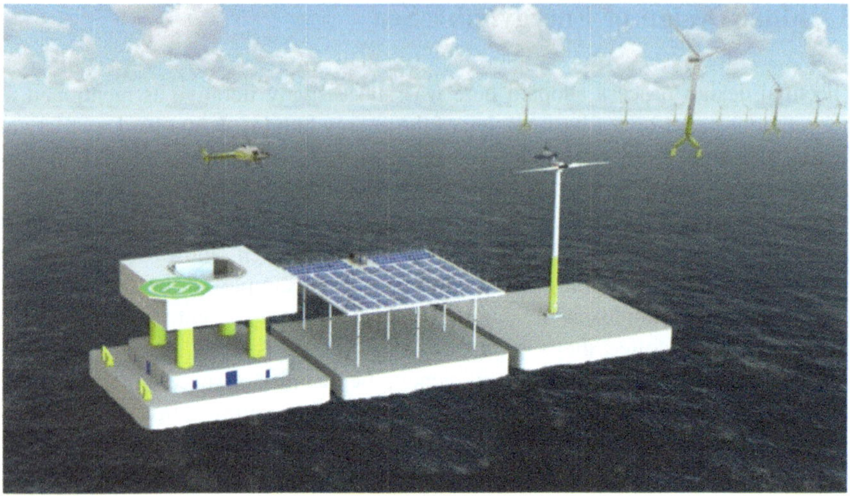

Fig. 1 3D-model of the final design of the Energyhub@Sea concept for the North Sea site

air conditioning is probably not needed. Other minor changes in design may be required when it comes to harsher environmental conditions. For example, other storage concepts could be developed as detailed solutions or other logistic measures could be better organized the processes on board. The area on the rooftop of warehouse and at the quayside make sport activities and other free time activities accessible to crews during the warmer months. However, year-round planting on the module will be difficult due to the salty spray and low winter temperatures.

Regarding energy supply, photovoltaic (PV) system and a medium-sized wind turbine are mainly considered. However, underpinning information concerning design of wave energy converter (WEC) had shown that it is economically unfeasible to integrate in the O&M hub, and therefore, it has not been included in this basic design layout at the North Sea site. As can be seen in Fig 1, PV system and 125 kW direct drive wind turbine mounted on Space@Sea floaters are integrated in the O&M hub.

3.2 *Energyhub@Sea at the Mediterranean Sea Site*

Essentially, the Energyhub@Sea was designed to be scalable depending on the region of operation. The weather conditions of the Mediterranean Sea are benign in terms of wave height and wind speed as compared to those of the North Sea. Therefore, unlike the North Sea site, the supporting columns can be omitted and the accommodation building can be placed directly on the storage hall. Another obvious element in the design is the leisure area. The milder conditions allow more frequent outdoor stays throughout the year. It is therefore imperative to consider

Fig. 2 3D-model of the final design of the Energyhub@Sea concept for the Mediterranean Sea site

outdoor recreation areas with running tracks, plants and garden furniture. The rooftop of the accommodation building and parts of the quay surface are available for this purpose bringing spectacular ocean views as bonuses. Figure 2 depicts the render of an Energyhub@Sea configuration for the Mediterranean Sea site.

Similar to the North Sea site's version, required energy are supplied mainly from PV system and the wind energy turbine. Additional PV systems can be mounted on the roof of the accommodation building. A compact 125 kW wind turbine mounted on Space@Sea floater is installed nearby O&M hub. Since WEC are economically unfeasible to integrate in the O&M hub, it has not been included in this final design at the Mediterranean Sea site.

4 Energy Storage System

Ensuring uninterrupted power supply is of paramount importance in the energy hub design. For this purpose, an energy storage system which is coupled with both the wind turbine and PV system was developed. Within this process a special focus was put on the evaluation of a fitting electrical energy storage unit.

4.1 Determination of Load Profiles

The behavior of the developed system was simulated by using the software Top-Energy. As a fundament for this simulation, an approach for the automated determination of flexible load profiles in time resolutions of up to ten minutes was developed. This load profile generator creates load profiles on basis of environmental conditions, technical characteristics and expected behavior of the inhabitants. In the scope of this work two locations were considered—one in the Mediterranean (Golfe du Lion) and one in the North Sea (Helgoland).

Figure 3 compares the electrical power consumption of the Mediterranean and the North Sea hub. It is clear that the annual total energy demands are supposed to be higher in the North Sea site than in the Mediterranean Sea. This behavior can be explained with the high heating demands in the North Sea site where frequent extreme weather conditions are expected to occur. Knowing this, it is interesting to see that the total amount of electrical energy, which is needed in the end, is higher at the Mediterranean location. Reasons for this are the higher occupation of the Mediterranean Sea hub and the utilization of heat pump technology as a source of heat and cold. Due to the high efficiency of water-based heat pumps two thirds of the heating and cooling demands can be met through the transfer of thermal energy with the surrounding sea water.

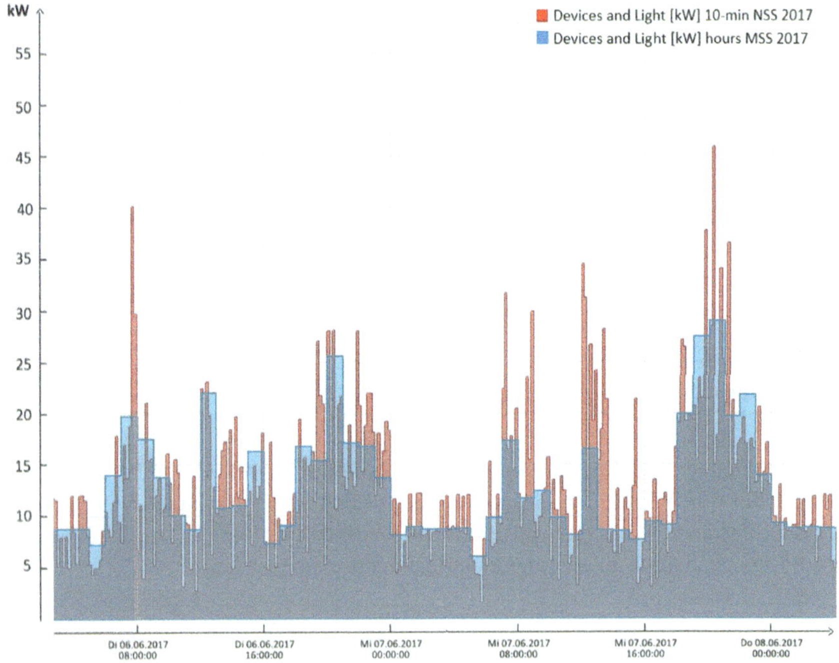

Fig. 3 Visualization of the differences between the North Sea Scenario (NSS) (red) in ten-minute-resolution and the Mediterranean Sea Scenario (MSS) (blue) in hour-resolution on basis of the electrical demands for 2017. The demands of the MSS are straightened through the application of load factors

4.2 Multi-Use of Wind, Solar and Wave

A system configuration consisting of a middle-sized wind turbine generator, a PV-system and a wave energy converter was simulated as a baseline scenario for the energy system in the first run. For this scenario, the aim was to determine a multi-use solution. Therefore, the nominal capacities for heating, cooling, hot water preparation and refrigeration at both scenarios were chosen big enough to meet the maximum demands occurring. The same applies for the energy storage systems. As a result, it was possible to develop this system but it has to be pointed out that the construction of a one-size-solution is neither energetically nor financially efficient.

The simulation results show that the wind turbine delivers the biggest share of electrical energy to cover the hub's energy consumption. Solar and wave power each supply a much smaller share. Each of the three energy sources has a relatively low utilization ratio. The wave energy converter exhibits the smallest utilization ratio of the considered energy sources. It was designed as solution for the whole Space@Sea project. Therefore, it aims to harvest as much energy as possible. This output structure does not fit the electrical demands of the energy hub. The total utilization ratios of both locations lie at about 10% which is one indication that the generation side of the developed Baseline Scenario is oversized.

Through analysis of the output structure, it was evaluated that wind and wave energy are the main energy sources in the given case. Solar power can be used to support both of them. The combination of wind and wave power is not efficient since the two are mostly operating at the same time. A combination leads to load peaks, which cannot be used and load valleys which cannot be filled.

In order to enhance the efficiency of the system, three optimization scenarios were assessed on the basis of a financial point of view both locations. A combination of wave and solar power and two scenarios consisting of wind and solar power were evaluated. The simulations showed, that the most promising system is the combination of a wind turbine and a PV system on top of the roof of the energy hub's accommodation building. Furthermore, it must be noted that especially the number of floaters and the dimensions of the battery have a major influence on the financial efficiency of the system. Figure 4 visualizes a situation at the North Sea location in January 2017. It can be seen that the electrical output of the wind turbine generator has the main impact on the balance of the system, while the PV system only supports. In order, to be able to compare the photovoltaic system on the rooftop of the accommodation building (PVr) and the PV system of the Combi-Module, the output of the Combi-Module was added to the graphic. The figure shows that even the much bigger PV system of the Combi-Module does not generate a high output enough to cope with the output of the wind turbine generator.

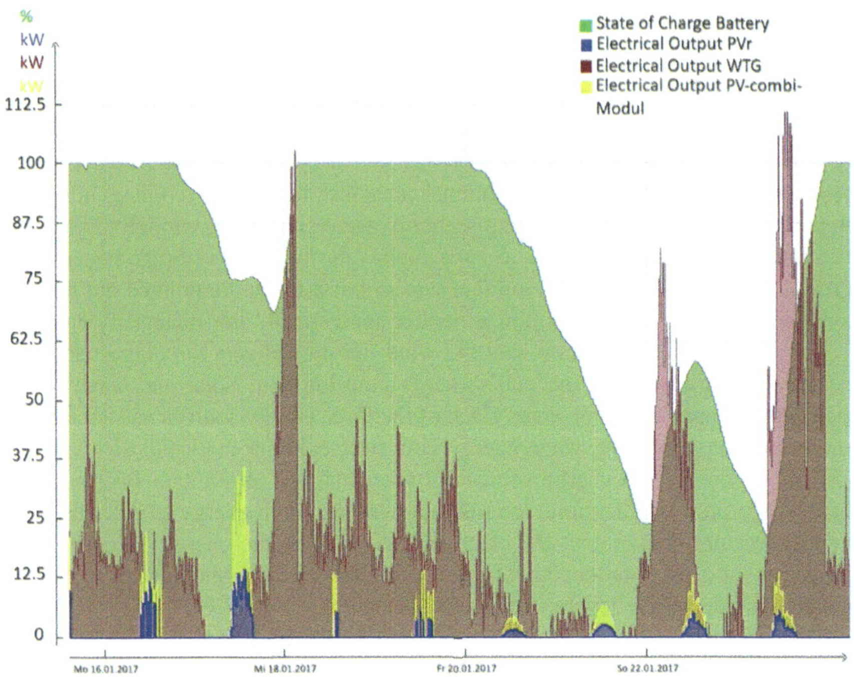

Fig. 4 Visualization of a situation and the resulting storage discharge. The graphic is the result of the simulation of the WTG-PVr scenario of the year 2017 at the North Sea location. The yellow graph represents the output of the PV-system of the Combi-floater at the same time

4.3 Detailed Design of the Storage System

The final dimensions of the baselines scenarios storage unit are 700 kWh storage capacity and 215 kW charge/discharge. In addition to the baseline scenario three optimization approaches of the system have been assessed. The resulting best-case scenarios consists of the wind turbine generator and a PV System while the wave energy converter gets excluded. The best-case scenario for the North Sea location has a battery with 880 kWh storage capacity and 155 kW charge discharge. The Mediterranean Sea scenario has a battery with capacity of 840 kWh and 165 kW.

Figure 5 shows a principle diagram of the electrical storage system and illustrates its functions. The battery storage system will be supplied from the grid through a power management system. The distribution board contains circuit breakers for voltage supply of the charger and inverter but also sends feedback to power management system regarding health status of the battery skid (including alarms, e.g. earth fault monitoring, under voltage relay). The skid will have local instrumentation (gauges) for the important parameters to be visible and easy to read.

The power management system will be able to switch from battery charging (ca. 155–215 kW) to battery supply to grid. The command will be given automatically

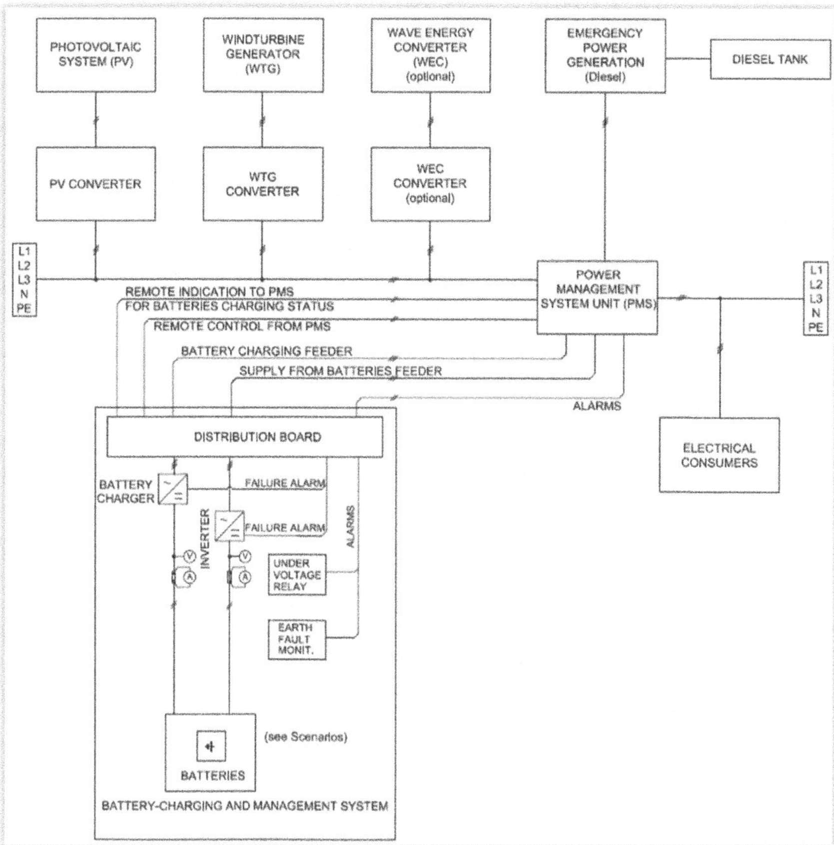

Fig. 5 Principle diagram of the storage system

through the distribution board from power management system. In case of any blackouts together with the batteries discharged, the emergency diesel generator will start and will feed the consumers from emergency distribution board. Even if the emergency diesel generator will be started from the power management system, it will run almost independent after that. The batteries' schedule of charging and discharging (e.g. to not decrease less than 40% of their Ah) capacity will be indicated by the battery manufacturer and vendor to achieve a maximum performance for their designated lifecycle [5].

To have less cables to run from the power management system to the main switchboard and emergency switchboard the serial communication profile will be considered. From a safety point of view the critical alarms and feedbacks from the distribution board and the power management system will be considered hardwired.

5 Overview of the Structural Design of the Accommodation

The accommodation building was designed with ordinary-strength hull structural steel, or mild steel and higher-strength hull structural steels such as H32 and H36 according to American Bureau of Shipping (ABS) rules and guidance [6–9]. The former has a minimum yield of 235 MPa and the latter have 315 MPa and 335 MPa respectively. It is envisioned that all steels can be grade A for the North Sea site. The mild steel is generally considered for the actual structure of the Building, as it is foreseen a deflections-driven design. HT32 is retained for possible areas requiring strengthening out of solely yielding criteria. The HT36 steel is only retained for the correspondence to the actual steel selected for a set of pillars, the sections of which will actually be provided in S355 steel according to the standard [10].

The selected solution is a typical marine/offshore steel structure comprising stiffened plate panels, retaining the caisson design only for a base caisson and the main upper decks supporting pillars, as possible. Also typical to marine structures, individual plate panels including within caisson were designed with single direction ordinary stiffener systems, which required change of system between several neighboring panels and thus broke the modularity of the design by a certain degree. Figure 6 depicts finite element model of the accommodation building. The structural design analyses were conducted by means of finite element methods to comply

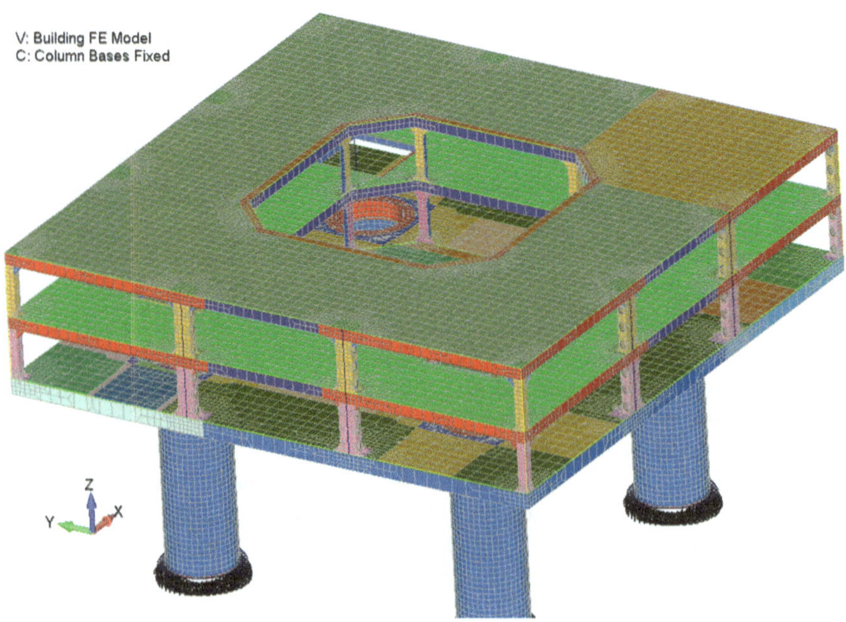

Fig. 6 Finite element model overview and constraints at supporting column bases

with ABS rules and guidance. Details of the structural analyses are not presented in this paper. The results show that the overall design is compliant with the provisions prescribed in [8] for all load cases and limitations considered in the Energyhub@Sea application. The bolt connections between the building and the supported columns are also deemed feasible above TRL5.

6 Conclusions and Recommendations

The study presented herein, was the final design of the Energyhub@Sea concept. The concept not only brings the possibility to generate renewable energy from the energy hub components in a sustainable way but also provides affordable workspace for service and maintenance of offshore wind parks. In compared to other existing offshore wind maintenance bases, it offers a solution with both low capital and operational expenses and provides the highest possible wind park availability [2]. It is clear that the quality of the development work made within the framework of Energyhub@Sea is very high. This well-designed super structure of offshore onsite-based O&M hub is beneficial in many aspects for offshore wind parks. It enhances the quality of work and living environment and reinforces O&M services for the offshore wind farms. On the basis of the designs, constructions, calculations and the adopted cost structure, a timely projecting of a prototype is possible. The virtue of robustness in the design will allow the concept to be applicable to other usage concepts like a combination of food (seaweed, mussel and fish farms) and energy production. With the parameters derived from the Energyhub@Sea concept, adjustments can be made accordingly with the intended functionalities and extended applications as the scope of the project or business grows. The overall discussion covers the possible use of the basic structure for other offshore applications in areas of service and logistics, aquaculture, research, tourism and living at sea. Thus, the innovation brought about by the Energyhub@Sea solution and its positive impacts on offshore wind activities strengthen the goal of Space@Sea concept.

References

1. Adam F, Dierken P, Aye MM, Flikkema M (2018) Energyhub@ Sea-One technical solution for offshore wind maintenance
2. Adam F, Dierken P, Aye MM, Zarnckel T, Wittmann F, Schmitt C (2019) Cost benchmark between three different offshore operations and maintenance solutions
3. Siemens SWT-8.0–154 (2017) https://en.wind-turbine-models.com/turbines/1465-siemens-swt-8.0-1544. Accessed 9 Feb 2017
4. Offshore-Arbeitszeitverordnung (2013) https://www.ilo.org/dyn/natlex/docs/ELECT RONIC /95550/112532/F63014703 2/bgbl113s2228_47220.pdf. Accessed 20 July 2013
5. Battery manufacturer Available via https://corvusenergy.com/energy-storage-system-html/
6. ABS (2019) Guide for building and classing accommodation barges, Jan 2019

7. ABS (2018) ABS notations and symbols – 8 Oct 2018
8. ABS (2019) ABS rules for building and classing steel barges, Jan 2019 (Barge Rules)
9. ABS (2019) ABS rules for building and classing marine vessels, Jan 2019 (Vessel Rules)
10. CEN, EN 10025–2: (2004) Hot rolled products of structural steels—part 2: technical delivery conditions for nonalloy structural steels

Nuclear Reactor Barge for Sustainable Energy Production

Mateusz Pater, Jonas Stampe, and Eirik Eide Pettersen

Abstract Emissions of greenhouse gases can be rapidly reduced if nuclear energy is deployed on a large scale. Conventional nuclear power plants are typically large (>1000 MW) and capital intensive, and thus undesirable or unfeasible in many regions including those characterised by numerous, disconnected power grids such as islands. Seaborg Technologies' Compact Molten Salt Reactor (CMSR) is an advanced, next-generation nuclear reactor based on molten salt reactor technology which cannot melt down, explode or be used for producing nuclear weapons. The reactor produces 250 MW of thermal energy in the form of superheated steam and will be installed on a Reactor Barge vessel that can be equipped with several reactors. The produced steam can be used to generate electricity, desalinate seawater, or facilitate other industrial processes requiring high temperature. The reactor is of a modular construction to facilitate shipyard assembly, and the barge is towed to a site where the reactor can operate for up to 12 years without refueling. Due to its small size, inherent safety features, and modularized construction, the Reactor Barge is expected to produce low-carbon energy which is cost competitive with fossil fuels. The lifecycle of a Reactor Barge is discussed in this paper, with a special focus on sustainable energy production, applications of the CMSR, and market prospects for nuclear power barges.

Keywords Nuclear · Floating nuclear power plant · Molten salt reactor · Modularity · Power barge

M. Pater (✉) · J. Stampe · E. E. Pettersen
Seaborg Technologies, Titangade 11, 2200 Copenhagen, Denmark
e-mail: matt@seaborg.co

J. Stampe
e-mail: jsj@seaborg.co

E. E. Pettersen
e-mail: eirikep@seaborg.co

Ł. Piątek et al. (eds.), *WCFS2020*, Lecture Notes in Civil Engineering 158,
https://doi.org/10.1007/978-981-16-2256-4_11

1 Introduction

With the world's population nearing eight billion [1] and continuing economic growth, global energy needs are constantly on the rise. There are various ways to meet the increasing demand, but the climate crisis has proven that energy sources that are sustainable and produce considerably smaller amounts of greenhouse gases than currently are needed.

1.1 Market Need

The modern energy market is complex. To satisfy a growing demand with low-carbon energy, intermittent renewables such as solar and wind power are being deployed on a large scale around the world. In order to always meet electricity demand, full integration of power grids is required and involves coupling renewables with, usually, fossil fuels that provide base load and that can be used for system balancing (load following).

Economies in Asia are especially developing at an accelerated pace. For instance, countries in Southeast Asia saw an average economic growth of up to 7.4% per year in 2013–2017 [2] with a substantial population growth (from approx. 600 million in 2010 to 660 million in 2019, forecast to grow to nearly 800 million in 2050 [3]). The rapid population growth and technological advances have led to increased urbanisation and industrialisation, resulting in more than an 80% increase in primary energy demand since 2000—an average of 3.4% per year compared with the global annual average of 2.0% over the same period [4].

1.2 Current Solutions

For the majority of countries worldwide, it is predominantly fossil fuels that provide electricity [5]. There is approximately 200 GW of coal power plant capacity under construction and 300 GW being planned in 2020 (pre corona crisis) in the world [6], and hundreds of gas and oil projects being planned or under construction [7]. In Indonesia 41 new coal power plants are being installed with 85 new plants being planned, adding to an existing 150 coal plants already operating as of January 2020 [6]. See Figs. 1 and 2 for other examples from Southeast Asia.

In Southeast Asia, fossil fuels constitute a particularly important part of the electricity mix as geographical restrictions set an upper limit for renewable generation. During the monsoon season fluctuations of the production of renewable energy sources such as wind and solar can happen on an hourly basis, making the balancing of supply and demand very challenging [8]. The higher degree of fluctuations and variability require a greater share of non-intermittent energy sources to ensure energy security and stability.

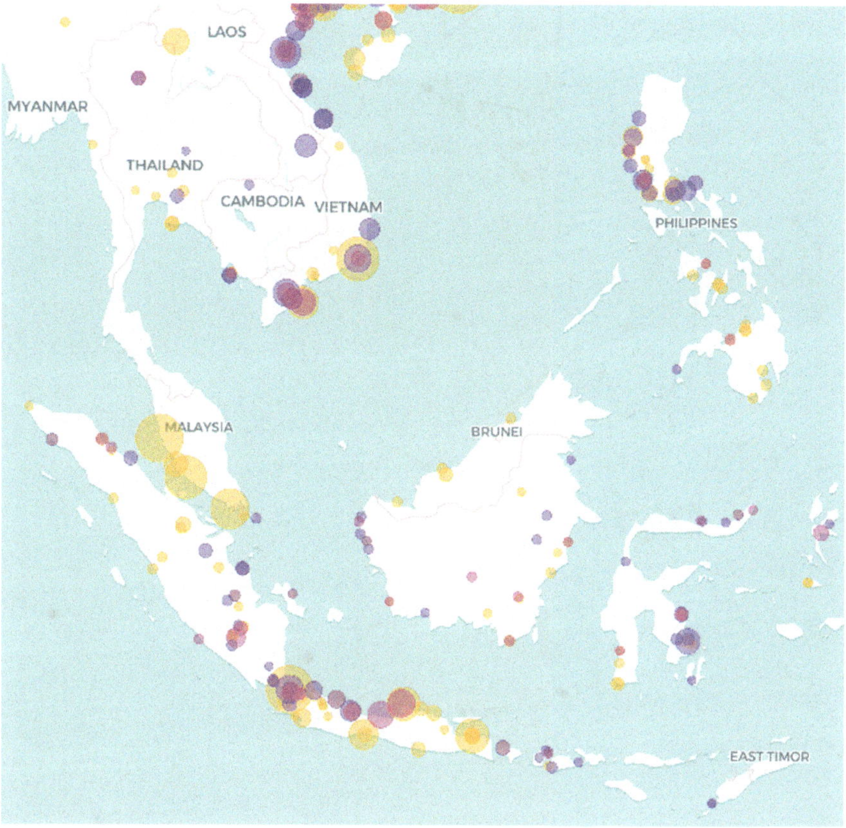

Fig. 1 Coal power plants in Southeast Asia in 2020 (yellow—operating, pink—under construction, violet—planned) [6]

At the same time, fossil fuels should be avoided due to their negative impact on the environment, both in terms of greenhouse gas emissions and local air pollution. An alternative to fossil fuels is nuclear power. Nuclear power plants (NPPs) are able to supply clean and reliable energy that can be used for both base load and load following [9]. According to the Intergovernmental Panel on Climate Change, nuclear is among the lowest carbon dioxide emitting energy technologies when taking into account the entire lifecycle of a facility (cf. Fig. 3). Consequently, replacing fossil fuel power plants and meeting growing energy demand with nuclear power would greatly reduce carbon emissions. Some countries in Southeast Asia are pursuing the development of nuclear power. With the large amount of coal power plants being built in Indonesia, the government has stated that 23% of the total power to be added to the national grid by 2027 should come from renewables and has concurrently categorized nuclear energy as a renewable energy source [10].

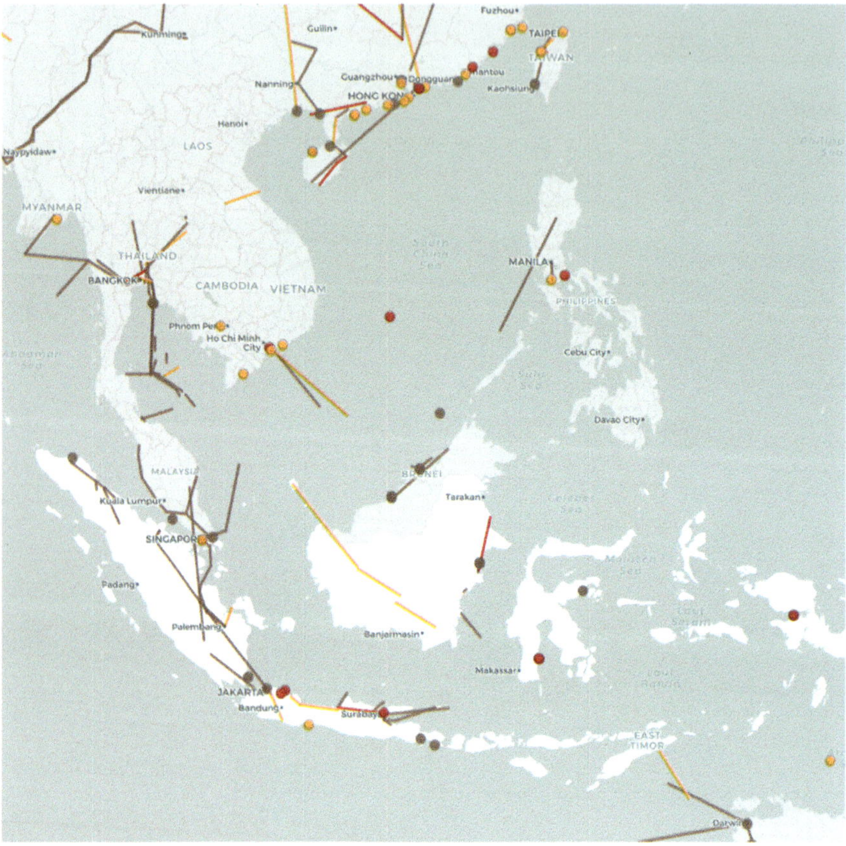

Fig. 2 Oil and gas projects in Southeast Asia in 2020 (brown—operating, red—under construction, yellow—proposed) [7]

Despite the attractive features of nuclear power, the conventional nuclear industry is facing difficulties. Conventional nuclear power plants, which are based on solid uranium fuel cooled by highly pressurised water, require large and complex systems to ensure reliable and safe operation. Consequently, and in order to leverage an economy of scale, modern nuclear power plants are large (typically above 1 GW of electric power) and require high upfront investments. Moreover, nuclear power lacks public support in many countries, some of which have initiated nuclear phase-outs following the Fukushima accident.

Small modular reactors (SMRs) are smaller and simpler nuclear reactors that aim to reduce the upfront investment cost compared to conventional nuclear plants [11]. To be competitive, SMRs rely on mass manufacturing, that is, an economy of multiples rather than an economy of scale. Several companies are working to develop SMRs. There are multiple start-up companies investigating novel designs,

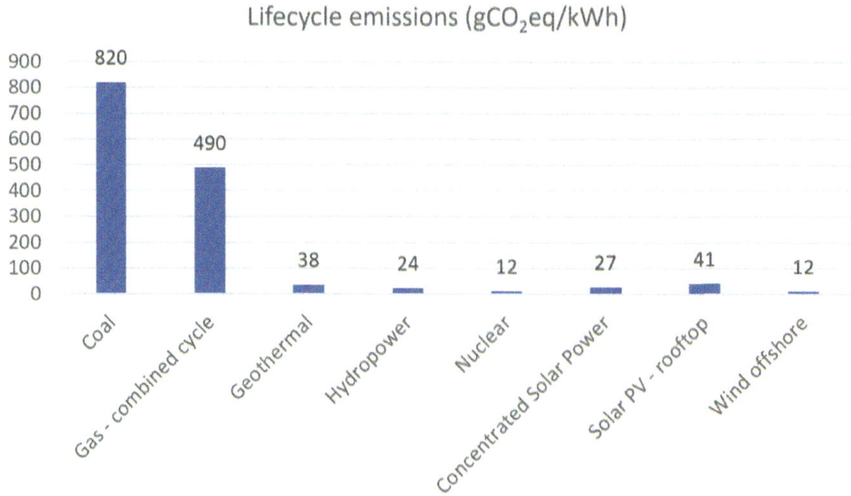

Fig. 3 Median lifecycle emissions of selected electricity supply technologies, based on IPCC [14]

as well as incumbents such as Rolls-Royce [12] or GE Hitachi [13] who are pursuing SMR concepts based on conventional reactor designs.

2 Compact Molten Salt Reactor

Seaborg Technologies, a private Danish company based in Copenhagen, is developing a next-generation nuclear reactor called the Compact Molten Salt Reactor (CMSR). One CMSR delivers 250 MW of thermal power which can be used for process heat and/or turned into 100 MW of electric power. It is installed into a floating vessel, the Reactor Barge, and has a 12 year service life time during which no refueling is required. The Reactor Barge is of modular construction and can host up to 10 active CMSRs. It has a 24 year service life time, during which it must be refueled once. The energy unit price is expected to be potentially lower than the one of currently operating coal power plants.

2.1 Molten Salt Reactor Features

The CMSR is a molten salt reactor (MSR [15]). The main difference between MSRs and conventional nuclear reactors is that the MSR fuel is a molten salt, i.e., is liquid during operation.

In the CMSR, the fuel is a molten fluoride salt which contains fissile uranium. In order to be in a liquid state, fluoride salts typically need to be kept well above 450 °

C [16]. There are various advantages of high temperature heat transfer media such as molten fluoride salts, both related to efficiency and nuclear safety.

The most important one is that radioactive material is retained within the molten salt. Fission reactions of uranium fluoride molecules produce highly reactive free fluorine. Most fission products in the fuel salt are chemically bound with this free fluorine. Also, as fluoride salts keep a low solubility in water and have low vapour pressure, resulting fluoride compounds are therefore bound within the fuel salt and remain there even in contact with air and water.

This means that in a hypothetical event where all the containment barriers are breached, nearly all solid and liquid radioactive elements will be contained within the fuel [17]. At the same time, any gases that are produced are constantly removed from the fuel. If a pipe channeling the salt breaks and the salt manages to escape, the leaked salt solidifies and turns into a rock after it cools to below the solidification temperature, thus posing minimum risk to operator personnel and the public.

Since molten salts do not require pressurisation to remain liquid at high temperatures, there are no dispersive forces such as in conventional NPPs.

Additionally, molten salts expand when their temperature increases [16]. If the fuel salt expands, the probability of fission is reduced which slows down the fission rate. Therefore, a power excursion that heats up the fuel salt causes a negative feedback. This effect in molten salt fuels makes MSRs inherently safe by alleviating unplanned power changes.

Seaborg Technologies is investigating a freeze plug which can act as a passive safety system for the CMSR. A salt flowing in a pipe can form a solid plug if the liquid is kept below its solidification temperature. This plug can be intentionally introduced in a pipe and maintained by means of active cooling. In the event that the cooling capability is lost (e.g. during an accident), the molten salt that is above the freeze plug causes the plug to melt, which in turn leads to the salt flowing to another part of the circuit. If such a freeze plug is placed above a drain tank having its geometry optimised for cooling, the salt can automatically drain into the safety container by gravity without any actuation or power.

Because of the high temperature of the molten salts, heating water in a steam generator and turning it into steam is more efficient [18] and brings the steam to a superheated state that is required for many industrial applications.

An approach to safety through physics considerations is preferred to the traditional safety by engineering which is achieved with complex, redundantly engineered safety systems. This approach allows for a simpler and more cost-efficient design.

2.2 CMSR Features

Seaborg's proprietary moderator—a material that increases the probability of fission reactions while lowering the amount of uranium needed—unlike in the majority of MSRs, is a liquid. The moderator is molten sodium hydroxide. Its properties enable the CMSR to be much more compact than if, for example, a graphite moderator was used. The first MSRs built in the past (e.g. the Molten Salt Reactor Experiment [19]) used solid graphite as moderator material, which also caused problems as graphite deforms and cracks under neutron irradiation. Those problems, however, are overcome with molten sodium hydroxide which is not structurally degraded by radiation.

See Fig. 4 for a model of the CMSR's primary circuit and surrounding systems, which includes a cut-away view of the reactor core vessel on the left, drain tank at the bottom, chemistry control unit on the right, and primary heat exchanger behind.

The reactor core is built of fuel tubes in which the fuel salt is circulated, and which are surrounded by sodium hydroxide. The fuel salt is moved by a pump and goes into a primary heat exchanger.

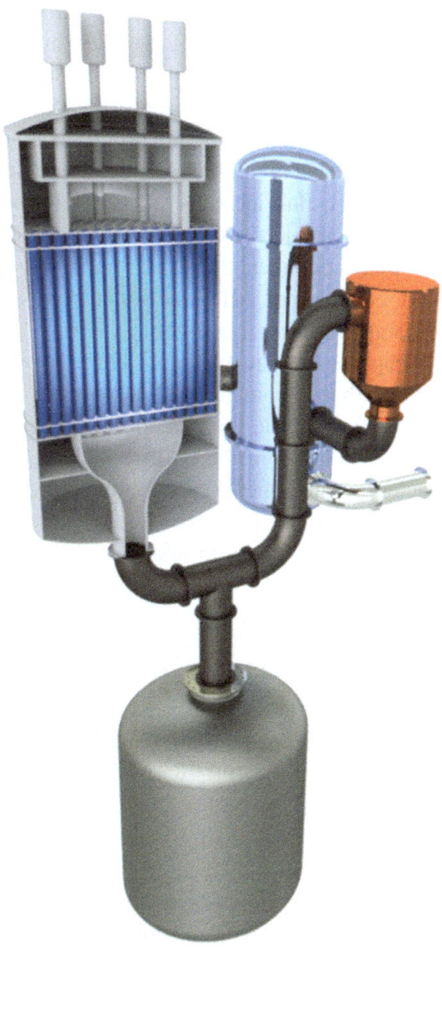

Fig. 4 Model of the Compact Molten Salt Reactor's primary circuit and surrounding systems

Before reaching the primary heat exchanger, the fuel salt enters the chemistry control unit, which, among other things, controls the corrosivity of the salt and extracts noble gases produced by fission. Gaseous fission products are collected and sent to gas processing systems.

The heat transport system in the CMSR is equipped with pumps and heat exchangers. The fuel salt transfers fission heat to the secondary salt in the primary heat exchanger. The secondary salt is a buffer between the radioactive fuel salt and steam. This salt is circulated through another heat exchanger (steam generator) in which the secondary salt transfers heat to water at high pressure which is then converted into superheated steam.

The drain tanks are the CMSR's main safety feature. The reactor is designed in such a way that energy production terminates and radioactivity drastically decreases once the draining procedure is initiated. The drained salts are cooled and can be allowed to solidify or kept in a liquid state for immediate reloading. Both active and passive (without the use of external power) methods of cooling can be used to remove the residual heat generated from radioactive decay.

The CMSR produces long-lived nuclear waste, that is, used nuclear fuel (traditionally referred to as spent fuel), which is similar to that produced by conventional nuclear reactors. In contrast to conventional nuclear waste, the used fuel from the CMSR will be in solid fluoride salt form and unpressurised since the fission gases have been continuously removed during reactor operation. Moreover, unlike oxide fuels, fluoride salts are generally non-soluble in water. The total volume of used fuel produced over the lifetime of the CMSR is roughly equivalent to a 10-foot shipping container; the total volume of radio-activated components roughly to a 40-foot container. All waste streams can and will be handled within existing nuclear decommissioning frameworks.

Importantly, the fuel cost represents a larger fraction of the total cost of the CMSR than a conventional light-water reactor because of its high capital expenditure. As a result, there is an important financial incentive to reprocess and re-use the used fuel from the CMSR. Indeed, Seaborg's goal is to fuel the CMSR with nuclear waste, chiefly plutonium, as soon as regulatory and supply chain obstacles can be overcome. This will result in a reduction of the lifetime of the waste of three orders of magnitude, to approximately 300 years.

3 CMSR Reactor Barge

Floating solutions for energy production are not new. Akademik Lomonosov [20], a Russian floating NPP, was connected to an isolated electricity grid in Pevek (Chukotka) in December 2019. It is designed to operate for 40 years and supply electricity to more than 100,000 households. The reactor is a PWR with a thermal capacity of 150 MW, requiring refueling every three years. With the first floating nuclear power plants being commercialised, Seaborg's Reactor Barge enters an emerging market with an improved technology.

3.1 Features

One Reactor Barge can house from two to ten CMSRs operating simultaneously, with a maximum power output of 1 GW electric (2.5 GW thermal). The Reactor Barge can deliver superheated steam and convert nuclear heat into electricity.

Table 1 shows the approximate dimensions of the Reactor Barge with two operating reactors (500 MW thermal, 200 MW electric) that constitute one Power Module (cf. Fig. 5). The barge's length changes with the increased number of Power Modules. Depending on the customer's needs, a desired number of Power Modules are built into the barge.

In each Power Module, two CMSRs are pre-installed in the shipyard during construction of the Reactor Barge. The limiting factor on the service life time of the CMSRs is the material degradation from ionising irradiation and corrosion by the molten salts. To extend the service life time of the Reactor Barge, a Power Module is therefore prepared with space for two additional CMSRs. After 12 years of operation, the two original CMSRs have depleted their fuel and are shut down. Prior to this, two new CMSRs have been delivered to the Reactor Barge, installed

Table 1 Approximate dimensions of the Reactor Barge with two operating CMSRs (one Power Module)

Overall length	Width	Main deck depth	Total height
98.5 m	32 m	22 m	39.5 m

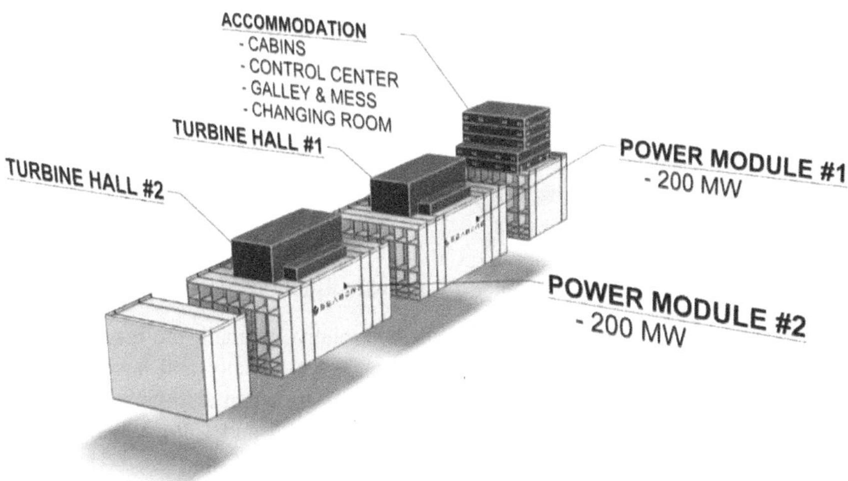

Fig. 5 Schematic representation of a two-module Reactor Barge. There are two active reactors in each Power Module which add up to an electrical power of 200 MW per module. The CMSRs are connected to turbines that produce electricity

next to the operating CMSRs, fueled, tested, and commissioned so as to take over power production. Consequently, the service life time of a Reactor Barge is extended to 24 years. The initial CMSRs are left within the Reactor Barge for the remaining 12 years. The radioactivity and radioactive decay heat production of the CMSRs and the spent fuel decrease with time [21], which simplifies decommissioning.

After another 12-year period of operation, the Reactor Barge is towed back to a dedicated shipyard for decommissioning. Only the radioactive components of the CMSR are required to undergo a nuclear-grade decommissioning scheme, whereas the rest of the Reactor Barge will be repurposed.

The Reactor Barge will be built in certified shipyards where integration with the Power Modules will take place. Shipyard standardisation towards construction, inspection, and testing results in enhanced production of CMSRs and Reactor Barges which allows for greater economy and efficiency compared to conventional NPPs.

Floating barges can be transported by sea from the construction site to the place of operation (the Reactor Barge is not propelled and it must be towed). This provides the barge with excellent siting flexibility, which subsequently reduces the financial risk incurred by asset owners and investors.

Installing CMSRs on floating Reactor Barges exerts a positive influence on safety: nuclear reactors that float cannot be nearly affected as much by earthquakes and tsunamis as on land. Additionally, the Reactor Barges can be placed on a single site for a chosen amount of time and then readily moved to another site if such a need occurs. This process of relocating a barge is simplified due to standardised environmental assessment criteria.

Due to the high temperature of the fuel salt, the heat transfer efficiency is increased [18], resulting in a range of new applications to which the CMSR and the Reactor Barge can be adapted. Seaborg Technologies has prioritised the following applications:

- Electricity,
- High-temperature hydrogen and ammonia production,
- High-temperature synthetic fuel production,
- Seawater desalination,
- District heating,
- Any combination thereof enabled by a customisable modular design.

Since the power produced by the CMSR is relatively small, the reactor can be used in standalone, remote area facilities that need electricity and have access to sea. Small islands or cities with separate electrical grids can also benefit from the CMSR. Additionally, the CMSR can adapt to demand and follow load. All the above features result in a great grid and siting flexibility of the Reactor Barges.

3.2 *Modularity*

The CMSR will be built in modules which will be assembled in a shipyard. The serial production of Power Modules is scalable and lowers the cost and construction time of the reactor.

The Reactor Barge is also of a modular design. The modular approach to constructing Reactor Barges provides customers with a wide array of flexible power solutions. To supply a requested amount of capacity, a Reactor Barge is assembled from the appropriate number of Power Modules. The practice of adding lengthwise modules to the Reactor Barge to increase capacity mirrors the traditional approach of constructing ships. It also resembles modern strategies employed for instance in the automotive industry in which a modular platform is equipped with battery capacity commensurate with the driving patterns of the particular customer [22].

3.3 *Current Status and Outlook*

In 2020 Seaborg successfully finished the conceptual design phase and commenced work on the basic design of the CMSR. Engineering and experimental validation activities in the laboratory and through prototyping, as well as regulatory licensing processes are ongoing. The first phase of regulatory approval has been successfully completed. The Reactor Barge design is under development in cooperation with industrial partners. The company intends to construct a first commercial prototype in the mid-2020s and has the ambition to scale up production within a few years thereafter.

The potential for Seaborg's business model to make a global impact is unprecedented. The company is focused on fast scaling, with the production of 1000 CMSRs per year by 2035 which would offset more than 2.5 billion tons of carbon dioxide. Under this plan, the first reactors would come online in 2025 in Southeast Asia followed by rapid scaling. This has a unique potential of transforming global energy markets with significant contributions towards meeting emission targets from the IPCC climate goals for 2050 [23] while ensuring great returns for investors.

4 Conclusions

Seaborg will sustainably power the growing world population while combating both the climate crisis and energy poverty. Several Compact Molten Salt Reactors, a next-generation modular nuclear reactor that produces up to 250 MW of thermal power and 100 MW of electrical power, are assembled in modular Reactor Barges.

The Reactor Barge has a capacity of 200–1000 MW of low-carbon electrical power and needs refueling only once over its 24-year service life time.

A simplified approach to safety in the CMSR and extensive modularity of the floating Reactor Barge are some of a number of characteristics that ensure a cost-competitive and attractive market fit. Highly flexible power solutions based on inherently safe, low-cost, and low-carbon next-generation nuclear reactors have a unique potential for replacing fossil fuels and reducing global greenhouse gas emissions.

References

1. Worldometer (2020) https://www.worldometers.info/. Accessed 1 June 2020
2. OECD (2019) Economic Outlook for Southeast Asia, China and India 2020: Rethinking Education for the Digital Era
3. Worldometer (2020) Population of South-Eastern Asia 2020. https://www.worldometers.info/world-population/south-eastern-asia-population/. Accessed 1 June 2020
4. IEA (2019) Southeast Asia energy outlook 2019. IEA, Paris
5. IEA (2019) World energy outlook 2019. IEA, Paris
6. CarbonBrief (2020) Mapped: The world's coal power plants in 2020. https://www.carbonbrief.org/mapped-worlds-coal-power-plants. Accessed 29 May 2020
7. Global Energy Monitor (2020) The global fossil project tracker. https://greeninfo-network.github.io/fossil_tracker/. Accessed 15 June 2020
8. Dunning CM et al (2015) The impact of monsoon intraseasonal variability on renewable power generation in India. Environ Res Lett 10
9. Lokhov A (2011) Load-following with nuclear power plants. NEA updates, NEA News 2011 29.2:18–20
10. PT. PLN (PERSERO) (2019) Rencana Usaha Penyediaan Tenaga Listrik 2019–2028 (Electricity Supply Business Plan 2019–2028)
11. World Nuclear Association (2020) Small nuclear power reactors. https://www.world-nuclear.org/information-library/nuclear-fuel-cycle/nuclear-power-reactors/small-nuclear-power-reactors.aspx. Accessed 1 June 2020
12. Rolls-Royce (2017) Small modular reactors. https://www.rolls-royce.com/media/our-stories/innovation/2017/smr.aspx. Accessed 2 June 2020
13. GE Hitachi (2020) BWRX-300. https://nuclear.gepower.com/build-a-plant/products/nuclear-power-plants-overview/bwrx-300. Accessed 2 June 2020
14. Schlömer S, Bruckner T, Fulton L et al (2014) Annex III: Technology-specific cost and performance parameters. In: Edenhofer O, Pichs-Madruga R, Sokona Y et al (eds) Climate Change 2014: Mitigation of Climate Change. Contribution of Working Group III to the Fifth Assessment Report of the Intergovernmental Panel on Climate Change. Cambridge University Press, Cambridge and New York, p 1335
15. World Nuclear Association (2018) Molten salt reactors. https://www.world-nuclear.org/information-library/current-and-future-generation/molten-salt-reactors.aspx. Accessed 28 May 2020
16. Sohal MS, Ebner MA, Sabharwall P, Sharpe P (2013) Engineering database of liquid salt thermophysical and thermochemical properties. Idaho National Laboratory, Idaho Falls
17. Cooper D (2020) Next-generation nuclear reactor. Materials World May 2020:36–38

18. Sunden B (2005) High temperature heat exchangers (HTHE). In: Shah RK, Ishizuka M, Rudy TM, Wadekar VV (eds) Proceedings of Fifth International Conference on Enhanced, Compact and Ultra-Compact Heat Exchangers: Science, Engineering and Technology. Engineering Conferences International, Hoboken, pp 226–238
19. Haubenreich PN, Engel JR (1969) Experience with the molten-salt reactor experiment. Nucl Appl Technol 8:118–136
20. NS Energy (2020) Akademik Lomonosov floating nuclear power plant. https://www.nsenergybusiness.com/projects/akademik-lomonosov-floating-npp/. Accessed 2 June 2020
21. Todreas NE, Kazimi MS (1993) Nuclear systems I: thermal hydraulic fundamentals. Taylor & Francis, New York, FL, USA
22. Volkswagen Danmark (2020) Den modulopbyggede eldrivlinjeplatform (MEB) (Modular electric drive matrix (MEB)). https://www.volkswagen.dk/da/id-og-e-mobilitet/id-viden/e-mobilitets-teknologi/meb.html. Accessed 13 July 2020
23. IPCC (2014) Summary for Policymakers. In: Edenhofer O, Pichs-Madruga R, Sokona Y et al (eds) Climate Change 2014: Mitigation of Climate Change. Contribution of Working Group III to the Fifths Assessment Report of the Intergovernmental Panel on Climate Change. Cambridge University Press, Cambridge and New York

Aquaculture in Multiple Use of Space for Island Clean Autonomy

Zoe J. Fletcher

Abstract Economic activities at sea are predicted to intensify and move further offshore as the global demand for food and energy continues to increase, this presents new technological and non-technological challenges. In order to ensure this expansion in a sustainable and ecological manner, different activities can be combined at the same location using multi-use offshore platforms. The Horizon 2020 funded MUSICA project aims to demonstrate the use of an innovative Multiple Use Platform (MUP) infrastructure for multiple Renewable Energy Sources, desalination and aquaculture services for small islands. Using the platform, the services can share the same space and work synergistically together, sharing Supply Chains and reducing costs associated with Operating and Maintenance. The platform will provide 3 forms of renewable energy. Electricity from the platform will be used to provide green support services for the island's aquaculture and the desalination unit on the platform will provide up to one thousand meters cubed of water for a water stressed island. The pilot MUP will be installed on site at Inousses Island to test and demonstrate the validity in a real operating environment. The project will also address key non-technical barriers to MUPs including those that are environmental, regulatory and legal.

Keywords Offshore · Aquaculture · Multi-Use · Renewable energy · Food security

1 Introduction

As the blue economy continues to expand, competition for the use of productive coastal space is increasing. Coastal resources are already heavily exploited, meanwhile global population and economic growth continue to increase demand for provisioning services such as food, power and water. By moving activities further

Z. J. Fletcher (✉)
AquaBiotech Group, Central Complex, Naggar Street, Targa Gap, Mosta 1761, MST, Malta
e-mail: zoejfletcher@icloud.com

Ł. Piątek et al. (eds.), *WCFS2020*, Lecture Notes in Civil Engineering 158,
https://doi.org/10.1007/978-981-16-2256-4_12

offshore and reducing the spatial foot-print through multiple use of space, these demands can be met in a more sustainable and ecological manner [1].

The global demand for high protein food sources, such as fish, is increasing. The planet has seen a 122% rise in food fish consumption in the last 28 years. Many of the world's wild capture fisheries are fully or over exploited, as a result the rise in production from global capture fisheries from 1990 to 2018 was just 14%. In the same period, global aquaculture production has increased by 527% to meet these demands [2]. European aquaculture has an annual production of 1.4 tonnes of live aquatic animals [3], however Europe is still heavily dependent upon imports from non-EU countries to satisfy its demand for fish. Increased production and competitivity of the sector are needed to reduce this dependency. By moving aquaculture offshore we can increase production and potentially decrease environmental impacts simultaneously.

On 18 May 2017 the European Commission, together with 14 EU countries signed the Valetta Declaration to launch the Clean Energy for EU islands initiative. The initiative is aimed at accelerating the clean energy transition on Europe's more than 2700 islands, this initiative will help islands reduce their dependency on energy imports by making better use of their own renewable energy sources and embracing more modern and innovative energy systems [4]. The Valletta Declaration calls for actions in sustainable mobility, water and waste management, citizens' empowerment, and sustainable tourism. Marine offshore renewable energy sources can offer a clean energy supply for islands that do not have the land capacity to accommodate sufficient renewable energy systems.

Moving activities offshore presents new challenges, both technologically and in terms of planning and permitting. Multi-use off offshore space can be cost saving by sharing the burden of communications, infrastructure and planning and permitting. The Horizon 2020 funded Multi-Use of Space for Island Clean Autonomy (MUSICA) project implements a valuable combination of activities, suited to the environmental characteristics of the Mediterranean. The project will demonstrate the use of an innovative Multiple Use Platform (MUP) infrastructure for multiple Renewable Energy Sources, desalination and aquaculture services for small islands. Using the platform, the services can share the same space and work synergistically together, sharing surveillance, energy, infrastructure and reducing costs associated with Operating and Maintenance.

2 MUSICA

The MUSICA project will produce a Multi-Use Platform (MUP) and Multi-use of Space (MUS) combination for the small island market. The MUSICA platform offers a decarbonising solution to water, food and power security for small islands, their marine initiatives (Blue Growth) and ecosystems. The solutions provided by the MUSICA MUP include:

- Three forms of renewable energy (RE) (wind, PV and wave) (total 870 kW) providing a non-correlated supply of competitively affordable electricity.
- Innovative energy storage systems hosted on the MUP, such as compressed water/air storage and batteries. These will be capable of providing all required storage for power on the island and platform, in addition to electrical output smoothening.
- A smart energy system, including demand response, modelling and forecasting.
- Renewable energy powered water desalination, capable of providing 1000 m^3 of water.
- Support services for islands' aquaculture, and a recharging station for electricity and water [5].

By combining these activities in one area (Fig. 1), it will reduce losses to the ecosystem whilst simultaneously accommodating a greater amount of economic activity in a small area.

The MUSICA MUP is a viable infrastructure not just for multiple provisioning services for Renewable Energy Solutions (RES), desalination and aquaculture, it will also provide energy to contribute to cultural services, such as boating, for small islands.

MUSICA is capable of supporting a small island with a population of up to 2000 people. Islands of this size typically do not have access to investment, expertise, infrastructure, and economies of scale that larger islands have. Of the 2700 EU islands referred to in the EU Atlas of Small Islands [6], 75% are 'small' islands where MUSICA can replicate its smart island MUP solution. MUSICA also contributes to decarbonising islands' Blue Growth initiatives by providing "green"

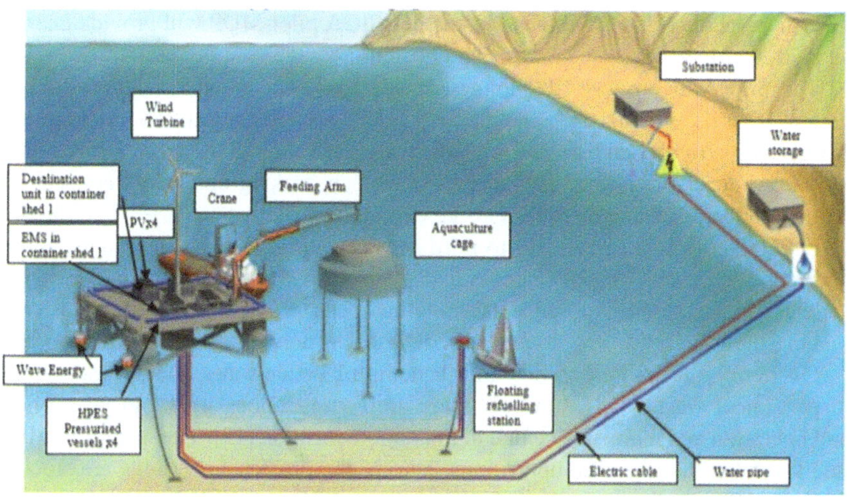

Fig. 1 Image demonstration of MUSICA Platform, RES, Aquaculture, recharging station services and shore connection (Static and Dynamic Cable combined)

support services for island's aquaculture, and electricity and water recharging station services for leisure boats [5]. The Smart Islands Initiative [7] has identified that many larger EU islands cannot have high island-based RES for several reasons including tourism, space, agriculture and local objection. The MUSICA smart MUP can therefore also offer a unique solution for these larger islands too. The pilot demonstration of the MUP and MUS technology will be located off the island of Innousses, a small island with a population of 800 people, located in the Aegean Sea.

Containerisation of MUP components following the initial planning phase will enable the MUP energy solution to be configured to an island's requirements and delivered complete.

For example, the pilot demonstrated on Innousses island will be configured to provide up to 61% RES. Innousses requires the MUSICA MUP for 2 key reasons:

1. Innousses has 0% RES, and due to it's remote location islanders pay 3 times the land rate for grid electricity. Moreover, the quality and reliability of the existing electricity supply is poor (20 kV grid connection to Chios island)
2. Innousses has no local water source. 30% of the island water is produced by a desalination plant powered by diesel.

2.1 Aquaculture in MUSICA

Floating multi-use platforms can combine several activities into one area; therefore, aquaculture can benefit from the use of multi-use platforms when moving into production at remote offshore sites [8]. In order to increase food security and blue growth aquaculture for small islands, the MUSICA pilot MUP will provide energy, support, storage and protection for an aquaculture cage.

The pilot will produce 200 tonnes of sea bass (Dicentrarchus labrax) or sea bream (Sparus aurata), in a single submersible cage located 50 m from the MUSICA platform. Sea bass and sea bream are native to the Mediterranean and are the most intensively farmed fish species in Greece, making up 99% of aquaculture sector production value in 2018. Greece exports approximately 72% of production to EU markets and a further 7% to third markets [9]. Globally, Greek aquaculture accounts for around 24% of sea bass and sea bream supply. In addition, rearing of sea bass in increased exposed conditions has been proven to be successful [10], making them the ideal species for MUSICA MUP demonstration in Innouses.

Offshore aquaculture has potential benefits of better water quality and greater dispersion of wastes [11], however there are complications with logistic services and operation and maintenance due to the remote location [12]. The MUSICA MUP offers solutions to these complications. During the project, fingerlings are stocked between 50 and 80 g and grown between 18 and 24 months [13, 14] to a harvest size of 400 g. The fish are fed on a formulated pellet feed, which is stored in a silo on the platform, and delivered to the cage via a feed pump located on the MUP. The feed pump will be powered using RES energy from the platform.

Fish feeding will be monitored in the submersible cage using underwater camera technology. Water parameters, such as oxygen, salinity and temperature will also be monitored. The power for feeding and monitoring equipment for the duration of the production cycle will be provided by the MUP RES. The platform will also offer some wind and wave protection to the aquaculture structure and equipment.

The cage and mooring design is based on existing environmental data including wind, waves, current, water depth and production goals. Based on available data, the cage chosen for this project is a six-leg submersible cage, with concrete mooring blocks. This design will also accommodate insertion of the cage into existing fish farm for initial licensing and permitting purposes, and also allow it to be moved out into open sea once permitting for the MUP is complete.

The performance trial will measure and evaluate structural performance of the cage and interactions with the MUP. Collection of environmental data with be assessed along with data gathered for the quantity and quality of fish produced and the amount exported, in order to assess the real economic benefit of the structure to the local aquaculture production potential.

2.2 MUSICA Concept and Design

The success of MUSICA will be achieved by advancing the MUP concept developed by the University of the Aegean (UoAeg) and EcoWindWater (EWW). The EWW floating platform was successfully trialled in Heraclea in 2010 for 2 years. This project will substantially modify and develop the EWW MUP to become the MUSICA MUP [4].

The MUSICA project is building on the findings of the MARIBE project. The MARIBE conducted 9 case studies on real MUS/MUP combinations, assessing their technical viability, business plans, financial viability, and risk [15].

The EWW MUP concept case study was a wind power floating desalination platform in the Aegean [16]. The EWW market and business case was the highest rated project overall in MARIBE and thus, was selected as the basis for the MUSICA innovation.

MUSICA will be 50% larger than EWW and will have advanced ICT including EMS as well as innovative energy storage and desalination equipment. Two multiple-use of space (MUS) customers, aquaculture and electricity and water recharging station services, will benefit from the services of the modular MUP platform in a 1.5 year trial.

The project has been designed to meet three key objectives. The first objective of the is to develop the pilot demonstrator of the MUSICA MUP solution and install on site at Innousses Island. This will test and demonstrate the validity of the MUP in a real operating environment. Therefore, the first objective will involve the completion of all technical innovations and improvements to the EWW MUP required, and the installation of renewable energy systems comprised of an 800 kw wind turbine, 60 kw solar panels and a 10 kw wave energy device.

The development will also include a complete smart energy system including a highly integrated and digitalised smart grid. This will accommodate additional modelling and forecasting based on high flexibility services from distributed generation, demand response, simulation software and innovative methods of storage of electricity. The integrated ICT system will be configurable to the island requirements and will be designed to ensure that upon commercialisation it is possible to be delivered with the MUP as a complete unit.

In addition, completion of the MUP will involve installation of 1000 m^3 desalination units and innovative energy storage facilities on the MUP for the island energy management. This will be completed using hydro-pneumatic energy storage (HPES) in combination with conventional batteries.

The second objective of MUSICA is to demonstrate, test and validate the MUSICA MUP, covering functional readiness, cost reduction, shared expenditure and value added to economic, environmental & social viability. The 1.5 year sea-based trial period will cover demonstration requirements for Technology Readiness Level 7 and validate the MUP in its operating environment. Performance indicators will be generated based on analytical predictions and real-data gathered during the trial. Results from the first test cycle will be used optimize design and implementation of required improvements. Non-functional readiness, including manufacturability, deployability, insurability and maintainability requirements will be validated. Further, the shared use of services and resulting cost-savings for MUP desalination, recharging, aquaculture and RES will be tested throughout the project term and optimised. A full evaluation, including detailed technological and economic, environmental and social viability, including changes made for optimisation, will be carried out at the end of the 1.5 year sea trial. The validation indicators and targets for the full evaluation include 61% RES penetration on Innousses (up from 0%), and demonstrated storage of 49 kWh from HPEs and 540 kWh from lithium ion batteries, on the MUP, 100% desalinated water requirements for Innousses from MUP (1000 m^3 per day), successful deployment and operation of aquaculture cage in open sea (including validation of quantity and quality of aquaculture produce), successful integration of wave energy into storage and RES supply, successful operation of floating electricity and water recharging station, supplied by MUP, for 18 months.

The third objective will support economically viable replication of the MUSICA smart MUP solution through Exploitation and Sustainability Plans for commercialisation beyond the project end. This will include market research, detailed business plans for Innousses and 5 other islands (Telendos, Heraclea, Chios, Gozo Malta, La Gomera Canaria Islands), skills training and workshops, and policy briefs on technical & non-technical barriers for mass market replication.

Health and safety (H&S) plans and improvement of professional skills will also be addressed in order to help maximise the societal acceptance and use of the

MUSICA MUP based RES, water supply and aquaculture support service. Successful completion of objectives will also align impacts of the project with Sustainable Development Goals.

3 Summary

Understanding and addressing barriers to multi-use of marine space will be beneficial to supporting the world's growing population. In response to the known challenges, innovative projects seek to find commercially viable solutions to the provision of services for humankind. Moving offshore can lead to reduced competition for space and provide better conditions for raising fish and generating renewable energy.

The MUSICA project is working to create a containerised product for commercialisation and use in offshore waters around small islands. The Multi-Use Platform will be able to host a number of essential services for small islands such as renewable energy generation through wind, wave and solar, desalination of seawater, and provision of power and shelter for offshore aquaculture activities. In addition, the platform can contribute to sustainable tourism by providing a recharging station for leisure boats.

The positive impacts of the MUSICA project include increased island energy autonomy, including the potential for energy export and revenue generation, reliable desalinated water supply for water stressed islands, increased sustainable aquaculture production, and reduced environmental impact in comparison to RES, desalination and aquaculture activities carried out in coastal areas or on land.

This project will demonstrate that the MUSICA MUP is a viable enabling infrastructure for multiple RES, desalination and Blue Growth aquaculture services for small islands. By sharing the same space, working synergistically together, sharing Supply Chains and reducing Operating and Maintenance costs, the MUSICA MUP offers both private and public benefits in addition to addressing challenges associated with the increasing demand for marine space. By 2050, a large percent of EU small islands will have MUSICA MUPs, contributing to their carbon neutrality and food and water security.

References

1. Kite-Powell (2017) Economics of multi-use and co-location. In: Buck BH, Langan R (eds) Aquaculture perspective of multi-use sites in the open ocean: the untapped potential for marine resources in the anthropocene. Springer, Cham, Switzerland, p 242
2. FAO (2020) The state of world fisheries and aquaculture 2020. Sustainability in action. FAO, Rome
3. European Commission (2019) Aquaculture in the EU. Eurostat. https://ec.europa.eu/eurostat/web/products-eurostat-news/-/EDN-20191015-2. Accessed 25 Jun 2020

4. European Commission (2017) Valletta 18 May 2017. Political declaration the clean energy for EU islands. European Commission, Brussels, pp 1–2
5. European Commission (2019) Multiple-use-of space for island clean autonomy. CORDIS. https://cordis.europa.eu/project/id/862252. Accessed 20 May 2020
6. ESPON (2010) Islands' atlas, the development of the islands—European islands and cohesion policy (EUROISLANDS), ESPON 2013 program. ESPON, University of the Aegean
7. Komninos K, Florou A (2018) Smart islands initiative. http://www.smartislandsinitiative.eu/en/index.php. Accessed 22 May 2020
8. Kaiser JB, Chambers MD (2017) Offshore platforms and mariculture in the US. In: Buck BH, Langan R (eds) Aquaculture perspective of multi-use sites in the open ocean: the untapped potential for marine resources in the Anthropocene. Springer, Cham, Switzerland, pp 375–391
9. Anatonopoulou V (2020) Greece, no. 1 fish farming producer in the EU! Greek Food News. https://greekfoodnews.com/greece-number-1-fish-farming-producer-in-the-eu. Accessed 25 Jun 2020
10. Sturrock H, Newton R, Paffrath S, Bostock J, Muir J, Young J, Immink A, Dickson M (2012) Prospective analysis of the aquaculture sector in the EU PART 2: characterisation of emerging aquaculture systems. European Commission, Joint Research Centre. Institute for Prospective Technological Studies, Seville, Spain
11. Holmer M (2013) Sustainable development of marine aquaculture off-the-coast and offshore—a review of environmental and ecosystem issues and future needs in temperate zones. In: Lovatelli A, Aguilar-Manjarrez J, Soto D (eds) Expanding mariculture farther offshore: technical, environmental, spatial and governance challenges. FAO technical workshop, 22–25 March 2010, Orbetello, Italy. FAO fisheries and aquaculture proceedings no 24. FAO, Rome, pp 135–171
12. Papandroulakis N, Thomsen C, Mintenbeck K, Mayorga P, Hernandez-Brito JJ (2017) The EU-project "TOPOS". In: Buck BH, Langan R (eds) Aquaculture perspective of multi-use sites in the open ocean: The untapped potential for marine resources in the Anthropocene. Springer, Cham, Switzerland, pp 355–374
13. Towers L (2010) How to farm European seabass. The Fish Site. https://thefishsite.com/articles/cultured-aquatic-species-european-seabass. Accessed 25 May 2020
14. Towers L (2010) How to farm gilthead seabream. The Fish Site. https://thefishsite.com/articles/cultured-aquatic-species-gilthead-seabream. Accessed 25 May 2020
15. Dalton G (2016) Case studies with companies. Marine investment for the blue economy. https://maribe.eu/blue-growth-reports/case-studies-with-companies/. Accessed 27 May 2020
16. Eco Wind Water (2016) Aegean offshore wind desalination. Final Report. http://maribe.eu/download/2514/. Accessed 15 Jun 2020

Building Floating Aquaculture Farms with Expanded Polystyrene in Singapore

Dominic Kang, Paul Ong, and Jan Roël

Abstract Domestic fish production is vital for Singapore to diversify its food sources. In 2018, local aquaculture farms produced up to 9% of the fish stock consumed in the country. Most of these farms operate along the Straits of Johor on floating wooden structures (also known as kelong). These structures require constant repair and maintenance, which could be costly and disruptive for farm owners. At present, there are no formal building regulations and guidelines available on these structures in Singapore. This paper proposes an engineered design of an offshore aquaculture farm that is compliant with building and maritime standards. The floating base of this new farm is constructed using expanded polystyrene and concrete based on the design concepts developed by FlexBase. Fish will be reared in a closed containment system, which will allow farm owners to safeguard their livestock against pollution and unfavourable environmental conditions. The new farm, designed to have a working life of 50 years, will require minimal maintenance over its service period.

Keywords Aquaculture · FlexBase · Polystyrene · Singapore · VLFS

D. Kang (✉) · P. Ong
Department of Civil and Environmental Engineering, National University of Singapore, 1 Engineering Drive 2, 07-03 E1A, Singapore 117576, Singapore
e-mail: dominic.kang@u.nus.edu

P. Ong
e-mail: paul.ong@nus.edu.sg

J. Roël
FlexBase BV, De Brink 436, 2553 HH The Hague, The Netherlands
e-mail: j.w.roel@flexbase.eu

Ł. Piątek et al. (eds.), *WCFS2020*, Lecture Notes in Civil Engineering 158,
https://doi.org/10.1007/978-981-16-2256-4_13

1 Introduction

Singapore is a city-state located in South-East Asia. Being a resource scarce country, it is heavily reliant on food imports to sustain its population. However, domestic food production remains vital as it helps Singapore to diversify its food sources and achieve food security.

In 2018, aquaculture farms produced up to 9% of the fish stock consumed in Singapore [1]. Most of these farms operate offshore along the straits of Johor in between Pulau Ubin and mainland Singapore and they rear a variety of saltwater fish such as snappers and groupers.

These aquaculture farms operate on wooden structures, also known as kelong, and they rely on wooden piles or air-filled plastic canisters to keep afloat at sea. However, these kelong structures are not built according to any structural design standards. As such, there is no formal process to assess the safety and integrity of these structures. Additionally, they require frequent maintenance and replacement to remain functional. At present, there is very little literature available on kelong structures in Singapore.

2 Proposed Design for Aquaculture Farming

There is a need for an engineered solution to address the safety concerns of these existing farms. Also, there is an operational need to build sturdy and resilient structures for aquaculture farming, as frequent repair work can be disruptive and costly for farm owners.

This paper proposes a new design for a floating aquaculture farm (see Fig. 1) that is compliant with existing building and maritime standards. The structure is designed for a service lifespan of 50 years to address the issue of frequent repair work in existing farms. The new farm will be moored at sea to prevent environmental forces from carrying it away.

The floating base of the farm is made using expanded polystyrene foam (EPS). Traditional floating bases are generally built with conventional air-filled pontoons to keep the entire structure afloat. In recent times, EPS has been used in place of

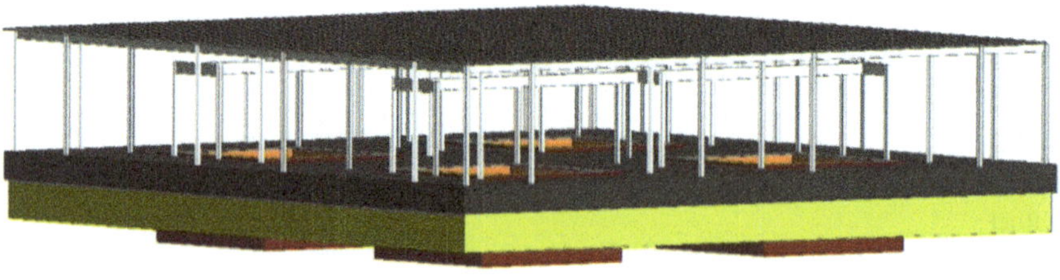

Fig. 1 3D view of proposed floating aquaculture farm design

air-filled pontoons for floating bases. EPS is less dense than water [2]. As such, it is able to provide structures with buoyancy just like conventional air-filled pontoons. Over the years, various builders have developed and patented numerous variants and configurations of EPS floating bases. These foams are readily available on the market and are relatively inexpensive to manufacture.

One of these builders is FlexBase, a Dutch company which specialises in the design and construction of floating and amphibious structures. FlexBase has developed a few types of floating bases with EPS and reinforced concrete, and the usage of their bases ranges from residential to industrial. These bases can be constructed directly on water. They require minimal maintenance during its entire service life. FlexBase floating modules have been used to construct numerous floating houses and large exhibition spaces such as the Rotterdam Floating Pavilion (see Fig. 2).

Currently, there is no precedent in using EPS as a structural material for floating bases in Singapore. Hence, the realisation of this farm with FlexBase will make it the first of its kind in the country.

For standard FlexBase modules, slots are cut in the EPS layers to cast concrete beams, as shown in Fig. 3. These concrete beams are designed to receive vertical actions from the top structure. A concrete floor slab is cast directly on top of the beams to spread the vertical actions. The base of the farm will be constructed according to this configuration.

Fig. 2 Rotterdam Floating Pavilion, The Netherlands

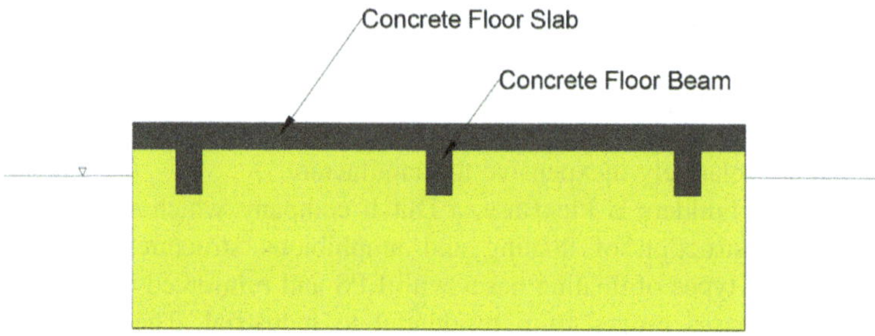

Fig. 3 Configuration of FlexBase module for farm

3 Floating Farm Dimensions

The farm consists of a single storey with no roof access, as shown in Figs. 4, 5 and 6. The allowable height in the storey is 4.5 m. The base is a square module with a length of 30 m, making it a very large floating structure (VLFS). Openings are engineered symmetrically in the base module to host 4 water tanks. The latter will be used to rear fish stock in a close containment system. For station keeping, the farm will be moored out at sea.

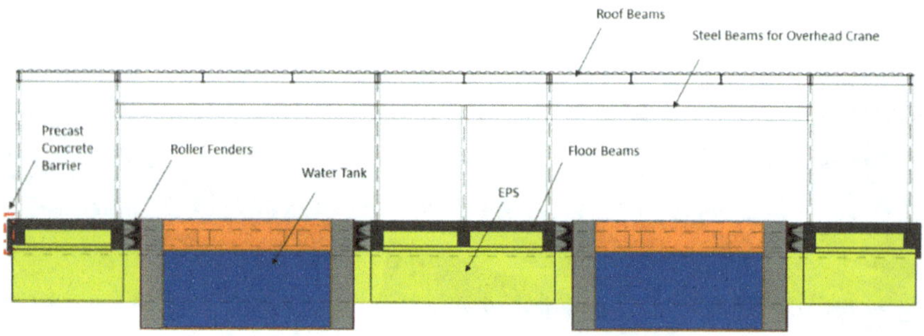

Fig. 4 Elevation view of farm

Fig. 5 3D view of farm base

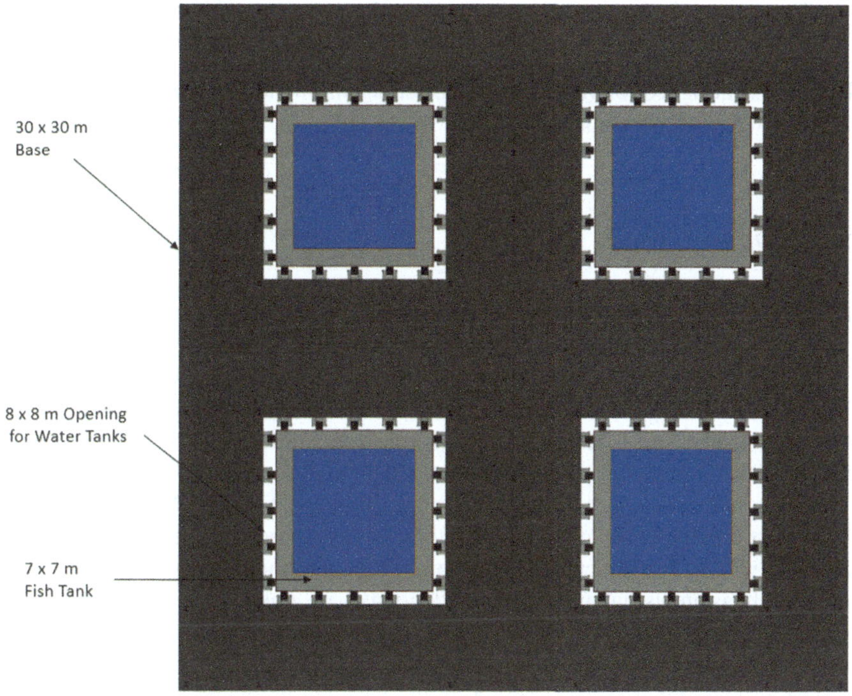

Fig. 6 Plan view of farm base

The structural frame components of the farm are sized based on the farm's floor area. For the structure above the floating base, the beams, columns, and roof are constructed with steel sections. Reinforced concrete is used for the floor slab and floor beams of the floating base. Lastly, EPS foam blocks are used for buoyancy.

4 Structural Design

In 2015, Singapore adopted Eurocodes as the prescribed structural design standards in replacement of British Standards [3]. The Singapore version of Eurocodes, denoted as "SS-EN", is used with the corresponding national annexes denoted as "NA to SS EN". The entire farm structure is designed to satisfy the ultimate and serviceability limit states, as prescribed by SS-EN and the corresponding national annexes.

5 Variable Actions Acting on Structural Components

The farm is designed for various variable actions throughout its service life, such as the imposed roof and floor loads, and wind loads.

5.1 Imposed Loads

The roof of the farm is designed as a shelter with no roof access. Based on guidelines in EN 1991-1-1, the imposed roof load shall be taken as 0.75 kN/m^2 [4].

The variable action on the floor is based on human traffic and the weight of industrial machinery. In this case, the relevant machinery for the farm would be the water purification units. These units will be placed symmetrically on the structure to prevent unwanted tilts at sea.

For the floor area on the farm, a design imposed load of 5.0 kN/m^2 is assumed. This imposed load is assumed to act on the centre of gravity of the people onboard, since the weight and dimensions of other entities contributing to the imposed floor load (such as industrial machinery and goods) are unknown at this stage. The average height of a person is taken as 1.71 m and his/her centre of gravity will be assumed at mid-height.

The various imposed loads acting on the farm are presented in Table 1.

5.2 Wind Action

Wind action is governed by guidelines set out in EN 1991-1-4 [5]. Based on past weather data in Singapore and guidelines in the Singapore National Annex, the wind parameters in Table 2 are assumed.

The farm is designed without external walls. However, it is possible for tarp sheets to be installed for shade along the perimeter in the future. The wind pressure on these surface areas could be significant. To be conservative, wind loads are calculated by taking into consideration the placement of these sheets on all sides.

Table 1 Imposed loads

Imposed roof load	0.75 kN/m^2
Imposed floor load	5.0 kN/m^2

Table 2 Wind parameters

Terrain	Sea (0)
Basic wind velocity (vb)	20.0 m/s
Mean wind velocity (vm)	23.2 m/s
Peak velocity pressure (qp)	0.65 m/s^2

Table 3 Wind forces acting on farm structure and tank

Wind force acting on farm structure	109.4 kN
Wind force acting on tank	9.0 kN

As the farm is symmetrical in both x and y axes, the wind loads acting on the farm in all 4 directions (North, South, East and West) are assumed to be the same. The maximum wind forces acting on the farm structure and the tank are presented in Table 3.

5.3 Environmental Actions

Waves and current exert forces on the farm throughout its entire service life. Singapore is located along the equator and it experiences 2 monsoon seasons (northwest and southwest) that are separated by consecutive inter-monsoon periods [6]. During the southwest monsoon season, squalls that developed over the Sumatra island are blown by south-westerly winds towards Singapore. The presence of squalls can result in higher than usual sea waves. A violent Sumatra squall in 2012 was modelled by a team of researchers to investigate its impact on the generation of sea waves [7]. The wave parameters of the study interval around the vicinity of Pulau Ubin are shown in Table 4. These values are used to determine the wave action on the farm.

5.4 Current Force

The force exerted by the water currents is determined by the Morison Equation. By using the significant wave height and period in Table 4, the parameters of the water waves are obtained and presented in Table 5. The maximum inertial forces acting on the farm and tanks are shown in Table 6.

Table 4 Wind and wave parameters along Pulau Ubin [7]

Significant wave height	0.50 m
Peak period	3.5 s

Table 5 Parameters of water waves

Maximum wave height (Hmax)	1.86 m
Wave radian frequency (ω)	1.80
Wave number (k)	0.32
Wavelength (L)	19.49 m
Maximum horizontal displacement (ζmax)	0.93 m
Maximum horizontal velocity (uHmax)	1.67 m/s
Maximum horizontal acceleration (aHmax)	3.0 m/s^2

Table 6 Maximum inertial forces

Wave propagation angle	Inertial force acting on farm	Inertial force acting on each tank
0°	220.0 kN	174.3 kN

5.5 *Wave Drift Force*

Wave drift forces acting on the farm and each tank are calculated and shown in Table 7.

5.6 *Total Environmental Forces on the Farm*

The total horizontal force acting upon the farm is 1089.7 kN, as summarised in Table 8.

6 Design of Structural Components

Steel materials are used for the roof sheets, beams, and columns, while reinforced concrete and EPS are used for the construction of the base module.

6.1 *Steel Corrugated Roof*

Steel corrugated sheets, LCP LYCORRIB® 28, will be used for the roof structure to shield the farm from the elements [8]. Made of high tensile steel, these sheets are lightweight, with a mass per unit area of 4.71 kg/m^2.

Table 7 Wave drift forces

Wave drift force acting on farm	13.0 kN
Wave drift force acting on each tank	3.0 kN

Table 8 Total horizontal force acting on farm

Wind force	109.4 kN
Current force	220.0 kN
Wave drift force	13.0 kN
Environmental loads acting on all tanks	746.3 kN
Total horizontal force	1089.7 kN

6.2 Steel Columns and Roof Beams

The structural frame above the floating base module will be constructed with S275 steel. A steel frame is preferred over a concrete frame for a lighter design. The chosen sizes of steel sections are presented in Table 9.

6.3 Concrete Barriers

Concrete barriers are installed along the perimeter of the farm to serve as formwork for the casting of concrete floor beams and slab, as shown in Fig. 4. Each concrete barrier is sized with a thickness and height of 0.15 and 1.4 m, respectively.

6.4 Concrete Floor Beams and Slab

The floor beams of the structure, which are reinforced as shown in Fig. 7, will resist all actions from the top structure. A concrete base slab of 300 mm thickness will be cast on these floor beams to receive and transfer the imposed floor loads.

For the concrete elements of the base module, corrosion is likely to be induced by chlorides from sea water. The exposure class designation for the concrete beams

Table 9 S275 steel profiles for columns and beams

Column	UKC 152 × 152 × 31
Beam	UKB 305 × 127 × 37

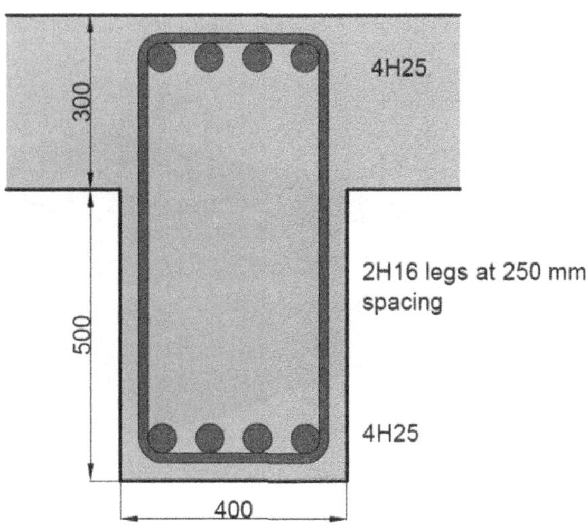

Fig. 7 Concrete beam design (reinforcement of slab omitted)

and slab is chosen to be XS3 in accordance with SS EN 206-1, which accounts for higher local ambient conditions in Singapore [9]. For this structure, the parameters in Table 10 are relevant for the concrete design.

6.5 Overhead Cranes

Overhead cranes will be placed over the openings to harvest the fish in the tanks as shown in Fig. 8. These beams will be placed at 4 m above the floor to provide sufficient clearance for the fish tanks.

6.6 Roller Fenders

Roller fenders will be installed along the sides of the openings to guide the vertical movement of the tanks under different loading conditions. These fenders are designed to absorb the horizontal environmental loads that act on the tanks.

Table 10 Parameters for EC2 Design

Concrete class	C45/55
Density of concrete	2500 kg/m^3
Characteristic compressive cylinder strength (fck)	45 MPa
Characteristic yield strength of reinforcement (fyk)	500 MPa
Nominal cover to top, bottom and side reinforcement (cnom)	40 mm
Standard fire resistance period (R)	60 min
Minimum width of beam	120 mm

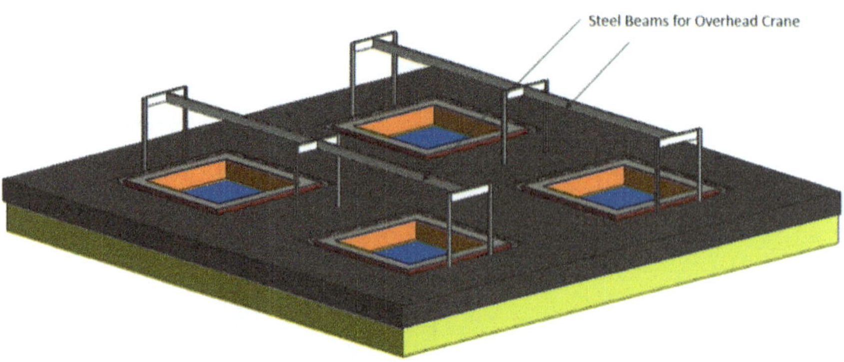

Fig. 8 Steel frame supporting overhead cranes

Fig. 9 Arrangement of fenders at the openings

A roller fender type, CR 35 [10], will be installed on the floating base. As shown in Fig. 9, a total of 10 roller fenders will be installed on each side of the opening, giving each side the capacity to resist an imposed loading of 3500 kN.

7 Stability and Buoyancy

The structure is designed against capsizing and sinking. Naval architecture principles are applied to determine the buoyancy and stability of the structure.

According to "Guidelines and Recommendations for Moored Floating Structures in Singapore", stability requirements shall be governed by DNV shipping and offshore standards [11].

The proposed farm structure has a long flat-bottomed base which makes it similar to a barge in terms of geometry. Based on DNV-OS-C301 and DNV Rules for Ships [12], the following conditions must be satisfied for the general stability criteria of a "ship shaped" structure [13]: Area under the righting lever curve (GZ curve) shall not be less than 0.055 metre-radian up to a heeling angle of 30° and not less than 0.09 metre-radians up to a heeling angle of 40° or the angle of flooding if this angle is less than 40°; The righting lever (GZ) shall be at least 0.20 m at an angle of heel equal or greater than 30°; The maximum righting lever should occur at an angle of heel preferably exceeding 30° but not less than 25°; The initial metacentric height, GM0, must not be less than 0.15 m.

It is important to ensure that the farm is buoyant and stable under normal operational conditions. Therefore, the farm is designed for equilibrium and initial stability, and statical stability.

7.1 Equilibrium and Initial Stability

Using Archimedes Principle, the total buoyant force and the initial stability of the farm can be determined. The coordinates of the C.O.G. of the structure are calculated and presented in Tables 11 and 12 for different loading conditions. Table 13 shows the range of free and submerged height for a total EPS volume of 1368 m^3.

The calculated values for the initial metacentric height GM in Table 14 show that the metacenter in both axes are above the centre of gravity. These values are well above the minimum height of 0.15 m recommended in the DNV standards.

7.2 Statical Stability

Statical stability refers to a floating body's ability to return to its upright position when it is acted upon by a destabilising force while stationary [14]. When a

Table 11 Centre of gravity of farm under dead loads only

Loading condition	Dead loads only		
Centre of gravity (Distance from origin)	x	y	z (KG)
	15.0 m	15.0 m	2.43 m

Table 12 Centre of gravity of farm under dead + imposed loading

Loading condition	Dead + imposed loading		
Centre of gravity (Distance from origin)	x	y	z (KG)
	15.0 m	15.0 m	2.77 m

Table 13 Submerged heights of floating base under various loading conditions

Dead loads only	Free height	Submerged height
	1.21 m	1.39 m
Dead + imposed loading	Free height	Submerged height
	0.58 m	2.02 m

Table 14 Metacentric heights of farm structure under different loading conditions

Loading condition	Dead loads only	Dead + imposed loads
Moment of inertia of floor area (m^4)	52906.3	52906.3
Volume of submerged region (m^3)	882.2	1287.9
BM (m)	60.0	41.1
KM (m)	60.7	42.1
GM	58.2 m (>0.15 m)	39.3 m (>0.15 m)

destabilising force acts upon the structure, a tilt will occur. This tilt is also known as a heeling angle and it is a variable in the computation of the righting and heeling arms. By comparing the righting and heeling arms, the statical stability of the structure can be determined.

To determine the effect of movement, the shifting of equipment onboard is simulated as shown in Fig. 10. Assuming that the total weight of the equipment is 500 kN, the heeling arm and moment are calculated by simulating movement from the centre of the floor area to the centre of 1-half of the floor area. This movement will result in a shift in the centre of gravity of the farm structure.

For statical stability, the heeling arms of these loading conditions are checked against the curve of statical stability (also known as the GZ curve) to determine the rotation of the structure when it is stationary. Following which, the angles of rotation are checked against DNV standards to ensure that the prescribed limits are not exceeded. An unfavourable factor of 1.5 is taken for all destabilising loads.

The heeling angles of various loading cases of heeling arms are listed in Table 15.

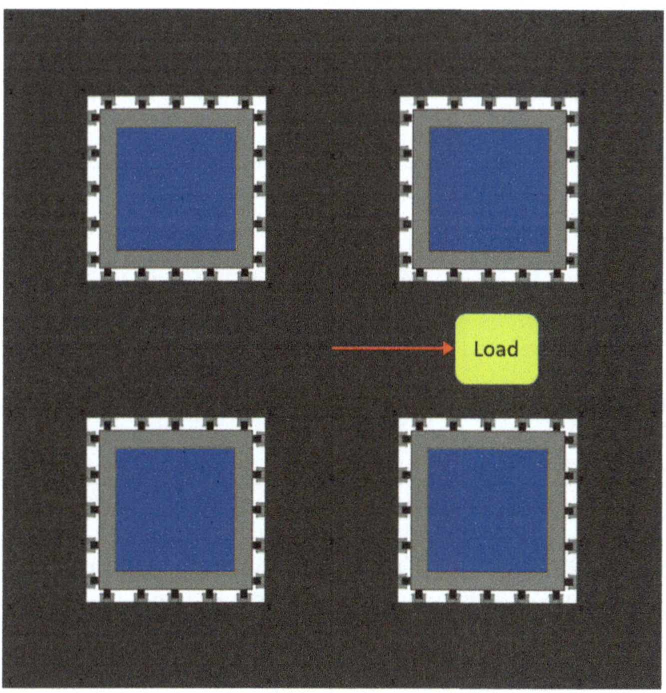

Fig. 10 Movement of load from centre of floor area to centre of 1 half of floor area

Table 15 Loading cases for statical stability analysis (angle and arms)

Loading case	x-x (or y-y)		
	Heeling angle	Righting arm (m)	Heeling arm (m)
(A): Wind action (basic wind)	0.07°	0.05	0.05
(B): Wind action (gust wind)	0.11°	0.07	0.07
(C): Movement of equipment from centre of the floor to centre of ½ of the floor	1.28°	0.44	0.44
(D): Combined loading of movement of equipment and gust wind	1.33°	0.51	0.51

7.3 Dynamic Analysis

A dynamic analysis can be performed to determine the farm's response to the motion of waves over time. Such analysis can be done by meshing the farm's model and by inputting parameters of the waves using hydrodynamic software. Due to limited resources, this analysis is not covered in this paper.

8 Tank Design

The fish tanks are designed using materials such as steel, EPS, and high strength plastic. As such, it is important that these tanks, remain buoyant and stable throughout its service life. Hence, the intact stability of these tanks is considered in the design process. When all 4 tanks are full and operational, a combined volume of 282 m^3 will be made available for the rearing of fish stock.

8.1 Tank Specifications

A 3.5 m high tank is designed to float independently without transferring vertical loads to the floating base. A plastic chamber will be contained within the outer steel chamber as shown in Fig. 11. In reference to Fig. 12, a plastic chamber will be contained within a steel chamber. In between both chambers, EPS will be used to fill the space to aid the tank in buoyancy. For a full tank, the maximum water level will be set at 2.40 m, measured from the base of the plastic chamber.

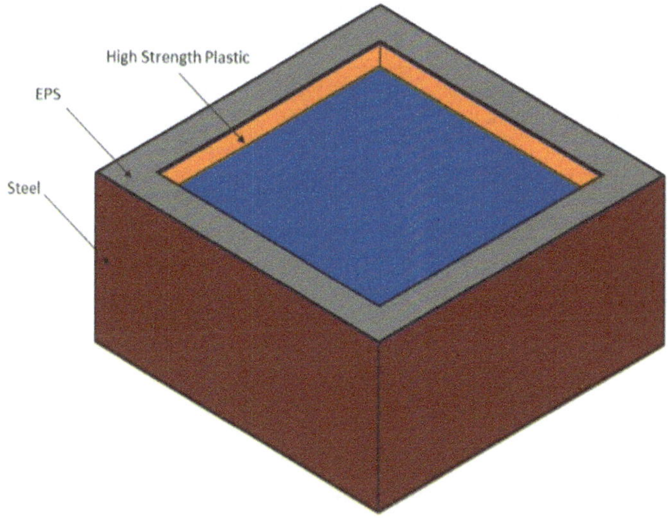

Fig. 11 Tank design (3D)

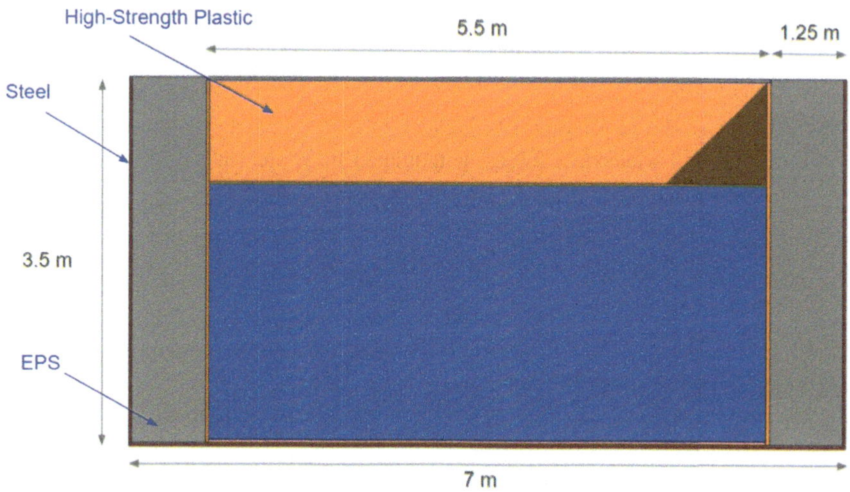

Fig. 12 Tank design (plan view)

8.2 *Intact Stability of Tank*

Calculations were done to determine the tank's free height when it is empty, half-filled, and full based on the maximum water level. This is summarized in Table 16. For all tank loading conditions, the free heights of the tank must not exceed the allowable height limit of the crane-supporting steel frames.

Table 16 Free heights of tank under different loading conditions

Loading condition	Empty tank (m)	Half-filled tank (m)	Full tank (m)
Free height	2.56	1.84	1.12
Submerged height	0.94	1.66	2.38

Table 17 Metacentric heights of tanks under different loading conditions

Loading condition	Empty tank	Half-filled tank	Full tank
KG	1.05 m	0.97 m	1.43 m
Moment of inertia	200.08 m^4	200.08 m^4	200.08 m^4
BM	4.33 m	2.46 m	1.71 m
KB	0.47 m	0.83 m	1.19 m
GM	3.75 m (>0.15 m)	2.32 m (>0.15 m)	1.48 m (>0.15 m)

Like the floating base, the initial stability of the tank is also determined when it is under different loading conditions. The various metacentric heights of the tank under different loading conditions are satisfactory as shown in Table 17.

8.3 *Forces on the Fenders*

The fenders, as shown in Fig. 13, are designed to resist the combined action of wind, current and wave drift forces as tabulated in Table 18.

9 Mooring System

Mooring lines are designed to keep the floating farm within its deployed position. The depth of waters along the Straits of Johor is taken to be 20 m [7]. Based on the shallow water profile, a catenary system of chain cables is proposed.

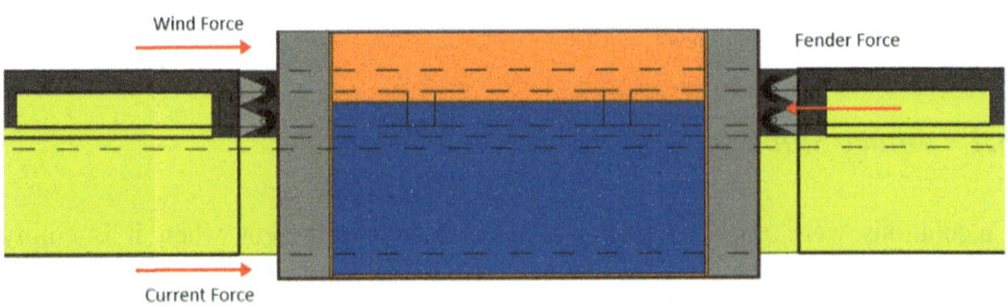

Fig. 13 Force experienced by fenders

Table 18 Total environmental loads acting on 1 tank

Wind force	9.3 kN
Current force	174.3 kN
Wave drift force	3.0 kN
Total environmental loads acting on 1 tank	186.6 kN

Table 19 Chain design

Force acting on floating base parallel to chain	1089.7 kN
Chain tension at floating base	1109.0 kN
Safety factor	2.0
Design tension of chain	2217.0 kN
Chain diameter	50 mm
Minimum breaking strength of chain	2740.0 kN

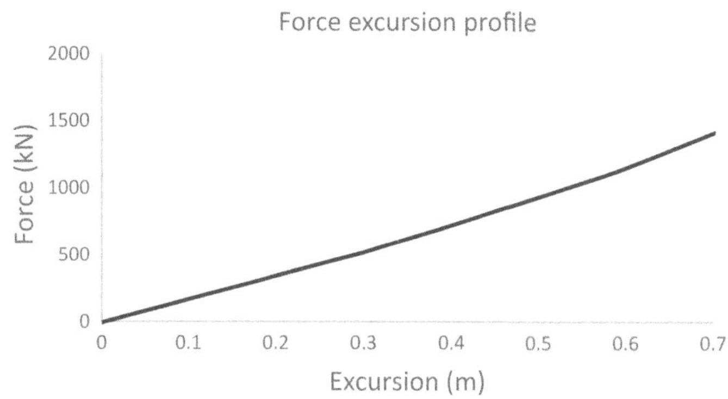

Fig. 14 Force excursion profile

Keeping in mind a safety factor of 2.0 for quasi-static analysis, a R4 chain of 50 mm diameter is recommended. As shown in Table 19, the minimum breaking strength of the chain satisfies the design tension of the cable.

As shown from the force excursion profile in Fig. 14, the expected offset from the original position is 0.56 m.

10 Conclusion

This paper has presented a floating aquaculture farm built on an EPS floating base developed by FlexBase. The farm is shown to satisfy the ultimate and serviceability limit states, as prescribed by SS-EN and the corresponding national annexes. It is also compliant to the maritime standards of DNV for equilibrium and stability.

The structure will be able to meet the service lifespan of 50 years and address the issue of frequent repair work in the existing kelong farms.

The realisation of this project will provide a safe working environment for farmers and there will be less disruption due to maintenance work. This will allow farmers to concentrate their efforts on fish rearing and increase domestic production.

As land is scarce in Singapore, the successful deployment of this EPS floating base could also generate ideas and possibilities in building more of these EPS floating bases for activities that are currently land-based. By relocating these activities offshore, precious land space can be freed up in the country for other meaningful purposes.

References

1. Singapore Food Agency (2019) The food we eat. In: Singapore's food supply. Singapore food agency. https://www.sfa.gov.sg/food-farming/singapore-food-supply. Accessed 01 Jun 2020
2. Horvath J (1994) Expanded polystyrene (EPS) geofoam: an introduction to material behavior. Geotext Geomembr 13(4):263–280
3. Building and Construction Authority of Singapore (2013) BCA introduces new international structural design standards for buildings. Building and Construction Authority of Singapore. https://www1.bca.gov.sg/docs/default-source/docs-corp-news-and-publications/media-releases/pr25032013_ec.pdf. Accessed 01 Jun 2020
4. European Committee for Standardization (2002) Eurocode 1: actions on structures—part 1-1: general actions—densities, self-weight, imposed loads for buildings. CEN, Brussels
5. European Committee for Standardization (2010) Eurocode 1: actions on structures—part 1-4: general actions—wind actions. CEN, Brussels
6. National Environment Agency (2009) Weatherwise Singapore. In: Weatherwise Singapore. National Environment Agency. https://www.nea.gov.sg/docs/default-source/resource/guidebook-on-climate-of-singapore.pdf. Accessed 01 Jun 2020
7. Ly D, Tan C, Ma Q (2015) Characteristics of Sumatra squalls and modelling of the squall-generated waves. In: Gourbesville P, Cunge J, Caignaert G (eds) Advances in hydro informatics. SimHydro 2014 conference, Sophia Antipolis, June 2014. Springer, Singapore, pp 157–174
8. LCP Building Products (n.d.) LCP LYCORRIB 28. In: LCP LYCORRIB 28. LCP Building Products. http://www.tsh.com.sg/wp-content/uploads/2016/07/LCP-LYCORRIB28-1.pdf. Accessed 01 Jun 2020
9. Singapore Standards Council (2014) Concrete: specification, performance, production and conformity. Springer, Singapore
10. ShibataFenderTeam (2019) Products. In: Products. ShibataFenderTeam. https://www.shibata-fender.team/files/content/Downloads/SFT_Product_Catalogue_A4_English.pdf. Accessed 01 Jun 2020
11. NUS, SINTEF (2019) Guidelines and recommendations for Moored floating structures in Singapore
12. DNV GL (2015) Offshore standard DNVGL-OS-C301, stability and watertight integrity. DNV GL, Oslo
13. DNV GL (2019) Rules for classification, ships part 3 chapter 15 stability. DNV GL, Oslo
14. Biran A (2003) Ship hydrostatics and stability. Butterworth-Heinemann, Oxford, pp 125–270

Floating Rice Fields, The Quest for Solutions to Combat Drought, Floods and Rising Sea Levels

Soon Heng Lim

Abstract Amazing as it seems, there is a case for growing rice on floating platforms in the sea. The capital expenditure to develop this is offset by the opportunity to repurpose the land for a number of commercially attractive activities. It would also eliminate crop failures due to droughts and floods soil degradation due to higher salinity and lower water table. This paper examines ways crops have been cultivated in the past. Crops are almost exclusively grown on land. Urban farming in recent years have led to the development of structures enabling crops to be produced in multi-tiers over the same plot of ground. Crops have also been grown over water surfaces on floating and semi floating structures on freshwater bodies. Apart from seaweeds, no vegetables have ever been successfully cultivated in the sea. While low-salt tolerant rice is being developed, rice has never been grown in the sea. The paper proposes a system where rice is grown on floating platforms with freshwater fed from a freshwater storage facility founded on the seabed. The key to success for the system is to develop a water storage system which is fed by the runoffs from the rice fields and supplemented by lakes or rivers on land.

Keywords Rice · Drought · Flood · Salinity · Climate-resilient · Sub-sea water storage

1 Rice Is Life

> Rice is the staple food of more than half of the world's population—more than 3.5 billion people depend on rice for more than 20% of their daily calories. Rice provided 19% of global human per capita energy and 13% of per capita protein in 2009. Asia accounts for 90% of global rice consumption, and total rice demand there continues to rise.
>
> Global rice consumption is projected to increase from 450 million tons in 2011 to about 490 million tons in 2020 and to around 650 million tons by 2050. However, supply projections

S. H. Lim (✉)
President Floating Solutions LLP, Society of Floating Solutions (Singapore), 7 Temasek Boulevard #12-07, Suntec City Tower 1, Singapore 038987, Singapore
e-mail: lsh@sixtrees.com.sg

Ł. Piątek et al. (eds.), *WCFS2020*, Lecture Notes in Civil Engineering 158,
https://doi.org/10.1007/978-981-16-2256-4_14

> seem to agree that the global rice supply is unlikely to be able to meet this 44% projected growth in demand. The global rice area harvested increased by 1.38% per year between 1961 and 1977, but since then slowed to just 0.33% per year.

Currently the world's rice fields cover a total area about the size of Saudi Arabia (2 million sq. km). With coastal erosion and rising sea levels the coverage will decrease rather than increase. This is exacerbated by the migration of millions of people from inland to the coast each year resulting in increasing pressure on agricultural land.

Many today still remember Elvis' hit "Are you lonesome tonight?" back in 1960. Incredulously, since then the world population has more than double from 3 billion to 7.8 billion. It is clear, growing rice the way, it has been done in the past three thousand years is hardly sustainable.

The migration of global population towards the coast puts huge pressure on land use. It also stresses the biodiversity of flora and fauna on the planet. The next leap is to go beyond where the land ends.

2 Growing Food in Multi-tier Farms

The past few decades see the emergence of cultivation of crops above ground. In cities from New York to Tokyo, residents are growing not only ornamental plants but vegetables and fruits on roof tops and balconies, more as a pastime than with economic objectives.

Bringing food from traditional farms to the consumer requires long and complex logistics which add costs to the product but not its value. It requires the use of trucks, railways, ships, and aircrafts, refrigerated containers and refrigerated warehouses before it reaches the grocery. To reduce the cost of that supply chain the idea of growing food on a commercial scale near to the city seems attractive.

In 1999 Columbia University professor Dr. Dickson Despommier developed the idea of a multi-story building in which layers of crops could be grown on each floor 5. He still champions vertical farms.

The idea is good, but it cannot be easily scaled up because there is too much competition for use of urban land. In many cities land is already too expensive for housing, let alone farming.

Vertical farming faces many challenges. Plants need sunlight. Growing them in the vertical direction reduces the amount of sunlight each plant receives. To overcome this artificial light is needed. Alternatively, plants can be grown in troughs that are rotated by a conveyor belt, but this is "dilutes" the share of sunlight for each plant as it is moved into the shaded side in each cycle.

Plants cannot be grown in the usual way with nutrients mixed in soil, so aquaculture is needed. The system requires nutrient loaded water to be circulated. Pumps and pipes need to be installed. The building structure to support the plants degrade with time. The entire system requires regular maintenance. Energy is needed to keep it running. The carbon footprint is naturally higher per unit of output.

In most cases, even with the cost of transportation factored into the equation, growing crops on land in remote locations, is still cheaper than growing them in vertical farms in urban areas nearer the consumer.

3 Challenges to Ground-Based Crops

Many crops such as rice, wheat, potatoes, corn, and soya beans are not adaptable for vertical farming. They are grown in vast quantities and require huge land masses.

The lands that they have been cultivating for the last two hundred years have been severely degraded. Natural nutrients on the topsoil are washed out by floods. Ground water is depleted by its extraction during times of drought. Coastal and riverine lands are eroded as sea levels rise. Rising levels leads to higher soil salinity which is detrimental to crops especially to rice.

In January 2020, the Thai ministry of agriculture declared that rice farmers in 22 provinces might not have enough water to grow rice in an area over 3600 km^2 in size (2.25 million rai). This is because the extraction of subterranean water to grow rice is prohibited as there is insufficient water for drinking and for industries.

In 2011, 20,000 km^2 of the Thai nation was inundated for six months affecting 13.6 million people and resulting in economic damage worth US $46.5 billion according to the World Bank.

A 41-month drought was recorded in India beginning in 2015.

According to Scientific American referencing Bangladesh, "A three-foot rise in sea level would submerge almost 20 percent of the entire country and displace more than 30 million people—and the actual rise by 2100 could be significantly more." Severe floods killed at least 61 people, displaced nearly 800,000 and inundated thousands of homes across a third of Bangladesh, after two weeks of heavy monsoon rains in July 2019 in that country.

> Salt stress often causes photosynthesis decrease, plant growth inhibition, biomass loss, and partial sterility, all of which lead to yield reduction.

Electrical conductivity (EC) of irrigation water is the attribute most often used for monitoring the salinity of irrigation water, by the practicability of its measurement and high correlation with the amount of soluble salts, since the EC is the measure of resistance passage of electric current between electrodes in a solution where ionic solutes (cations and anions) are present (Doneen 1975). Grattan et al. (2002) estimate a yield loss of 50% with an EC of around 7.4 dS m^{-1}. In some cases, however, the salinity of soil solution from 1.9 dS m^{-1} is already sufficient to significantly reduce the seedlings biomass and an EC of 3.4 dS m^{-1} compromises their survival according to researchers Zeng and Shannon.

Research into extraction of salt by mechanical and botanical means have not met with measurable success. It is costly and not a satisfactory solution (Fig. 1).

Fig. 1 The scourge of climate change, soil salinity, drought and floods can only be overcome by moving from land to sea. The technology to enable that is at hand

3.1 Salt Tolerant Rice

"Growing rice in swamps, bogs, and clay-like or salty coastal waters, which comprise about a third of the total arable land in China, has typically been impossible because salt stresses the plants. That makes photosynthesis and respiration a challenge for the stalks, causing them to stop growing and die. An increasing amount of land is expected to face this problem as sea levels rise".

China leads the world in research to genetically modify the DNA of natural rice to produce new strains that are salt water tolerant. More than 200 strains are experimented on with diluted salt water. The strains can tolerate up to a salt concentration of about "10% of the level in sea water"

To forestall undue exuberance Ren Wang, assistant director general for agriculture at the United Nations' Food and Agriculture Organization said, "It's still only maybe 10% the level of salt in sea water,"—"quite far" from any practical application.

China's legendary agronomist Yuan Longping works tirelessly to hybridize rice since the 1970s. He is the authority on genetic manipulation of rice cultivars and the creator of the Green Super Rice which has a yield of 17 tons per hectare, twice the national average. He reckons commercial saltwater-tolerant rice is eight to ten years away 10.

4 Hydraulic Versus Genetic Engineering to Increase Rice Yield

In the meanwhile we need to go forward with a way to grow proven strains of rice in floating rice fields in lakes where there is fresh water and in the sea by devising means of storing fresh water to overcome prolonged droughts and rising sea levels.

This is a mechanical solution, using hydraulic engineering rather than genetic engineering. It is a solution that applies the principles of buoyancy known to Archimedes more than 2000 years ago.

Rice would be a good crop for engineers, particularly naval architects, to put their knowledge and skill to the test, as it is under imminent threats from rising sea levels, costal erosion, droughts, and soil salinity. They should team up with agronomists specializing in rice cultivation, pest infestation, harvesting and storage to address all phases of the rice cycle.

4.1 The Economics of Relocating Rice Fields Away from Land

Is it worth the effort economically? Given that land where rice is grown is cheap, does it make sense? Will landowners obstruct such a move for fear of losing an income from farmers? How would it change the life of farmers and their families?

Current wholesale price of rice is about USD 600 per ton and a less than optimal yield is about 10 tons per ha per year. Over a ten-year period, ignoring inflation, the value of the output is USD 60,000 per hectare. It would seem that any system to be economically viable need to be address this as the ceiling-barrier to entry.

4.1.1 Cottage Industries

That however is an oversimplified benchmark. It does not account for the fact that every hectare of rice field on land that is replaced with a hectare of rice field floating in water, the vacated plot on land can be put to much higher economic use. The land can be used to keep livestock (goats, cattle, ducks, chicken), build factories to produce goods and keep many employed, even if these factories are only cottage industries producing clay pots or rattan furniture. The farmers would enjoy an additional income in the long months between sowing and harvesting.

4.1.2 Eco-Tourism

Other possible revenue generating ideas include eco-tourism. In Ubud, Bali, 5-star hotels with panoramic view of rice fields charge rates in excess of USD 1000/night.

Villagers benefit from finding employment as hotel staff, tourist guides and selling local handicrafts.

4.1.3 Compatible Crops

The land could also be used to grow other crops that are more climate resilient and drought tolerant. A suitable crop to cultivate would be coconut, as it grows naturally on the sea front with very little need for fertilizers or maintenance and copra can be turn into cooking oil. Another is mangrove plants (Rhizophora). This plant grows naturally in swamps, acts a defence against erosion, produces timber that is useful as piles and provides habitats for crabs, eels and other edible creatures.

When rice is grown on floating platforms, the space beneath the platforms can be put to good use for shellfish aquaculture to rear crabs, cockles, mussels and oysters. See Fig. 2.

4.1.4 Solar Energy

With plunging prices of solar panels "by 99% over the last four decades" according to MIT News, and the long-term escalating price of hydrocarbon fuels a point could be reach in the future when landowners in Bangladesh and other rice growing countries where the land is cheap may find it more financially rewarding to use the land to harvest solar energy than growing rice. Solar energy is unaffected by the vagaries of the weather, pests, and escalating prices of fertilizers (Fig. 3).

Instead of exporting rice landowners, in say Vietnam, may find it more profitable to export energy by undersea cables to Singapore. Australia is already planning to

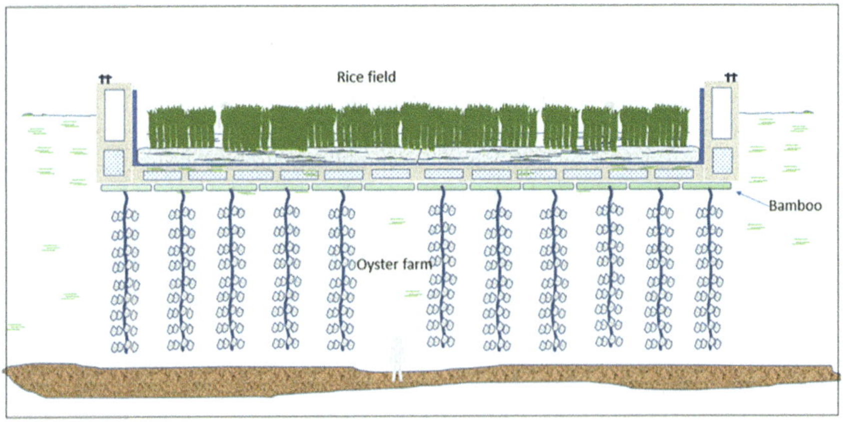

Fig. 2 Oysters and mussels grown under the floating rice field provide additional income for farmers

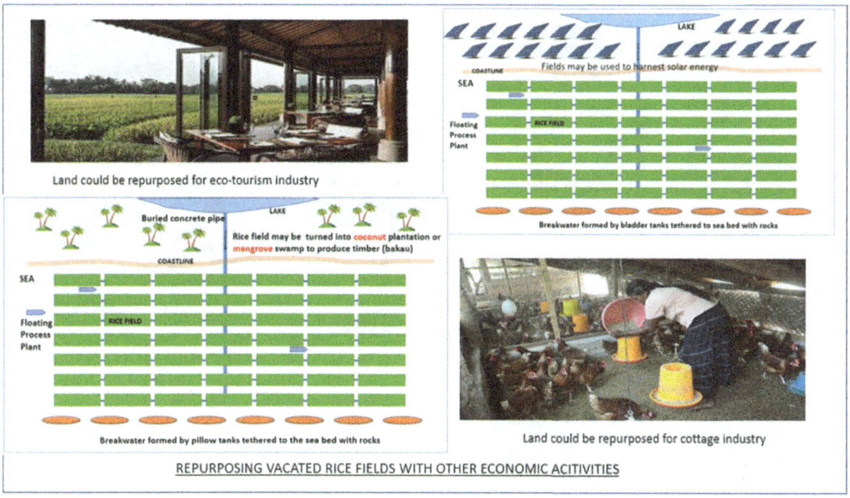

Fig. 3 Vacated rice fields may be repurposed to generate income in a variety of ways from eco-tourism to solar farms

do so. Vietnam would be more competitive because of its cheaper manpower cost and shorter distance to Singapore. Singapore has less diversity in its choice of energy sources than it has in procuring rice.

In summary if rice is grown in floating farms in lakes or in the sea, the freeing up of the land would bring about a higher level of wealth to the community as a whole.

5 Traditional Floating Farms

Artificial floating farms are not new. Fruits and vegetables have been cultivated successfully and organically for many years in various parts of the world.

Cultivation of crops on floating decaying vegetable matter makes enormous sense. It is less labour intensive, one need not plough the land, the thriving microorganic ecosystem till the soil every hour of the year. The water is always there so there is no need to construct irrigation systems. Boats move produce from farms to markets without a trace of greenhouse emission.

Inle Lake in Myanmar is well known for a community whose economy is built around the production of fruits and vegetables on floating platforms. These platforms are formed by bundling masses of water hyacinths and kept in position with bamboo stakes. There are 3000 hectares of such farms in the 260 km^2 lake. The hyacinths grow in huge profusion rapidly all year round, so the locals are never in short supply of "building material," or organic fertilizers. Tomato is the main crop owing to their demand, but other crops are also grown (Fig. 4).

Fig. 4 Traditional floating farms in freshwater lakes

The prize for the most amazing use of water bodies for growing food goes to the Aztecs of Mexico. They famously build the chinampas as early as the fourteenth century in the lakes in the Valley of Mexico14. Some of these chinampas were large with footprints measuring 90 m × 10 m. These plots of were constructed in marshlands and the plants are sowed on compost material and silt kept in place by timber stockade. The top of the soil is just high enough to keep the roots from being permanently soaked in the water beneath it.

They were laid out in a rectangular grid surrounded by a network of canals serving as "roadways." Boats were punt around to collect the produce during harvest time.

The chinampas still exist today. It seems chinampas ownership is legally recognized and can be passed on to succeeding generations, attesting to the durability of the structure and its engineering excellence. According to the FAO the "chinampas located in Xochimilco, Tláhuac and Milpa Alta comprise more than two thousand hectares in which about 12 thousand people work, mainly cultivating vegetables and flowers, including 51 domesticated agricultural species and 131 species of ornamental plants."

6 Floating Farms on the Sea

The floating farms mentioned above are constructed over fresh (i.e. non saline) water. No one has attempted to construct a floating farm over the sea. This easily explained by the fact that except for seaweeds, all plants for human consumption requires fresh water.

Storing freshwater in the sea was the main challenge. To build a floating reservoir up recent years would entail constructing a rigid steel or concrete box. That is easily done except that it is not easy to hold the box down when it is empty due to the uplift.

In 2018, College of Science and Technology, Nihon University's Dr. Shinji Sato, under the auspices of the Society of Floating Solutions (Singapore) presented a paper in Singapore on an experimental project integrating solar energy, effluent gas and algae in a floating structure in saline water. Unfortunately, we are unable to obtain any published results.

7 Water Conservation in Paddy Fields and Pest Control

The water balance in the rice field comprises the following processes:

Into the ponded field:

- Rain fall
- Irrigation from river and lake sources
- Ground water by osmosis

Out of the ponded field:

- Evaporation
- Transpiration
- Seepage into the soil
- Run offs overflow the embankments nutrients for crop washed away.

The above processes are largely uncontrolled. Heavy rainfall may cause flooding. Irrigation may not always be possible in dry seasons. Ground water table is lowered each year due to human activities and extraction of water for drinking (Figs. 5 and 6).

In land-based fields, enclosures are not often used (even as protection against birds and insects). Evaporation is therefore uncontrolled and exacerbated by wind. Seepage of water is exacerbated as ground water table recedes.

The above are the dilemma of rice cultivators. Floating rice fields offer opportunities to overcome these challenges. To reduce evaporation and insect and bird attacks plastic sheets are easily spread over bamboo pole supports slotted into sockets on concrete base. This simple solution reduce the need and cost for pesticides. The devastation by marauding locust can be prevented by covering the rice

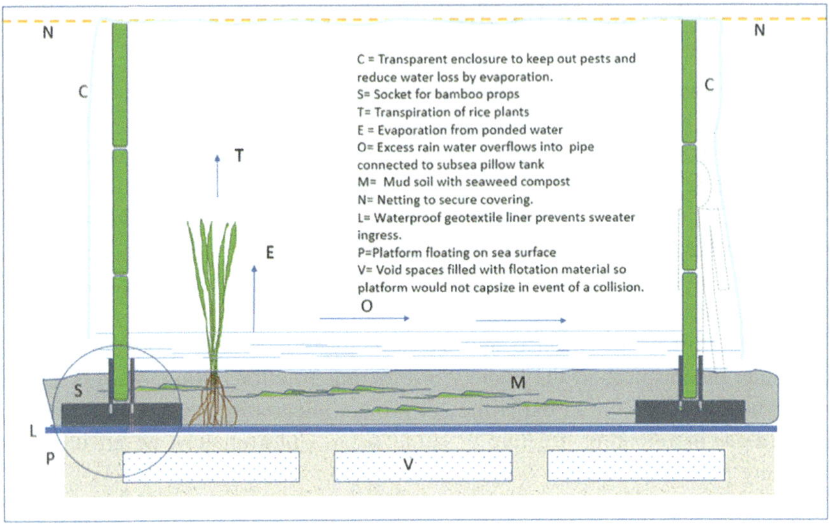

Fig. 5 The transparent enclosure reduces evaporation, keeps out pests, birds, and swarms of locust from devasting the field. A nylon netting secured to the sides of the floating dock prevents it from being blown away

Fig. 6 Ducks efficiently control pests and weeds in rice fields and offer their droppings as fertilizer to boot

fields with plastic sheets. When the plants are flowering, beehives may be placed in the enclosures for pollination. Ducks18 may be kept inside enclosure to keep snails and other under control without worry that they may fly away and never return.

The water in the rice pond and is in a close loop and does not get flushed out of the system. This the nutrients either organic or artificial is not wasted when a flash flood occurs.

8 Why Might It Be Worthwhile to Explore Growing Rice in the Sea?

As far as rice is concerned the demand will continue to grow but the land available to cultivate it is severely being marginalized. Vertical farms for rice is not a practical solution so the only option that remains is to use the space in the sea.

There are several challenges with using sea space:

- The forces of nature: wind, waves, current have to be overcome.
- A solution has to reduce water loss through run offs, and evaporation.
- Water in excess of immediate requirements has to be stored.
- Storing water on land (to support a floating farm) is difficult as there will be loss through evaporation and seepage and contamination by salt laden molecules of water. Siltation is another problem: it reduces the capacity of the reservoir over time.
- And any plot of land set aside as a reservoir quite obviously is no longer available to cultivate rice.

8.1 Challenges

Storing freshwater in the sea will overcome the challenges mention above.

The following are the considerations. The storage system has to fulfill the following criteria:

- It has to be protected from damage by passing boats or flotsam or willful acts and terrorism.
- It should not be carried away by wind, waves, or tidal currents.
- As the system is charged with water or water is extracted from it, it should remain substantially at the same elevation relative to the seabed. That is to say it does not float higher as it empties or lower as it is being filled.

A rigid tank is fine when it is full of water. It sinks to the seabed and remains there. However, when water is extracted from it will start to float at some point. It can be held down with cables. However, the tension in the cable will escalate from zero to some value and at some point, it may exceed its capacity or the holding capacity of its anchorage.

9 Using Bladder Tank Under the Sea

To resolve this, a "bladder tank" (sometimes referred to as a pillow tank) is proposed. This is a flexible collapsible containment bag shaped like a pillow, when inflated, used by the military for water storage 20. It can be folded away for transportation.

When used to store water the stress on the membrane skin is a function of the difference in densities of the fluid inside and outside of the tank. On land water in the tank exerts a considerable pressure on the wall of the tank as the density of water is about 900 times larger than air. The fabric is subject to a high tensile stress.

In the sea the situation is reversed; the denser fluid, seawater, is outside and the lighter fluid water, is outside. The water inside is incompressible.

If the tank is connected by a pipe so that the water inside has a free surface experiencing the same atmospheric pressure as the seawater, the pressure on both sides will equalize as the water rises. (The pressure in a liquid is the head x density.)

In seawater, the fabric of the tank experiences a low stress as the difference in densities of the two fluid is small. The fabric itself may experience an uplift if its density is smaller than seawater. If connected to a pipe to above the waterline the fluid inside will automatically adjust itself so that the pressure at any given elevation is equal inside and outside the tank (Fig. 7).

While pillow tanks have been used to transport water over great distances on the sea, as storage facilities for freshwater under the sea it has never been done (at least

Fig. 7 Bladder or pillow tanks are inexpensive ways of conserving water on land and in the sea

there is no documentary evidence of this application.) The potential of this innovation has yet to capture the imagination of engineers and agricultural economists.

Flexibladder in South Nowra, Australia, manufactures bladder tanks up to 2000 m^3 (larger than an Olympic size swimming pool.) For subsea application, we believe the capacities can be much higher if properly protected from mechanical damage and from forces when empty due to its self-buoyancy. Other manufacturers include several from Europe, the US, China, Japan, and South Korea.

9.1 Material and Fabrication

There is a whole range of polymers that may be considered for the fabrication of these tanks: HDPE, LDPE, LLDPE are some of the common ones. The good news is that the prices of these material is likely to decline.

Linear low-density polyethylene (LLDPE) has superior tensile strength, superior impact, frictional, and puncture resistance. It is also UV resilient and recyclable. It has a density of 0.910–0.925 g/cm^3. It will therefore float making it simple for recovery if necessary, for maintenance or repair. It has good durability and will easily last ten years. The tensile strength at yield and breaking are respectively 10–30 MPa and 25–45 MPa, which compares favourably with rubber.

It is probably an appropriate choice for our purpose. A 2000 m^3 tank would require approximately 2500 m^2 of material. Assuming a wall thickness of 1.2 mm, and a density of 0.92 kg/cm^3, it would weigh 2760 kg (0.0012 m × 2500 m^2 920 kg/m^3). The material will cost USD 6900. (According to Alibaba's website the price is about USD 2.5 per kg.) The labour cost to fabricate the tank depends on the local labour cost. See properties in Table 1.

The methods of joining polymers are well documented and readily available online and in the references mentioned at the end of this paper 21. They range from application of heat, induction current, ultrasonic, laser, and microwave. The skill necessary to perform the bonding of the LLDPE can be easily acquired (unlike welding steel structures.) With the supervision of the manufacturer's engineer they may be performed by local technicians on site.

9.2 The Undersea Water Storage Facility

The Water Storage Facility is preferably placed directly beneath the floating rice field if there is sufficient depth to accommodate it, or a distance away if necessary if the water is shallow. (A floating bladder tank is also an alternative. Its disadvantage is it occupies space on the surface.)

The bladder tank is kept enclosed in a concrete chamber which is secured to a concrete base plate as shown in Fig. 8. To prevent damage by large marine creatures (crabs, sting rays, sharks etc.) the chamber has hinged gates. As the tank

Table 1 Properties of common polymers

Polymer	Flexibility	Color	Impact strength	Chemical resistance	Water vapor barrier	Gas barrier	Stress crack	Impact resistance	Density
HDPE	Flexible but more rigid than LDPE	Natural color is milky white, semi-translucent depending on density	Good	Good	Good	Poor	Good		0.963
LDPE	Very flexible	Natural milky color, translucent	High	Good	Good	Poor	Good	Good	0.910–0.940 g/cm^3
LLDPE	Very flexible	Natural milky color, translucent	High	Good	Good	Poor	Good	Good	0.93
Polypropylene	Rigid for containers	Opaque, natural grayish yellow in natural form.		Good	Excellent	Poor	Excellent	Excellent	0.905 g/cm^3
Polyvinyl chloride- PVC	Flexible to rigid	Transparent to yellowish color in natural state	Low	Good	Good		Some		1.35–1.45
MDPE	Good	Translucent	Good	Excellent		Poor	Average	Good	0.926–0.940 g/cm^3

LLDPE would perform well as a subsea water storage tank. *Source* Global Plastic

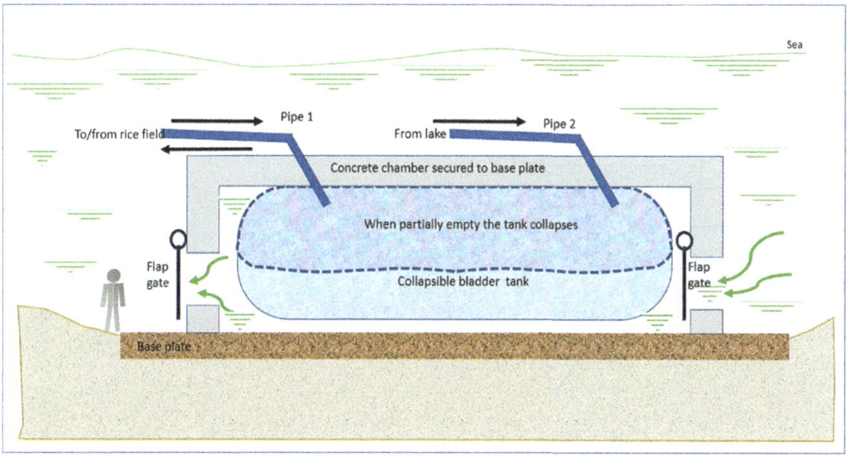

Fig. 8 Storing water under the sea allows space above to be used. The water pressure inside and outside the bladder is equal

inflates, the displaced seawater leaves the chamber through one gate. As it deflates seawater enters via another gate. The gates to be oriented in such a way as to induce a scouring effect in the chamber to flush out silt. These gates can be remotely locked if necessary.

The chamber is of concrete construction. Little reinforcement is needed as the structure is not subject to any significant bending. The main force it will experience is the uplift due to the lower density of the water in it relative to seawater.

It is important to consider the effect of tidal currents. Sites experiencing strong tidal currents should be avoided.

Studies need be carried out to determine the impact of tidal forces at the selected site, as this will affect the design of the station keeping system to keep the chamber and tank from floating.

The base plate is sized to provide sufficient mass to ensure the chamber is not lifted when the bladder is full.

Data relating to the permeability of LLDPE is not readily available. It is not likely that there will be significant water loss by osmosis.

There are two pipes, one connects the bladder tank to the rice field and the other to a source of fresh water (which may be a river or a lake) on land. Pipe 1 supplies the rice field with water during dry periods and drains away surplus run offs when it rains. This is a virtually close loop system, both for water and any nutrients dissolved in the water.

Pipe 2 replenishes the bladder tank with water from the land source during a prolonged drought. If the land water source as depleted the crop can still be safe by importing water from elsewhere and pumping it into the rice pond which then may be transferred in the bladder tank.

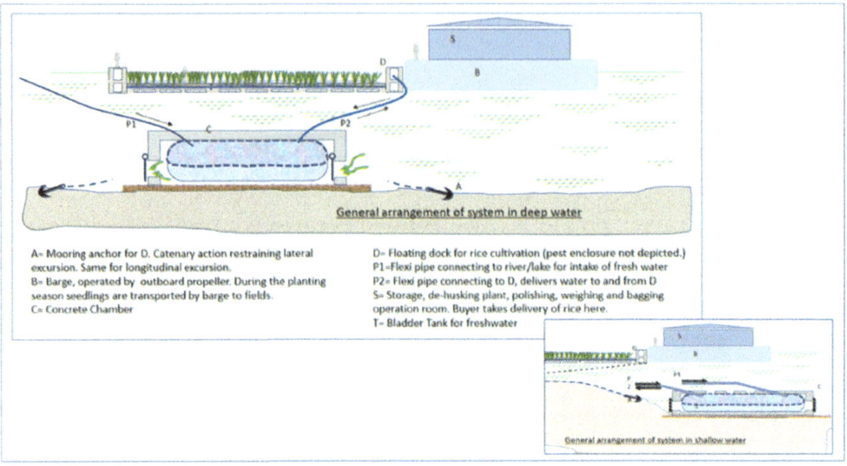

Fig. 9 Upper graphic shows preferred layout of the field and chamber. Inset shows layout that may be adopted if water is shallow

If water depth permits the chamber may be placed directly beneath the floating field. If not, it may be placed at a distance away. See Fig. 9.

Depending on the weather pattern, the regularity of rain, severity of drought, the storage capacity can be designed accordingly, bearing in mind that waste due to evaporation and transpiration of the plant is reduced with the field being under cover. In a prolonged abnormal drought, it is possible to barge water to replenish the bladder tank if the water resources nearby dries up. Charging the storage capacity is easily done by topping the field with water which is then drained into the bladder tank beneath the sea which is not exposed to evaporation.

10 The Floating Field

Just as vegetables grow on roof tops of urban buildings, so can rice on floating docks and be free of droughts, floods, or the salt in the ground and pestilence in the air. Precious runoffs can be stored, and evaporation can be curtailed. All varieties and strains of rice, long grains, short grains, medium grains, Jasmine, Basmati, Japonica or Arborio that suit any palate can be grown in the sea all that is needed is a dock to keep seawater and no-saline water separate.

The water level in the fields may need to be adjusted over the growing cycle from planting seedlings to harvesting. This can easily be achieved with a floating pond. Water is either pumped into or out off the ponded field into the undersea storage tank. No plant nutrient is lost in the process.

Attempts to develop new salt resistant strains of rice take years for results to show and more years for the taste buds to accept. Even though the new strains may tolerate some salinity in the soil, rising sea levels and coastal erosion are too costly to fight against.

> To address seawater intrusion and salinity ingress, every year villagers create an embankment, but the structure is unable to keep the sea at bay and breaches regularly. "The rate at which the sea is coming close to us makes us believe that soon it will engulf our entire village. Our farming has already gone kaput," laments K. Kannan, 55-year-old secretary of the local farmers' association. According to him, villagers have been demanding a permanent cement embankment from the public works department (PWD) to keep the village safe from the sea. But no action has been taken on that front.

10.1 *Construction*

The dock for floating field may be of any practical size. 50 m wide and 100 m long (1/2 hectare) would be a handy side from the viewpoint of construction, launching, towing, and mooring at its designated site. The moulded depth has to ensure that its freeboard is large enough to overcome the splash of the sea in a storm. The section is shown in Fig. 10. (It does not show the full width due to the aspect ratio.) The same design can be replicated and of course there is economy of scale if more are manufactured.

10.1.1 Dock Body

Of what material can the dock be fabricated? It all depends on the environment where the dock is to be located. It can be of timber, or HDPE (high density

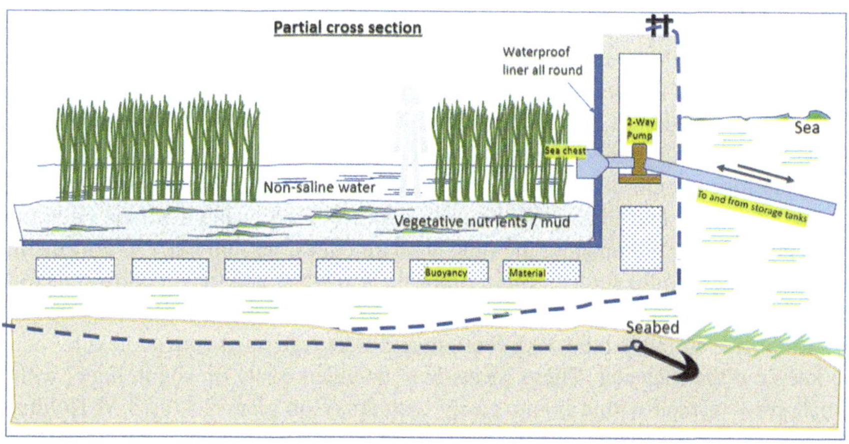

Fig. 10 Water in the pond of the floating field is maintained at a desired level by draining rainwater to the bladder or replenishing it from it by pumping. Marine nylon ropes are used to maintain position but allows field to float up and down with tidal change

polyethylene) or light weight reinforced concrete, respectively in order of cost, strength, rigidity, and durability. In the case of concrete, and HDPE void spaces are filled with buoyancy material so that seawater will not fill them when a leak occurs.

The dock may not have the rigidity of a steel dock that is used to lift ships for maintenance but there is no need for such a high standard of structural integrity in the interest of affordability. Under sheltered conditions where waves are less than 1.5 m and wind speeds less than 20 m/sec, the simple construction will suffice. In the worse case scenario, the dock can be submerged so that it is filled with more water from bladder tank to provide better stability and refloated after the storm has abated. By submerging, the sides are less exposed to wind, the mass is increased, and the center of gravity is lowered which all converged to keep the dock safe.

The design of the dock should be simple, basic, and intuitive. It need not be subject to the many rules and regulation that governs the design and construction of ocean-going vessels. It is a sustainable structure. The cultivation of rice does not pollute the sea. Under heavy weather no one will be working so no lives will be in danger.

The load on the dock consists of the soil to support the roots, the water, and the rice plant when fully grown and bearing rice. The level of water in the dock is kept constantly at the desired level. When water rises beyond the set level it overflows into a sea chest which is connected by a flexible Pipe 1 as illustrated in Fig. 8 in the previous section. When it falls to a set level, the pump automatically cuts into transfer water from the bladder tank via Pipe 2. The pipes are of inexpensive HDPE material, flexible and corrosion free and easily replaced if damaged. A 150 mm diameter size would be adequate. The length and flexibility allow for some degree of motion without unduly straining the pipe.

To keep the dock in position against tidal currents, waves and wind, some form of mooring system is needed. The system should allow the dock to rise and fall with the tide. This can be achieved with tubular steel pipes, but it would require a piling machine for its installation. We recommend for simplicity the anchor and cable catenary system. The anchor can be replaced by a large boulder if it is difficult to procure. Instead of steel cables, nylon marine mooring ropes may be just as effective. Ropes should be class approved. The mooring rope runs beneath the dock to the opposite side at the forward and after end of the dock. To prevent longitudinal drift, the rope should be inclined at about 30° to the longitudinal center line of the dock. The upper end of the rope is lashed to a bollard or bitt with the necessary precaution to avoid damaging the rope at sharp edges. 60 mm dia. (50 ton) marine mooring ropes would suffice. These ropes float in water and may require sinkers to configure the catenary geometry.

The inside surface of the dock (bottom and sides) are lined with geotextile sheets to keep out the seawater. These sheets may be glued easily on site in hours with a small crew of men within hours, easily with spray-on glue such as 3 M HoldFast 70. The bonding strength of the glue is as good as the strength of the textile. The liner further enhances the strength of the bonding between the modules (timber or HDPE) used to form the geometry of the dock.

No painting is necessary on the exterior of the dock. Rubber tires or timber may be used as fenders to protect the side against barges that may come alongside. Some void spaces may be used as stores for tools and ploughing machines, shelters for farmers in bad weather, as well as stores for seeds and fertilizers.

10.1.2 Hydrostatics of the Rice Floating Dock

Hydrostatically, the floating rice field is similar to a floating swimming pool but with one main difference. The swimming pool is filled with water to a level that is close to the level of the water outside the pool. It means that each of the four walls of the pool is experiencing similar (though not necessarily identical) pressure distribution inside and outside. This is not the case with the floating rice field.

Another floating structure that has somewhat similar shape is the floating dock commonly used to lift ships out of water for maintenance. It consists of two "wing walls" erected on a flat top pontoon. By ballasting its tanks, the dock is submerged so that a ship can be maneuvered into it between the walls. By subsequently deballasting the same tanks the dock is floated lifting the ship in it above water. In both cases (afloat or submerged) the wing walls experience the same pressure distribution. Because of this identical water pressure distribution on each side, each wall experiences zero tipping moment.

The situation in the floating rice field is rather different. The water pressure distribution acting on each wall is unsymmetrical as shown in the diagram C in Fig. 11. The wall needs a reaction, R to keep it in equilibrium. That reaction is provided by the pontoon.

A good location for developing a cluster of floating fields would be in a sheltered bay. Breakwaters if necessary, can be constructed with concrete tetrapods.

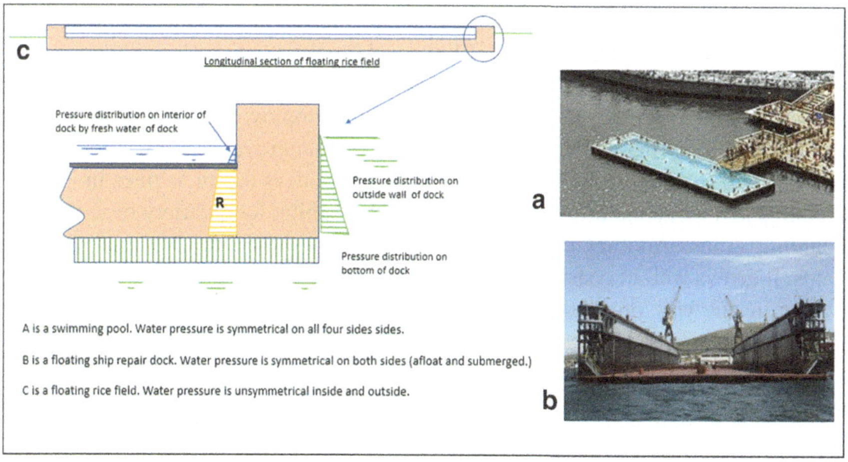

Fig. 11 The pressure distribution diagram on the sides and bottom of the field

However, it is worthwhile considering floating breakwaters with pillow tanks tethered by marine ropes to rocks. These floating breakwaters double up as additional water reservoirs.

All the hydrostatic forces for the floating field are moderate as the structure and it load are light. The water in the pond may vary unlikely to overstress the structure. This of course can be confirmed by FEM (finite element method).

11 Financing

Finding the necessary capital is of course crucial for small time farmers. However, if he owns the land on which he grows his rice he can pledge that as a collateral for a loan and if that is not sufficient to cover the cost of the floating rice field the field itself may also be collateralized until the loan is redeemed. Even if the land is under threat of sinking under rising sea levels, it can be economically productive for instance as fish or seafood farms or as mangrove forest. A floating structure for planting rice can be designed to last for several decades and can be easily repossessed, relocated, repurposed, and sold so it is a low risk collateral.

12 Concluding Remarks

There are two unrelated but fundamental reasons why floating rice fields may lead to the betterment of marginalized farmers. The first is an existential one for many farmers as well as consumers of rice. Vast areas of coastal land are so degraded by ground water extraction, increased levels of salinity and drought that its viability for rice cultivation is undermined. Most rice fields are not short of water but short of reservoirs to harvest and store water for use during the dry seasons. That problem has never been adequately addressed but can be now.

The second reason is that relocating rice farms to the sea frees up land that can be used to produce goods and services and provide employment for thousands more people benefiting the local economy. Growing mangrove swamps for example protects land from erosion and the trunk (bakau wood) is useful as piles in low rise buildings and monsoon drains. It is also a natural habitat for a variety of shell fishes including crabs. Another crop worth considering is coconut, the mother of all plants. It grows well in coastal areas, requires little maintenance. Every part of it has a value in one way or another.

The ways to repurpose the land are only limited by the imagination. From rearing animals for meat to producing bamboo furniture the land can provide more employment for villagers to supplement income from rice.

Suggested Reading/Viewing Material, Available Online

1. Ricepedia, http://ricepedia.org/rice-as-food/the-global-staple-rice-consumers
2. Rice farming: complete guide from seeds to harvest. https://www.youtube.com/watch?v=J_mMS3EkHok
3. This is how rice grows in Japan. https://www.youtube.com/watch?v=joLWIEd2bPM
4. Rice farmers move towards sustainability, UN Environment Program. https://www.unenvironment.org/news-and-stories/story/rice-farmers-move-towards-sustainability
5. The vertical farm: feeding the world in the 21st century, Dr. Dickson Despommier
6. Off-season rice crop in 22 provinces threatened by drought, The Nation Thailand 13 January 2020
7. India's longest drought: 41-month-long dry spell from 2015–18 was longest in 150 years. https://www.news18.com/news/india/indias-longest-drought-41-month-long-dry-spell-from-2015-18-was-longest-in-150-years-finds-study-2358355.html
8. Rising sea forces villagers to abandon rice farming, by Nidhi Jamwal. https://indiaclimatedialogue.net/2019/01/02/coastal-villagers-abandon-farming-as-sea-level-rises/
9. The unfolding tragedy of climate change in Bangladesh, Scientific American, Robert Glennon on April 21, 2017
10. An 87-year-old scientist may have just unlocked the secret to growing rice in saltwater. https://www.businessinsider.com/yuan-longping-chinese-rice-scientist-growing-saltwater-rice-2017-11
11. Agrisea ocean farms will grow rice in saltwater this year, Andrea D. Steffen. https://www.intelligentliving.co/agrisea-ocean-farms-grow-rice-saltwater/
12. Inle lake and the floating gardens—it's all about tomatoes. https://www.youtube.com/watch?v=f8p0sx4xRd4
13. Chinampa, https://en.wikipedia.org/wiki/Chinampa
14. Chinampas of Mexico city were recognized as an agricultural heritage system of global importance. http://www.fao.org/americas/noticias/ver/en/c/1118851/
15. Chinampas, The floating gardens of Mexico. https://www.ancient-origins.net/ancient-places-americas/chinampas-floating-gardens-mexico-001537
16. Water balance of flooded rice in the tropics by Siva Sivapalan. https://www.intechopen.com/books/irrigation-and-drainage-sustainable-strategies-and-systems/water-balance-of-flooded-rice-in-the-tropics
17. Improvement of salt tolerance using wild rice genes, Ruidan Quan et el. https://loop.frontiersin.org/people/463320/overview
18. The use of ducks in rice fields in the control of weed and pest. https://www.permaculturenews.org/2016/09/15/use-ducks-rice-fields-control-weed-pest/
19. Filling the world's rice bowl—with less water. https://www.voanews.com/east-asia-pacific/filling-worlds-rice-bowl-less-water
20. Smarter liquid storage for industry. https://www.labaronne-citaf.com/fields-of-use/
21. Global Plastic, https://www.globalplasticsheeting.com/hdpe-vs-lldpe-vs-ldpe
22. Geotextile seaming solutions using 3 M™ HoldFast 70 cylinder spray adhesive. https://www.youtube.com/watch?v=d-J0GRSG2ns
23. Coconuts, Plantation International. https://www.plantationsinternational.com/coconuts/

Motion Control Strategies for Smart Floating Cranes

Wouter Bentvelsen, Guus Jonathan Gorsse, Niels Bouman, Vincent Bashandy, Vittorio Garofano, and Jovana Jovanova

Abstract Floating structures have raised interest in the recent years for different applications, from living and farming at sea to renewable energy production. To support the logistics on the floating structures, floating cranes are necessary and their designs are constantly improved. Increasing developments in the automation industry paved the way for automated crane operations. In this work, motion control of a smart crane is presented with particular attention to the performance under wave motion. In this research, a scaled down, two-dimensional mathematical model of a gantry crane is derived using Lagrangian mechanics and DC motors dynamics. This results in a nonlinear system that is capable of simultaneous traversing and hoisting a container. The system is simulated in MATLAB Simulink environment and a proportional-derivative control and a state feedback control are designed and implemented. Their robustness is explored by modelling sensor behavior, external disturbances and floating platform dynamics. Both control strategies were able to keep stability in a disturbed system. During simulation, the sway angles never exceed 10°. Smaller oscillations occurred using the state feedback control. Therefore, it creates a smoother response compared to the proportional derivative control, which ultimately translates to increased safety, turnover rate and durability of the crane.

W. Bentvelsen (✉) · G. J. Gorsse · N. Bouman · V. Bashandy · V. Garofano · J. Jovanova
TU Delft, Mekelweg 4, 2628 CD Delft, The Netherlands
e-mail: W.Bentvelsen@student.tudelft.nl

G. J. Gorsse
e-mail: guusgorsse@hotmail.com

N. Bouman
e-mail: N-Bouman-1@student.tudelft.nl

V. Bashandy
e-mail: V.A.Bashandy@student.tudelft.nl

V. Garofano
e-mail: V.Garofano@tudelft.nl

J. Jovanova
e-mail: j.jovanova@tudelft.nl

Ł. Piątek et al. (eds.), *WCFS2020*, Lecture Notes in Civil Engineering 158,
https://doi.org/10.1007/978-981-16-2256-4_15

Keywords Motion control · Smart cranes · Floating platform · State feedback control · PD control

1 Introduction

The off-shore engineering industry is quickly growing and innovating to meet current societal demands. Transport of goods utilizes cranes to move containers from or to ships in off-shore fixed and floating platforms. The challenges in the logistics can be addressed by adding automated crane portals on the floating platforms.

One of the bottlenecks in the process of shipping containers is the handling rate which is why multiple ship-to-shore (STS) gantry cranes are usually used to (un) load a containership [1]. The trend in "smart" technology and automation has been adopted in port equipment design [2, 3]. Currently, some autonomous STS gantry cranes exist however, most are still operated by a high skilled operator who is responsible for minimizing container sway as this is the cause for almost half of the accidents involving containers [4, 5]. The sway angle is the rotation angle of the hanging container due to the inertia of the container and the effects of side-to-side and upward-downward motion as depicted in Fig. 1.

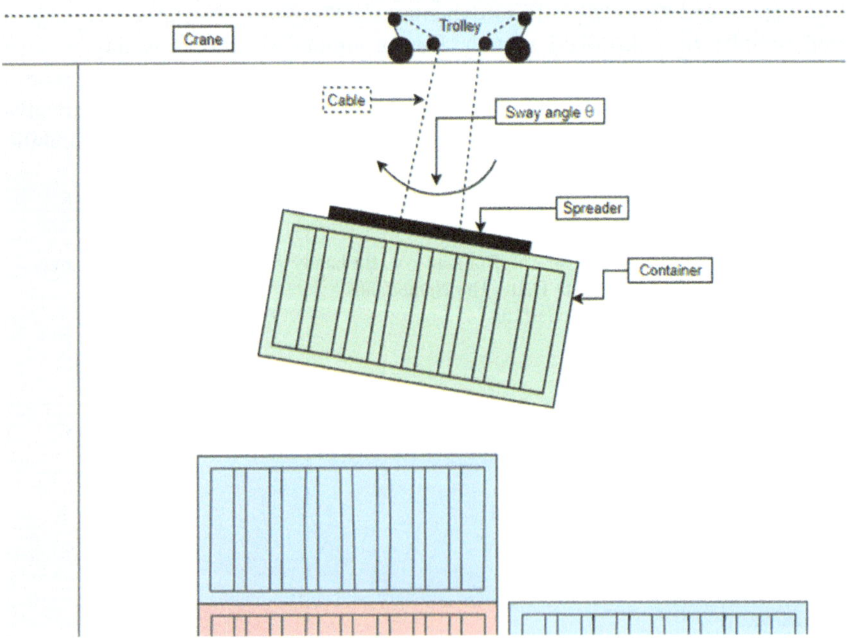

Fig. 1 A schematic depiction of container sway

Sway control can result in a higher harbor efficiency due to lower handling times and improved safety. Out of the quay, on an offshore platform, the sway effects of a crane will play an even bigger role when handling containers due the motion of floating platforms and ships. This paper explores how sway control strategies for STS gantry cranes perform under stress and when they are moved on a floating platform.

In order to smartly reduce sway of STS gantry cranes, multiple control strategies currently exist [6]. However, the performance of such systems when they are disturbed with floating platform dynamics is not yet explored in detail. This research investigates two sway control strategies, PD (proportional derivative) and state feedback, on their performance under disturbances. Both feedback control strategies are known to handle disturbances well [7]. In order to do so, a nonlinear dynamic system with simultaneous hoisting and traversing is derived for the crane. Motor dynamics, sensor effects, floating platform dynamics and disturbances such as wind are taken into consideration in order to test the robustness of the proposed control strategies. These models are simulated using MATLAB Simulink environment. The results comparison is discussed in detail, leading to conclusions and future work.

2 Modelling of the General System

The proposed system is a two-dimensional scale model of a STS gantry crane. The movement space of a container is confined by a 2×0.5 [m] plane, with a 1.5 [kg] container. A full-sized container crane moves in a plane of up to 150 [m] $\times$ 40 [m], which can have a container weight up to 25,000 [kg]. Since the model is relatively small, it is actuated by 2 DC motors, one for traversing, (x-movement) and one for hoisting (L-movement). The system is modelled in such a way that actuators and gearboxes will be fixed to the frame, rather than to the moving cart. The following Fig. 2 shows a simplified CAD model of the crane and pulley system. Here, the angle of sway in the lower image is 8°. Therefore, if the sway is confined during movement to a maximum $\pm 10°$, safety of the scale model can be guaranteed. For a real-size STS gantry crane this value might be lower due to stricter safety and quality constraints.

The dynamic model of the system has two components, the crane and the DC motors. Since the actuators are non-ideal, they have certain constrains. Therefore, they need to be modelled in order to simulate a realistic actuated system.

In order to derive the crane dynamics, the Fig. 3 has been used. It can be noted that this system is a simplified overview of what is going on in the actual system. Most important is that the system represents a pendulum on a cart model with variable pendulum length.

Fig. 2 Simplified 2 × 0.5 [m] CAD model of a scaled version STS gantry crane

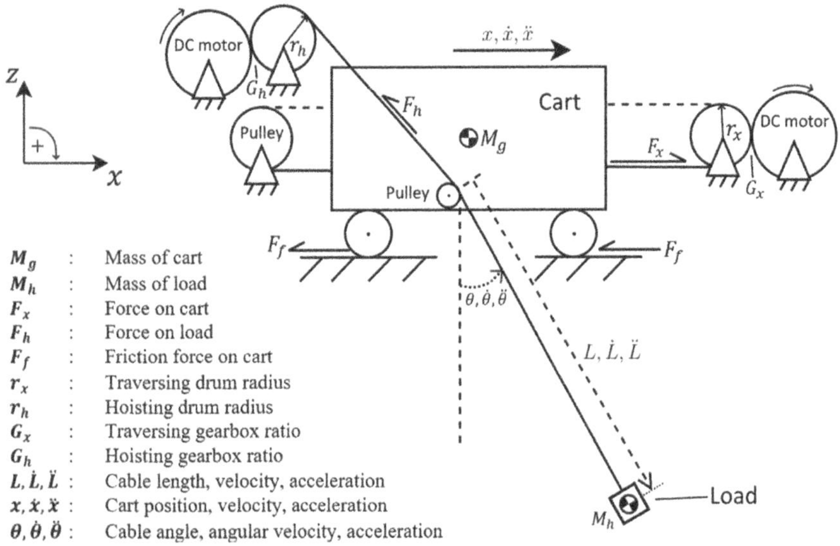

Fig. 3 Simplified diagram of the gantry cart including DC motors and pulleys systems

For example, the container and cart are simulated as point masses and friction as well as damping is initially assumed to be zero. This system is defined by three states, x, θ, L. To derive the equations of motion for this system, Lagrangian mechanics was used to derive Eq. 1:

$$\mathcal{L} = \frac{1}{2}\left(M_g\dot{x}^2 + M_h\left(v_{2x}^2 + v_{2z}^2\right) + M_h\dot{L}^2\right) + M_h g L \cos\theta \tag{1}$$

where v_{2x} and v_{2z} represent the velocities of the load. When the Lagrangian process is followed, the equations of motion are derived as below.

$$\begin{aligned}\ddot{x} &= \frac{F_x + F_h \sin\theta}{M_g} \\ \ddot{\theta} &= -\frac{\cos\theta \frac{F_x + F_h \sin\theta}{M_g} + 2\dot{L}\dot{\theta} + g\sin\theta}{L} \\ \ddot{L} &= -\frac{F_h}{M_h} - \left(L\dot{\theta}^2 + g\cos\theta - \frac{\sin\theta(F_x + F_h\sin\theta)}{M_g}\right)\end{aligned} \tag{2}$$

In order to use these equations in the final controller models, they need to be put to state space. For the PD controller they are first partly linearized under the assumptions: $\sin(\theta) = \theta, \cos(\theta) = 1, \dot{\theta}^2 = 0$.

In state space form this yields the following model:

$$\begin{aligned}\dot{\mathbf{q}} &= A\mathbf{q} + B\mathbf{F} \\ \mathbf{y} &= C\mathbf{q} + D\mathbf{F}\end{aligned} \tag{3}$$

where $\boldsymbol{q} = \left[\dot{x}x\dot{\theta}\theta\dot{L}L\right]^T$ & $\boldsymbol{F} = [F_x F_h g]^T$

This yields

$$A = \begin{bmatrix} 0 & 0 & 0 & 0 & 0 & 0 \\ 1 & 0 & 0 & 0 & 0 & 0 \\ 0 & 0 & \frac{-2\dot{L}}{L} & \frac{-g}{L} & 0 & 0 \\ 0 & 0 & 1 & 0 & 0 & 0 \\ 0 & 0 & 0 & 0 & 0 & 0 \\ 0 & 0 & 0 & 0 & 1 & 0 \end{bmatrix}, B = \begin{bmatrix} \frac{1}{M_g} & \frac{\theta}{M_g} & 0 \\ 0 & 0 & 0 \\ \frac{-1}{M_g L} & \frac{-\theta}{M_g L} & 0 \\ 0 & 0 & 0 \\ \frac{-\theta}{M_g} & \frac{-1}{M_h} & -1 \\ 0 & 0 & 0 \end{bmatrix} \tag{4}$$

The C matrix becomes the identity matrix as all states can be measured, although when modelling the results, non-ideal sensors shall be taken into account. The D matrix becomes the zero matrix as there is no direct output term in the equations of motion. In order to implement the crane dynamics into a state space controller, the complete non-linear dynamics can be used.

The actuators that control the traversing and hoisting movements are simulated as armature-current controlled DC-motors. In case of such DC-motors, the field current is held constant and the current is controlled through the armature voltage [8, 9]. The transfer function from the input armature current to the resulting motor torque is, with K_t the motor torque constant.

$$\frac{T_m(s)}{I(s)} = K_t \tag{5}$$

where

$$I(s) = \frac{U(s) - K_e \dot{\omega}(s)}{R + Ls} \quad (6)$$

Combining these equations allows for a voltage input to result in a torque output with a certain rotational speed depending on the load attached to the motor.

To use these motors most efficiently, a gearbox was implemented into the system. The gear ratio is calculated knowing the optimal point of the motor through simulating the motor characteristics. The nominal power of the motor can be simulated for every angular velocity of the motor and by dividing this through the losses (heat generation and motor friction) an optimal point can be found. These gear ratios differ for each load applied to the motor.

Other elements that are in the system model are trolley friction (rolling resistance), pulley efficiency (rolling resistance), pendulum friction (caused by wind resistance), sensor delay, sensor errors and noise disturbances. The system model is implemented in MATLAB & Simulink and solved using a variable step method.

The floating platform dynamics implemented to check the overall robustness to oscillating disturbance, is a simplified way to simulate real waves which gives an effect of an added sway angle and translation of the cart. The waves are modelled as sinusoidal sources with addition of random oscillations smoothed out by a low-pass filter as in Eq. 7. This way, the wave source has the oscillatory, and unpredictability properties of sea waves. This is not sufficient for modelling real waves due to disregarding properties like angular momentum and drag [10], but it is suitable for exploring the robustness of the floating crane subjected to a 'random' oscillatory disturbance.

$$\varphi(t) = R_{LPF}(\sigma, \mu, t)\sin(\omega_R t) + \sin^{-1}\left(\frac{C}{P}|\sin(\omega_s t + \varphi_B) - \sin(\omega_s t + \varphi_A)|\right) \quad (7)$$

With $\varphi(t)$ being the calculated wave angle and φ_A and φ_B being the initial phase angles at point A and B which are both endpoints of the floating platform. ω_R and ω_s are the angular frequencies of respectively the random wave and the sinusoidal wave. $R_{LPF}(\sigma, \mu, t)$ is a low-pass filtered random number depending on the mean σ, variance μ and time t. The size of the platform is given by P. C is the amplitude of the sinusoidal wave.

3 Motion Controller Design

In order to minimize "stress" on the controller, the input is shaped using a low pass filter. The effect of this is that the input will not be a step, rather a smoothed out curve. When a step is the input, overshoot from the controller generally occurs, which is in a real-life STS gantry crane could cause damage and accidents and is therefore to be avoided. By smoothing out the input curve to the desired (reference) state, the feedback error becomes smaller initially and stays smaller throughout the movement, therefore the controller is better able to follow the reference. Which results in little to zero overshoot.

3.1 PD Control

The DC motors are controlled by the error of the desired rope length and the container location, $x + L\sin(\theta)$. Most importantly, to be able to use PD control on this system, a separate gravity compensation is necessary. Because as the system is reaching its target location the proportional and derivative error is zero, so no voltages are supplied to the motor anymore, however the gravity is still pulling on the cable, resulting in the load to descend. Clearly the motor needs an offset minimum amount of voltage to compensate for gravity, the so called 'holding torque'. By adding a minimum voltage (later called holding voltage) to the motor corresponding to this force so the rope length does not change. By modeling the motor model with different opposing moments, the 'holding voltages' linked to this torque can be found using fixed point iteration. Resulting in the following characteristic Eq. 8:

$$V_{hold} = \frac{12}{0.457} T \tag{8}$$

3.2 State Feedback Control

For the state feedback control, full nonlinear equations of motion can be used in the process part. In the process block, the input $\boldsymbol{u} = -K\boldsymbol{q} + k_r\boldsymbol{r}$ is a voltage that is applied to the motors, where $\boldsymbol{q}$ is the state vector and $\boldsymbol{r}$ is the reference signal containing the desired x-position of the gantry and the desired cable length.

$$\boldsymbol{r} = \begin{bmatrix} x_{ref} & L_{ref} \end{bmatrix}^T \tag{9}$$

The process is separated in two parts: the motor dynamics and the crane dynamics. The inputs to the motor are the voltages $\boldsymbol{u}$, the outputs are the forces $\boldsymbol{F}$ that act on the crane.

For this system, the K matrix will be of size 2 × 6, thus consisting of 12 individual values, making it impossible to tune it by hand. Instead, the method of Linear Quadratic Regulation (LQR) will be used. With LQR, the costs of errors in the states $\boldsymbol{q}$ are described by a 6 × 6 matrix Q matrix and the costs of usage of the inputs $\boldsymbol{u}$ is described by a 2 × 2 matrix R. Here each Q_i and R_i represent the cost factor associated with i. These can be varied to tune the systems behavior.

$$Q = \begin{bmatrix} Q_{\dot{x}} & 0 & 0 & 0 & 0 & 0 \\ 0 & Q_x & 0 & 0 & 0 & 0 \\ 0 & 0 & Q_{\dot{\theta}} & 0 & 0 & 0 \\ 0 & 0 & 0 & Q_\theta & 0 & 0 \\ 0 & 0 & 0 & 0 & Q_{\dot{L}} & 0 \\ 0 & 0 & 0 & 0 & 0 & Q_L \end{bmatrix}, R = \begin{bmatrix} R_x & 0 \\ 0 & R_h \end{bmatrix} \tag{10}$$

To determine the controller gains, the A, B and C matrices are required [11]. To obtain these, the entire process from the voltages $\boldsymbol{u}$ to the states, including the motors, gearboxes, pulley efficiencies etc. must be implemented into one model. For this, the equations of motion, Eq. 2, are the starting point. A substitution for the forces needs to be found that is expressed in terms of the applied voltages. The DC motors however have nonlinear dynamics, which need to be linearized. When the sum of moments around the motor shaft is taken, a formula for the motor force F_m can be obtained.

In this formula, the current i can be substituted by the inverse Laplace transform of Eq. 6 (rewritten for i). The angular velocity and acceleration of a motor can be expressed in the corresponding state, using dummy q (for traversing, $q = x$; for hoisting, $q = L$), with $\dot{\omega} = \frac{\dot{q}}{r_q G_q}$ and $\ddot{\omega} = \frac{\ddot{q}}{r_q G_q}$. On top of that, because of the pulley system with efficiency η_p, the force that actually actuates the load is $F_{load} = \eta_p F_m$. Incorporating these and translating F_{load} to F_x for the cart and to F_h for the spreader results in.

$$F_x = -\eta_{px} \frac{K_t^2 + bR}{r_x^2 G_x R} \dot{x} - \frac{\eta_{px} J}{r_x^2 G_x} \ddot{x} + \frac{\eta_{px} K_t}{r_x R} U_x - \frac{\eta_{px} K_t L_i}{r_x R} i_x' \tag{11}$$

$$F_h = -\eta_{ph} \frac{K_t^2 + bR}{r_h^2 G_h R} \dot{L} - \frac{\eta_{ph} J}{r_h^2 G_h} \ddot{L} + \frac{\eta_{ph} K_t}{r_h R} U_h - \frac{\eta_{ph} K_t L_i}{r_h R} i_h' \tag{12}$$

These can then be plugged into the equations of motion. The resulting equations have to be linearized further. To solve this, four measures are taken:

- The motor inertia J is relatively small, and therefore set to zero.
- The controller gains are calculated for a constant L (and therefore also $\dot{L} = 0$). The negative effect of this is reduced by the use of gain scheduling (explained later).
- The controller gains are calculated for a zero degree sway angle (and therefore also $\dot{\theta} = 0$). This is not considered to be a problem since the desired maximum sway angle is 8°. Which in radians is just 0.14 rad. Furthermore, this is only maximum, and most of the time it will be lower.

The voltages receive priority to be taken into the state vector. In cases where θ, $\dot{\theta}$, L or $\dot{L}$ can be taken into the state vector, this should be done instead of setting them constant/to zero. When these measures are all applied, the equations of motion can be converted to state space.

$$A = \begin{bmatrix} -\eta_{px}\frac{K_t^2+bR}{M_g r_x^2 G_x R} & 0 & 0 & 0 & 0 & 0 \\ 1 & 0 & 0 & 0 & 0 & 0 \\ \eta_{px}\frac{K_t^2+bR}{M_g r_x^2 G_x RL} & 0 & -\zeta\dot{\theta} & -\frac{g}{L} & 0 & 0 \\ 0 & 0 & 1 & 0 & 0 & 0 \\ 0 & 0 & 0 & 0 & \eta_{ph}\frac{K_t^2+bR}{M_g r_h^2 G_h R} & \frac{g}{L} \\ 0 & 0 & 0 & 0 & 1 & 0 \end{bmatrix} \tag{13}$$

$$B = \begin{bmatrix} \frac{\eta_{px}K_t}{M_g r_x R} & 0 \\ 0 & 0 \\ -\frac{\eta_{px}K_t}{M_g r_x RL} & 0 \\ 0 & 0 \\ 0 & -\frac{\eta_{ph}K_t}{M_g r_h R} \\ 0 & 0 \end{bmatrix} \tag{14}$$

The C matrix needs to be of a 2×6 size, and is taken to give $\boldsymbol{y} = [x \quad L]^T$, thus becomes

$$C = \begin{bmatrix} 0 & 1 & 0 & 0 & 0 & 0 \\ 0 & 0 & 0 & 0 & 0 & 1 \end{bmatrix} \tag{15}$$

As said in section earlier, the cable length L is taken to be constant. However, in the real world, the cable length is variant and never zero during operation. Choosing a single value for L is thus improper. Therefore, controller gains have been calculated for the following cable lengths.

$$L = \{0.10 \quad 0.15 \quad 0.20 \quad \dots \quad 0.95 \quad 1.00\} \tag{16}$$

During operation, the gain scheduler monitors the current cable length using sensors and applies the controller gain corresponding to the ones calculated for the cable length closest to the current length.

4 Simulation Results and Discussion

To obtain useful results, tests have been performed where the task is to do one full cycle of container transshipment. This has been tested both on a stationary and on a floating platform with the PD controller and the state feedback controller. Each test includes a random external disturbance on the actuator forces with a maximum magnitude of 10% of the highest possible actuator force as seen in Fig. 4.

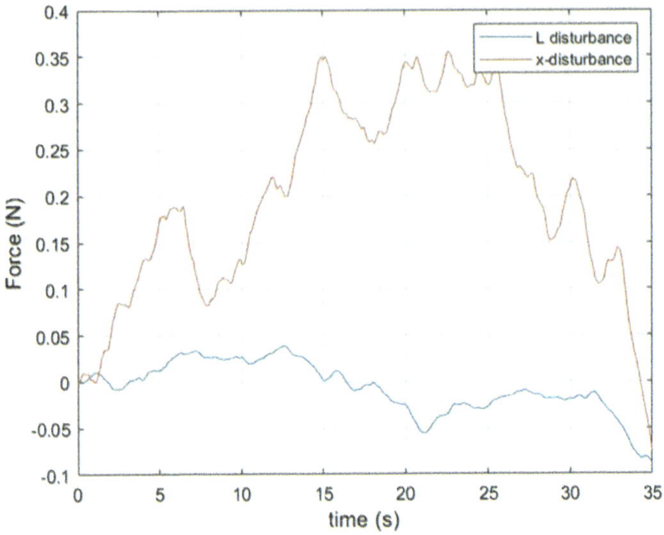

Fig. 4 The 10% random disturbance size plotted against time

A full cycle can be divided in four steps:

- Step 1: The traversing from the crane center to the container pickup location on the ship.
- Step 2: The hoisting and traversing with the container.
- Step 3: The descending and traversing with the container and
- Step 4: The hoisting and traversing without the container.

These points are shown as dots in the results. All tests include sensors that have a delay of 10 ms and an error of 2%. The load sensor has an error of 5%. In the tuning process of both control strategies the intention is to reach the target position as fast as possible while keeping the maximum sway angle within ±10°. On top of that, oscillating behavior has to be kept to a minimum. In addition, an animation of the system in real-time was made as seen in Fig. 5. This is for the case of a crane with PD controller on a stationary platform with 2.5% disturbances. To show how well the performance is, the containers are on a moving ship, as they would in real life. The pulley corresponding pulley efficiency (η_p) have been roughly estimated at 75 and 86% for the hoisting cable and respectively for the translation of the cart. The rolling resistance of the cart (steel wheels on steel rails) is estimated to be 0.0020. The pendulum damping factor of the rope is estimated to be 0.02.

In this paper, only the results of the case of the crane on a floating platform are shown. From the results of the case of the crane on a stationary platform it became evident that both systems performed very well and kept the sway angles within 10°. However the state feedback executed the tasks slightly faster, with less shocks and less oscillations. Therefore it is regarded as the better performer when the crane is stationary.

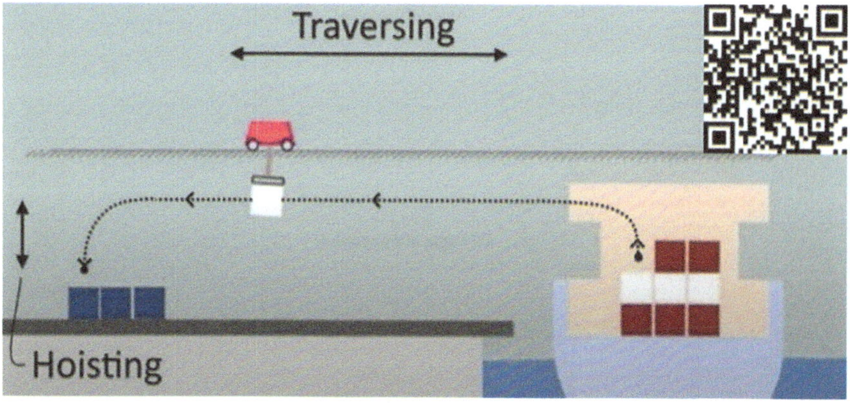

Fig. 5 The animation of the simulated crane system with PD control

Figures 6 and 7 show results for the case for a crane on a floating platform with the PD controller and the state feedback controller respectively. The systems work with input shaping, but in the graphs this is not shown.

For the crane on a floating platform, the PD control gives really good results. Execution of tasks is fast and the sway induced by the waves is successfully counteracted. Also, the sway angles are still suppressed within $\pm 10°$, though there is a lot of oscillation. The state feedback also performs well, but has more trouble. Performance is about 10 s faster, but the waves cannot seem to be counteracted. Rather, as can be seen in the cable angle plot, the exact wave pattern is present. Still, when looking at the container location plot, the accuracy of the state feedback is pretty good. Yet, the accuracy of the PD is even better, almost the same as for the case of the crane on a stationary platform. Therefore, it seems like the PD is more able to adapt to the waves, and therefore more robust than the state feedback.

The performance of the control strategy is highly related to how the system is tuned. This is iterative work for both the PD and state feedback, though it can be concluded that the state feedback is easier to tune. This is due to being able to assign more importance to certain states and/or actuators, which is much more intuitive than tuning a PD.

The system in this research is modelled as a load hanging with a single infinitely rigid cable, whereas in a full-size STS gantry crane this would be with multiple elastic cables. However, it is expected that the spreader would be less prone to swaying in such a model. Assuming elastic cables do not make a big difference, modelling the system with a single cable may therefore be regarded as a worst-case scenario.

The noise and disturbance in the model is made by a random generator, whereas in real life this may not be such randomly distributed. However again, the randomness may be regarded as a worst-case scenario, as this is the furthest away from monotony behavior.

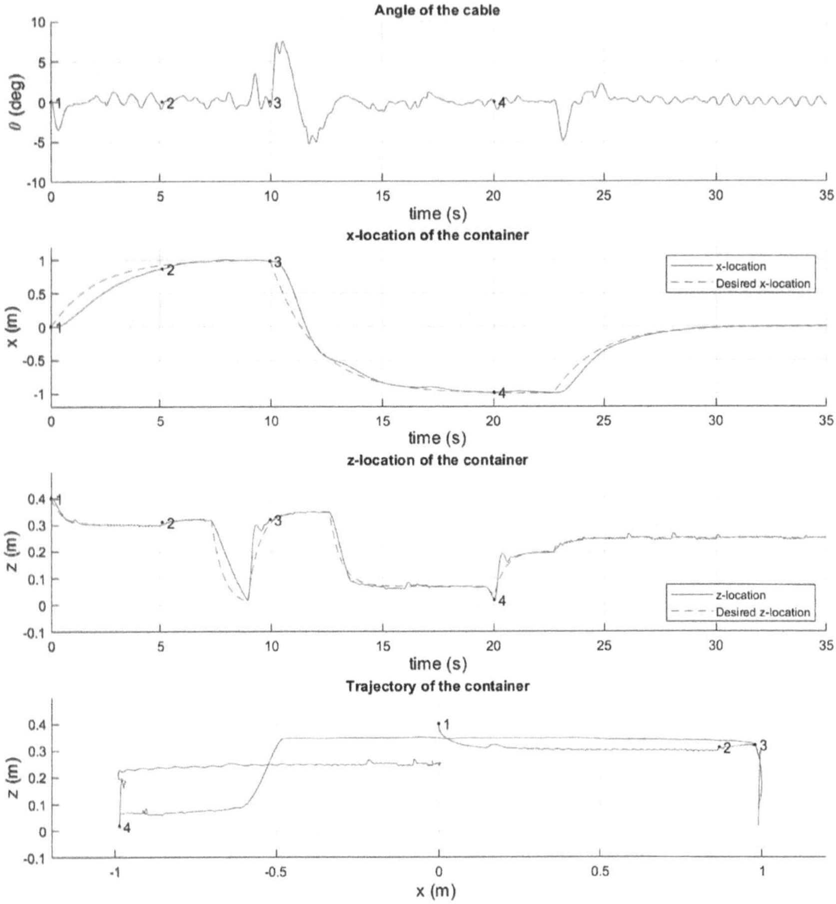

Fig. 6 The results of the PD controller with floating platform dynamics

The PD uses a partly linearized system model. Therefore, if the linearisations are not justified, the simulation results may not represent what would happen in reality.

Since the sway angles always stay within $\pm 10°$, the linearisations $\sin(\theta) \rightarrow \theta$ and $\cos(\theta) \rightarrow 1$ are acceptable. On the other hand, the linearisation $\dot{\theta}^2 \rightarrow 0$ is not justified, as the highest angular velocity is 0.8 rad. However for about 95% of the trajectory, it does not exceed 0.5 rad. The state feedback controller does not contain linearisations.

The system is a scale model of a crane. Due to scale-effects, the results could be somewhat different for a real size crane. Nevertheless, these results show the overall robustness of a floating crane, which should be at the same level for real size crane.

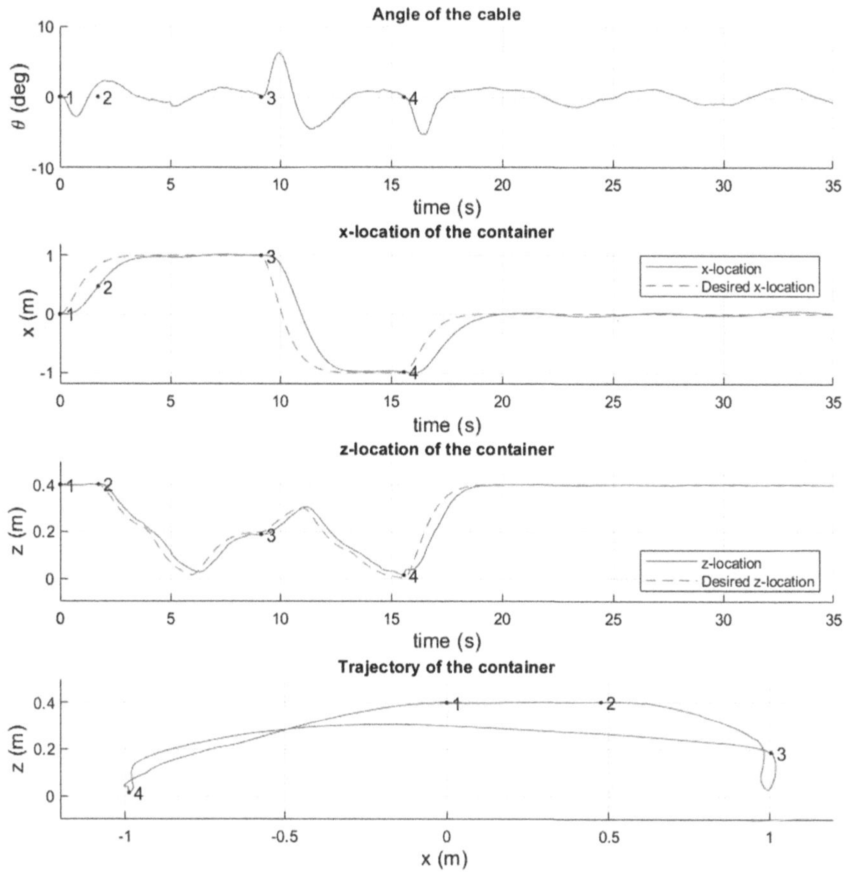

Fig. 7 The results of the state feedback controller with floating platform dynamics

5 Conclusion

This research resulted in a scale model of a ship-to-shore gantry crane system. Crane and motor dynamics, frictional forces, external disturbances, sensor effects, pulley efficiencies as well as any physical constraints such as maximum motor voltages were all implemented. Next, this model was simulated and controlled by two different motion controllers, a PD controller and a state feedback controller. Multiple tests are performed with both controllers.

From the tests done for cases with the crane on a stationary platform it can be concluded that both controllers handled the tasks well, however the state feedback performed slightly better than PD.

From the tests done for cases with the crane on a floating platform it can be concluded that still both controllers fulfilled the tasks, however the state feedback had more trouble to compensate for the waves than the PD. Therefore, the PD seems to be more robust than the PD. Still, more research has to be done to be able to conclude which is better.

Naturally, to draw conclusions for a real life and full scale gantry crane, more research and testing has to be done. To this end, topics that could be investigated in a follow up research could include: other types of input shaping as this noticeably influences controller responses, a full-scale model incorporating elastic behavior, third dimension effects and safety measures such as collision prevention. More research in optimal trajectory determination should also be done for increased efficiency. The implemented floating platform dynamics were simplified and therefore, more realistic floating platform dynamics should be researched in the future as well.

References

1. Holmes Jellett J (2019) Harbours and sea works. https://www.britannica.com/technology/harbor. Accessed 2 Jun 2020
2. Andritsos F, Perez Prat J (2000) The automation and integration of production processes in shipbuilding, state-of-the-art
3. Witschge L (2019) Rotterdam is building the most automated port in the world. https://www.wired.co.uk/article/rotterdam-port-ships-automation. Accessed 2 Jun 2020
4. Roberto G, Edmundo R (1975) On bang-bang control policies. In: Proceedings of the 7th IFIP conference on optimization techniques: modeling and optimization in the service of man, part 2. Springer, Berlin, Heidelberg, pp 587–602
5. PEMA, TT Club, ICHCA International (2019) Recommended minimum safety features for quay container crane. Available via TT Club. https://www.ttclub.com/fileadmin/uploads/tm/Loss_Prevention/191118_PEMA_BP03_Recommended_Minimum_Safety_Features_For_Quay_Container_Cranes.pdf. Accessed 8 Jun 2020
6. Ramli L, Mohamed Z, Abdullahi A, Jaafar H, Hazriq IM, Lazim I (2017) Control strategies for crane systems: a comprehensive review. Mech Syst Signal Process 95C:1–23. https://doi.org/10.1016/j.ymssp.2017.03.015
7. Belanger NM (1995) Control engineering: a modern approach. Saunders College Publishing, Philadelphia
8. Control Tutorials for Matlab (2017) Simulink. DC motor speed: system modeling. http://ctms.engin.umich.edu/CTMS/index.php?example=MotorSpeed&section=SystemModeling. Accessed 20 Apr 2020
9. Dorf RC, Bishop RH (eds) (2001) Modern control systems. Pearson Education Inc, Upper Saddle River
10. Jonkman JM (2009) Dynamics of offshore floating wind turbines—model development and verification. Sci Am 5(12):459–492
11. Åström KJ, Murray RM (2008) Feedback systems: an introduction for scientists and engineers. Princeton University Press, Princeton

Research

Modular Multi-purpose Floating Structures for Space Creation

Jian Dai, Øyvind Hellan, Arnstein Watn, and Kok Keng Ang

Abstract Modular multi-purpose floating structures (MMFS) are an innovative approach for space creation on the sea. The basic idea is to create "land on sea" by connecting a number of standardized modular units to form the desired size and shape for generic applications. The research presented in this paper was part of the multi-purpose floating structure (MPFS) project funded by the Land and Liveability National Innovation Challenge (L2 NIC) Directorate and JTC Corporation in Singapore. This paper presents an overview of the concept development and evaluation of the modular units and inter-modular connectors. Results from detailed structural and hydrodynamic analyses as well as scaled model tests show that the proposed solution is technically feasible. The construction methodology and preliminary cost estimate are also presented and discussed.

Keywords Modular floating structures · Connector · Hydrodynamics · Model test · Construction

J. Dai
Department of Civil Engineering and Energy Technology,
Oslo Metropolitan University, 0166 Oslo, Norway
e-mail: jiandai@oslomet.no

Ø. Hellan (✉)
SINTEF Ocean AS, 7052 Trondheim, Norway
e-mail: Oyvind.Hellan@sintef.no

A. Watn
SINTEF/NTNU, 7465 Trondheim, Norway
e-mail: Arnstein.Watn@sintef.no

K. K. Ang
Department of Civil and Environmental Engineering,
National University of Singapore, 117576 Singapore, Singapore
e-mail: ceeangkk@nus.edu.sg

Ł. Piątek et al. (eds.), *WCFS2020*, Lecture Notes in Civil Engineering 158,
https://doi.org/10.1007/978-981-16-2256-4_16

1 Introduction

Singapore is a condensed city-state. With a land space of only 710 km^2, it is home to more than 5.8 million residents. In order to sustain the development growth, Singapore has been making efforts to create useable space for different purposes through various means such as high-rise buildings, land reclamation, underground utilization, etc. Among these approaches, the very large floating structure (VLFS) technology is also proposed as one viable way to create space on the sea. Construction of large floating structures is technically feasible in Singapore because of its benign sea state condition and strong offshore construction capability. Towards exploring and implementing large floating structure applications in Singapore, the Land and Liveability National Innovation Challenge (L2 NIC) Directorate and Jurong Town Corporation (JTC) funded the project Multi-Purpose Floating Structures (MPFS). The research work was carried out by the National University of Singapore (NUS), SINTEF and the Norwegian University of Science and Technology (NTNU) and included collaborative multidiscipline research including theoretical analyses, laboratory testing and evaluation of construction and maintenance costs. Besides the research institutions, Shimizu Corporation has also been invited to participate in this project to evaluate the engineering method for construction, installation and marine operations of the proposed floating structures, together with the cost estimation.

The MPFS project aims to develop innovative and optimal structural and foundation solutions, lightweight concrete recipes as well as construction methods for (1) hydrocarbon storage, (2) floating bridges and (3) modular multi-purpose floating structures. The research on the hydrocarbon storage focuses on the development of the world's first floating prestressed light-weight concrete storage facility with a capacity of 300,000 m^3 [1, 2]. The research on the floating bridge focuses on developing new design concepts for the ASEAN's first floating bridge spanning over 500 m of shallow waterbody [3, 4]. An overview of the research activities on these two studies was given in [5].

The objective of the modular multi-purpose floating structures (MMFS) is to develop innovative solutions for the creation of "land on sea" of a desired shape and size for multiple applications. Some of the identified applications include floating aggregate storage facility, floating flatted factories, floating houses (dorms) and floating fish farms. Space creation is to be achieved by connecting a suitable number of modular floating units. These floating units and connectors should be standardized and optimized to reduce costs by ease mass production and optimized installation.

This paper presents the development of innovative design concepts for modular multi-purpose floating structures. Innovative concept designs and evaluation of the modular units and inter-modular connectors are presented. Results from structural and hydrodynamic analyses as well as scaled model tests are presented and discussed. In addition, this paper also presents the construction methodology and preliminary cost estimate of the proposed design concepts.

2 Design Concepts

As its name suggests, the design of modular multi-purpose floating structures should be generic with focuses on some specific applications. The specific applications are aggregate storage, flatted factories and housing properties. The concept should however be generic and not limited to these applications. Besides, the design shall meet a few requirements, including optimal basic shapes, sizes and connector designs, optimized use of sea space, fulfilment of classification rules and safety measures for floating structures, cost-effective solutions for logistics, construction and installation. The footprint of the deployment is 10 ha. The design working life shall be 60 years with minimum maintenance.

The design permanent and imposed loads vary with the application of the superstructures. It may be appropriate to categorize the design payload scale into three groups. In group 1, the average payload is smaller than 25 kPa representing greeneries, public space and 2-storey dorms/apartments. For group 2, the average payload ranges from 25 to 45 kPa representing typical light industrial applications such as 2-storey flatted factories and low-rise housing properties such as 3-4 storied dorms/apartments. Group 3 is for heavy industrial applications with average payload beyond 45 kPa but limited to be less than 80 kPa. This limit is imposed in view of the loss of the advantage of the floating structure due to large draft to seawater depth ratios resulting in non-economical designs. Aggregate storage is one example of heavy industrial applications.

2.1 Shape and Size of Modular Unit

The modular multi-purpose floating structures are formed by connecting several basic modules to form the required size and global shape. The research team has considered three options in basic shapes. Option 1 consists of rectangular modules connected in a staggered configuration with square and triangular modules at the edges and corners, as illustrated in Fig. 1. For convenience, this basic shape option shall be collectively termed "RECT". Option 2 consists of only square modules, and thus termed "SQUARE". Option 3 comprises hexagonal modules. It is termed "HEXA". For all design options, the modular units are to be made of prestressed concrete with internal bulkheads serving as stiffeners.

From a preliminary comparison, design option 1 may provide wider choices in connecting modules, wider choices in inter-modular connection stiffness and natural straight edges for berthing purposes. The construction and fabrication are comparatively easier and cheaper. In addition, the regular shape enables easier planning of the superstructure. On the other hand, design option 3 provides a single standardized shape and size with corresponding optimal use of structural material and rigid inter-modular connection. Both design options also face their challenges. For example, there are three types of modular units needed in design option 1.

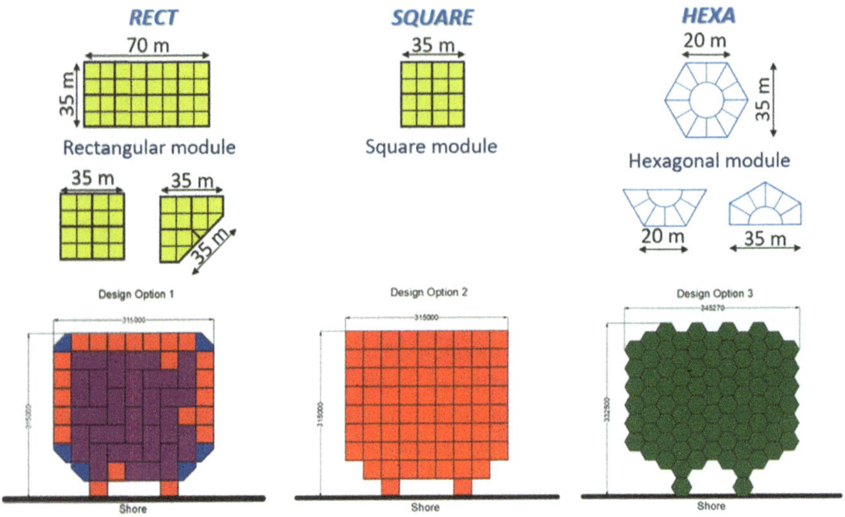

Fig. 1 Design options for modular units

Design option 3 lacks natural flat edges in the global layout for easy berthing unless half hexagonal modules are engaged. Design option 2 is considered in between in terms of the pros and cons.

2.2 *Inter-modular Connector*

The design of the module to module connection is critical in the development of the MMFS. The connection should ideally have the capabilities such as self-alignment, impact attenuation, easy engagement and adequate strength [6]. Besides, it is desirable that to allow for a certain degree of rotational flexibility such that the moment developed due to environmental loads could be greatly reduced. For concrete floating modular units, the conventional design is based on the use of prestressing tendons/bars and shear keys. Figure 2a illustrates one example of connecting two concrete floating modules. The connection engages the modules through prestressed tendons both at the top and bottom of the modules. The tendons form a force couple that is capable of resisting moments developed at the connection interface. The shear to be transferred at the connection is taken by the shear key. Due to the prestressing of tendons, the connection is always engaged, thereby forming a rigid connection. Figure 2b gives another example by Hyundai based on prestressing bars that is used in the Incheon Concrete Floating Quay project [7].

It is essential to prevent the potential of water leakage and to ease the offshore activities by having the operation only at the top surface of the floating units. The research team brainstormed and proposed an innovative inter-modular connection

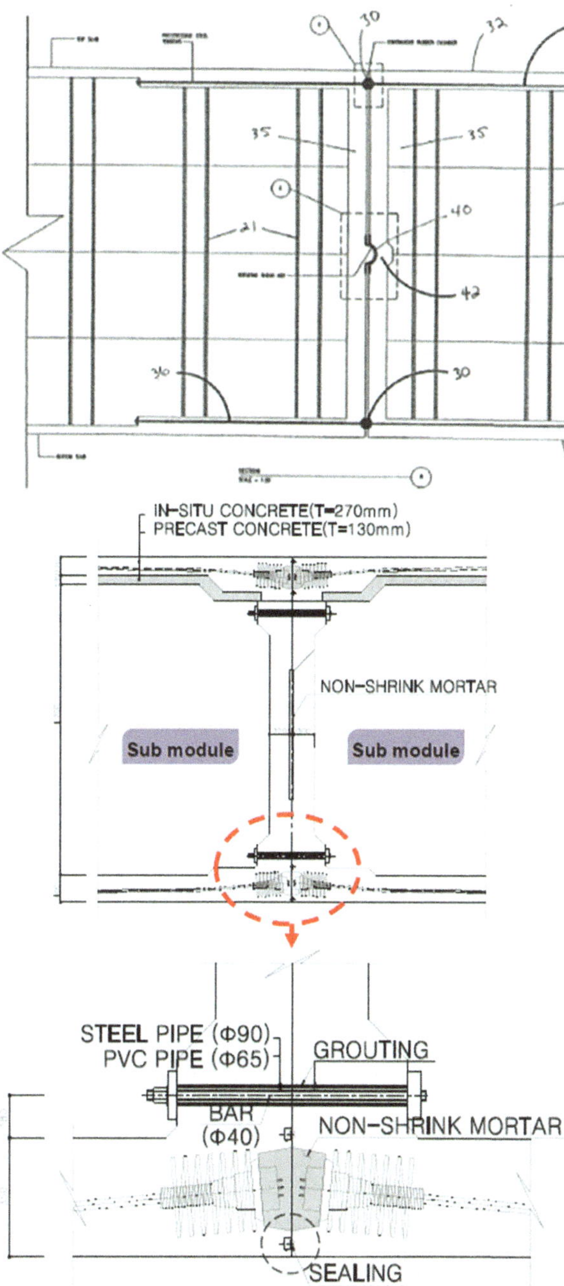

Fig. 2 Conventional connection designs: design by Dr. Alfred Yee (*top*) and design by Hyundai (*bottom*)

(see Fig. 3). This connector design comprises a top part with a tension member and a shear key. The top connector resists tension/splitting forces between connected modules and does not transfer moment. It can be used as a winch during the alignment of the modules. The shear key comprises a movable male part that can be adjusted up-and-down and in-and-out to cater for the construction tolerance. The up-and-down adjustment of the shear key is made possible through a jack. The in-and-out adjustment can be achieved by fitting in a shim plate. An alternative design with concrete shear keys is also proposed (see Fig. 4). Once the positioning of concrete shear keys is achieved, they can be grouted to the module using high-strength rapid-hardening grout.

2.3 *Station-Keeping System*

The development and evaluation of station-keeping solutions for floating structures involve several factors, including soil conditions, water depth, sea environments, sea-bed erosion and construction and maintenance costs. Based on the previous experiences on the hydrocarbon storage facility and floating bridges, it is concluded that mooring dolphin is a feasible foundation solution for floating structures in Singapore. This is in view that the soils could vary from layered soil to soft marine clay, the water depth nearshore is fairly shallow, and the sea conditions are relatively calm. It is also worth noting that the intended modular multi-purpose floating

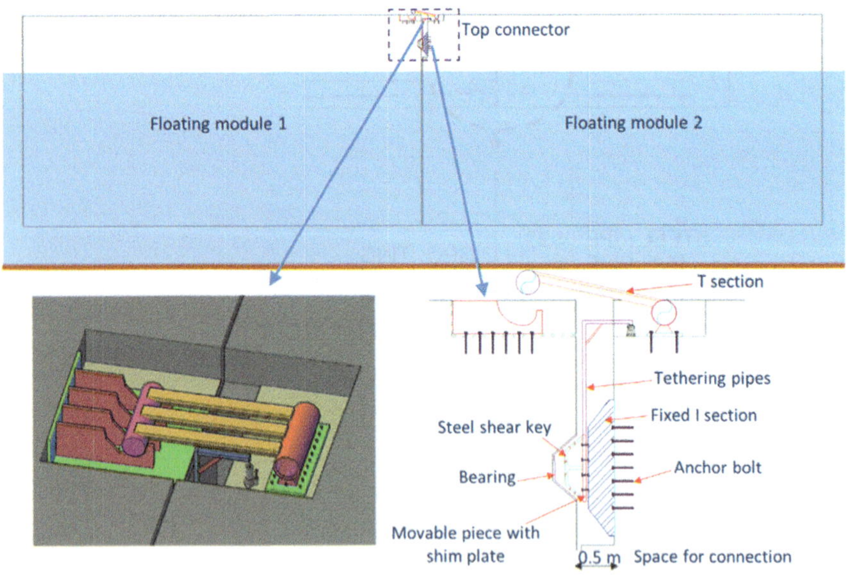

Fig. 3 New connection design for modular multi-purpose floating units with steel shear key

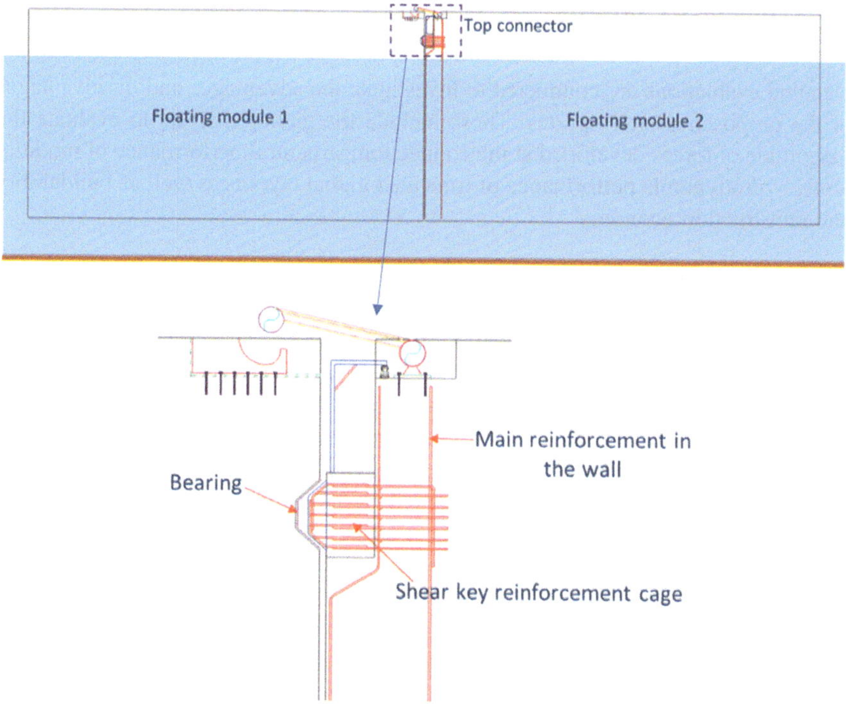

Fig. 4 Alternative connection design with concrete shear key

structure has a very similar overall geometry and thus environmental loads to the hydrocarbon storage facility. Thus, the mooring dolphin designed for the hydrocarbon storage facility is adapted here. An illustration of a mooring dolphin with inclined piles is shown in Fig. 5.

Fig. 5 Illustration of mooring dolphin

3 Concept Evaluation

Detailed evaluations are conducted to investigate the advantages and disadvantages of the proposed design options. These include the global analysis to evaluate the magnitude of forces developed at the connection, structural performance of modular units, hydrodynamic performance of truncated global layouts as well as buildability and construction economy.

3.1 Global Analysis of Connection Force

The global performance of the three proposed design options is compared through detailed finite element simulations. A segment of the global layout of the MMFS is selected for the evaluation of the modules under various possible load patterns. Figure 6 shows the selected segments of the global layouts under four different loading patterns corresponding to the three design options. Finite element models are constructed accordingly based on plate theory with the properties idealized from the actual 3D modular model. Inter-modular connection stiffness is assumed to be rigid. The modules are supported by area springs to represent the hydrostatic pressure. Two different loading levels are considered, namely 25 and 80 kPa, which represent the loading corresponding to a floating residential application and heavy-duty aggregate storage, respectively. This study aims to evaluate the shear forces and moments developed at the inter-modular connectors and how they are affected by the modular shape and size as well as the global configuration.

Different loading patterns are considered and the one inducing the highest connection loads for each design option is identified. This load pattern is then employed in a detailed study on the global connection loads. The total connection shear force, bending and twisting moments are obtained by scaling up the truncated numerical models to an equal global footprint. The results from the analyses are, listed in Table 1. The results show that the difference in shear force and twisting moment between RECT and HEXA is relatively small. However, the connection

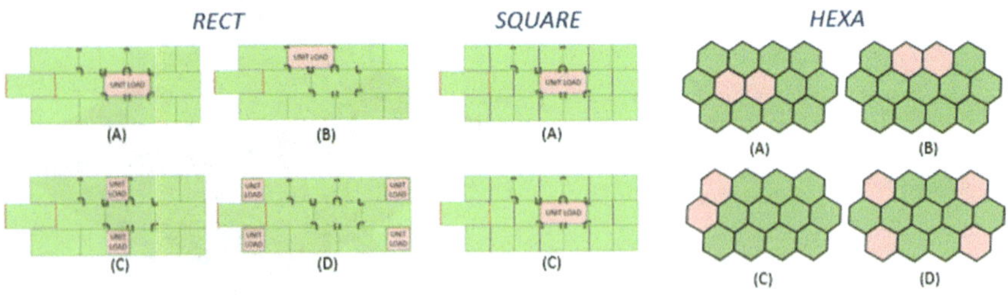

Fig. 6 Truncated global models and examples of possible load patterns

Table 1 Comparison of connection loads of different design options

Total connection load	RECT	SQUARE	HEXA
Connection length	805 m	1,120 m	980 m
Shear (25 kPa)	432 MN	570 MN	450 MN
Shear (80 kPa)	1.3 GN	1.7 GN	1.3 GN
Bending (25 kPa)	11.5 GNm	14.4 GNm	8.4 GNm
Bending (80 kPa)	36.8 GNm	48.0 GNm	26.8 GNm
Twisting (25 kPa)	6.8 GNm	6.3 GNm	5.3 GNm
Twisting (80 kPa)	16.2 GNm	22.5 GNm	16.8 GNm

bending moment with HEXA is found to be 27% lower than RECT. SQUARE is also found to have the highest connection loads due to its long total connection length.

3.2 *Structural Performance of Modular Units*

This section compares the structural performance of various modular units. In the comparative study, the highest loading level, i.e. 80 kPa representing heavy-duty aggregate storage, is considered. Finite element models are developed for the different shapes using the commercial software ABAQUS. Solid brick element (C3D20R) is used to discretize the main concrete structural components. The floating modules are supposed to remain in the linear elastic range of behavior at the service state. A linear elastic model is defined and the lightweight concrete density (including steel reinforcement) is assumed as 2,000 kg/m^3. The material input for the concrete model has a Young's Modulus, E, of 30 GPa and a Poisson's ratio, υ, of 0.17. Through iterative structural analysis using the finite element method with the consideration of a minimum freeboard of 2 m to ensure operational purposes, the total height of the floating modules is found to be 18.5 m. Note that such a heavy-duty application is meant for long-term strategic storage of aggregate. Therefore, frequent loading and unloading which may cause significant change in the draft of the floating modules are not expected.

Figure 7 shows the sectional view of a floating unit with considered loadings. The load of 80 kPa is applied as a uniformly distributed loading on the top slabs of the modular unit. The hydrostatic pressure due to seawater is taken into account. Springs are modelled beneath the base slab in the FE models to simulate the upward buoyancy effects. Draft values for each floating module are determined by balancing the upward buoyancy force with downward load effects.

Table 2 summarizes the required amount of concrete and steel for the three different design options. More details regarding the finite element modelling and analysis results can be found in [8, 9]. As it can be seen from Table 2, the three design options have similar structural performance. More specifically, RECT and

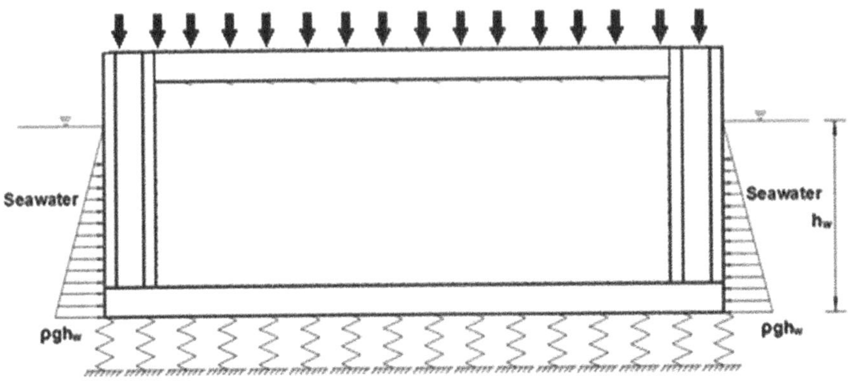

Fig. 7 Schematic diagram of loadings applied on arbitrary cross section

Table 2 Concrete and steel usage for different design options

Design	Draft (m)	Freeboard (m)	Total concrete volume per planar area (m^3/m^2)	Total steel weight per planar area
RECT	16.1	2.4	3.2	0.16
SQUARE	16.38	2.12	3.3	0.19
HEXA	16.5	2	3.2	0.16

HEXA have the same amount of concrete and steel, while SQUARE requires a slightly higher amount of the material.

3.3 *Hydrodynamic Performance of Connected Units*

Both physical model tests and corresponding numerical simulations are carried out to evaluate the hydrodynamic performance of the proposed design options. In the numerical model, a one-line configuration made of connected square or hexagonal units is considered for the sake of simplicity, as illustrated in Fig. 8. In view of their geometric difference, 7 square modules and 8 hexagonal units are considered, respectively. Each modular unit is considered as a rigid body and the potential flow theory is applied to model the fluid. In each model, hinge connectors are used for the floating unit in connection to the quayside. Note that the model made of square units can also be used to represent the design option of "RECT" in view of the fact that each rectangular unit can be considered as two rigidly connected square units.

Model tests of the proposed design concept were also carried out at the wave basin in the National University of Singapore during November 2018. The tests focused on the hydrodynamic behavior of the modular system in waves. Due to the limited area in the wave basin, it was not impossible to test an entire system.

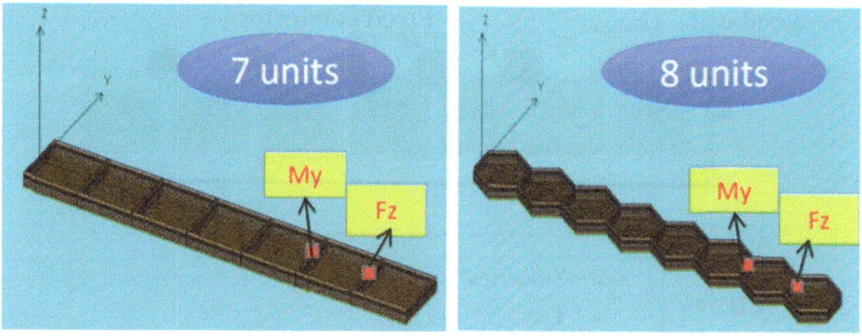

Fig. 8 Numerical models of connected modular units

Fig. 9 Experimental model of one-line system

Therefore, tests were performed only on systems comprising square units. Figure 9 shows the setup of an experimental model of 7 connected modular units at the wave basin.

Figure 10 shows the comparison of the bending moment developed at the connectors from the numerical and experimental models. In the comparison study, regular waves with a wave height of 2 m and a period of 7 s in the longitudinal direction of the system are considered. Good agreement is found between results obtained by using the numerical model and experimental model. Although the numerical results tend to be slightly larger than the model test results. This slight overestimation is probably due to the uncertainty occurred during the experiment,

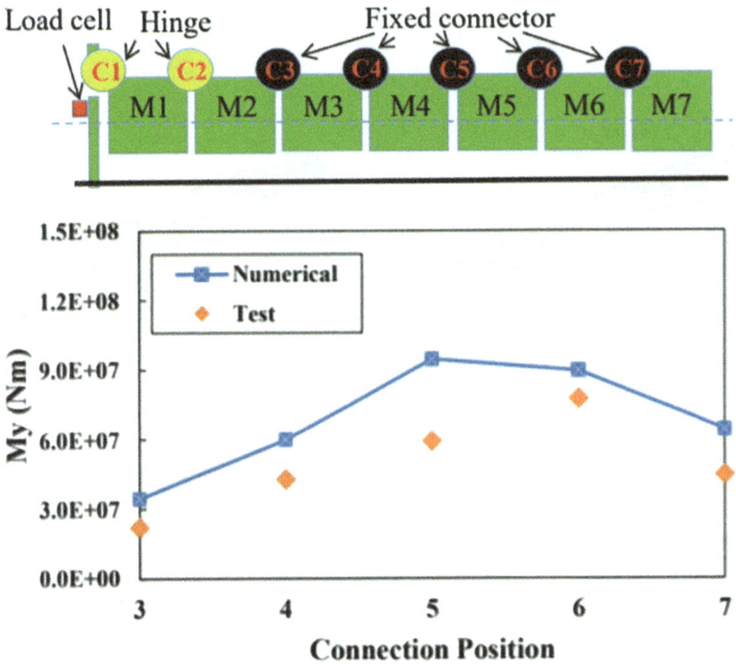

Fig. 10 Moment at inter-modular connections

such as the viscous effects or shallow-water effects. Nevertheless, the numerical model is verified and validated.

The numerical model is next applied to investigate the difference in the hydrodynamic performance between square and hexagonal modular units. Table 3 lists the motion of floating units and connection forces developed under a 100-year sea state with a significant wave height of 1.8 m and peak period of 7 s. In general, the design option HEXA has slightly smaller motions and forces under wave

Table 3 Hydrodynamic performance of one-line modular systems

Parameter		SQUARE	HEXA	HEXA/SQUARE (%)
Pitch (°)	Max	0.196	0.187	95.4
	Min	−0.227	−0.212	93.4
	Std	0.062	0.058	05.6
Shear (kN)	Max	3.20E+03	2.90E+03	90.6
	Min	−2.97E+03	−2.84E+03	95.6
	Std	9.11E+02	8.40E+02	92.2
Moment (kNm)	Max	1.34E+05	1.23E+05	91.8
	Min	−1.30E+05	−1.20E+05	92.3
	Std	3.82E+04	3.48E+04	91.1

actions. However, the difference between SQUARE, RECT and HEXA is rather small. For more detailed information of the hydrodynamic study, the reader may refer to [10, 11]. It is worth highlighting that there was a large-scale model test conducted at SINTEF Ocean to examine the hydrodynamic performance of the proposed hydrocarbon storage facility [12]. The experimental results are also of important value in view of the fact that both applications have similar overall dimensions.

3.4 Buildability and Cost Comparison

Integration of superstructure and floating substructure may help to reduce the cost of a floating facility. Normally the design of floating structure starts with superstructure planning, and then design the floating structure corresponding to superstructure's size, height, column span. Following this process, it is possible to optimize the floating structure design and reduce the cost. However, since the current study focused only on the floating structure, the buildability and cost evaluation based on the schematic design are indicative only.

It is assumed that the footprint of superstructures covers 60% of the floating structure area with a plot ratio of 2.5, i.e. the ratio of gross floor area is 250%. The superstructure is 4-storied and a total payload of 40 kPa is assumed to be applied to the floating units.

For the construction of the modular units, it is recommended that they are fabricated using dry docks. There are only a few dry docks available in Singapore and the region which have the capacity of construction a few units concurrently. In general, all modular units can be constructed using standard methods. However, the hexagonal units have oblique sides which tend to introduce complications of the construction. An evaluation by Shimizu corporation shows that the buildability of HEXA design option is about 20% lower than RECT and SQUARE [13].

For the cost comparison, only indicative overall cost is examined. This is due to the fact that there are many uncertainties in the plan and details at the conceptual design stage. However, this indicative cost is aimed to be utilized as the basis for the purpose of further study of design and construction methods to reduce the cost. It should be noted that the estimation was conducted by Shimizu corporation using Japanese rates due to the lack of locally available information. The cost estimate includes concrete floating work and the landing/towing work. For the latter, the outfitting at quayside and offshore connection work as well as dock rental fee based on Japanese rates are considered. Also note that although the Japanese rates are used in the cost estimate, the study focuses on the relative cost between different design options. For the detailed analysis of the construction cost, other available fabrication facilities elsewhere may be consulted. However, this is outside the scope of the study.

The evaluation estimates that the total construction costs for RECT and SQUARE are virtually the same as they have a very similar geometry. HEXA has

Table 4 Comparison between design options

Selection criteria	RECT	SQUARE	HEXA	Weightage
Global static performance	1	0.78	1.37	40%
Hydrodynamic performance	1	1	1.07	10%
Payload carrying capacity	1	0.96	0.99	50%
Efficiency of structural system	1	0.89	1.15	100%
Buildability	1	1	0.8	–
Cost-effectiveness	1	0.95	0.87	–

more modular units and the construction with oblique walls increases the cost for each module by about 20–30%. Overall speaking, the total construction cost for HEXA is about 15–20% higher than RECT and SQUARE [13].

3.5 *Comparison of Design Options*

With the studies on the global performance, structural behavior, hydrodynamic performance and buildability and cost-effectiveness, a quantitative comparison is made possible by assigning appropriate indices to the performance of the design options in each of the identified selection criteria with appropriate weightage, as shown in Table 4. Note that for all selection criteria, the performance of RECT is set to 1 and chosen to be the basis for comparison. Higher indices refer to better performance. In general, all design options are technically feasible. HEXA is found to perform the best in term of the global static performance owing to the natural interlocking mechanism with oblique connectors, followed by RECT and SQUARE in sequence. RECT scores the best in terms of ease of cost effectiveness, structural performance, construction, ease of marine operations and planning of super-structure. In view of this, it may be reasonable to choose RECT for further detailed research analysis and design.

4 Conclusions

This paper presents the development of innovative design concepts for a modular multi-purpose floating structure. Studies on the global static behavior, structural and hydrodynamic performance, station keeping system as well as buildability and cost economy are carried out. The results show that construction of large floating structures for generic applications in Singapore coastal waters is very doable because of the benign sea state and strong offshore industry capability. The study also reveals that all proposed design options are technically feasible. Considering both the efficiency of the structural system and cost economy, the design option comprising mainly rectangular modular units appears to have the best performance and thus can be chosen for future detailed research, analysis and design.

Acknowledgements This research is supported in part by the Singapore Ministry of National Development and the National Research Foundation, Prime Minister's Office under the Land and Liveability National Innovation Challenge (L2 NIC) Research Programme (L2 NIC Award No L2 NICTDF1-2015-2). Any opinions, findings, and conclusions or recommendations expressed in this material are those of the author(s) and do not reflect the views of the Singapore Ministry of National Development and National Research Foundation, Prime Minister's Office, Singapore. The support from JTC Corporation is also deeply appreciated by the authors.

References

1. Dai J, Ang KK, Jin J et al (2019) Large floating structure with free-floating, self-stabilizing tanks for hydrocarbon storage. Energies 12:3487
2. Jiang D, Tan KH, Dai J et al (2018) Structural performance evaluation of innovative prestressed concrete floating fuel storage tanks. Struct Concrete 20:15–31
3. Wan L, Magee AR, Hellan Ø et al (2017) Initial design of a double curved floating bridge and global hydrodynamic responses under environmental conditions. In: Proceedings of the 36th international conference on ocean, offshore & arctic engineering, Trondheim, Norway, June 2017
4. Wan L, Jiang D, Dai J (2020) Numerical modelling and dynamic response analysis of curved floating bridges with a small rise-span ratio. J Mar Sci Eng 8:467
5. Ang KK, Dai J, Hellan Ø et al (2020) Design and potential applications of floating structures in Singapore. In: Wang C, Lim S, Tay Z (eds) WCFS2019. lecture notes in civil engineering, vol 41. Springer, Singapore
6. Boldbaatar T, Yoon D-G (2013) A study on the connector of floating platform based on concrete structures. J Korean Soc Mar Environ Saf 19(1):37–44
7. Lee SH (2017) Modular floating structures. Workshop on multi-purpose floating structures, 1 March, Trondheim, Norway
8. NUS, SINTEF (2019) Modular multi-purpose floating structures: structural analysis and design of MMFS modules. Technical report no. 21 submitted to JTC
9. NUS, SINTEF (2019) Modular multi-purpose floating structures: structural analysis and design of MMFS connectors. Technical report no. 22 submitted to JTC
10. Ren N, Zhang C, Magee AR et al (2019) Hydrodynamic analysis of a modular multi-purpose floating structure system with different outermost connector types. Ocean Eng 176:158–168
11. NUS, SINTEF (2019) Hydrodynamic response of modular multi-purpose floating structures (MMFS). Technical report no. 20 submitted to JTC
12. SINTEF, NUS (2019) Floating fuel storage facility model tests. Technical report no. 29 submitted to JTC
13. Shimizu Corporation (2019) Construction methodology and cost estimate for MMFS. Technical report no. 28 submitted to JTC

A Study on Stability of Floating Architecture and Its Design Methodology

Toshio Nakajima, Yuka Saito, and Motohiko Umeyama

Abstract Herein, the authors describe an overall approach to the architectural design of floating structures such as floating houses. The primary aim of this study is not to present a method for stabilizing floating structures, but rather to provide a design synthesis method for use when designing such structures. More specifically, we propose an integrated procedure for use at the preliminary design stage of such structures that systematically facilitates their overall design. As an inclining platform could endanger the people on board, it is necessary to determine an adequate metacentric height in order to prevent such occurrences. This measurement, which is defined as the distance between the center of gravity of a floating structure and its metacenter, quantifies the initial static stability of a floating body. Based on this idea, we consider the associated problems as well as the methods used in practical procedures, and combine them to introduce a unique approach called the "required GM" method. We also discuss the different and various aspects used in basic configuration determinations of floating architectural structures, such as the aspect of static stability and the overall process used at the conceptual design stage. In addition, illustrative examples of an idealized floating platform embodying the simplest possible structures are provided to illustrate these points.

Keywords Floating house · Sustainable · Natural hazard · Flood house · Climate change

T. Nakajima (✉)
Waterfront Real Estate Co., Ltd., 7-20-402, Takatsuka-cho, Nishinomiya, Hyogo 6620066, Japan
e-mail: tmunakajima@gmail.com

Y. Saito
Building Department, Design Engineering Division, OIL&GAS PROJECT COMPANY, JGC CORPORATION, 2-3-1, Minato Mirai, Nishi-ku, Yokohama 2206001, Japan
e-mail: saiyk89@gmail.com

M. Umeyama
Department of Civil and Environmental Engineering, Tokyo Metropolitan University, 1-1, Minamiosawa, Hachioji, Tokyo 1920397, Japan
e-mail: tbbtp647@gmail.com

Ł. Piątek et al. (eds.), *WCFS2020*, Lecture Notes in Civil Engineering 158,
https://doi.org/10.1007/978-981-16-2256-4_17

1 Introduction

In spite of widespread movements to reduce greenhouse gas emissions, substantial reductions have yet to be achieved, and it may already be impossible to halt the acceleration of rising sea levels caused by the influence of global warming [1]. The situation is worsened when we consider that many of the world's cities are located in low-lying areas that are in dangerous proximity to the sea and/or large rivers. A novel remedy against such serious conditions is to modify such cities by incorporating floating solutions into architectural structures as a way to achieve safety and security against flooding hazards [2, 3]. Examples of such dangerous conditions can be seen in the island nation of Singapore, which must be constantly vigilant against flooding because most of the island is flat and very close to sea level. Since Singapore also suffers from both shortages of both freshwater and available land, we have previously explored a unique floating solution that involves extensive, large-scale excavations of flat areas in order to create numerous artificial reservoirs, on top of which pontoon-type buoyant structures with supertall and high-density residential buildings can be constructed [4]. This method, which is designed to transform water surfaces into floating cities, also provides potential solutions to increased population levels (Fig. 1), and it is expected that many other cities around the world could potentially benefit by utilizing similar structures to avoid future flooding crises.

At the other end of the size spectrum, tremendous numbers of floating houses have already been constructed on seaside areas and lakes all over the world [5]. However, since these floating houses are small in size, they are relatively unstable,

Fig. 1 Perspective view of "Singapore Water City"

which means they frequently move under the influence of the winds and waves. This can cause many people to feel uncomfortable. In contrast, large floating hotels and/or urban condominiums built on floating foundations can provide more stable, and thus more comfortable, alternatives to small floating houses. In all such cases, to ensure occupant comfort and safety, the most important factors related to such floating structures are static stability and need to prevent unfavorable listing that could result from various loadings. Nevertheless, there will likely be an increasing need for even more floating structures in the near future, since climate change and rising sea-levels due to global warming will become rapidly more serious.

In a previous study, Takarada et al. [6] introduced a unique scheme called the "Required GM" method, which can be used to minimize the large-scale listing of semi-submersible oil rigs that might result from various environmental loads. Herein, that method is modified for the purpose of designing the floating architectural configurations by minimizing the inclining angles that can be generated. We believe that our proposed method is practically applicable for determining the structural elements of pontoon-type floating structures, and can be expected to have more practical applications when used by architects and/or designers at the preliminary design stages of such structures.

2 On the Stability of Floating Architectures

2.1 General

The need for stability plays a key role in the design of floating structures. Some items and/or elements of static stability for a floating structure are shown in Fig. 2. Indeed, its impacts on safety and comfort make it a fundamental consideration for architects and/or designers in their preliminary design stages. Therefore, at the beginning of the design process, it is important to determine the size of some principal elements, such as the length, width, and depth of the floating foundation. For a floating structure, the most important aspect of static stability is the location of the center of gravity (G). In this section, we discuss some fundamental studies on static stability for a pontoon-type floating structure as well as the basic theory of static stability. Here, the wind force is considered to be the most influential force, since it has generally been regarded to be the main force that must be counteracted according to the existing rules and/or regulations of classification societies. More details on wind force as well as wind intensity change with respect to height are shown in Appendix A.

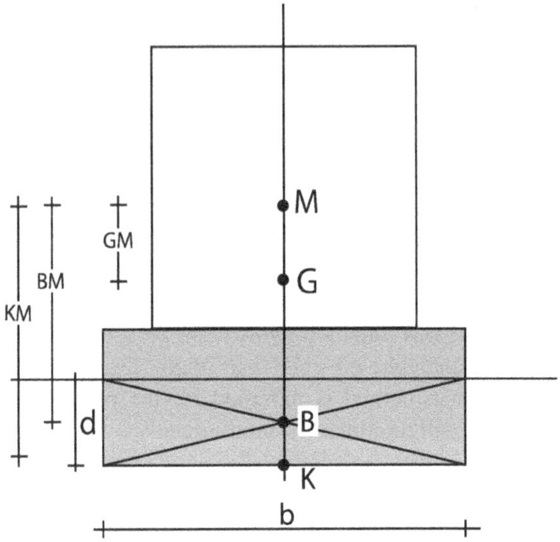

Fig. 2 Items of static stability

2.2 Fundamentals on Static Stability for Floating Architecture

When external forces such as wind act on a floating body, the free-floating body tilts due to the heeling (or overturning) moment introduced by forces such as that caused by wind, with the location of the center of buoyancy (B) shifting leeward and then the platform tilting in the inclining angle (ϕ), as is shown in Fig. 3.

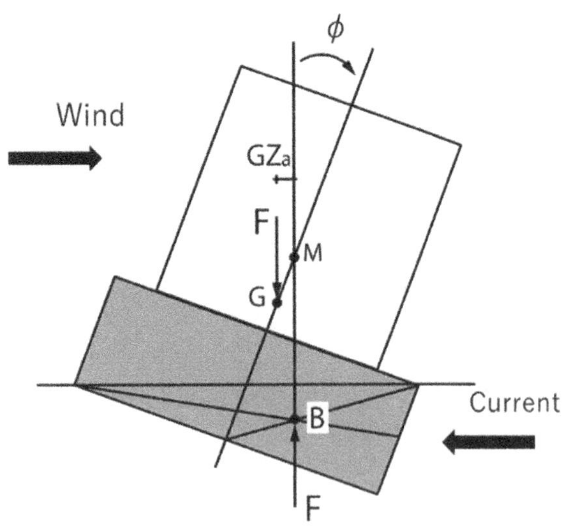

Fig. 3 Inclined condition of a platform due to winds

At that point, the corresponding righting (or restoring) moment (M_R) is created to cancel out this heeling moment.

The righting moment (M_R) is given by

$$M_R = F \cdot \overline{GZ} \tag{1}$$

$$(F = W = \rho g V)$$

where F is buoyancy, which is equal to the weight of the floating structure (W), g is the gravity, and ρ is the water density. The $\overline{GZ}$ is the horizontal distance (or lever arm) and is expressed by the following equation as in Ref. [6].

$$\overline{GZ}(\phi) = \sin\phi\left(\overline{GM} + \frac{1}{2}\overline{BM}\tan^2\phi\right) \tag{2}$$

When ϕ is very small, the second term of this equation can be ignored as follows.

$$\overline{GZ(\phi)} = \sin\phi \cdot \overline{GM} \tag{3}$$

This term also appears in the righting moment M_R as

$$M_R = \overline{GZ(\phi)} \cdot F = F \cdot \sin\phi \cdot \overline{GM}. \tag{4}$$

If the vertical distances of both the center of buoyancy (KB) and KG are known (see Fig. 2), both the KM and $\overline{GM}$ values are estimated easily by calculating the $\overline{BM}$ value obtained from the following equation.

$$\overline{BM} = \frac{I_y}{V} \tag{5}$$

Here, I_y is the moment of inertia in relation to the y axis, and is calculated by the following equations in the case of a rectangular pontoon (see Fig. 4).

$$I_y = \frac{1}{12}BL^3 + l_x^2 \cdot B \cdot L \tag{6}$$

Additionally, assuming that ϕ is sufficiently small, it can be computed by the following equation from Appendix B:

$$\varphi(\text{in degrees}) = \frac{180.}{\pi \cdot GM} \times \frac{\sum_i M_i}{W} \tag{7}$$

where $\sum_i M_i$ indicates the sum of possible overturning moments caused by various combined loadings while W indicates the total weight of the floating structure.

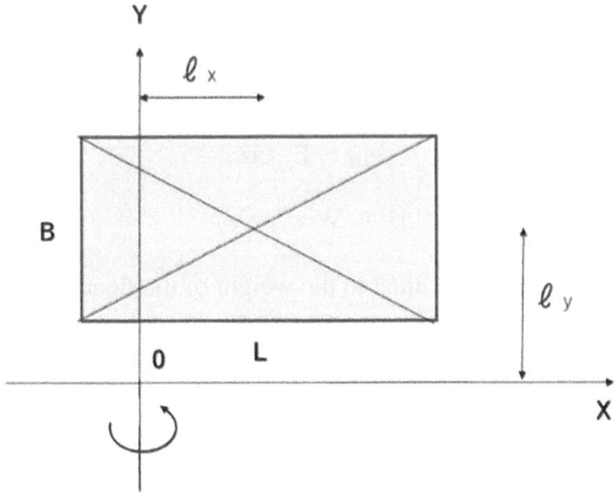

Fig. 4 Right-handed cartesian coordinate system of rectangular pontoon

2.3 *Static Stability Comparisons of Floating Architectures*

Several computational studies were conducted on the static stability of small-sized floating structures. In those studies, comparisons of the inclining angles that result from the same wind intensity (20 m/s) were carried out to show that platform inclining angle differences occur when the KG is at different locations. The drag coefficient (C_D) for the upper structure on the floating pontoon was assumed to be 2.0 for these computations. The configuration and dimensions of a floating house are shown in Fig. 5 and Table 1, respectively.

Figure 6 shows a comparison of inclining angles (in degrees) resulting from wind imposed on the non-dimensional length of pontoons (L/Lx) for four different KG values. In each case, the inclining angle of the floating structure is calculated when 20 m/s wind blows against the upper structure.

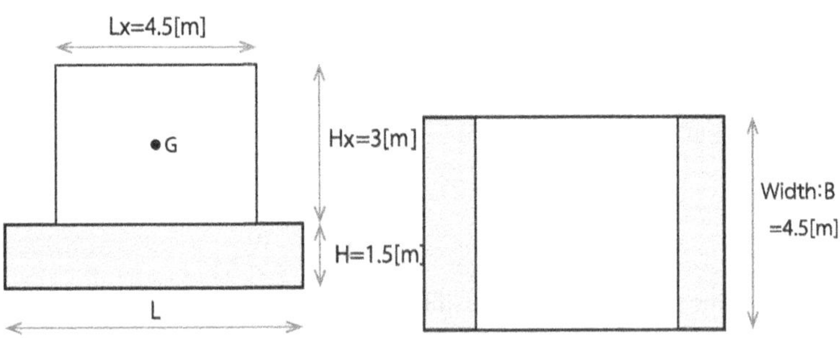

Fig. 5 Vertical and horizontal configuration of a floating house

Table 1 Principal particulars of a floating house

Upper structure (house)			Pontoon		
Length: Lx	4.5	(m)	Length: L	4.5–9.0	(m)
Width: B	4.5	(m)	Width: B	4.5	(m)
Height: H	3.0	(m)	Height: H	1.5	(m)
Total mass	24,300	(kg)			

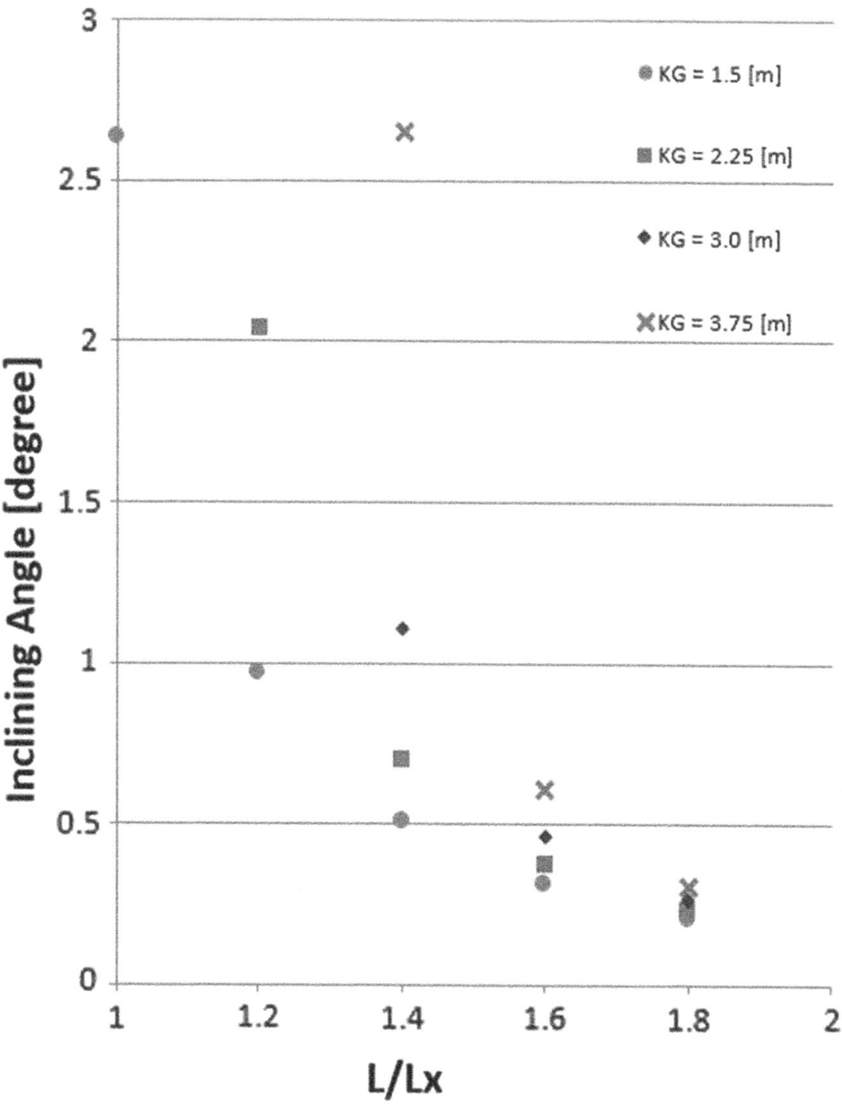

Fig. 6 Inclination due to wind for different KG values

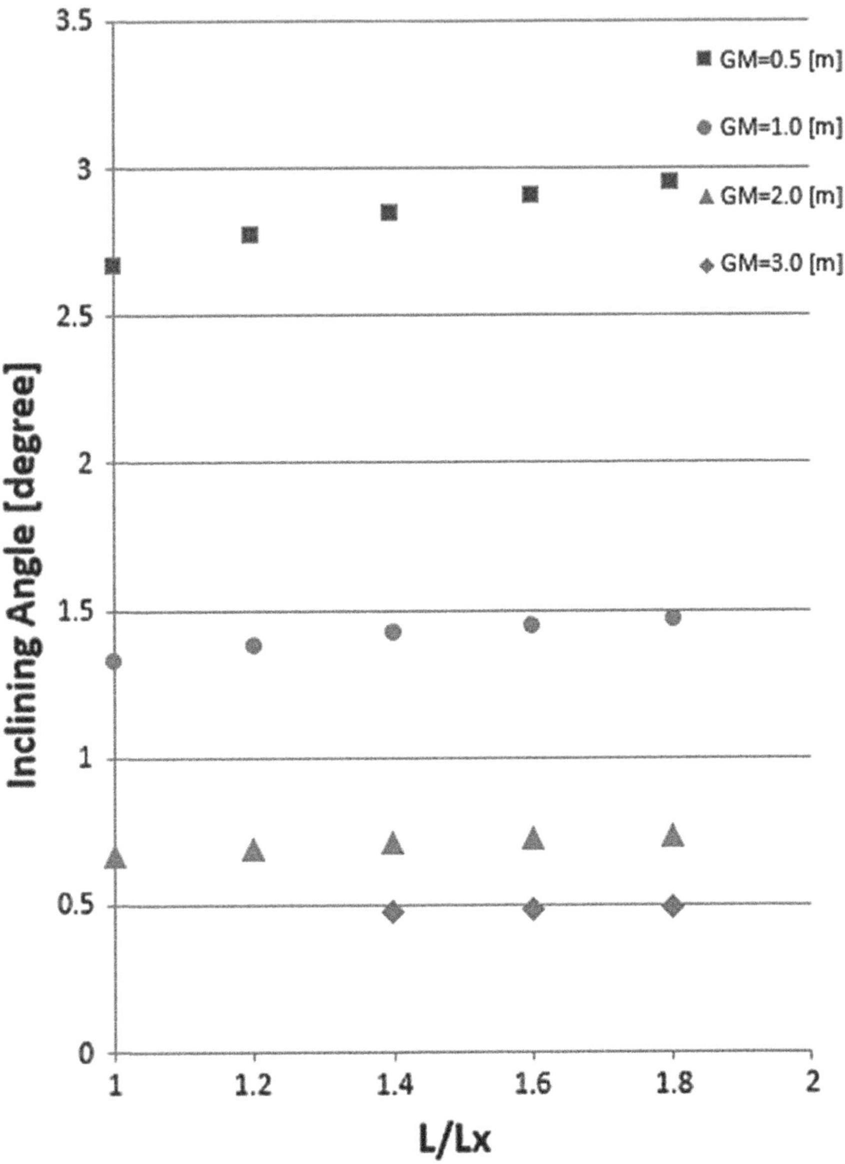

Fig. 7 Inclination due to wind for different GM values

Generally, it is known that the KGs of most floating structures are located in higher positions and it is largely believed that higher KG values result in platform instability. However, this is often not the case. In fact, Fig. 7 shows that a floating structure with a large enough $\overline{GM}$ value can be very stable, even if their KG locations are in high positions.

By comparing the inclination between the different KG value locations, the following conclusions can be made:

- Inclining angles decrease along the size of non-dimensional length (L/Lx) in all four KG cases.
- Smaller pontoon lengths (L/Lx) result in larger inclining angles for higher KG values.
- It is obvious that larger pontoon lengths (L/Lx) significantly increase stability.
- The inclining angle remains almost the same for the longest length (L/Lx), even if the KG differs in height, when compared with those for smaller pontoon lengths.

On the other hand, Fig. 7 shows a comparison of wind-driven inclinations for the different $\overline{GM}$ values of 0.5, 1.0, 2.0, and 3.0 m. Here, it can be seen that when the $\overline{GM}$ is same, the inclining angle is unaffected even if the pontoon length changes. Thus, since we can conclude that structure stability depends primarily on the $\overline{GM}$ value, it can be said that static stability is the dominant factor for floating structures.

3 Determination of Principal Size of Floating Architectures

3.1 General

From previous studies, it is clearly evident that the $\overline{GM}$ value is the influential factor in the static stability of a floating structure. Accordingly, when designing floating structures for civil purposes, decisions regarding the appropriate dimensions are derived using the $\overline{GM}$ value to ensure comfort as well as safety. The present method reported in this paper is effective for determining the principal size of these structures easily and thus can be utilized by architects and/or designers aiming at creating superb floating architectural designs.

In this paper, the "Required GM" method [6] is used as a tool for determining the principal elements of a floating architecture such as small houses, floating hotels, and similar structures. Furthermore, since the floating structures considered here are general platforms for civil purposes in moderate water environments, the inclining angle is assumed to be sufficiently small. Based on this idea, a new practical method of estimating of the $\overline{GM}$ value is presented in the following section.

3.2 The "Required GM (Req. GM)" Method

To begin with, some definitions and assumptions are made here. The simplest yet most fail-safe approach is where the axis is fixed and the floating unit tilts around this axis with zero trim.

In addition, assuming a wall-sided vessel, the following assumptions are made:

- The steady forces in the vertical direction are small and there are no draft changes.
- The time averaged righting moment is the same in waves and in calm water.
- The lower edge of the platform does not raise free of the water when the unit is in an inclined state.

The required GM value is determined in the following equations derived from Appendix B. The $\overline{GM}$ value (Req. GM) necessary to avoid a serious inclining angle is provided in Eq. (8). When the ϕ is sufficiently small, the required GM is derived as follows [6]:

$$\text{Req. GM}(\phi) = \frac{\sum_i M_i}{W \cdot \sin \phi} = \frac{180.}{\pi \cdot \Phi d} \times \frac{\sum_i M_i}{W} \tag{8}$$

where Φd is inclining angle in degrees.

Accordingly, the $\overline{GM}$ value depends upon the sum of the overturning moments. If there are other items to be considered, they may be added as required.

3.3 Computational Approach for a Floating Architecture

Present design methodology is based on the fact, outlined in the previous section, that the static stability of a floating structure is mostly dependent upon its $\overline{GM}$ value. However, since a correct KB value is not defined at the beginning, a certain tentative value should be assigned. The design approach for a floating architecture is shown in the computation flowchart of Fig. 8.

First, the underwater volume (V) of a floating architecture and its KG value are computed in advance. Next, all the overturning moments due to wind, current, and other factors are estimated while also determining the ϕ value for the floating architecture.

- First, the "Req. GM" value is calculated using Eq. (8).
- Then, the dimensions of principal elements such as the length (tentative), draft (tentative) and so forth for the floating structure are assumed as the first trial.
- The KB value is then tentatively calculated from these design parameters.
- Next, the moment of inertia is obtained from the calculated $\overline{BM}$ value.
- The $\overline{GM}$, KG, and tentative KB values are used for this computation.

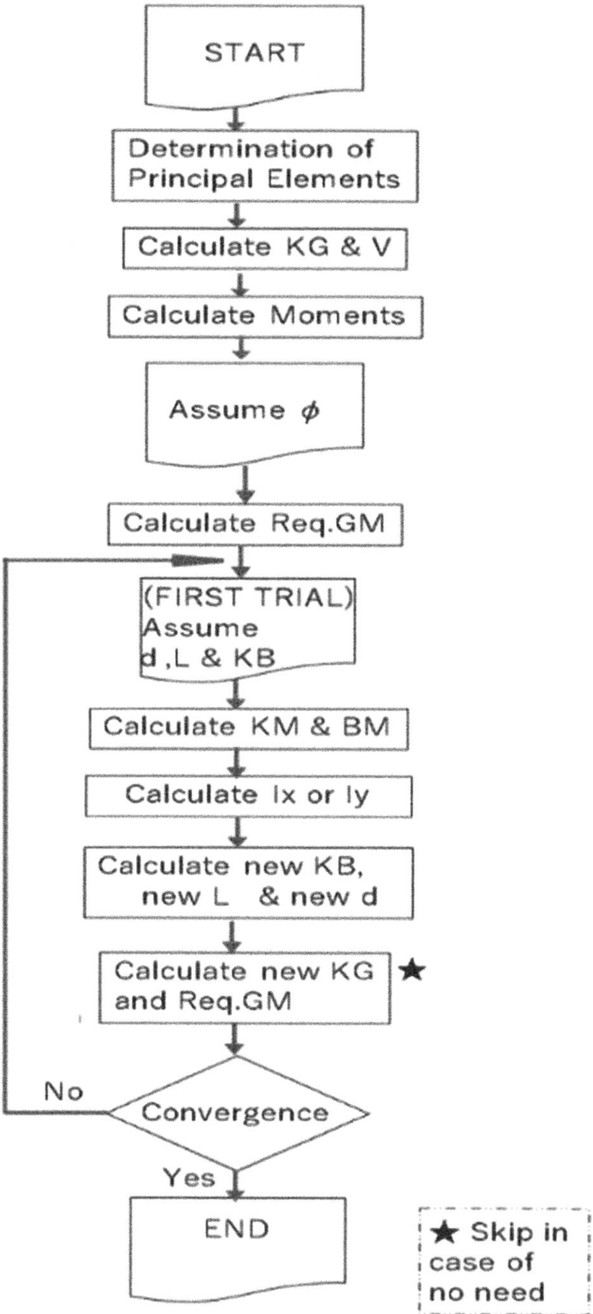

Fig. 8 Flowchart of computation

- Finally, tentative values for the length (L), draft (d), and other factors are obtained.
- If necessary, a new KG value is computed.
- If the KG value changes, the "Req. GM" value is calculated again.
- At that point, the process returns to the previous original step (as shown by the backward arrow on the left side in the flowchart) and the newly obtained L and d values are used for the next trial.
- This step is repeated until the final design values of L, d, and other related factors have converged.

4 Design Approach for Floating Architectures

4.1 Case Study of a Floating House

4.1.1 Problem

To find an appropriate length (L) and draft (d) for the pontoon of a floating house, as shown in Figs. 9 and 10, assuming the ϕ value becomes 1.5° under the combined conditions of wind force, pile reaction, and the simultaneous movement of eight persons toward the leeward side.

Conditions

- The wind velocity is 20 m/s (wind force and moment are given as 6,000 N and 4,200 N m, respectively).
- The reaction force on the pile on the leeward side is given as 6,000 N.
- Eight persons (64 kg/person) shift 2.26 m towards the leeward side simultaneously.

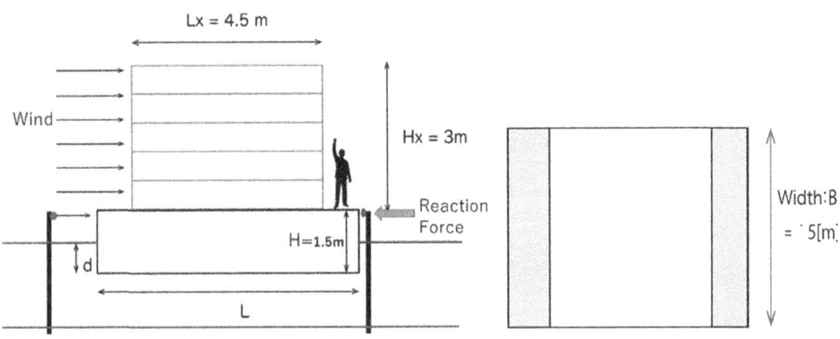

Fig. 9 Vertical and horizontal views of a floating house

Fig. 10 Photo of a floating house

Other Given Data

- The building mass is 16,500 kg and the pontoon mass is 7,888 kg.
- The building height, length, and width values are 3.0, 4.5, and 5.0 m, respectively.
- The pontoon height and width are 1.5 and 5.0 m, respectively.
- The KG location is given as 2.3 m (fixed).
- Other data: water density (ρ); 1,000 kg/m^3, air density; 1.205 kg/m^3.

Solution

Since the total mass is calculated as 24,900 kg (= 16,500 kg + 7,888 kg + 64 kg × 8), the underwater pontoon volume (V) becomes 24.9 m^3. Then, the total weight (W) is 244,020 N (24.9 m^3 × 1,000 kg/m^3 × 9.8 m/s^2). The schematic elevational view of the floating house is shown in Fig. 11.

Calculation of the Total Overturning Moment $\left(\sum_i M_i\right)$

- The overturning moment around the KG due to the movement of eight persons is

$$M_P = 64.0\,\text{kg} \times 9.8\,\text{m/s}^2 \times 8\ \text{ persons} \times 2.26\,\text{m} = 11,340\,\text{N m} \tag{9}$$

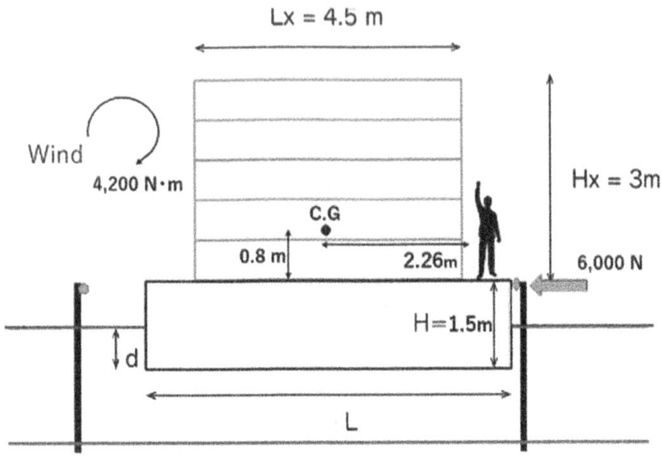

Fig. 11 Schematic elevational view of a floating house

- The wind heeling moment is (Mw) = 4,200 N m (given)
- Since the total horizontal force on the platform (wind) is 6,000 N (given), the reaction force of the pile has the same value. Then, the moment around the KG due to the reaction force of the pile at the leeward side (M_R) becomes

$$M_R = 6,000\,\mathrm{N} \times 0.8\,\mathrm{m} = 4,800\,\mathrm{N\ m} \tag{10}$$

where the vertical distance between the KG and the pile reaction location (lever) is 0.8 m (= KG − 1.5 m).

- Finally, the total overturning moments $\left(\sum_i M_i\right)$ are calculated as follows:

$$\begin{aligned}\sum_i M_i &= 4,200\mathrm{N\ m\ (wind)} + 4,800\mathrm{N\ m\ (pile)} + 11,340\mathrm{N\ m\ (people)}\\ &= 20,340\mathrm{N\ m}\end{aligned} \tag{11}$$

- The "Required GM" value can be obtained by using Eq. (8) as follows: Here, the ϕ value is given as 1.5° from the beginning.

$$\mathrm{Req.\ GM}\ (1.5^\circ) = \frac{20,340\,\mathrm{N\ m}}{244,020\,\mathrm{N}\cdot\sin\,(1.5^\circ)} = 3.18\,\mathrm{m} \tag{12}$$

Parameter Update Process

- First trial:
  The initial length L1 (tentative) is set to 5 m and thus the draft (d1) becomes 0.996 m.
  (Here, the value of underwater volume (V) is given as 24.9 m^3.)
  Then, the KB location 1 becomes 0.498 m.
- The KM value is then obtained by summing the KG (2.3 m) and the $\overline{GM}$ (3.18 m), and is thus 5.48 m.
  Here, for simplicity, the KG value is assumed not to change.
- Accordingly, the $\overline{BM}$ value (= KM − KB) becomes 4.982 m, and the moment of inertia can then be calculated as follows:

$$I_y = \overline{BM} \times V = 4.982\,\text{m} \times 24.9\,\text{m}^3 = 124.0518\,\text{m}^4 \quad (13)$$

- Since the moment of inertia (I_y) is 124.0518 m^4, L2 becomes 6.2115 m instead of 5.0 m using the relation shown in Eq. (6) while the draft (d2) becomes 0.8017 m and the new KB location becomes 0.4009 m (see Table 2).
- The trials are repeated until convergence is obtained.

Calculations of Wind Velocity and Force

Since the wind velocity U_{10} is given as 20 m/s at the height of 10 m for this case study, the wind velocity at 2.25 m high (= 1.5/2 m + 3.0/2 m) becomes 16.16 m/s from the following computation, which is described in Appendix A.

$$U_{2.25} = 20\,\text{m/s} \times (2.25/10)^{1/7} = 20\,\text{m/s} \times 0.808 = 16.16\,\text{m/s} \quad (14)$$

Next, the wind force (F_W) is calculated assuming that the drag coefficient (C_D) is 2.034.

$$\begin{aligned} F_W &= \frac{1}{2} \times 1.205\,\text{kg/m}^3 \times (16.16\,\text{m/s})^2 \times 2.034 \times (3\,\text{m} + 0.75\,\text{m}) \times 5.0\,\text{m} \\ &= 6,000\,\text{N} \end{aligned} \quad (15)$$

Table 2 Repeated trials for length (L) and depth (d) for a floating house

Trial	L (m)	d (m)
1	5.0000	0.9960
2	6.2115	0.8017
3	6.2416	0.7979
4	6.242	0.798

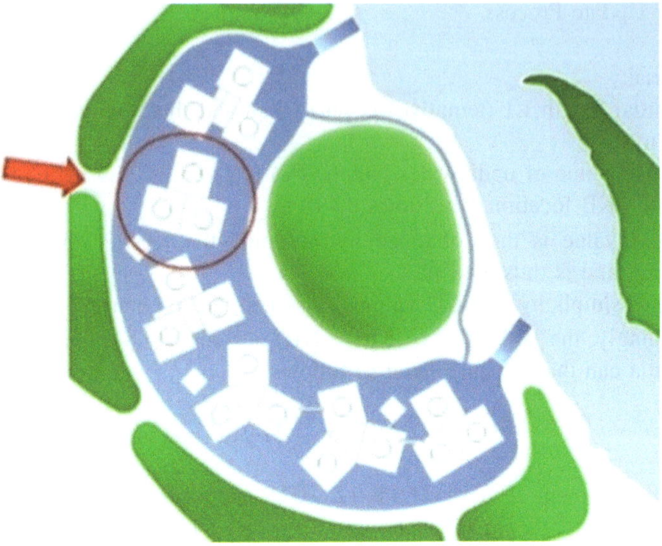

Fig. 12 Plan view of artificial reservoir

4.1.2 On the Design of Singapore Water City

If constructed, the Singapore Water City project would cover approximately 105.8 ha, have a length of 2,190 m, a width of 483 m, and an average water depth of about 25 m (see Figs. 12 and 13). As part of this project, three 51-story high-rise towers, in which 48 floors of each building would be residential space, would be constructed on the same floating foundation composed of three square-shaped pontoons (or floating modular units) (see Fig. 14).

The total number of households and the population of the three residential towers would be nearly 2,260 and 8,100, respectively. The total living space in the three residential towers would be approximately 372,960 m^2. We firmly believe that building semi-floating water cities of this type, which could house numerous high-rise towers, would effectively mitigate Singapore's ongoing land shortage while also alleviating the pressures of the city's growing population. An accurate assessment of the stability of a floating water city is a fundamental requirement for ensuring its safety, especially when considering the construction of high-rise towers on a floating foundation. As high-rise towers will be subjected to the highest levels of wind intensity, calculations of the static stability and inclination of these towers are crucially important.

The total weight of three 51-story high-rise towers that are 210 m high and 60 m wide is estimated at approximately 529,200 ton-force (tf) by assuming the unit weight of 1.33 tf/m^2. The deadweight sum of the three pontoons (or floating modular units) is 146,880 tf based on the specific weight of 0.167 tf/m^3, which was obtained from measurements taken during the "Mega-float Project" [7, 8]. It follows

Fig. 13 Bird's eye view of "Singapore Water City"

Fig. 14 Elevational view of high-rise towers on a floating foundation

that the weight of the pontoon of the Singapore Water City project would be 676,080 tf, plus variable loads of approximately 39,420 tf, while the total underwater volume (V) would be 715,500 m^3. The $\overline{BM}$ and $\overline{GM}$ values for this project are calculated as 348.2 m and 257.5 m, respectively, while the vertical location of the center of gravity (KG) for these floating systems is 103.1 m.

Next, assuming a drag coefficient of 2.0 and a maximum wind intensity of 60 m/s at 105 m above the water level, the total wind force on the three high-rise towers is calculated to be 163,976.4 kN [4]. Furthermore, the corresponding total wind heeling moment is calculated as 17,217,522 kN m [4].

The coordinate system is shown in Fig. 15 while the inclined condition of a floating foundation is shown in Fig. 16. The pontoon freeboard is designed to be 2.0 m while the ϕ value of the floating foundation is set to be 0.55° under serious conditions involving a wind velocity of 60 m/s. Here, the ϕ value of 0.55° is considered to be the flooding angle at which the upper deck end of the floating foundation reaches the maximum calm water level, and is thus chosen for safety to be the inclination limit.

As is shown in Fig. 17, height of a super high-rise tower (condominium) is 210 m while the width is 60 m. On the other hand, the underwater volume (V) for 3 pontoons (715,500 m^3) is expressed as follows:

$$\text{V (underwater volume)} = (\text{L} \times \text{B} \times \text{D} + \text{Vadd}) \times 3 \text{ pontoons} \tag{16}$$

$$\text{Vadd} = 75\,\text{m} \times 75\,\text{m} \times 10\,\text{m} \tag{17}$$

Here, Vadd is the volume of the additional bottom buoyancy unit (see Fig. 17), and B = L.

Fig. 15 Coordinate system of a floating foundation

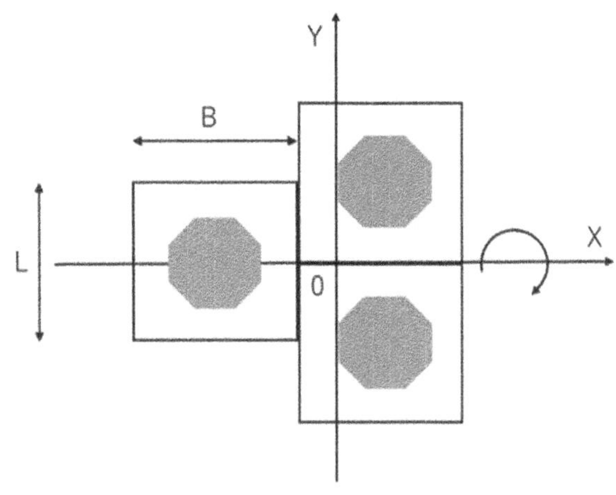

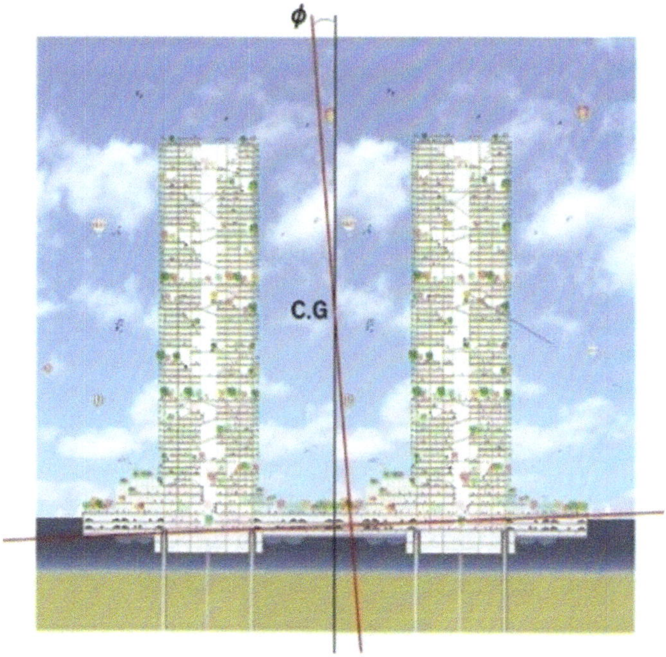

Fig. 16 Inclination of a floating foundation

Problem

To find the length (L) and the draft (d) of each pontoon when the ϕ value is 0.55° under a wind velocity of 60 m/s with the following given values:

- KG: 103.115 m (fixed)
- Underwater volume (V): 715,500 m^3 (fixed)
- Wind heeling moment (Mw): 17,217,522 kN m (fixed)

Solution

- The "Required GM" value can be obtained by Eq. (8) as follows:
  Here, the ϕ value is given as 0.55°.

$$\begin{aligned} \text{Req. GM } (0.55^\circ) &= \frac{17,217,522\,\text{kN m}}{715,500\,\text{m}^3 \times 9.8\,\text{m/s}^2 \cdot \sin 0.55^\circ} \\ &= 255.800\,\text{m} \end{aligned} \tag{18}$$

Then, the KM $\left(= \text{KG} + \overline{\text{GM}}\right)$ becomes 358.915 m (= 103.115 m + 255.8 m).

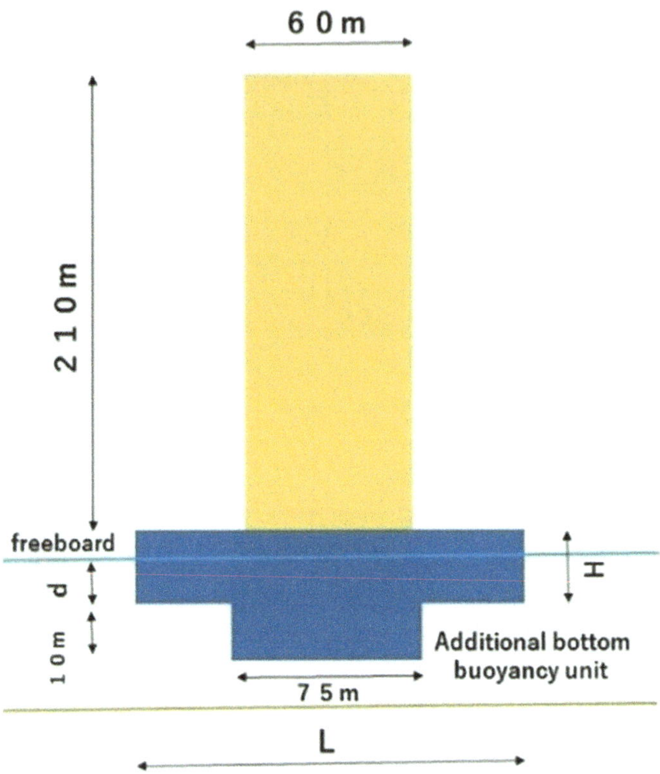

Fig. 17 Principal dimensions of 51-story high-rise tower on a pontoon

Parameter Update Process

- First trial: The initial length L1 (tentative) is set to 100.0 m with a underwater volume (V) of 715,500 m^3, and the draft d1 becomes 18.225 m.
  Then, the KB location 1 becomes 15.7840 m.
- Accordingly, the $\overline{BM}$ (= KM − KB) becomes 343.131 m (= 358.915 m − 15.7840 m).
- Then, the moment of inertia Ix can be calculated as follows:

$$Ix = \overline{BM} \times V = 343.131\,m \times 715,500\,m^3 = 245,510,230.5\,m^4 \tag{19}$$

- Since the moment of inertia (Ix) is 245,510,230.5 m^4, the second estimation of length L2 becomes 134.5092 m using the relation shown in Eq. (6), and the draft (d2) becomes 10.0731 m. Then, the KB location 2 changes to 12.6694 m (see Table 3).
- The trials are repeated until convergence is obtained.

Table 3 Repeated trials for length (L) and depth (d) for "Singapore Water City"

Trial	L (m)	d (m)
1	100.0	18.225
2	134.5092	10.0731
3	134.8134	10.0277
4	134.8151	10.0274

From these computational results, the final dimensions of the L and d of a pontoon are determined to be 135 m and 10 m, respectively.

5 Concluding Remarks

Looking a century or more into the future, the most important keyword used in relation to urban areas will undoubtedly be "sustainability," which is a concept that covers all the actions of humankind and the necessity of making efforts to build urban areas that are capable of withstanding global environmental hazards [8]. The authors believe that the adoption of floating solutions in urban areas can provide solutions to most water-related environmental hazards, and thus has the potential to create a future style for cities worldwide. However, because there are major differences in the purposes and/or missions of floating platform types, it will be difficult to design floating platforms for civil purposes if the process is restricted to applying current technology and established designs, such as those used in the offshore floating platforms operated by the oil industry [9]. Therefore, to minimize these difficulties, the authors have introduced a practical and useful process for designing floating structures destined for civil use. In terms of safety, the most important part of this process is avoiding large inclinations and establishing adequate static stability under the various simultaneous multi-loading conditions that floating structures can be expected to experience.

To develop a practical method for calculating the static stability of a floating structure while aiming to ensure the comfort and safety of its residents, we formulated a process based on tools of the so-called "Required GM" method, which was originally developed for semi-submersible type oil rigs in the harsh environments of the open sea [6]. In this paper, the floating foundation problem was considered in a general fashion, and methods used in the process were presented and discussed along with some useful studies on static stability. Finally, we proposed a computation design process that adequately considers principal dimensions (such as length, width, and draft) at the initial planning stage.

This paper presented two parts with different aspects. The first was basic knowledge on static stability in relation to the overall design process at the conceptual design stage, while the second discussed practical trials for determining the basic configurations of two floating structure types. In the first part, the importance of both the $\overline{GM}$ and KM values in maintaining appropriate stability levels for floating structures was confirmed via comparisons among calculated results. From these results, it was found that the KG location has a considerable effect on the inclining angle and that the $\overline{GM}$ value is the critical factor governing the inclination of a floating platform. In the second part, two study cases, one involving small-sized floating houses and the other showing large high-density floating residences, were presented as practical trial examples. The presented results show that our proposed basic planning and design methodology is imminently applicable to various floating structure types. The authors hope that architects and/or engineers will find our proposed design procedure, as well as the floating architectures studies set forth in this paper, to be of practical use.

Acknowledgements The authors are grateful to Business Strategy Promotion Center at Chodai Co., Ltd., for their financial support. Some of the content within this paper was quoted from work conducted by Ms. Yuka Saito, which was carried out as research in support of her thesis presented at Tokyo Metropolitan University during the 2015–16 period.

Appendix A

In general, steady wind force (ΔF_Z) at z m can be obtained by the following equation:

$$\Delta F_Z = P_Z \cdot A = \frac{1}{2}\rho \cdot U_Z^2 \cdot C_D \cdot A \tag{20}$$

where P_Z is the wind pressure (kg/m^2) at z m, C_D is a drag coefficient, A is the projected area (m^2), and ρ is the air density.

In general, accurate C_D values are obtained via wind tunnel tests.

Here, it should be note that wind velocity is measured at the height of 10 m and the average value over a period of 10 min is used. The wind velocity (U_Z) changes along the vertical location are shown in Fig. 18, and can be estimated by the following equation. Note that wind velocity is lower near the ground due to friction.

$$U_Z = U_{10}\left(\frac{z}{10}\right)^{\alpha} \tag{21}$$

where U_{10} is the wind velocity at a height of 10 m, and α is the surface roughness.

It is known that the value of α is 1/7 on the sea surface and 1/4 in an urban area.

Appendix B

According to Ref. [6], the static righting lever GZ at an inclining angle φ of a wall sided vessel is expressed as follows:

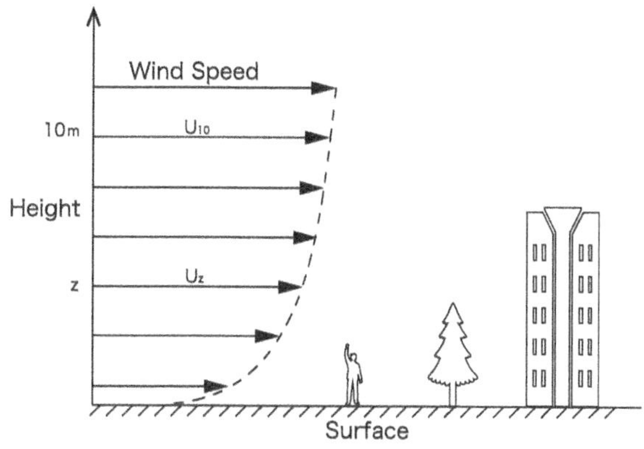

Fig. 18 Schematic view of wind speed

$$\overline{GZ}(\phi) = \sin\phi\left(\overline{GM} + \frac{1}{2}\overline{BM}\tan^2\phi\right) \tag{22}$$

Supposing an overturning moment due to the sum of various components $\left(\sum_i M_i\right)$ and the righting moment (M_R), the following expression is established:

$$\overline{GZ}(\phi)\cdot W = M_R(\phi) = \sum_i M_i(\phi) \tag{23}$$

Equation (22) is then multiplied by the weight of a floating foundation (W) to give

$$W\cdot\overline{GZ}(\phi) = \sum_i M_i(\phi) = W\cdot\sin\phi\left(\overline{GM} + \frac{1}{2}\overline{BM}\tan^2\phi\right) \tag{24}$$

After rearranging, we have

$$W\cdot\sin\phi\cdot\overline{GM} = \sum_i M_i(\phi) - \frac{1}{2}W\cdot\sin\phi\cdot\overline{BM}\cdot\tan^2\phi \tag{25}$$

Dividing Eq. (25) by $W\cdot\sin\phi$ on both sides gives

$$\overline{GM}(\phi) = \frac{\sum_i M_i(\phi)}{W\cdot\sin\phi} - \frac{1}{2}\overline{BM}\times\tan^2\phi. \tag{26}$$

When ϕ is small, the second term of Eq. (26) can be ignored. Thus, the value of the $\overline{GM}$ which is required (Req. GM) for a small inclination is given by the following equation:

$$\begin{aligned} \text{Req. GM }(\phi) &= \frac{\sum_i \mathrm{M_i}}{\mathrm{W} \cdot \sin \phi} \\ &= \frac{180.}{\pi \cdot \Phi \mathrm{d}} \times \frac{\sum_i \mathrm{M_i}}{\mathrm{W}} \end{aligned} \tag{27}$$

or

$$\Phi \mathrm{d}\ (\text{in degrees}) = \frac{180.}{\pi \cdot \mathrm{GM}} \times \frac{\sum_i \mathrm{M_i}}{\mathrm{W}} \tag{28}$$

where Φd is the inclining angle in degrees.

Thus, a "required GM" method that evaluates and compensates for various heeling moments under a variety of combined environmental loadings is proposed.

References

1. Intergovernmental Panel on Climate Change (2014) IPCC fifth assessment report—climate change 2014. Intergovernmental Panel on Climate Change
2. Nakajima T, Umeyama M (2019) An integrated floating community based upon a hybrid water system: toward a super-sustainable water city. Lecture notes in civil engineering, vol 41. Springer, Heidelberg, pp 309–327
3. Nakajima T, Umeyama M (2015) A new concept for the safety of low-lying land areas from natural disasters. J Ocean Eng Mar Energy 1(1):19–29 (Springer)
4. Nakajima T, Yamashita Y, Harada S, Andoh (2020) Sustainable water city in Singapore. In: Oceans'20 MTS/IEEE international symposium Singapore, 11–14 August (to be published)
5. Maarten K (2010) Connecting modular floating structures. TU Delft M.S. thesis, A7-3 A9-8
6. Takarada N, Nakajima T, Inoue R (1986) A phenomenon of large steady tilt of a semisubmersible platform in combined environmental loadings. In: 3rd international conference on stability of ships and ocean vehicles, 22–26 September 1986, Gdansk
7. Sato C (2003) Result of 6 years' research project of mega-float. Ocean space utilization technology. Ministry of Land, Infrastructure and Transport of Japan, pp 436–442
8. Nakajima T, Umeyama M (2013) Water city as solution to escalating sea level rise in lower-lying areas. In: Oceans'13 MTS/IEEE international symposium, San Diego, California, 23–26 September 2013
9. Hendriks TM, Mendonca Santos J (2018) Challenges in stability assessment of offshore floating structures. In: Guedes Soares, Teixeira (eds) Maritime transportation and harvesting of sea resources, October 2017. Taylor & Francis Group, London, p 13. ISBN 978-0-8153-7993

Towards an Understanding of the Stability Assessment of Floating Buildings

Artur Karczewski

Abstract One of the most important aspects of the design of floating facilities such as ships, floating offshore structures or floating houses is stability. Its impact on both general safety and operational aspects renders it a fundamental consideration already in preliminary design stages. Usually, the concept of sufficient stability of floating buildings is associated with the ability to keep an allowed heel angle and residual freeboard, despite the action of the heeling moment. Once in the water, a floating object has to withstand different environmental conditions. It is always acted upon by forces from various factors. The main sources of load are wind, waves and the shift of inhabitants. However, the challenges in the assessment of stability are also connected to how a floating structure responds to these sources of load and also to the method of performing a stability analysis. This paper focuses on wind load and revisits several challenges encountered in the stability calculations and in the prediction of the behavior of floating buildings. A review of the current regulations was also performed in this respect. The obtained results indicate the necessity and also direction of further considerations related to the safety of stationary floating objects.

Keywords Stability · Method · Stationary floating object

1 Introduction

The requirements for the stability of stationary floating buildings focus on assessing the size of the freeboard and the maximum angle of heel due to the action of the heeling moment.

The minimum freeboard is determined for floating on an even keel. The residual freeboard is the distance between the waterline and the deck of a floating object in

A. Karczewski (✉)
Faculty of Mechanical Engineering and Ship Technology, Gdańsk University of Technology, Gdańsk, Poland
e-mail: artkarcz@pg.edu.pl

Ł. Piątek et al. (eds.), *WCFS2020*, Lecture Notes in Civil Engineering 158,
https://doi.org/10.1007/978-981-16-2256-4_18

inclination. The heeling moment is usually caused by the action of wind or a concentrated force (e.g. movement of the people inside). It should be noted, that the allowed heel angle is very small and that generally, the safe range for heeling is several degrees, and it does not extend more than 4–5° in most circumstances. This is a completely different situation from the case with ships for which large heel angles are considered first and foremost and the initial range of stability is not a cause for concern.

However, just as for ships, stability assessment for floating buildings is carried out under the following assumptions:

- the object is in static equilibrium,
- external forces are static (wind strength and direction are constant over time),
- external forces act perpendicular to the long axis and parallel to the water surface,
- the float is watertight and the center of mass of the object is fixed.

In practice, this means that only transverse stability (rotation along the longer axis) is ultimately considered. In addition, it is also assumed, that the object floats on an even keel, regardless of tilt. Indirectly, this means that the buoyancy center is fixed so as to not translate longitudinally during the object's heel. This is true when the rotation is along an axis of symmetry. However, the real movement of a floating object is three-dimensional and the buoyancy center can move longitudinally due to the asymmetrical distribution of buoyancy relative to the plane of rotation [1].

This is especially important for objects, which are arbitrarily placed relative to the wind and can neither change their position nor their orientation to it. This problem has been noticed in ships and offshore structures [1–5]. Stability calculations should be made for free floating units regardless of the orientation of their heeling axis. From the results thus obtained, the most critical axis direction should be selected. A stability analysis should be carried out for this weakest case.

This is the first point of concern taking into consideration the technical regulations currently in force (Table 1). At present, the angular movements of a free floating object are not taken into account, and this can significantly affect the safety of houses on the water.

Secondly, if the main factor which causes tilting of a floating building is, for example, wind, then it can be assumed that the direction of its action relative to the main axes of the building is usually neither perpendicular nor parallel to them. For this reason, the external heeling moment is distributed into components with respect to these individual axes, and the movement of the object is multidirectional.

In addition, floating buildings differ in shape, size, various dimensions and in the relationship between these dimensions. There are cuboids (Fig. 1a, b), spheroids (Fig. 1d) and buildings which can change their topology (Fig. 1c). They are not always symmetrical. This, in itself, poses a challenge to the computational methods used in stability analysis.

This article will consider the behavior of a stationary floating object with regard to changing wind direction. Reducing the stability problem to a response to wind

Table 1 Freeboard, heel angle and design wind pressure requirements according to different guidelines

Title	Region	Min. freeboard	Max. heel angle	Min. residual freeboard	Wind pressure
		Fb	*φ*	*Fbr*	*q*
		(m)	(°)	(m)	(Pa)
Building code for float homes [6]	Alaska, USA	0.36 (14″)	4	1/2 Fb	[a]
Regulation of the construction and maintenance of floating homes [7]	California, USA	0.38 (15″)	4	1/3 Fb	480
British Columbia float home standard [8]	British Columbia, Canada	0.40	5	1/2 Fb	275
Technical regulation on the stability, buoyancy, etc. of houseboats and floating structures [9]	Denmark	0.50	4	1/3 Fb	500
Queensland development code. MP 3.1 floating buildings [10]	Queensland, Australia	0.40	–	0.25	[a]
NTA 8111 Drijvende bouwwerken [11]	Netherlands	0.30	4	0.15	[a]

[a]Referenced to building standards

Fig. 1 Examples of stationary floating objects: **a** floodule floating house, **b** the floating house in the Czerniakowski Port in Warsaw, **c** Watervilla Middelburg, **d** Rotterdam Floating Pavilion

pressure, excluding other loads, can be justified by the relatively higher influence of this factor on stability and the aforementioned variability in relation to the main axes of the structure. Therefore, such a reduction is assumed to be acceptable during the theoretical research of this investigation into the relationship between safety and the calculation method used for the assessment of stability.

The case of a free floating object has been considered. To investigate the basic phenomena, the calculations have been carried out without taking anchoring into consideration.

2 Stability Calculation

The basic method for determining the size of the angle of heel, φ, and the residual freeboard, *Fbr*, is the righting moment arm, *GZ*, as a function of the heel angle. The effect of external forces on the object are angular displacements, corresponding to the point of equilibrium between the heeling moment and the righting moment. *GZ* is understood as the distance between the action of the buoyancy force, *B*, and the force of gravity, *G*, in still water at a given heel angle. As the floating object is inclined through small angles of heel, the lines of buoyant force intersect at a point called the metacentre, *M*, which is the instantaneous point of rotation (Fig. 2).

For a ship, calculations are carried out on the assumption that the heeling moment acts on the longitudinal axis of the object, while the righting moment is related to the longitudinal balance axis. This discrepancy has little effect on the final results with regard to ships. The reason is the size of the object itself and the large *L/B* ratio, which is about 6. For a floating building, for which the *L/B* ratio tends towards 1, the calculation technique is important. It significantly affects the size of the forces that want to heel and want to trim the object at the same time. For a square building and an oblique wind, the trimming and heeling forces are equal to each other. The building moves in space within two degrees of freedom: it trims and

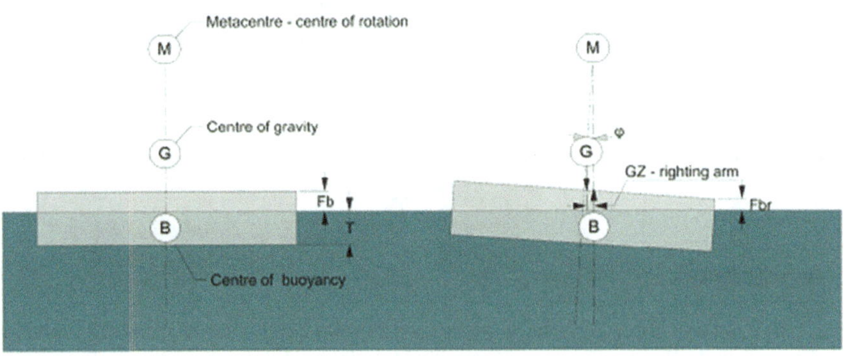

Fig. 2 The parameters using to assess a stability of stationary floating buildings

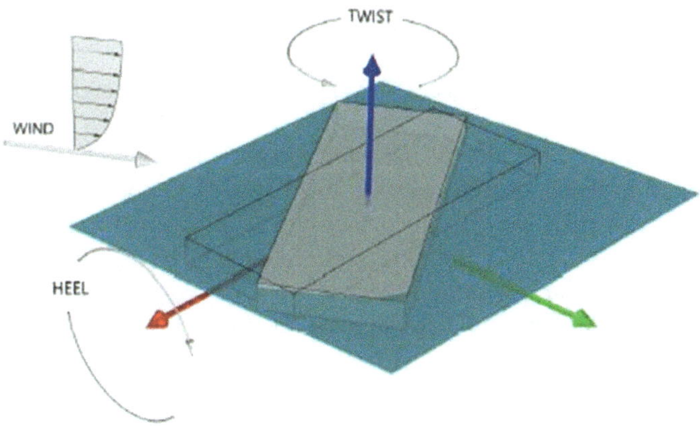

Fig. 3 Definition of twist and heel

heels at the same time. The wind moment is parallel to the wind pressure plane which is perpendicular to the wind direction in an upright position, and the axis of rotation is the intersection of the waterplane and the wind pressure plane. In other words, the object positions itself in such a way that the final potential energy is minimal, i.e. so that the rotation work is minimal [1, 3, 4]. This movement is illustrated by the so-called minimum stability curve. There are several methods for determining this curve. Each of them is quite complicated and laborious. Since this work does not discuss purely theoretical issues related to the stability of a free-floating object and focuses mainly on safety aspects, it has been assumed that the curves of the righting arms of objects rotating around an axis parallel to the axis of the heeling moment are to be used for analysis; therefore, before heeling, the object is initially twisted in the plane of floating by the angle corresponding to the direction of the wind (Fig. 3).

3 Wind Heeling Moment

The wind causes forces on the walls of an object. In shipbuilding, only the forces on the walls exposed directly to wind are considered. The value of the force depends on the shape of an object, on the wall area and on the wind velocity. The generalized relationship for this force can be written as follows (based on [12]):

$$F_W = C \cdot q \cdot A \tag{1}$$

where *C*—shape coefficient; *q*—wind pressure; *A*—projected area of the ship normal (perpendicular) to the direction of the force.

The wind heeling moment for floating buildings is expressed by the following formula (based on [13]):

$$M_H = F_w \cdot (D + 0.5 \cdot (H - T)) \tag{2}$$

where D—depth of floating system (m), H—superstructure height (m), T—draft (m).

It is assumed that the value of the wind moment does not change as the heel angle changes, because such changes are small enough to not affect the results.

The amount of wind pressure for which the stability criteria are checked is a contractual matter. It is most often associated with the local conditions of the geographical region for which the given regulations are in force. Therefore, the value of wind pressure used in a design varies between 275 and 500 Pa (Table 1).

4 Shape Coefficient

One of the basic issues in calculating the total value of wind stress on the exposed faces of an element is determining the shape coefficient. This is not a new problem, and it has been studied for many years in many scientific papers [14, 15]. For design purposes, the shape factors for different structures are given in Eurocode [16]. Based on this publication, the shape coefficient for smooth members with rectangular cross-section with sharp edge may be taken as (Fig. 4):

$$\begin{cases} C_{S1} = 2 \times \cos\alpha & \\ C_{S2} = \left(\frac{L+B}{L}\right) \times \sin\alpha & \text{for } 1 \leq L/B \leq 2 \\ C_{S2} = 1.5 \times \sin\alpha & \text{for } L/B > 2 \end{cases} \tag{3}$$

where α—angle between the direction of the wind and the axis of the exposed member or surface.

The shape coefficient of oblique direction can be calculated using the following relationship:

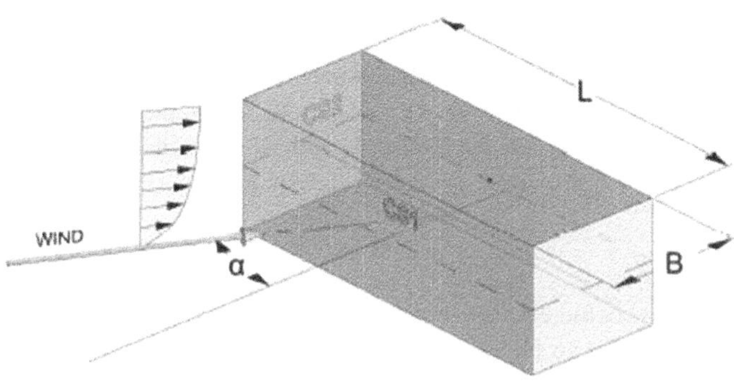

Fig. 4 Shape coefficient on rectangular cross-sections

$$C = \sqrt{C_{S1}^2 + C_{S2}^2} \tag{4}$$

5 Wind Area

The wind can blow in different directions relative to a floating building. The situation in which it is from a direction perpendicular to the longer axis is a special case, although it is accepted in commonly used technical regulations. The projected surface of the exposed elements of a floating building is treated as the wind pressure surface. This approach comes from the shipbuilding industry where it makes sense because the ratio of length to width of the ship is relatively large. However, it is different in the case of floating buildings. Firstly, they don't have to be built on a rectangular plan which is generally a slender shape. Secondly, their proportions are different and can vary between 1 and 6. Therefore, if the direction of the wind is changing, it can easily be seen that the side projection area, *A*, will be different. Figure 5 shows the correlation of the area, *A*, and the ratio *L*/*B* (for constant ratio *H*/*L* equal to 1). The ratio *L*/*B* = 1 is typical for off-shore structures, while the ratio *L*/*B* = 6 represents a typical seagoing vessel. The *L*/*B* ratio for floating buildings

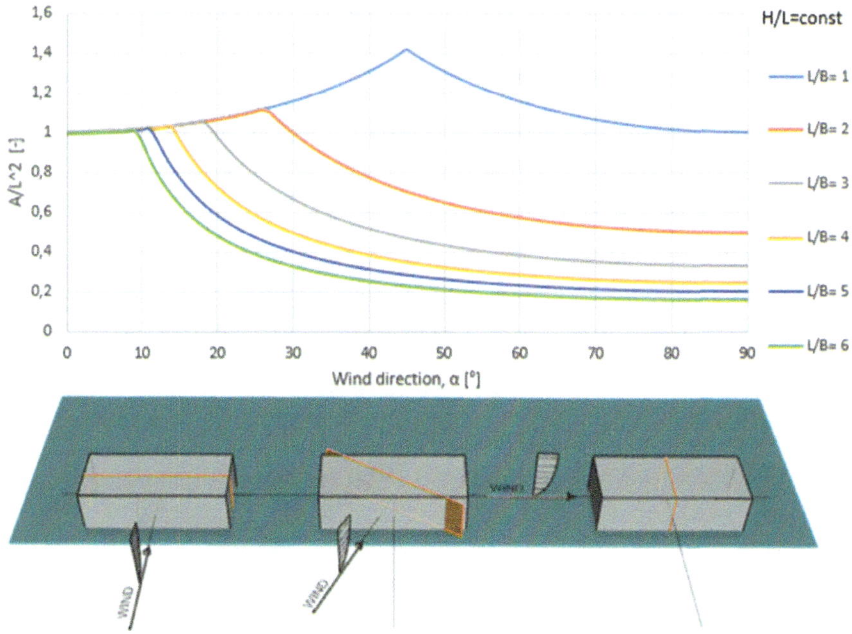

Fig. 5 Increase of wind area relative to twist

typically falls between these values. For the considered, rectangular object, depending on the *L*/*B* ratio, the wind area is assumed to be the maximum value at the angle of rotation corresponding to the diagonal.

6 Stability Evaluation

To assess the impact of the selection of the heeling axis on the results of the stability assessment, research scenarios were adopted for the selected theoretical model of a floating object, including its twist to the direction of the wind in a range of horizontal angles from 0° to 90° with a step of 10°.

In the research, the object consists of a rectangular floating system (floating device, float) and an open superstructure (Fig. 6). Viewed simply, the upper part applies loads (gravity and wind pressure), while the lower part generates reactions (buoyancy and righting moment) in the system.

For such an object, the choice of the stability calculation method will have no effect on the results obtained for extreme twist angles. However, it will be noticeable at intermediate angles where heeling waterplanes are not symmetrical.

General characteristics of the object are presented in Table 2. It should be noted that this is a relatively low draft facility, intended for very shallow waters. Due to its relatively light weight, this construction also appears to be susceptible to external loads such as wind, which is especially interesting in this study. At the same time, such objects are becoming more and more popular with the expansion of both water and amphibious construction in new geographical locations [17]. The ratio of the main dimensions *L*/*B*, which equals 2.67, can also be regarded as representative for this type of building [18]. This value is also close to the midpoint of the range 1–6. Thanks to this, it is possible to examine the behavior of a typical object in the wind.

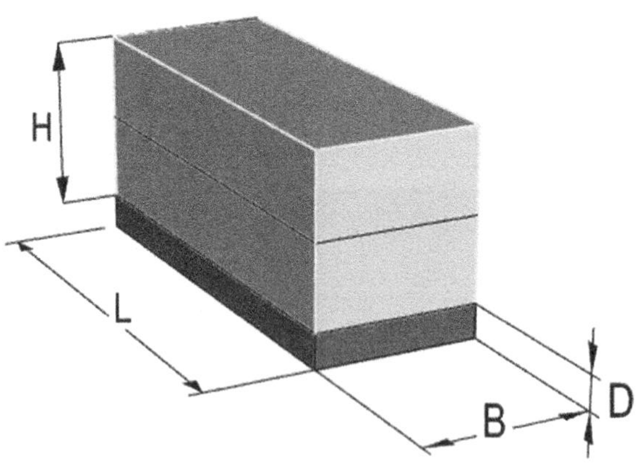

Fig. 6 Main dimensions of examined object

Table 2 General characteristics of examined model

Characteristic	Sym. (m)	Value
Float and superstructure length	L	16.0
Float breadth	B	6.0
Float depth	D	1.2
Superstructure height	H	6.0
Draft	T	0.75
Vertical center of gravity	VCG	2.04

7 Results

The righting arm curve for individual cases of twisting is presented in Fig. 7. In each case, the curves vanish at the angle where the extreme point of the float enters the water. It has been assumed that flooding of the deck is not acceptable because its immersion poses a threat to the people and things onboard. The minimum range of stability occurs for a twist angle of 80°, which is consistent with the results of other studies for objects with similar geometric characteristics, such as barges [3]. On the other hand, for the initial angles (up to 30°), although the range of the allowable angle of heel varies (for example at 0° it is 8.5 and at 30° it is 6.8), the increases in the value of the righting arms are small. It is worth noting that at angles of 0° and 10°, the values of the righting arms are practically the same; only the safety range changes. For 10°, the value drops by about 0.4°.

The value of wind pressure corresponding to the safety criteria for selected regulations has been subsequently determined based on the stability characteristics

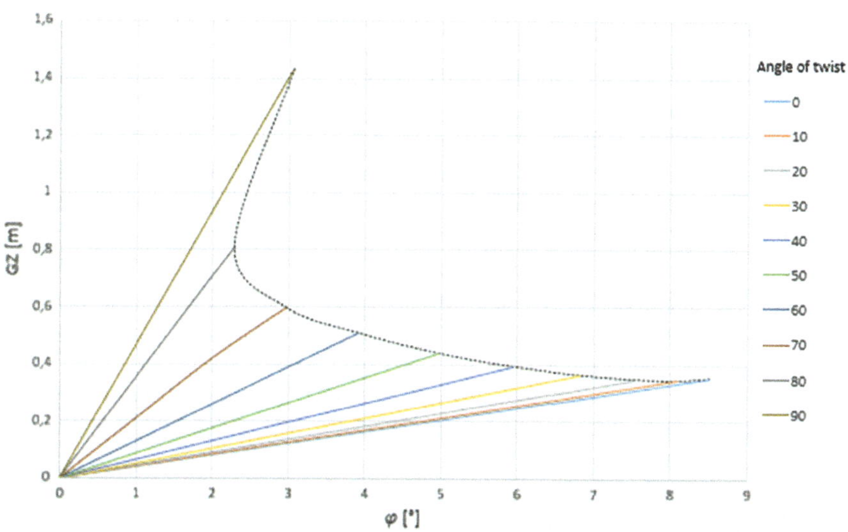

Fig. 7 Righting arm curves for different angle of twist

Table 3 Wind pressure

Twist (°)	Criteria				
	Vanish stab.	Heel—4°	Heel—5°	1/2 *Fb*	1/3 *Fb*
	Wind pressure (Pa)				
0	458	213	266	293	195
10	433	212	265	284	189
20	400	210	264	259	172
30	556	325	407	345	230
40	722	484	606	415	277
50	902	727	–	465	310
60	1,123	1,123	–	499	332
70	1,377	–	–	521	347
80	1,917	–	–	533	355
90	3,410	–	–	537	358

obtained for individual twist angles. In a few cases, the obtained results are lower than the values indicated in individual provisions (Table 3). This is the case even if the wind is perpendicular to the long axis of the building. However, this is more of a design problem than a formal one and as such, it requires design changes. With regard to regulations, it can be noted that the minimum values of the computational wind pressure occur for the angle of 20°. This value is very close to the diagonal angle of the tested object, which is 20.5°.

8 Discussion

This article discusses the stability of a floating buildings, taking into account the change in the wind direction and the possibility of building rotation along an axis which is not orthogonal to the main axes of the structure. The free-floating object method has been adopted to assess the stability. This is a relatively new issue, which has been studied mainly in works dealing with ships and offshore facilities by such authors as Pawłowski [1], van Santen [2–4], Vassalos et al. [5], and others. However, this issue has not been discussed before in relation to floating buildings. This is most likely due to its apparent triviality (uselessness) to the maritime industry. Often, angles of heel, that are less than 5°, are disregarded because they coincide with the swing range of a ship (they overlap with the period of the ship's own sway). This phenomenon does not take floating buildings into account because they are most often sited in shallow and sheltered waters and additionally, such facilities are stationary.

The results obtained show that the calculation method significantly influences the stability characteristics of a tested object, and thus its safety. The methods of calculating the values of the wind force, which are adapted to the current regulations, differ significantly from those used to obtain the calculations in this article. This indicates that, as in the case of offshore structures, the problem is not negligible and may significantly affect safety with respect to living on floating objects.

9 Conclusions

- Taking into account the possibility of free positioning of an object in relation to external load in the calculations gives less favorable results than the currently adopted heeling procedure relative to the longitudinal axis.
- Because the stability assessment is based on the principle of a free-floating object, it significantly reduces the range of permitted movements of the object, which means there is a reduction in the safety range.
- The current technical regulations regarding the stability of floating objects follow the regulations for ships. They should be carefully assessed and possibly refreshed to take into account the current state of knowledge and computational possibilities.
- The design of stationary floating buildings is primarily the responsibility of architects. Because the issue of safety, as has been shown in the article, is neither simple nor straightforward, it is advisable that they utilize the help of naval architects in this work. At the same time, for the same reasons, any new stability regulations and procedures should be accompanied by an explanation of how to perform the calculations and how to meet the various demands encountered. This is a recommendation that has also appeared in works relating to offshore constructions.

References

1. Pawłowski M (2016) Stability of free-floating ship. Finally report. Polish Register of Shipping, Gdansk
2. van Santen JA (2009) The use of energy build up to identify the most critical heeling axis direction for stability calculations for floating offshore structures. In: Santos Neves MA, Belenky VL, de Kat JO, Spyrou K, Umeda N (eds) Contemporary ideas on ship stability and capsizing in waves. Springer, Dordrecht, pp 193–216
3. van Santen JA (2013) Problems met in stability calculations of offshore rigs and how to deal with them. In: Proceedings of 13th international ship stability workshop, Brest, pp 9–17
4. van Santen JA (2018) Challenges in stability of offshore floating structures. In: Guedes Soares C, Teixeira (eds) Maritime transportation and harvesting of sea resources, Taylor & Francis Group, London, pp 13–21
5. Vassalos D, Konstantopoulos G, Kuo C, Welaya Y (1985) A realistic approach to semisubmersible stability. In: SNAME transactions, vol 93. Society of Naval Architects & Marine Engineers, pp 96–128

6. City and Borough of Sitka (2019) Building code for float homes. In: Home rule charter of the city and borough of Sitka. https://codepublishing.com/AK/Sitka/html/Sitka19/Sitka1915.html . Accessed 22 Nov 2020
7. Marin County (2016) Regulation of the construction and maintenance of floating homes. In: California municipal code. https://library.municode.com/ca/marin_county/codes/municipal_code. Accessed 22 Nov 2020
8. Float Home Standard (2003) British Columbia float home standard. https://www2.gov.bc.ca/assets/gov/farming-natural-resources-and-industry/construction-industry/building-codes-and-standards/guides/2003_float_home_standard.pdf. Accessed 22 Nov 2020
9. Danish Maritime Authority (2007) Technical regulation on the stability, buoyancy, etc. of houseboats and floating structures. https://www.dma.dk/Vaekst/Rammevilkaar/Legislation/Pages/Technical-regulations.aspx. Accessed 01 Aug 2020
10. Department of Housing and Public Works (2007) Floating buildings. In: Queensland development code. https://www.hpw.qld.gov.au/__data/assets/pdf_file/0011/4313/mp-3-1-floating-buildings.pdf. Accessed 01 Aug 2020
11. Nederlands Normalisatie-Instituut (2011) NTA 8111 Drijvende Bouwwerken. NEN, Delft
12. Det Norske Veritas (2010) Recommended practice, DNV-RP-C205, environmental conditions and environmental loads. https://rules.dnvgl.com/docs/pdf/dnv/codes/docs/2010-10/rp-c205.pdf. Accessed 1 Aug 2020
13. Polish Register of Shipping (2019) Rules for the classification and construction of Inland waterways vessels, part IV, stability and freeboard. https://www.prs.pl/uploads/srl_p4.pdf. Accessed 01 Aug 2020
14. Lindsey WF (1938) Drag of cylinders of simple shapes, report no. 619—National Advisory Committee for Aeronautics. https://ntrs.nasa.gov/search.jsp?R=19930091694. Accessed 1 Aug 2020
15. Haddara MR, Guedes Soares C (1999) Wind loads on marine structures. Mar Struct 12(3): 199–209. https://doi.org/10.1016/S0951-8339(99)00023-4
16. Comité Européen de Normalization (2005) EN 1991-1-4. Eurocode 1: actions on structures—part 1-4: general actions—wind actions, Brussels
17. Ostrowska-Wawryniuk K, Piątek Ł (2020) Lightweight prefabricated floating buildings for shallow Inland waters. Design and construction of the floating hotel apartment in Poland. J Water Land Dev 44(I–III):118–125. https://doi.org/10.24425/jwld.2019.127052
18. Piątek Ł (2018) Architecture of floating buildings: conditions and directions of development in Poland after 2000. Dissertation, Warsaw University of Technology

Numerical Investigation of Hydroelastic Effects on Floating Structures

Changqing Jiang, Ould el Moctar, Thomas E. Schellin, and Yan Qi

Abstract Hydroelasticity effects of an offshore floating structure comprise the combined motions and deformations of the floating body responding to environmental excitations. The review of research on hydroelasticity of very large floating structure shows that understanding the physical phenomenon has increased, but discussions of practical implications of hydroelasticity on offshore structure design are rare. Conventionally, floating structure designs are based on a rigid quasi-static analysis, meaning that the hydrodynamic loads are estimated under rigid assumption and then applied to the elastic structure regardless of structural inertia. Here, the hydroelastic behavior of a standard floating module designed within the scope of the Space@Sea project was numerically investigated, and the role of hydroelasticity in the practical assessment of a large floating structure was demonstrated. The fluid dynamics relied on a Computational Fluid Dynamics (CFD) code, and the structural responses were computed by a Computational Structural Dynamics (CSD) solver. The CFD-CSD solver was coupled using an implicit two-way coupling approach, computing the nonlinear 6-DoF rigid body motion separately from linear elastic structural deformations. First, the numerical model was validated against benchmark test data, and then a standard floating module in waves was assessed in terms of structural integrity and motions. Maximum stresses and bending moments obtained by the coupled CFD-CSD approach and the traditional rigid-quasi-static approach were compared, and the implication of hydroelasticity on the floating module was assessed. The hydroelastic criterion and the validity of a rigid a quasi-static analysis determined the effects on dynamic responses.

C. Jiang (✉) · O. el Moctar · T. E. Schellin · Y. Qi
University of Duisburg-Essen, Duisburg, Germany
e-mail: changqing.jiang@uni-due.de

O. el Moctar
e-mail: ould.el-moctar@uni-due.de

T. E. Schellin
e-mail: thomas.schellin@uni-due.de

Y. Qi
e-mail: yan.qi@stud.uni-due.de

Ł. Piątek et al. (eds.), *WCFS2020*, Lecture Notes in Civil Engineering 158,
https://doi.org/10.1007/978-981-16-2256-4_19

Keywords Hydroelasticity · Space@sea · Floating structure · CFD · CSD

1 Introduction

The majority of the world's population lives in coastal areas where space has always been a premium commodity because available land space is limited. On the other hand, the waterfront may reclaim or change the utilization of large parts of existing land spaces as the sea level rises with global warming. With the increasing need for affordable deck space at sea, various concepts of man-made islands were proposed. While sandy islands are based on land reclamation technology, they are limited to shallow waters. Floating concepts have the advantage of being suitable for deeper waters, and they can be relocated if needed. Within the Space@Sea project, a standardized floating solution for offshore space was developed, and modularity was one of the key elements of this project. By following the analogy of standardization to enlarge a floating structure using a multitude of smaller structures, this project intends to significantly reduce building and installation costs and to provide the desired flexibility of additional deck space at sea. An overall discussion of this project can be found in [1].

Because of the relatively large sizes of very large offshore floating structures (VLFS) and their deployment in near shore areas, they face extreme environmental conditions that may differ significantly from those of smaller floating structures. These extreme conditions involve waves, winds, and currents, which can induce significant hydrodynamic loads on the floating structure leading to its elastic deformation. For the linear hydroelastic response of VLFS subjected to waves, a finite element-boundary element hybrid method is generally adopted. The fluid dynamics are usually solved based on the potential flow theory, using the Green function implemented in a boundary element method (BEM) to calculate the wave-induced loads and motions. The structural dynamics can be solved using either a dry-mode method [2] or a wet-mode method [3]. Additionally, a generalized mode method proposed by Newman [4] can also be used to calculate the structural dynamics, and an application of this approach can be found in [5]. For thin-plate type VLFS, its hydroelastic responses can be solved using an analytical approach [6] or a semi-analytical approach [7]. The hydrodynamics of floating bodies are based on the potential flow theory, which is unable to implicitly consider viscous effects or complex free-surface topologies arising from steep or overturning waves, appendages emerging from the water, and water on deck. A literature survey of the research on hydroelastic analysis of pontoon-type VLFSs can be found in [8]. An alternative is the computational fluid dynamics (CFD) approach, which can reveal detailed information of viscous flow features. The different effects of wave loads from potential flow and CFD solvers on fixed and floating offshore structures were benchmarked in [9]. For moored floating offshore structures, the predicted motions and loads from potential flow and CFD solvers were compared and discussed in

[10]. Diverse linear and nonlinear fluid-structure interaction models are here described from a structural, a quasi-static, and a hydrodynamic point of view.

There are several approaches to perform a structural analysis. They require the transfer of the predicted hydrodynamic loads to the computational structural dynamics (CSD) solver. Two approaches, namely, the one-way coupling and two-way coupling techniques, were evaluated herein to incorporate predicted hydrodynamic loads from a BEM solver or a CFD solver. One-way coupling required computing wave-induced loads based on rigid-body motion using a fluid dynamics solver and then applying these forces on a wet elastic model to predict structural responses, whereby the structural deformations computed by the CSD solver are not fed back to the fluid dynamic solver. Two-way coupling allows the elastic deformations of the structure to be fed back to the fluid domain, which can be important for the computation of pressures when the structure deforms during a slamming event or when large deformations influence the flow field. Our purpose was to determine the ability of different modeling tools (i.e., viscous and potential flow solvers) to accurately predict the hydrodynamic loads on a standard floating module as well as its hydroelastic deformations. The effects of body motions, structural deformations, and nonlinearities caused by steep waves, viscosity, and green water are discussed by comparing these approaches under moored and fixed conditions. We scrutinized the differences between a viscous flow solver and a potential flow solver in terms of accuracy and computational effort for assessing the hydroelasticity of the floating module. This enabled us to better understand the applicability of these two methods.

2 Numerical Approach

To investigate effects of hydroelasticity, our numerical approach included the calculation of hydrodynamic loads using a potential flow and/or a CFD solver. The fluid dynamics solvers were coupled with structural solvers to calculate the elastic deformations.

2.1 BEM-CSD Solver

First, the hydrodynamic analysis was conducted using the BEM solver ANSYS AQWA [11], a commercial suite of hydrodynamic programs widely used in the offshore industry. According to potential flow theory, the total potential of the forces acting on a floating body can be expressed as follows:

$$\Phi(x,y,z,t) = \Phi_I(x,y,z,t) + \Phi_D(x,y,z,t) + \Phi_R(x,y,z,t) \quad (1)$$

where Φ_I is the incident wave potential representing the incident waves, Φ_D is the diffraction potential representing the disturbance of the incident waves diffracted from the body, and Φ_R is the radiation potential representing the velocity potential of the body oscillations in the absence of incident waves. In linear theory, all of the above problems are assumed to be independent of each other if the floating body motions are small. The total unsteady potential has to be satisfied in the fluid domain, on the free surface, on the submerged body surface, on the sea bed, and for far-field radiation conditions at infinity. For an incompressible, inviscid, and irrotational fluid flow, its velocity potential satisfies the Laplace equation everywhere in the fluid domain:

$$\frac{\partial^2 \Phi}{\partial x^2} + \frac{\partial^2 \Phi}{\partial y^2} + \frac{\partial^2 \Phi}{\partial z^2} = 0 \tag{2}$$

The boundary condition at the sea bed of water depth h is:

$$\frac{\partial \Phi}{\partial z} = 0 \quad \text{for: } z = -h \tag{3}$$

For the free surface dynamic boundary condition is:

$$\frac{\partial^2 \Phi}{\partial t^2} + g\frac{\partial \Phi}{\partial z} = 0 \quad \text{for: } z = -h \tag{4}$$

The boundary condition at the surface of the rigid body, S, plays a very important role. The velocity of a water particle at a point at the surface of the body is equal to the velocity of this body point itself. The outward normal velocity, v_n, at a point $P(x, y, z)$ at the surface of the body is given by:

$$\frac{\partial \Phi}{\partial n} = v_n(x, y, z, t) \tag{5}$$

The radiation condition states that as the distance, R, from the oscillating body becomes large, the potential value tends to zero:

$$\lim_{R \to \infty} \Phi = 0 \tag{6}$$

The forces $\vec{F}$ and moments $\vec{M}$ follow from an integration of the pressure, p, over the submerged surface, S, of the body:

$$\begin{aligned} \vec{F} &= -\iint_S (p \cdot \vec{n}) \cdot dS \\ \vec{M} &= -\iint_S p \cdot (\vec{r} \cdot \vec{n}) \cdot dS \end{aligned} \tag{7}$$

The hydromechanical forces $\vec{F}$ and moments $\vec{M}$ can be split into four parts:

$$\begin{aligned}\vec{F} &= \vec{F}_R + \vec{F}_W + \vec{F}_D + \vec{F}_S \\ \vec{M} &= \vec{M}_R + \vec{M}_W + \vec{M}_D + \vec{M}_S\end{aligned} \tag{8}$$

where, $\vec{F}_R$ and $\vec{M}_R$ denote waves radiated from the oscillating body in still water; $\vec{F}_W$ and $\vec{M}_W$, approaching waves on the fixed body; $\vec{F}_D$ and $\vec{M}_D$, diffracted waves of the fixed body; $\vec{F}_S$ and $\vec{M}_S$, hydrostatic buoyancy in still water. The wave-induced hydrodynamic pressures resulting from the hydrodynamic analysis (AQWA) are directly mapped to the structural analysis model (ANASYS STRUCTURAL) using Workbench Interface to simulate the realistic structural behavior of a floating body subject to wave loading. The process of the integrated analysis approach is schematically shown in Fig. 1.

2.2 *CFD-CSD Solver*

Although potential flow solvers are widely used in marine hydrodynamics because of their robustness and computational efficiency, they fail to accurately predict oscillatory motions and high impact scenarios of floating structures at their natural frequencies because viscous effects are not implicitly considered. Additionally, they are also challenging to simulate complex free-surface topologies, such as steep or overturning waves and green water flows. A reliable assessment of structural dynamics of floating offshore structures requires an accurate prediction of hydrodynamic loads and nonlinear effects. An alternative is to simulate the free surface flows by coupling the Navier-Stokes equations with a Volume of Fluid (VOF) method. A finite volume method (FVM) discretizes the solution domain, using a finite number of arbitrarily shaped control volumes. Navier-Stokes equations describe the dynamics of a viscous and incompressible flow. They draw upon

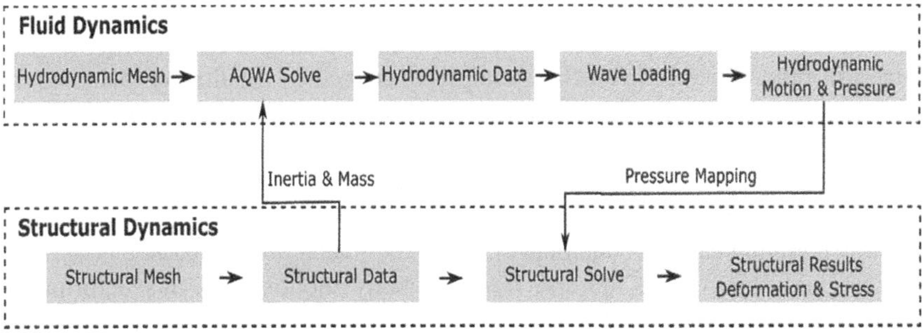

Fig. 1 The coupling procedure of the hydrodynamic and structural analysis for the BEM-CSD solver

the formulation of mass and momentum conservation. In integral vector notation, they are written as follows:

$$\frac{\partial}{\partial t}\int_{V}\rho dV+\int_{S}\rho\vec{u}\cdot\vec{n}dS=0 \tag{9}$$

$$\frac{\partial}{\partial t}\int_{V}\rho\vec{u}dV+\int_{S}\rho(\vec{u}\vec{u})\cdot\vec{n}dS=\int_{S}\vec{T}\cdot\vec{n}dS+\int_{V}\rho\vec{b}_{f}dV \tag{10}$$

where $\vec{u}$ is the fluid velocity field vector, $\vec{n}$ is the normal vector of S representing the surface area of the control volume V, $\vec{T}$ is the stress tensor, and $\vec{b}_f$ is a vector describing a body force per unit mass. The transport of turbulent momentum is described by time averaged and fluctuating terms for the flow quantities. The instantaneous local density, ρ, and viscosity, μ, are expressed in terms of the water volume fraction, α, as follows:

$$\rho=\alpha\rho_{w}+\rho_{a}(1-\alpha) \tag{11}$$

$$\mu=\alpha\mu_{w}+\mu_{a}(1-\alpha) \tag{12}$$

where subscripts a and w represent air and water, respectively. A linearized momentum component is first solved using an existing pressure and mass transport through the cell faces. Then an algorithm (SIMPLE) is used to correct the pressure. The solid domain is discretized into finite elements, and solution to the partial differential equation is obtained by imposing a set of boundary conditions. The governing equations for the dynamic response of an elastic body is idealized using finite elements, as follows:

$$\mathrm{M}\ddot{\vec{u}}+\mathrm{C}\dot{\vec{u}}+\mathrm{K}\vec{u}=\vec{R} \tag{13}$$

M, C and K are the mass, damping and stiffness matrices; $\vec{R}$, the external load vector applied; $\ddot{\vec{u}}$, $\dot{\vec{u}}$, and $\vec{u}$, the acceleration, velocity and displacement vectors of the finite element nodes. The equation is solved using an implicit integration scheme, whereas the equation is integrated over discretized time step Δt. The numerical fluid-structure interaction analysis is created by the fluid domain and the solid domain through one-way coupling or two-way coupling approaches. One-way coupling problems indicate that there is a small interaction between fluid and solid. This means that the fluid can impose some response to the structure, but the response is not significant enough to change the fluid flow. A sample analysis using one-way coupling can be found in [12]. Two-way interaction indicates that the fluid pressure and shear forces deform the solid body motion, and this deformation changes the fluid flow. This involves iteration to achieve a satisfactory solution. A nonlinear computational method for hydroelastic effects in extreme seas using a two-way coupling approach can be found in [13].

3 Validation of the CFD-CSD Solver

First, the CFD-CSD solver was validated against available benchmark experiments to confirm its capability and accuracy to calculate the fluid dynamics as well as the structural dynamics of a floating body.

3.1 Drop Test of a Rigid Cylinder

The accuracy of the fluid solver was verified using the water impact experiment of a circular cylinder performed by Greenhow and Lin [14]. Some simplifications to the experimental model were made. In the experiment, the cylindrical cylinder was a rigid body of radius $r = 0.055$ m, the size of the tank was 1.168×0.902 0.102 m, the water depth of the tank was 0.3 m, and the cylinder fell freely into the water from a drop height of $H = 0.5$ m. The initial water entry speed $V = \sqrt{2g(H - r)}$ was designed as the impact speed. The rigid cylinder was neutrally buoyant. This meant that the cylinder's weight equaled the buoyancy force acting on the totally submerged cylinder [14]. The parameters for the numerical setup are listed in Table 1, where d is the diameter, m is the mass, V_{entry} is the body entry velocity, ρ_f is density of the fluid, and ν_f is kinematic viscosity of the fluid.

To determine the sensitivity of the grid, simulations were performed on three different grids of constant *CFL* (Courant–Friedrichs–Lewy) number, which meant that the same refinement factors were applied in all dimensions and that spatial and temporal refinements were done simultaneously [15]. A refinement factor of $c = 2$ was uniformly specified for all spatial directions. Ratios specifying the number of cells for the coarse grid (G1), the medium grid (G2), and the fine grid (G3) were well matched by the factor $c^2 = 4$. The three grids used for the convergence study are summarized in Table 2.

Figure 2 shows the comparative penetration depths of the simulated rigid cylinder and experimental measurements. We see that decreasing the mesh size

Table 1 Parameters of the numerical setup for the rigid cylinder

d (m)	m (kg)	V_{entry} (m/s)	ρ_f (kg/m^3)	ν_f (m^2/s)
0.110	0.950	2.955	1.0×10^3	1.0×10^{-6}

Table 2 Summary of grids G1, G2, and G3

Grid	Δx	Δz	Δt	No. of cells
G1	Δx_1	Δz_1	Δt_1	36,352
G2	$\Delta x_1/c$	$\Delta z_1/c$	$\Delta t_1/c$	1,45,408
G3	$\Delta x_1/c^2$	$\Delta z_1/c^2$	$\Delta t_1/c^2$	5,81,632

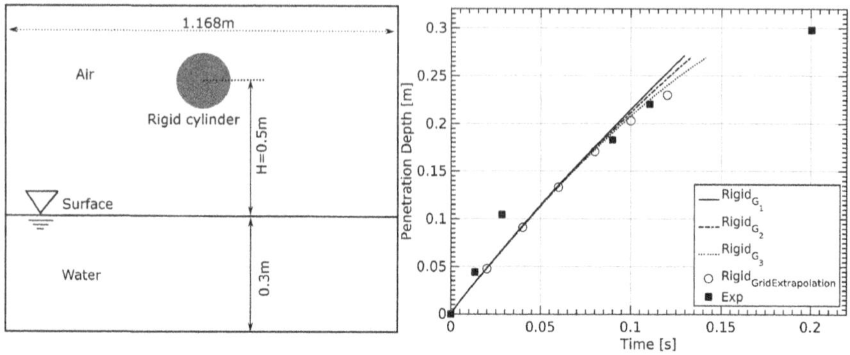

Fig. 2 Experimental setup (left) adopted from [14], and comparative time histories of simulated and measured penetration depths (right)

obtained a better agreement with experimental measurements and that the capability and accuracy of the adopted CFD method to simulate a floating body impacting a complex free surface was confirmed.

3.2 *Drop Test of a Deformable Cylinder Shell*

Then we simulated a drop test of a deformable cylinder shell using the coupled CFD-CSD solver, which was based on the experiment by Arai and Miyauchi [16]. The parameters of the numerical setup for the deformable cylinder shell are listed in Table 3, where d is diameter, t is shell thickness, ρ_s is structure density, E is Young's modulus, ν is Poisson's ratio, V_{entry} is body entry velocity, ρ_f is density of the fluid, and ν_f is kinematic viscosity of the fluid (Fig. 3).

We used an implicit two-way coupling approach, and Fig. 4 shows the comparison of strain at the bottom of the elastic cylindrical shell. We see that there was a good agreement for the lower frequency strain response although small deviations of higher frequency strain responses occurred. Refinement of the fluid domain and solid domain might have been necessary. From a practical standpoint, the low frequency response was more critical, and the favorable agreement between simulation and experiment confirmed the capability and accuracy of the coupled CFD-CSD solver to analyze the hydroelastic problem of a floating body in complex free surface flows.

Table 3 Parameters of the numerical setup for the deformable cylinder

d (m)	t (m)	ρ_s (kg/m^3)	E (Pa)	ν	V_{entry} (m/s)	ρ_f (kg/m^3)	ν_f (m$_2$/s)
0.306	0.003	3.04×10^3	1.0×10^{10}	0.340	2.955	1.0×10^3	1.0×10^{-6}

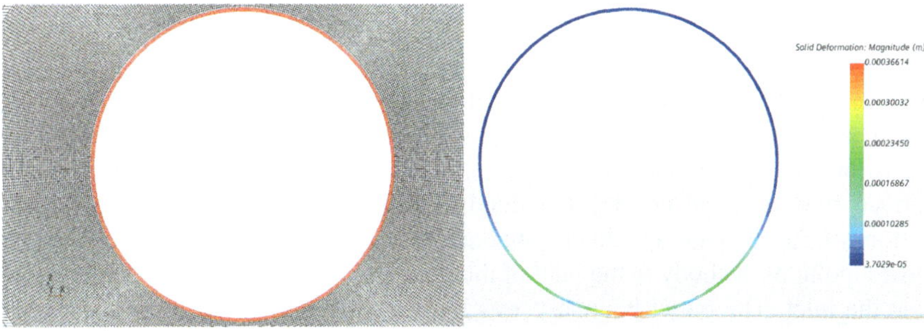

Fig. 3 Grid topology (*left*) of the fluid (*black*) and solid domain (*red*), and the deformation of the bottom point when the body entry into the water (*right*)

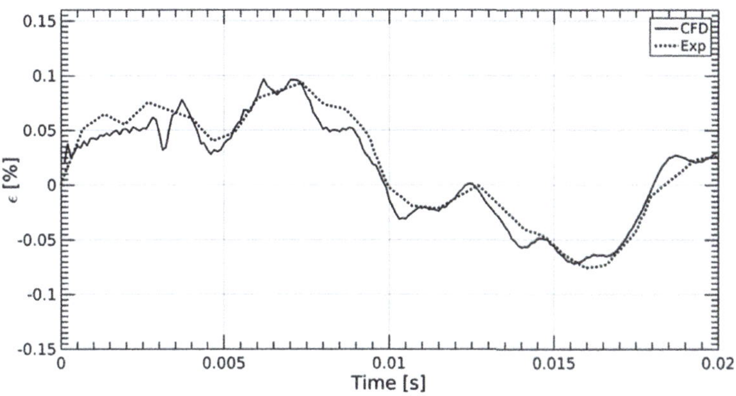

Fig. 4 Comparative time histories of strains at the bottom of the elastic cylindrical shell

4 Hydroelastic Assessment of a Standard Floating Module

We performed hydrodynamic and hydroelastic analyses for a standard floating module, developed within the scope of Space@Sea project. The material chosen for the modules is concrete. The driving consideration for concrete over steel were longevity, resistance to corrosion, fatigue life and production costs. The height of the floating module is H of 6 m, and its length equals its width, which is L of 45 m. The service draft is a half of body height λ. It was assumed that the greatest load will be caused by a wave whose height completely submerges the ends of the body while causing zero draft amidships. This deformed the module in sagging. This extreme wave was specified by a wave length of L and wave height of H. For additional details, see [17].

4.1 Finite Volume Model

A rectangular box defined the computational domain, as shown in Fig. 5. It consisted of three layers, namely, a uniform high-resolution middle layer extending above and below the calm water level to enclose the minimum and maximum surface elevation, and two layers extending from the middle layer to the top and bottom of the domain. The inlet boundary was located a distance of $2L$ from the closest point of the body to the inlet of the domain, including the wave forcing zone L at the inlet. The outlet boundary was located a distance of $2L$ from the closest point of the body to the outlet, including the wave forcing zone L at the inlet. To form an equilibrium state with the given position of the floating body, the moorings were modelled in a symmetric way. That meant that the z-coordinate of the anchor points close to the tank wall were situated symmetrically. All moorings were modeled as linear springs with a constant stiffness of 13.1 N/m and pretension of 6.8146 N. The x-coordinate of the associated anchors close to the tank walls is 2.226 m. The tests were performed at a geometrical scale of 1:70. Froude scaling laws were applied. Additional details can be found in the model test report of TU Delft [18] and a previous numerical study [19].

For regular waves of height of H and wave length L, grid and time step studies were performed on $2D$ grids without the presence of the floating body. The same approach discussed above was used, which stipulated that exactly the same refinement factors be applied to all dimensions and that spatial and temporal refinements be done simultaneously, thereby keeping the Courant number constant. A refinement factor of $c = \sqrt{2}$ was uniformly specified for all spatial directions. Ratios specifying the number of cells for the coarse grid (G0), the medium grid (G1), the fine grid (G2), and the very fine grid (G3) were well matched with the factor $c^2 = 2$. The four grids used for the convergence study are summarized in Table 4. The number of cells per wave height (c.p.w.h.) was varied from 15, 21, 30, and 42 to obtain the refinements in the wave propagation domain. On the coarse grid G0, smaller wave troughs η were observed compared to the three remaining

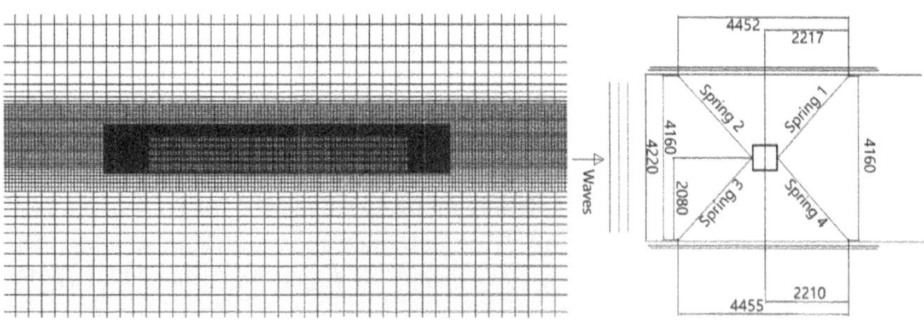

Fig. 5 Mesh topology of the computation domain (*left*) and the mooring set-up (*right*)

Table 4 Summary of grids G0, G1, G2, and G3 used for the wave study

Grid	Δx	Δz	Δt	No. of cells	c.p.w.h
G0	Δx_0	Δz_0	Δt_0	9600	15
G1	$\Delta x_0/c$	$\Delta z_0/c\backslash$	$\Delta t_0/c$	18,974	21
G2	$\Delta x_0/c^2$	$\Delta z_0/c^2$	$\Delta t_0/c^2$	37,950	30
G3	$\Delta x_0/c^3$	$\Delta z_0/c^3$	$\Delta t_0/c^2$	75,684	42

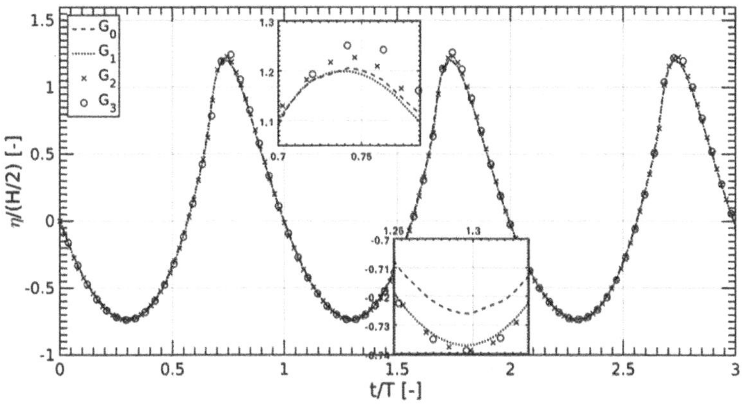

Fig. 6 Wave elevations obtained on four different grids of constant CFL number

grids, as shown in Fig. 6. It indicated that the grids refined from G1 to G2 and G3 contributed negligibly to the simulation accuracy while increasing the computational time. Therefore, we maintained the medium grid G1 for the rest of the simulations.

4.2 Finite Element Model

The material chosen for the floating modules was concrete. The driving considerations for concrete over steel were longevity, resistance to corrosion, increased fatigue life, and lower production costs. The 3D-FE model comprised a tetrahedral mesh of the appropriate thickness for the inner and outer surfaces. For additional details, see the design report [17]. The floating module comprised 86836 elements, and its grid topology is shown in Fig. 7.

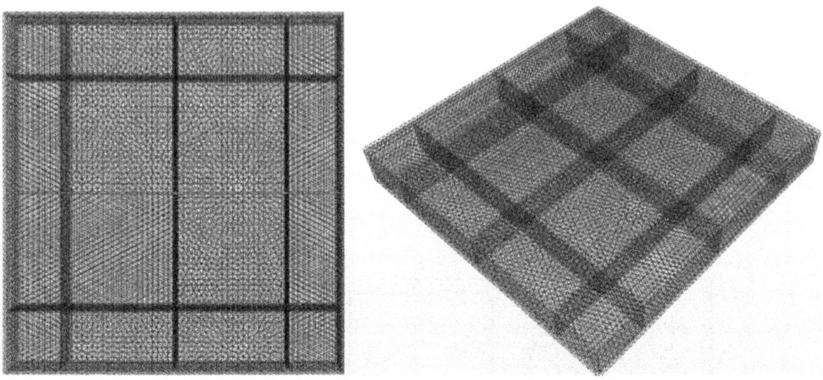

Fig. 7 Mesh topology of the solid domain, top view (*left*) and perspective view (*right*)

4.3 Numerical Results

The hydrodynamic and hydroelastic responses of a standard floating module in regular waves under moored and fixed conditions were assessed in this section. It starts with the investigation of the effects of rigid body motions on the wave-induced loads, continuing with a discussion of module's structural dynamics. Then the nonlinearities arising from complex free surface topologies are studied.

4.3.1 Effects of Body Motions on Wave-Induced Loads

Within the Space@Sea project, various floating island configurations were proposed, and each of these configurations was to be moored with different mooring setups. Thus, the motions and loads on a floating module depend on its location and moored condition. To demonstrate the effects of motion amplitude on wave-induced loads, we analyzed a floating module under fixed and moored conditions. Figures 8, 9, 10 and 11 plot the comparative results of the floating module under fixed and moored conditions. Figure 8 plots wave elevations; Fig. 9, wave induced forces in surge; Fig. 10, wave-induced loads in heave; Fig. 11, wave-induced moments in pitch. We see that wave elevations were identical and that the fixed module experienced higher wave-induced forces in heave and higher moments in pitch. For the wave-induced forces in surge, the crests are comparable. However, a moored module experienced a higher force directed opposite to the wave propagation. From a practical standpoint, although a fixed module experienced a smaller "trough" of wave-induced forces, the maximum wave-induced forces in the direction of wave propagation were likely to be critical rather than minimum wave induced forces. Additionally, a fixed module suffered a slightly longer wave impact duration caused by peak wave-induced forces, which would tend to increase structural fatigue. Consequently, to cover various motion conditions of a floating module in different

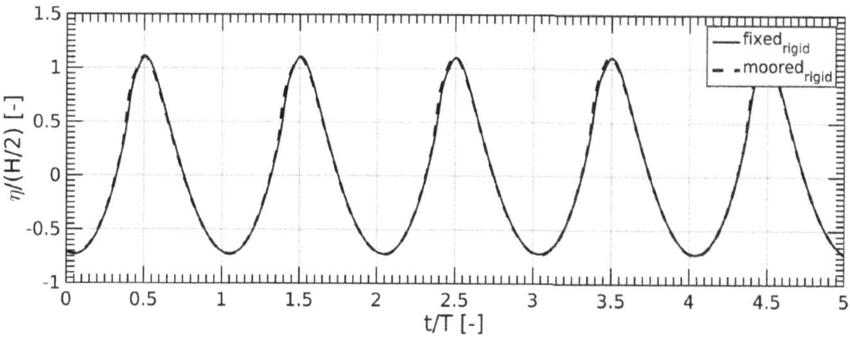

Fig. 8 Wave elevation of the module under fixed and moored conditions

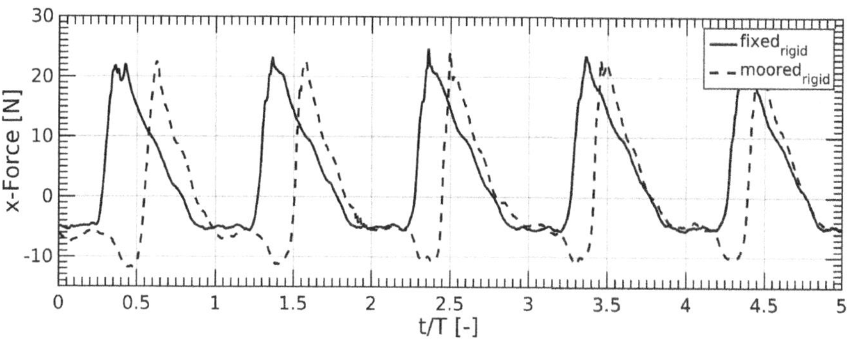

Fig. 9 Wave induced forces in surge of the module under fixed and moored conditions

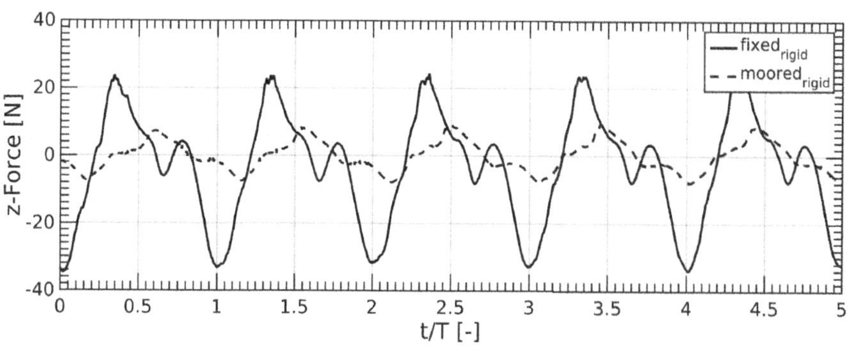

Fig. 10 Wave induced moments in heave of the module under fixed and moored conditions

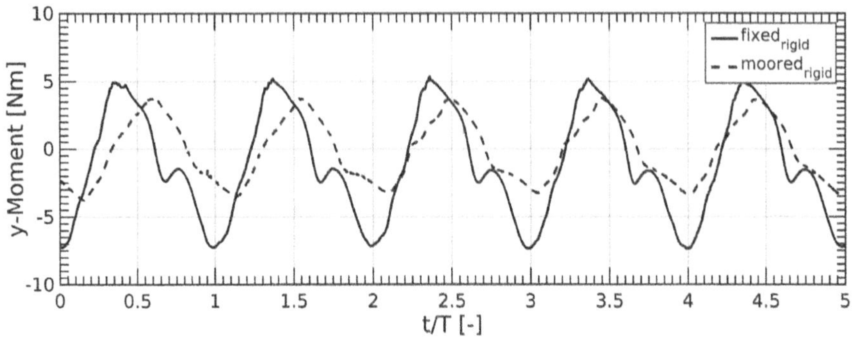

Fig. 11 Wave induced moments in pitch of module under fixed and moored conditions

island configurations with different mooring setups, the hydroelasticity analysis would have to be conducted for the fixed condition as well.

Figure 12 shows a sample screen shot of distributed hydrodynamic pressures and free-surface topologies around the floating module. In these figures, the top graph was obtained for the fixed condition; the bottom graph, from computations based on the moored floating module.

4.3.2 Effects of Structural Dynamics on Wave-Induced Loads

To investigate the effects of structural dynamics on wave-induced loads, we compared results obtained from the one-way and two-way coupling approaches. One-way coupling required computing wave-induced loads based on rigid-body motions using CFD and then applying the forces on a wet elastic model to predict structural responses, during which the structure deformations computed by the CSD solver was not fed back into the CFD solver. For the two-way coupled approach, the elastic deformations of the structure were fed back into the CFD solver, which was necessary for the computation of pressure changes when the structure deformed during a slamming event or when large deformations effectively influenced the flow field. Figures 13, 14, 15 and 16 plots comparative results of a rigid module and a flexible module under the fixed condition. Figure 13 plots the associated wave elevations; Fig. 14, the associated wave-induced loads in surge; Fig. 15, the associated wave-induced loads in heave; Fig. 16, the associated wave-induced moments in pitch. We see that the resulting wave elevations and wave induced forces and moments of the rigid module were almost identical to those of the flexible module. The module's structural deformation was small during the wave impact, and the deformation hardly affected the flow filed. Consequently, the one-way coupling approach was also sufficient for the hydroelastic assessment of the floating module.

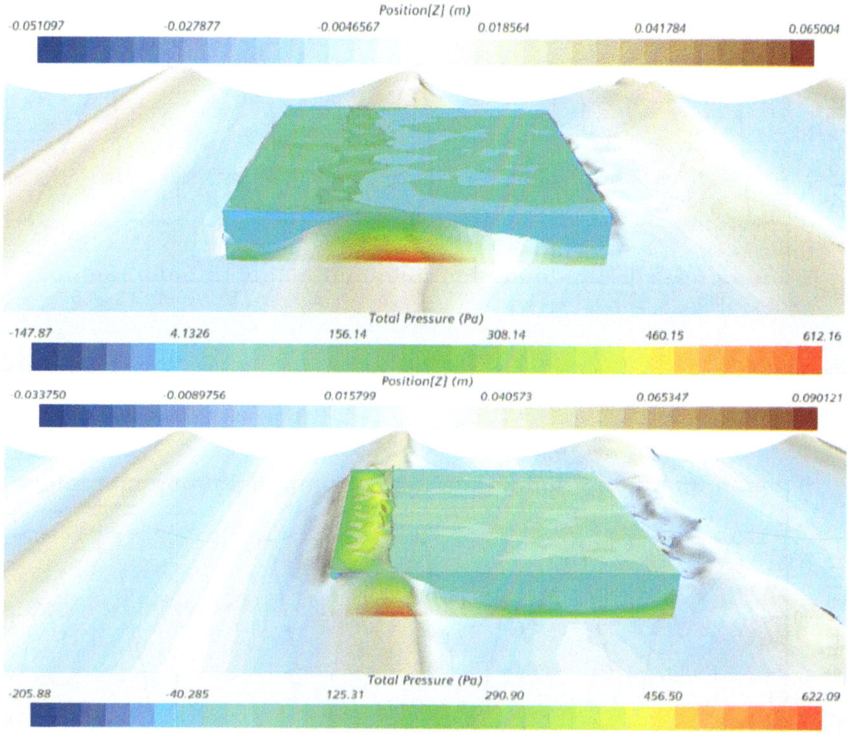

Fig. 12 Distributions of hydrodynamic pressures of the module obtained from computations based on the fixed condition *(*upper*)* and the moored condition *(*lower*)*

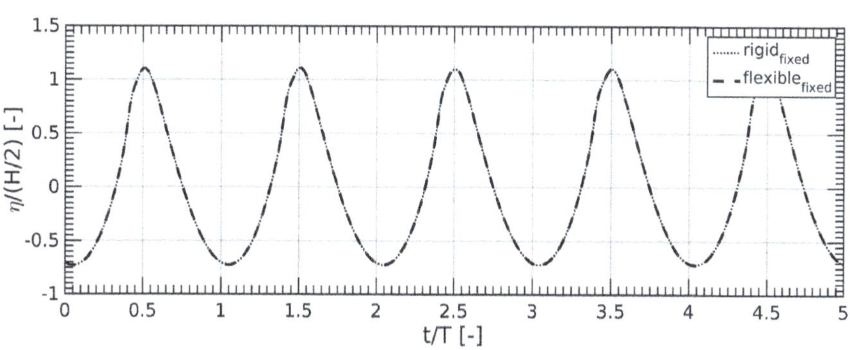

Fig. 13 Wave elevation of a rigid module and a flexible module in the fixed condition

Figures 17 shows a sample screen shot of distributions of hydrodynamic pressures (top) and structural deformations (bottom). The hydrodynamic pressures (top) were obtained from a rigid module; the structural deformations (bottom), from a flexible module at the same time instance.

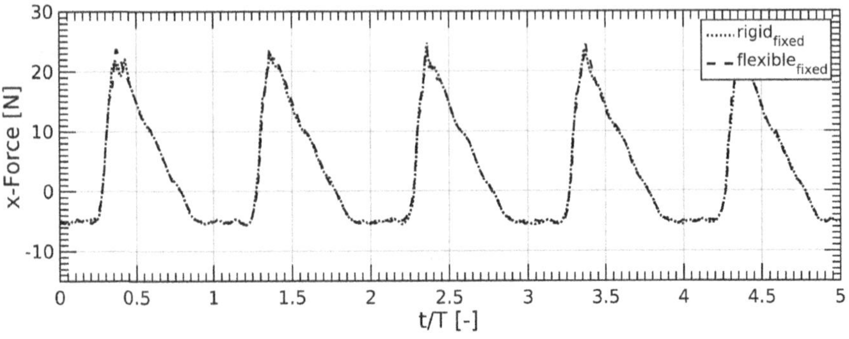

Fig. 14 Wave induced forces in surge of a rigid module and a flexible module in the fixed condition

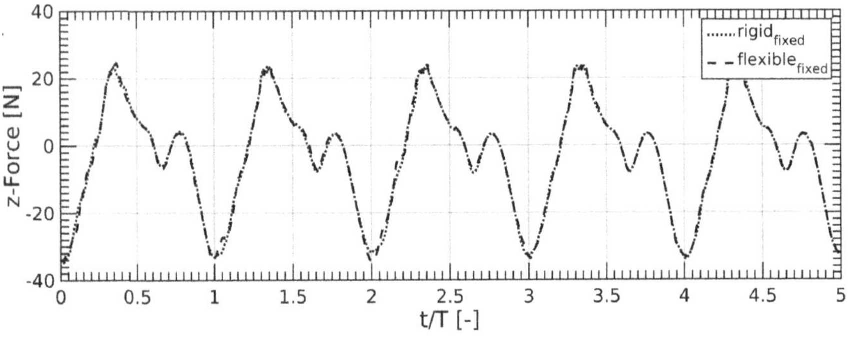

Fig. 15 Wave induced forces in heave of module under fixed and moored conditions

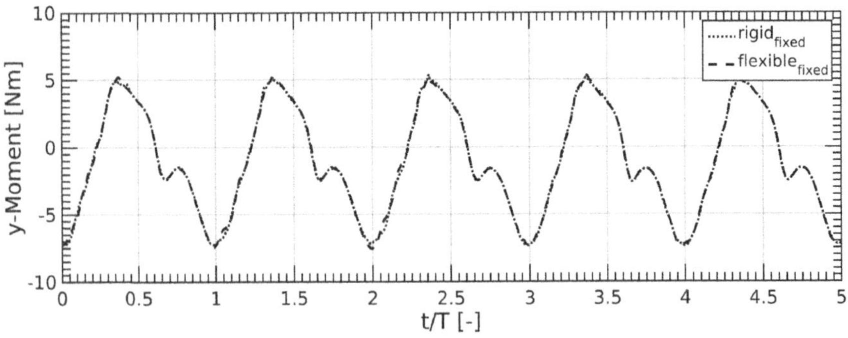

Fig. 16 Wave induced moments in the pitch of module under fixed and moored conditions

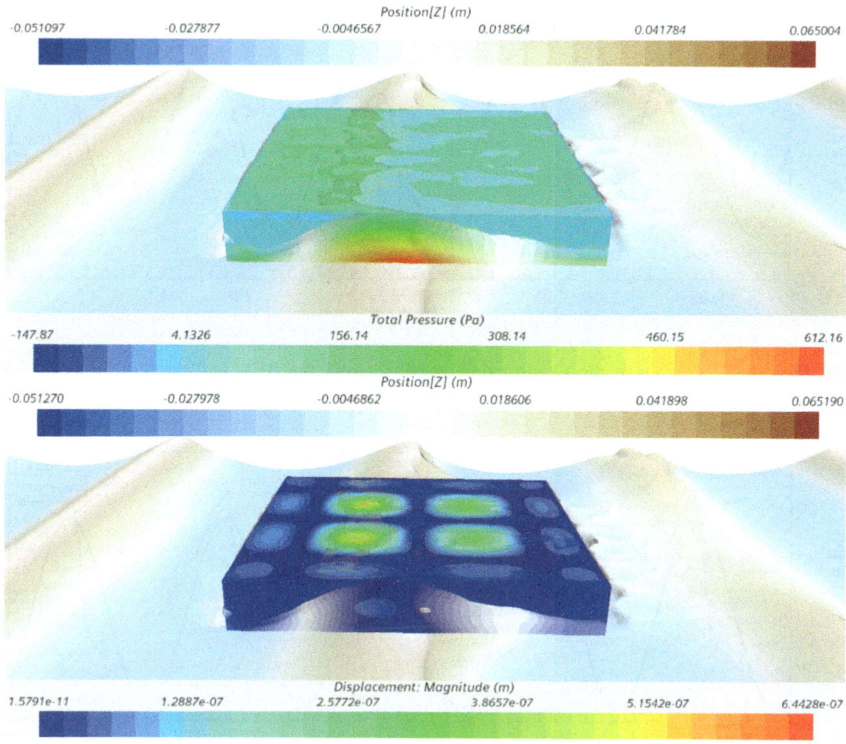

Fig. 17 Distribution of hydrodynamic pressures (upper) of a rigid module, and the structural deformation (lower) of a flexible module at the same time instance

4.3.3 Effects of Nonlinearities on Wave-Induced Loads

To investigate the effects of nonlinearities on wave-induced forces and moments, the hydroelastic assessment of the fixed module was also conducted using the BEM-CSD solver. The nonlinearities mainly arose from nonlinear waves, viscosity, and green water on desk, which we simulated using the CFD solver instead of the BEM solver. The BEM-CSD solver assumed a rigid module, and the stress and bending moment were computed by integrating forces acting on the module surface. The wave-induced hydrodynamic pressures resulting from the hydrodynamic analysis were directly mapped to the structural analysis model through an interface. Figures 18, 19, 20 and 21 plot the comparative results of a fixed module obtained from the BEM solver and the CFD solver. Figure 18 plots the associated wave elevations; Fig. 19 the associated wave induced loads in surge; Fig. 20, the associated wave-induced loads in heave; Fig. 21, the associated wave-induced moments in pitch. For the BEM solver, we modeled the wave forces in time domain using linear wave theory based on the linear coefficients obtained from the frequency domain analysis. We see that the nonlinear waves from the CFD solver had higher

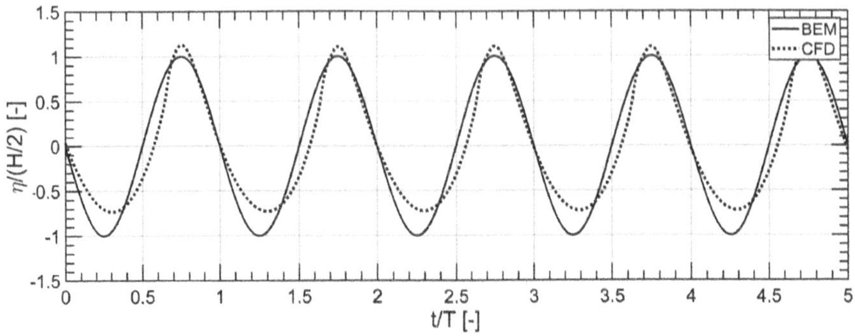

Fig. 18 Wave elevation of the fixed module obtained from the BEM and the CFD solvers

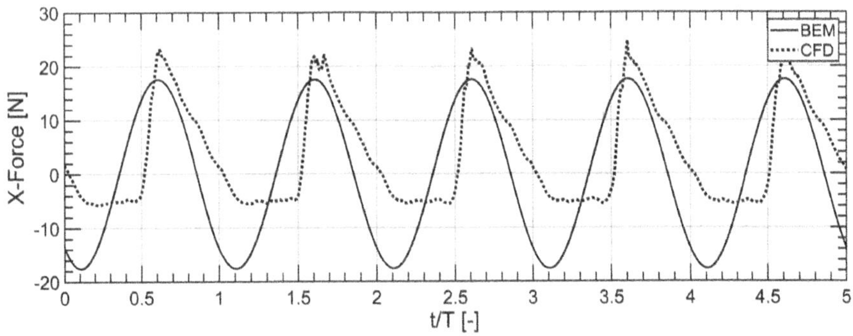

Fig. 19 Wave induced forces in surge of the fixed module obtained from the BEM and the CFD solvers

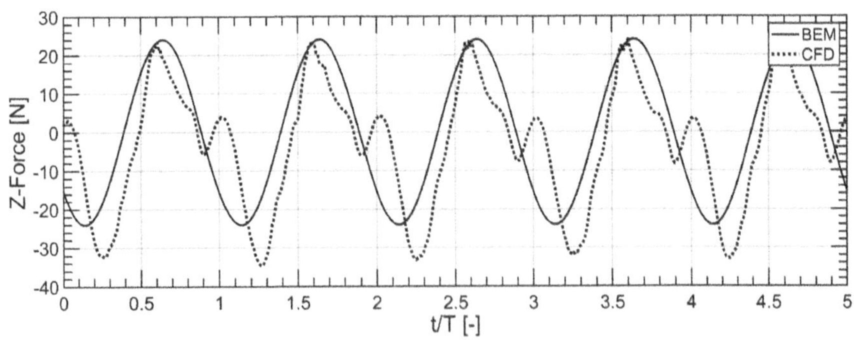

Fig. 20 Wave induced forces in heave of the fixed module obtained from the BEM and the CFD solvers

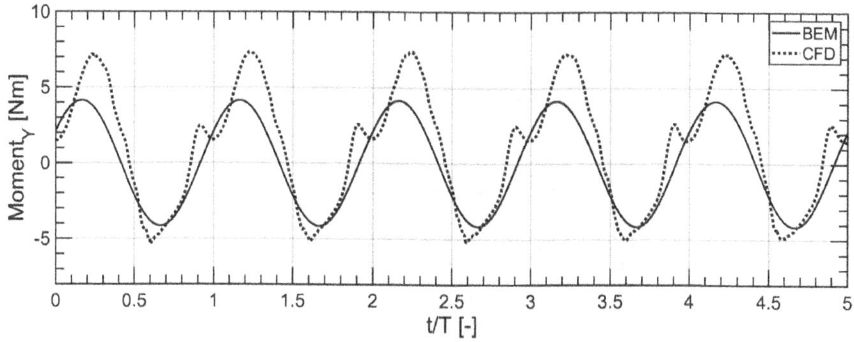

Fig. 21 Wave induced moments in pitch of the fixed module obtained from the BEM and the CFD solvers

wave crests and smaller wave troughs. In general, results obtained from the BEM solver had smaller maxima than those obtained from the CFD solver, especially for wave-induced moments in pitch. Asides from the limitations arising from the adopted linear potential theory, the BEM-CFD coupling approach could not generate a transient result for the structural analysis. Figure 22 plots the maximum deformation of the flexible module obtained from the BEM-CSD solver and the time history of the maximum deformation obtained from the CFD-CSD solver. We see that the maxima of deformations from the BEM-CSD solver were underpredicted, which was due to the limitations of potential solver as it is based on linear

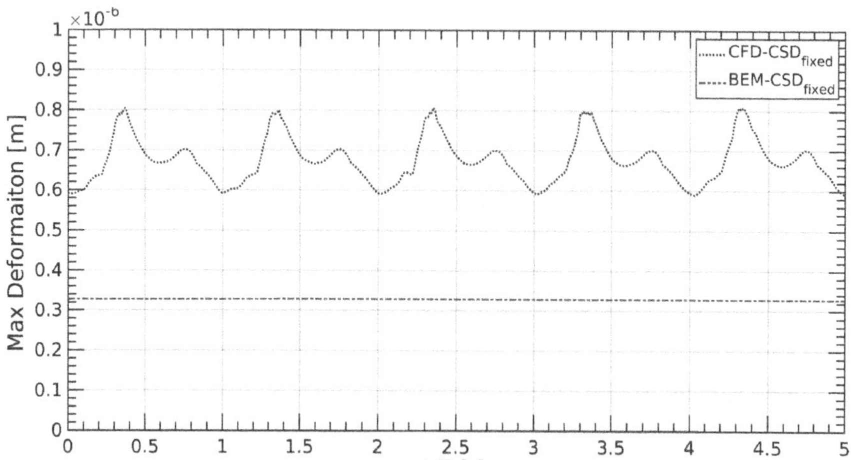

Fig. 22 Maximum deformations of the fixed module

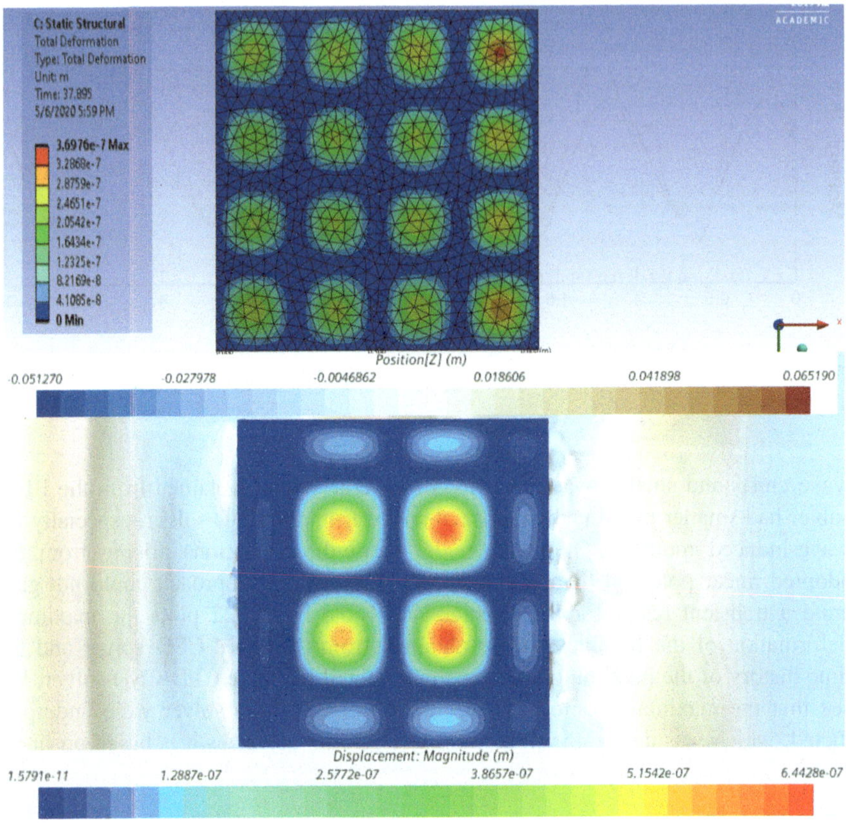

Fig. 23 Maximum deformations of the fixed module obtained from the BEM-CSD solver (*upper*) and the CFD-CSD solver (*lower*)

theory. To include nonlinear wave effects, viscosity, complex free surface topologies, and green water impacts, the flow dynamics computations would have had to be based an advanced CFD approach (Fig. 23).

5 Conclusions

We considered the hydroelasticity of a standard floating module in regular waves under moored and fixed conditions to assess the capabilities of the coupled BEM-CSD solver and CFD-CSD solver to predict wave-induced forces and overturning moments.

The effects of body motions and wave-induced forces and moments for a rigid floating module under fixed and moored conditions were compared. We found that the fixed module experienced greater wave-induced forces and overturning moments than the moored module. We analyzed effects of structural deformations on wave-induced forces and moments and compared the results for a flexible module in the fixed condition obtained from the one-way and the two-way coupling approaches. For this floating module, constructed of concrete, its deformation hardly affected the fluid dynamics and, consequently, the one-way coupling approach was adequate for its design.

Finally, to demonstrate the effects of nonlinearities caused by finite amplitude waves, flow viscosity, and complex free surface topologies, such as like green water effects on wave-induced forces and overturning moments, we adopted an integrated approach based also on potential flow theory. The underpredicted maxima of wave induced forces and moments indicated that, to assess the structural integrity of the module, we had to account for the fluid dynamic nonlinearities, and only the CFD based solver was then able to obtain reliable predictions.

Acknowledgements This work is the outcome of project Space@Sea as part of the European Union's Horizon 2020 research and innovation program under grant agreement No. 774253. The authors gratefully acknowledge the computing time on the supercomputer magnitUDE of the Center for Computational Sciences and Simulation (CCSS), provided by the Center of Information and Media Services (ZIM) of the University of Duisburg-Essen under DFG grants INST 20876/209-1 FUGG and INST 20876/243-1 FUGG.

References

1. Flikkema M, Waals O (2019) Space@Sea the floating solution. Front Mar Sci 6:553. https://doi.org/10.3389/fmars.2019.00553
2. Fu S, Moan T, Chen X, Cui W (2007) Hydroelastic analysis of flexible floating interconnected structures. Ocean Eng 34(11–12):1516–1531. https://doi.org/10.1016/j.oceaneng.2007.01.003
3. Humamoto T, Fujita K (2002) Wet-mode superposition for evaluating the hydroelastic response of floating structures with arbitrary shape. In: The twelfth international offshore and polar engineering conference
4. Newman JN (1994) Wave effects on deformable bodies. Appl Ocean Res 16(1):47–59. https://doi.org/10.1016/0141-1187(94)90013-2
5. Kashiwagi M (1998) A B-spline Galerkin scheme for calculating the hydroelastic response of a very large floating structure in waves. J Mar Sci Technol 3(1):37–49. https://doi.org/10.1007/BF01239805
6. Ohkusu M, Namba Y (2004) Hydroelastic analysis of a large floating structure. J Fluids Struct 19(4):543–555. https://doi.org/10.1016/j.jfluidstructs.2004.02.002
7. Ohmatsu S (2000) Numerical calculation method for the hydroelastic response of a pontoon-type very large floating structure close to a breakwater. J Mar Sci Technol 5(4):147–160. https://doi.org/10.1007/s007730070001
8. Watanabe E, Utsunomiya T, Wang CM (2004) Hydroelastic analysis of pontoon-type VLFS: a literature survey. Eng Struct 26(2):245–256. https://doi.org/10.1016/j.engstruct.2003.10.001
9. Nematbakhsh A, Gao Z, Moan T (2017) Benchmarking of a computational fluid dynamics-based numerical wave tank for studying wave load effects on fixed and floating offshore structures. J Offshore Mech Arct Eng 139(3). https://doi.org/10.1115/1.4035475
10. Jiang C, el Moctar O, Paredes GM, Schellin TE (2020) Validation of a dynamic mooring model coupled with a RANS solver. Mar Struct 72: https://doi.org/10.1016/j.marstruc.2020.102783

11. AQWA A (2014) Theory manual, release 16.0
12. Ley J, el Moctar O (2014) An enhanced 1-way coupling method to predict elastic global hull girder loads. In: International conference on offshore mechanics and arctic engineering, vol 45431. https://doi.org/10.1115/OMAE2014-24199
13. el Moctar O, Ley J, Oberhagemann J, Schellin T (2017) Nonlinear computational methods for hydroelastic effects of ships in extreme seas. Ocean Eng 130:659–673. https://doi.org/10.1016/j.oceaneng.2016.11.037
14. Greenhow M, Lin WM (1983) Nonlinear free-surface effects: experiments and theory. Massachusetts Institute of Technology, Department of Ocean Engineering
15. Oberhagemann J, Shigunov V, el Moctar O (2012) Application of CFD in long-term extreme value analyses of wave loads. Ship Technol Res 59(3):4–22. https://doi.org/10.1179/str.2012.59.3.001
16. Arai M, Miyauchi T (1997) Numerical simulation of the water impact on cylindrical shells considering fluid-structure interaction. J Soc Naval Architects Japan 182:827–835. https://doi.org/10.2534/jjasnaoe1968.1997.182_827
17. Breuls M, Blumer S, Iacob N, Blümel B (2019) Detail design report—deliverable 4.4
18. Thill RPC (2018) Model tests report—deliverable 4.3
19. Seithe G, Moctar OE (2019) Wave-Induced motions of moored and coupled multi-body offshore structures. 11th International workshop on ship and marine hydrodynamic. Hamburg, Germany, pp 22–25

Effects of Clearance Between Seabed and Bottom of a VLFS on Hydroelastic Responses

Tomoki Ikoma, Shoichiro Furuya, Yasuhiro Aida, and Lei Tan

Abstract Possibility to utilize sea areas by using very large floating structures (VLFSs) has remained when we consider impact to ocean environment, a change of uses, decommission and lifecycle. We can consider installing VLFSs in very shallow seas as well as deep sea areas. In Japanese case, it is, for instance, in Tokyo bay areas. Although added-mass increases very much due to a huge horizontal area, studies in which effects of variation of the added-mass on hydroelasticity are summarized are not a lot. Besides, how the elastic deformation changes has not also been investigated when clearance between seabed and a bottom plan decreases because of draft increasing in shallow sea areas. This study used the linear potential theory bases prediction method for hydroelasticity problems. The paper calculated pontoon type VLFSs. Then it was investigated and summarized how the elastic motion characteristics were affected by water depth, mass of a VLFS and the clearance and in taking into account of variation of the added-mass. From the results, effects of physical draft of VLFSs was not large when deep water depth and we could calculate the elastic response only using suitable mass of a VLFS even if the draft of the VLFS to be calculated was a bit different. In addition, when the clearance decreased, the added-mass increased very much and characteristics of the elastic motion were affected very much.

Keywords VLFS · Added-mass · Water depth · Clearance

T. Ikoma (✉) · Y. Aida · L. Tan
Department of Oceanic Architecture and Engineering, CST, Nihon University, 7-24-1 Narashinodai, Funabashi City, Chiba Prefecture 274-8501, Japan
e-mail: ikoma.tomoki@nihon-u.ac.jp

Y. Aida
e-mail: aida.yasuhiro@nihon-u.ac.jp

L. Tan
e-mail: tan.lei@nihon-u.ac.jp

S. Furuya
Department of Oceanic Architecture and Engineering, School of Science and Technology, Nihon University, 7-24-1 Narashinodai, Funabashi City, Chiba Prefecture 274-8501, Japan
e-mail: cssh19031@g.nihon-u.ac.jp

Ł. Piątek et al. (eds.), *WCFS2020*, Lecture Notes in Civil Engineering 158,
https://doi.org/10.1007/978-981-16-2256-4_20

1 Introduction

It is very useful to adapt a very large floating structure (VLFS) comparing with making an artificial island when we would like to expand areas to be developed toward sea areas. Because, we would be able to reduce environmental impact and cost. Besides, structures on a floating foundation are affected little by earthquakes. Generally speaking, a landfill construction method is better than to construct a floating system in shallow water less than 50 m. However, we should consider to adapt a floating system even in such as shallow seas depending on uses. We have been able to consider designing a VLFS in very shallow seas as well as deep sea areas. In Japan, there are investigations, for example, considered in Tokyo Bay [1]. The use of VLFSs in Tokyo Bay has been proposed since the 1960s. In 1994, the mega float technology research association had been established, and in 1995, a floating test structure of 300 m in long, 60 m in wide, 2 m in depth was constructed and installed in Tokyo bay. After that, from 1998 to 2000, the world's first floating airport with a runway of 1000 m had been installed and tested. This was called the Phase II model with 5 m in depth, see Fig. 1. The draft was very small compared to the horizontal scale of the mega-float, and it was a pontoon type like a membrane. In addition to the demonstration of elastic deformation, design technology of mooring equipment, and control function as an airport, it concluded that the evaluation method of the marine environment impact was proper and impacts of waves were not so significant.

Fig. 1 Phase II mega-float model

By the way, although we can predict hydroelastic deformation of such as a pontoon type VLFS with the linear potential theory. In addition, there is no issues to calculations of elastic motions within the theory. However, we know detail of hydrodynamic characteristics little when VLFSs are installed in too shallow water and effects of distance between bottom of VLFS and sea bottom which is here the clearance. When the clearance decreases, the added-mass increases and it is well known. Then, elastic motion of a VLFS with a condition of very small clearance may become small very much. However, we can hardly understand if it is correct or not. Besides, recently although VLFSs are designed to propose a new utilization concept, basic investigations of VLFSs have been a few in a long time.

An objective of this present study is to investigate detail of effects of the water depth very shallow and the clearance very small on hydroelastic motion of a VLFS. This study adapts the linear potential theory-based prediction method for calculations of hydroelasticity problems. The paper focus on a pontoon type VLFS. The present study investigates how the elastic motion is affected by water depth especially shallow water depth and the clearance very small.

2 Formulation

The section shows the theoretical formulation to calculate wave exciting forces and radiation forces with the linear potential theory. This paper does not focus on aircushions on a floating structure but the study uses multi-functional programme code which can predict hydroelasticity problems with effects of aircushions and/or oscillating water column type wave energy converters equipped on a floating structure.

2.1 Basic Assumptions and Boundary Conditions

This section reviews the formulations associated with a prediction method incorporating the hydrodynamic forces acting on an aircushion-type floating body. The basics of these formulations have been shown by Ikoma et al. [2, 3]. However, some equations have been changed in order to make them simpler and to modify the computer program code for analysis of velocity potentials in accordance with the present method.

Herein, velocity potentials are defined assuming a perfect fluid and the irrotational flow. In this paper, all problems are also linearized. The velocity potential Φ is defined as follows:

$$\vec{v} = \mathrm{grad}\Phi(x, y, z; t) \tag{1}$$

where t is time. In addition, separation of the time variable in the velocity potential is defined as follows:

$$\Phi_t(t) = \mathrm{Re}\left[-i\omega t\phi \cdot e^{-i\omega t}\right] \tag{2}$$

where ω is the angular frequency and i is a complex number which is $\sqrt{-1}$. The velocity potential ϕ is the sum of all wave components, which include the incident wave, the diffraction wave and the radiation waves. This can be expressed as follows:

$$\phi = a\phi_I + a\phi_D + \sum_{r=1}^{\mathrm{D.O.F}} q_r\phi_r \tag{3}$$

where a is the complex amplitude of an incident wave, q_r is the motion amplitude of the r-th mode, ϕ_I is the velocity potential of the incident wave, ϕ_D is the velocity potential of the diffraction wave and r is the velocity potential of the radiation wave due to the r-th motion. If the corresponding structure is a rigid body, there are six degrees of freedom (D.O.F.) which are reflected in q_r (surge, sway, heave, roll, pitch and yaw). We can consider the elastic motion by using eigenfunctions of elastic modes, in which case q_r represents the principal coordinate, which is also termed the generalized coordinate.

The coordinate system is shown in Fig. 2. The vertical axis is positive upward. The normal vector on the boundaries is defined as positive toward the hull from fluid region.

The government equation is the following Laplace's equation:

$$\nabla^2\phi(x, y, z) = 0 \tag{4}$$

The boundary condition on free water surface is given as follows:

$$K\phi - \frac{\partial\phi}{\partial z} = 0 \tag{5}$$

where K is ω^2/g, thus ω is the angular frequency and g is the acceleration of gravity. These formulations follow the ordinary linear potential theory and are well known.

In this study, it is assumed that geometrical deformation of an aircushion is caused only by vertical motion. When the structure in question is an elastic body, the vertical displacement can be expressed as follows:

$$\zeta(x, y) = \sum_{r=1}^{\infty} q_r\zeta_r(x, y) \tag{6}$$

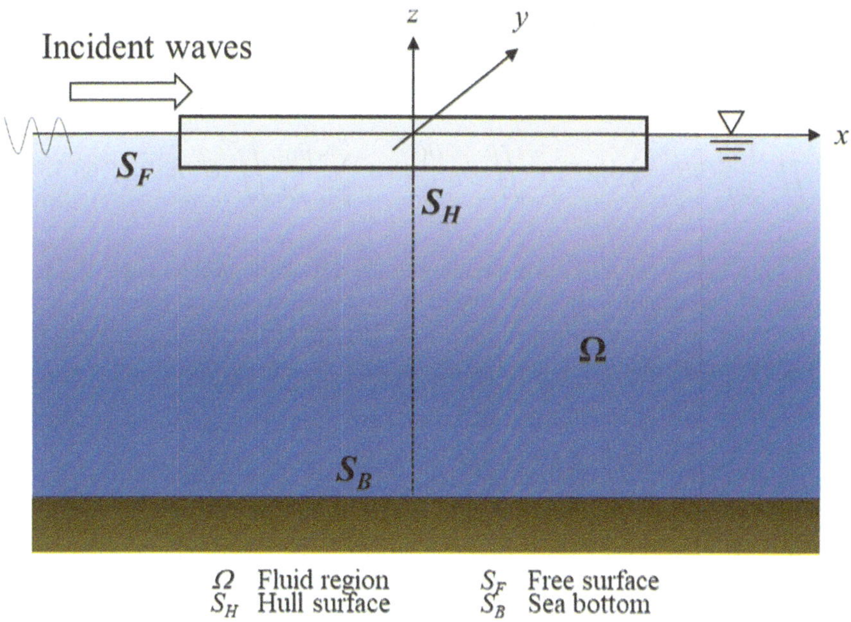

Fig. 2 Coordinate system used in this work

where ζ_r is an eigenfunction, which is at the same time a mode function.

$$-C_P\phi(P) = \iint\limits_{S_H} \left(\phi(Q)\frac{\partial G}{\partial n_Q} - \frac{G\partial\phi(Q)}{\partial n_Q}\right)ds + \iint\limits_{S_{FA}} \left(\phi(Q)\frac{\partial G}{\partial n_Q} - \frac{G\partial\phi(Q)}{\partial n_Q}\right)ds \tag{7}$$

The constant panel method is applied to the discretization in this study. When $\phi(P)$ is set at S_H, C_P is 0.5. When $\phi(P)$ is set at S_{FA}, C_P is 1.0. In this study, there is not an area of S_{FA} so that we take into account of only hull surfaces S_H.

2.2 Integral Equations

The boundary element method with Green's function, satisfying the condition of a free water-surface, is the method often called singular points distribution method. The potential distribution method is applied here.

The boundary integral equation is expressed as follows using Green's function G.

2.2.1 a. For Diffraction Problem

The velocity potential of diffraction component ϕ_D at S_H is given as follows:

$$-\iint_{S_H} \frac{\partial \phi_D}{\partial n} G ds = \frac{1}{2}\phi_D(P) + \iint_{S_H} \phi_D \frac{\partial G}{\partial n} ds + \sum_{n=1}^{N_{AC}} \frac{p_{D_n}}{\rho g} \iint_{S_{FA}} G \cdot n_z ds, \ P \text{ on } S_H \quad (8)$$

where p_D is the air pressure corresponding to the diffraction problem and N_{AC} is the number of aircushions. In this study, it is assumed that a plane of a free water-surface within aircushion is on $z = 0$. The boundary values of a right hand term are given as follows:

$$\frac{\partial \phi_D}{\partial n} = -\frac{\partial \phi_I}{\partial n} \quad (9)$$

Water elevation of the scattering component η_S is now defined as follows:

$$\eta_S = \eta_I + \eta_D \quad (10)$$

and thereby,

$$\eta_D = \eta_S - \eta_I \quad (11)$$

When η_D is set to S_{FA}, the integral equation is transformed into the equation regarding the water elevation within the aircushions as follows,

$$-\iint_{S_H} \frac{\partial \phi_I}{\partial n} G ds + \frac{\eta_I(P)}{K} = \iint_{S_H} \phi_D \frac{\partial G}{\partial n} ds + \frac{\eta_S}{K} + \frac{p_{D_m}}{K\rho g} + \sum_{n=1}^{N_{AC}} \frac{p_{D_n}}{\rho g} \iint_{S_{FA}} G \cdot n_z ds, \ m = 1 \text{ to } N_{AC}, \ P \text{ on } S_{FA} \quad (12)$$

where p_{Dm} represents the air pressure in an aircushion including position P.

In addition, the integral equation of air pressure within an aircushion is expressed as follows:

$$0 = \alpha_0 \iint_{S_{FA}} \eta_S ds + p_D \quad (13)$$

where α_0 can be:

$$\alpha_0 = \frac{\gamma P_S}{V_0} \quad (14)$$

2.2.2 b. For radiation problem

The boundary values on hulls are given as follows using mode shape functions:

$$n_r = \vec{n} \cdot (0, 0, \zeta_r) \tag{15}$$

and thereby the equation of the r-th mode can be expressed as follows:

$$-\iint_{S_H} n_r G ds = \frac{1}{2}\phi_r(P) + \iint_{S_H} \phi_r \frac{\partial G}{\partial n} ds + \sum_{n=1}^{N_{AC}} \frac{p_{r_n}}{\rho g} \iint_{S_{FA}} G \cdot n_z ds, \ P \text{ on } S_H. \tag{16}$$

If the observation point P is set on the free surface within an aircushion, the integral equation is expressed as follows:

$$\iint_{S_H} n_r G ds = \iint_{S_H} \phi_r \frac{\partial G}{\partial n} ds + \frac{\eta_r}{K} + \frac{p_{rm}}{K\rho g} + \sum_{n=1}^{N_{AC}} \frac{p_{rn}}{\rho g} \iint_{S_{FA}} G \cdot n_z ds, \ P \text{ on } S_{FA}. \tag{17}$$

The equation expressing the air pressure and the wave elevation η_r can be given as follows,

$$\alpha_0 \iint_{S_{FA}} \zeta_r ds = \alpha_0 \iint_{S_{FA}} \eta_r ds + p_r \tag{18}$$

2.2.3 Hydrodynamic Forces

The wave exciting force of r-th modes is obtained as follows,

$$f_{e,j} = \rho\omega^2 \iint_{S_H} (\phi_I + \phi_D) n_j ds + \iint_{S_{FA}} p_D n_j ds \tag{19}$$

The radiation force of k-th mode due to j-th motion mode is obtained as follows,

$$f_{jk} = \rho\omega^2 \iint_{S_H} \phi_j n_k ds + \iint_{S_{FA}} p_j n_k ds \tag{20}$$

In addition, the coefficients of added mass $A_{m,jk}$ and the wave making damping $D_{m,ik}$ are expressed as follows:

$$A_{m,jk} = \frac{1}{\omega^2} \mathrm{Re}\left[F_{jk}\right] \tag{21}$$

$$D_{m,jk} = \frac{1}{\omega} \mathrm{Im}\left[F_{jk}\right] \tag{22}$$

where Re means an real part and Im means an imaginary part of complex variables. This added mass includes the effect of hydrostatic pressures in itself, as shown by Ikoma et al. [2].

3 Results and Discussion

This section shows results of radiation forces and wave exciting forces in comparing effects of the gap between a floating structure and the sea bottom. Besides, the effect on elastic deformations is discussed.

3.1 Effects of Variation of Water Depth

For the investigation of effect of changing clearance between seabed and bottom of floating body, calculation models that 800 m long, 600 m wide in Figs. 3, 4 is used. Draft of the models was set at 2.0 m, and water depth is from 10 to 4 m, clearance

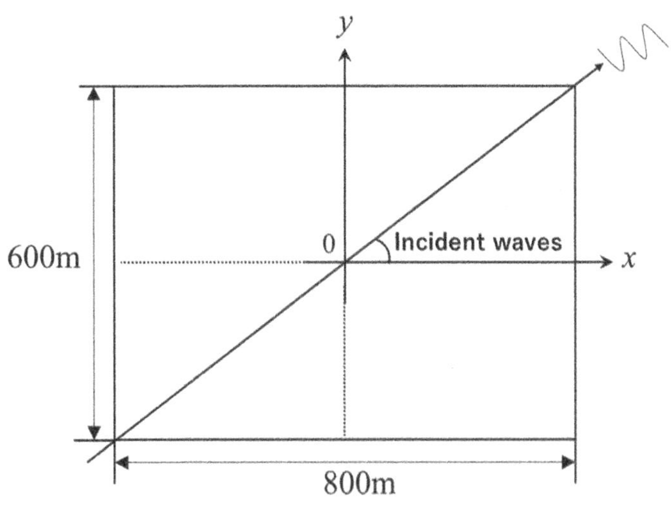

Fig. 3 Horizontal plane calculation model

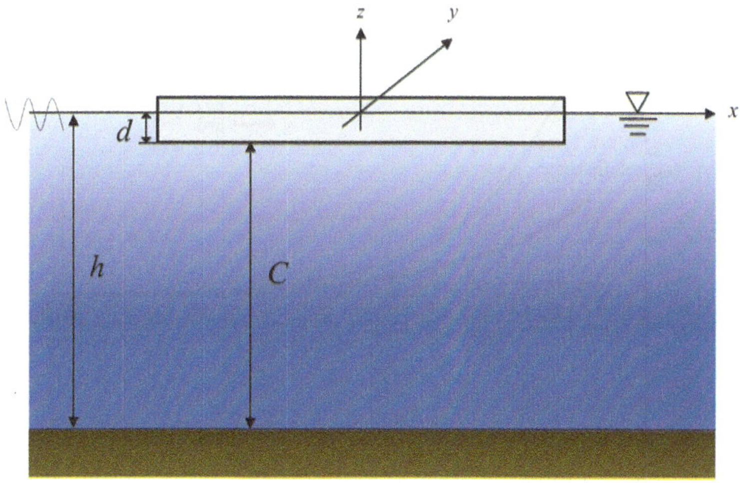

Fig. 4 Cross section of calculation model with clearance

Table 1 Calculation condition for constant draft

Length of body	800 m
Breadth of body	600 m
Draft of body	2 m
Water depth	150, 10, 7, 5, 4 m
Clearance	8, 5, 3, 2 m
Weight of body	9.4e + 9 N
Stiffness per unit (EI/B)	2.04e + 11 Nm^2/m

is from 8 to 4 m. Each condition is shown in Table 1. The incident wave angle was 0°. To compare with deep water waves, depth set to 150 m at least.

3.1.1 Results of Fluid Forces

From Fig. 5, The vertical axis is dimensionless added mass coefficient. ρ is fluid density, V is submerged volume. horizontal axis is dimensionless angular frequency. B is breadth of body g is acceleration of gravity. As the water depth becomes smaller, the effect of the water mass under the floating body becomes larger, and therefore the added mass coefficient increases. Also, in the high frequency area, the difference between each case becomes smaller and approaches zero. In deep water wave condition, added mass coefficient is almost zero in all frequencies. It is because there are many areas under the floating body, it is not compressed.

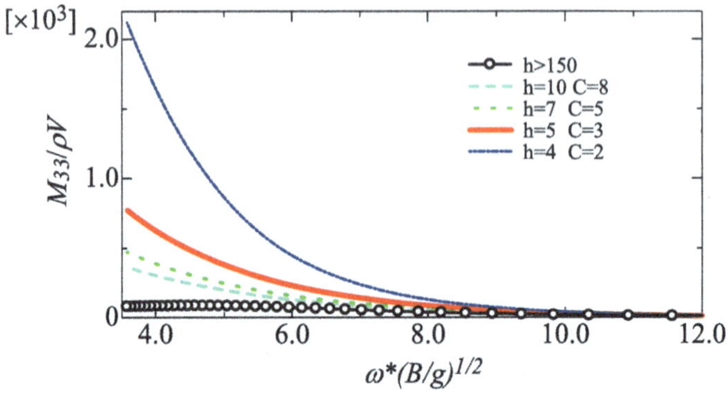

Fig. 5 Added mass coefficients of heave

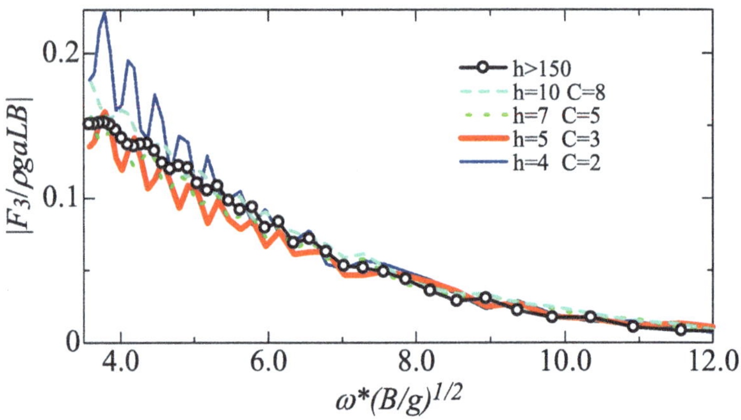

Fig. 6 Wave exciting forces on heave

From Fig. 6, vertical axis is absolute value of wave excitation force coefficient of heave. All of these is close in each case. However, the amplitudes of the excitation force increase as the clearance decreases. This value may depend on length of body, because of evanescent wave.

Wave damping coefficient was calculated, but values of that was not accurate. This is because the size of the calculation mesh was not enough for the clearance. Until now, it was necessary to divide the wavelength ratio sufficiently, but it is necessary to create a mesh considering clearance. The result of fluid force was not stable in the case of h = 3 C = 2.

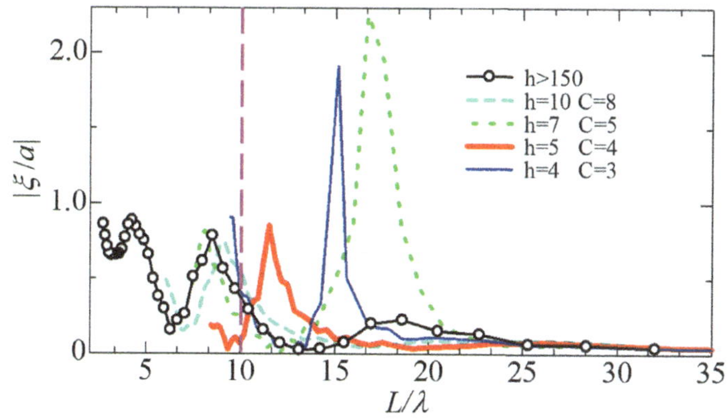

Fig. 7 RAOs of vertical displacement at end of floating structure

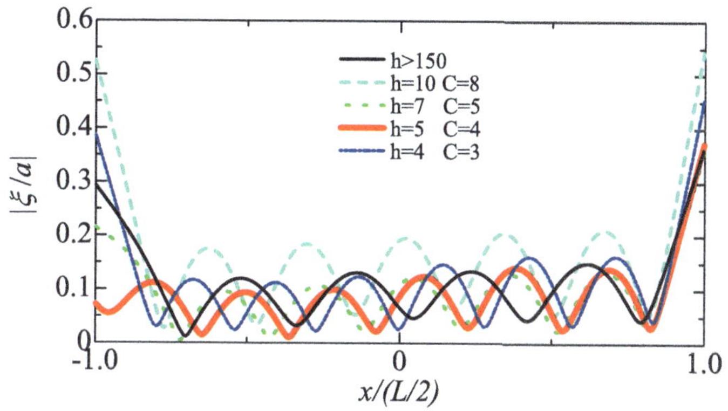

Fig. 8 Distributions of vertical displacement on y = 0 in case of L/λ = 10

3.1.2 Results of Elastic Motions

From Figs. 7, 8, since the wave frequency was calculated with the same value, the wavelength ratio changes when the water depth is changed. Therefore, the vertical response distributions are compared with L/λ = 10 where the wavelength ratio is the same value. Vertical axis is vertical displacement of the point of (−400, 0). It was expected that the response could be reduced because the added mass coefficient increased as the clearance decreased, but the response did not decrease significantly and there was not much difference. The effect of the added mass coefficient on the elastic motion of the floating body is not so large.

Table 2 Calculation condition for constant water depth

Length of body	800 m
Breadth of body	600 m
Draft of body	2, 4, 5 m
Water depth	10 m
Clearance	8, 6, 5 m
Weight of body	9.4e + 9 N, 1.9e + 10 N, 2.4e + 10 N
Stiffness per unit (EI/B)	2.04e + 11 Nm^2/m

3.2 *Effects of Variation of Clearance*

For the investigation of effect of changing clearance between seabed and bottom of floating. Draft of the models was set from 2.0 to 5 m and water depth is 10 m. Clearance is from 8 to 5 m. Each condition is shown in Table 2. The incident angle was 0°.

3.2.1 Results of Fluid Forces

From Fig. 9, as in the case of changing the water depth, as the draft increases, the effect of the water mass under the floating body increases, and the added mass coefficient increases. Also, it is slower to approach a zero compared to when the water depth changes.

From Fig. 10, there is a large difference in wave excitation force coefficient in this condition. Since there was not much change when the water depth was changed, the magnitude of wave excitation force coefficient is proportional to the submerged volume. The amplitudes of the excitation force increase as the clearance decreases.

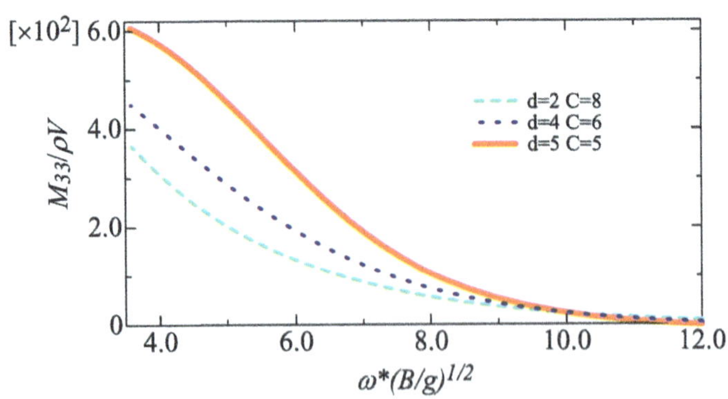

Fig. 9 Added mass coefficients of heave

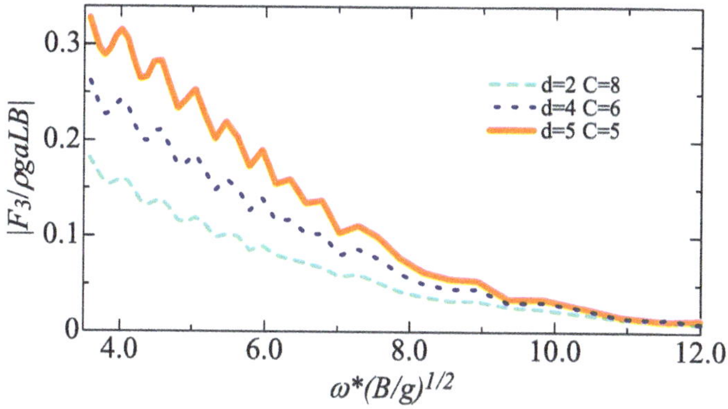

Fig. 10 Wave exciting forces coefficient of heave

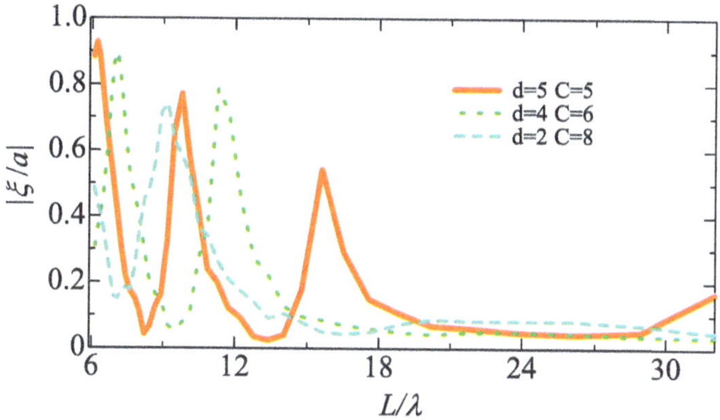

Fig. 11 Response function of vertical displacement at End of floating body

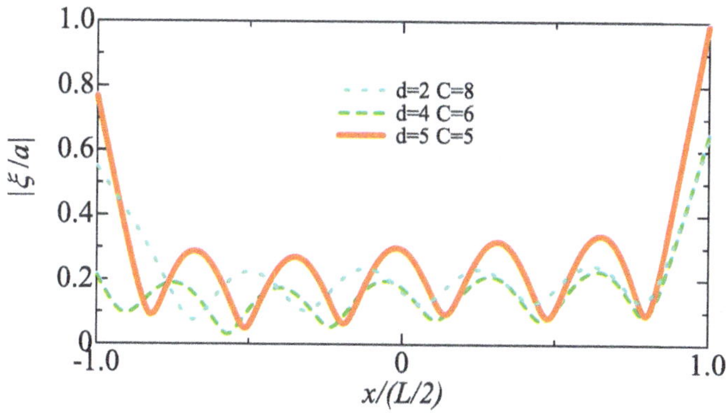

Fig. 12 Distribution of vertical displacement on y = 0 in case of L/λ = 6

3.2.2 Results of Elastic Motions

From Figs. 11, 12, the vertical response distributions are compared with $L/\lambda = 10$ where the wavelength ratio is the same value. Vertical axis is vertical displacement of the point of (−400, 0). The effect of the added mass coefficient on the elastic motion of the floating body is not so large. If the water depth is the same, the response peak shifts to the high frequency side as the draft is deepened.

The result of fluid force was not stable from d = 6 C = 4. Although the clearance was the same as when the water depth was changed, incorrect results were obtained when the draft was large.

4 Conclusions

In the paper, the effects of variation of shallow water depth and the clearance between bottom of a VLFS and sea bottom on motion response, wave exciting forces and the added-mass coefficient were discussed. From the investigation, it can be concluded as follows:

- The added mass increases according to the decrease of the clearance. The result is standard. The reasons would be that the fluid (water) can hardly move because of thin water region under the floating structure.
- The elastic motion is sensitively affected by variation of the clearance in very shallow water. (When water depth is relatively deep, ratio between length of a structure and wavelength significantly dominates rather than difference of water depth. From past my researches)
- The elastic motion is not reduced even if the water depth or the clearance is small. Then, resonance periods of elastic motion shifted to long period range. I think that the reason why the elastic motion cannot be decreased is large mass effects and the resonance in long periods.
- It is important to consider the clearance when we conduct numerical calculations of such as hydroelastic motions. However, the calculation becomes very sever because size of the clearance is close to that of calculation meshes to be distributed on the structure surface in order to predict the velocity potentials. It is necessary to appropriately change the size of the calculation mesh according to the clearance as well as against wavelengths.

References

1. Shuku M, Horiba S, Inoue S, Kobayashi E, Simamune S (2001) Overview of mega-float and its utilization, vol 38, No.2. Mitsubishi Heavy Industries, Ltd. Technical Review

2. Ikoma T, Masuda K, Rheem CK, Maeda H (2006) Response reduction of motion and steady wave drifting forces of floating bodies supported by aircushions in regular waves. In: Proceedings of the 25th international conference on OMAE'06, ASME, CD-ROM, OMAE2006-92072
3. Ikoma T, Kobayashi M, Masuda K, Rheem CK, Maeda H (2008) A prediction method of hydroelastic motion of aircushion type floating structures considering with draft effect into hydrodynamic forces. In: Proceedings of the 27th international conference on OMAE'08, ASME, CD-ROM, OMAE2008-57185

Measuring Hydroelastic Deformation of Very Flexible Floating Structures

Sebastian Schreier and Gunnar Jacobi

Abstract For Offshore Floating Photovoltaics (OFPV) applications, thin-film PV panels on lightweight floating support structures gain increasing scientific and commercial interest. Over the past years, several different concepts of thin-film OFPV have been proposed, with the common denominator of floating mattress or blanket-like support structures with very little draft in the order of centimeters compared to their width and length in the order of several tens to hundreds of meters. Mostly made from polymer foam materials, these floating support structures are more flexible than the conventional Very Large Floating Structures (VLFS) investigated in 1990s. The flexibility of a floating structure is expressed by the characteristic length derived from the ratio of the structural bending stiffness and the hydrostatic stiffness of the support. For conventional VLFS, this characteristic length is usually longer than the dominant wavelength of the ocean waves, resulting in only moderate structural deflections of the order of 1/10 of the wave height and the total thickness of the structure. The newly proposed structures have characteristic lengths of less than the wavelength of ocean waves. This allows the structures to move with the waves and follow the wave elevation like a floating blanket. Therefore, these structures are classified as Very Flexible Floating Structures (VFFS). Despite the growing interest in VFFS, little is still known about their hydroelastic deformation and their influence on the surrounding wave field. To start the experimental VFFS research at Delft University of Technology, Digital Image Correlation (DIC) measurements were carried out in this study to investigate the vertical deflection of a VFFS at model scale in a small towing. The model's characteristic length was 1/3 of the shortest wavelength and it was tested in long-crested regular longitudinal waves. The wavelength varied between 1/10 and 1/5 of the structure length. The measurements showed that the structure indeed

S. Schreier (✉) · G. Jacobi
Faculty 3mE, Department M&TT, Section SHS, Delft University of Technology, Mekelweg 2, Delft 2628 CD, The Netherlands
e-mail: s.schreier@tudelft.nl

G. Jacobi
e-mail: g.jacobi@tudelft.nl

Ł. Piątek et al. (eds.), *WCFS2020*, Lecture Notes in Civil Engineering 158,
https://doi.org/10.1007/978-981-16-2256-4_21

mostly followed the wave elevation and revealed 3D effects across the structure, which require deeper investigation into wave scattering of VFFS.

Keywords Offshore floating PV · Digital image correlation (DIC) · Hydroelasticity · Wave scattering · Very flexible floating structures (VFFS)

1 Introduction

Over the past decennia, investigation of hydroelastic response of floating structures was mainly motivated by research on sea ice and Very Large Floating Structures (VLFS), see the reviews of Karmakar et al. [1] as well as Squire [2, 3], Chen et al. [4], and Lamas-Pardo et al. [5]. As Squire [6] points out, there are modelling parallels between VLFS and sea ice investigations. On the other hand, Lamas-Pardo et al. [5] observe that none of the designed VLFS have ever been built, with the sole exception of the Mega-Float floating runway [7]. However, these projects did spark the scientific interest and progress in the field of hydroelasticity.

More recently, hydroelasticity receives renewed attention with the rise of (offshore) floating photovoltaics (FPV). Large modular floating structures for rigid PV panels are envisaged in various projects as summarized by Trapani et al. [8]. Flexible structures for FPV based on thin-film PV modules are demonstrated to have technical and economic potential [9, 10]. Jamalludin et al. conclude "that floating solar could be one of the most important ocean structures in the future" [11].

From the aforementioned reviews, it becomes apparent that there is an abundance of theoretical studies on hydroelasticity. However, experimental investigations and data are much less reported. Several experiments related to the Mega-Float VLFS project are reported in the late 1990s. Yago and Endo investigated a 9.75 m long, 1.95 m wide model with a draft of 16.6 mm [12]. Yago et al. conducted 2D experiments with a 50 m long, 5 m wide model with a draft of 10 mm [13]. In both studies, Yago et al. used models made of composite material to tune the model stiffness to match the scaled-down bending stiffness of the full-scale structure, which had horizontal dimensions of 300 m $\times$ 60 m and 5000 m 1000 m, respectively. They measured the structure elevation along the centerline of the model with a string-and-pulley system attached to potentiometers at intervals of 0.4 m. Kagemoto et al. employed optical tracking of 4 LEDs along the centerline of their model with length of 2 m and width of 0.5 m [14]. Their model consisted of segmented floating blocks that were attached to a flexible upper deck of 5 mm acrylic glass. Ohta et al. reported on a model test with a 15 m long model related to a full-scale structure of 1200 m in length [15]. Ohmatsu discussed several experiments related to the Mega-Float project [16]. He advised the use of potentiometers for the measurement of the structural deflection and elaborated on the issue of scaling the bending stiffness of the models.

In the field of sea-ice interaction, Squire conducted experiments with an ice-covered wave flume of 2 m $\times$ 1 m using an accelerometer to measure the vertical deflection of the ice cover [17]. Meylan employed 40 pressure transducers along the bottom of a wave flume of 20 m long and 1.83 m wide to investigate the effect of floating flexible sheets on the waves. He used polypropylene (PP) sheets over the full width of the flume with length of 1.22 m and thickness of 3.175 mm as well as 6.35 mm [18]. Sakai and Hanai carried out experiments with polyethylene (PE) sheets of 5 and 20 mm thickness to simulate sea ice in waves in a 2D setup [19]. They measured the vertical displacement of the sheet at several points along the centerline of the model using ultrasonic sensors. Montiel et al. investigated the behavior of compliant floating circular discs in waves [20]. Their models had a radius of 0.72 m and were made from expanded Polyvinylchloride (PVC) material of 3, 5, and 10 mm thickness. They used optical markers distributed over half of the model surface that were individually tracked by a stereoscopic system. Bennetts et al. conducted experiments in a 3D setup using PVC and PP material for their 1 m wide models, which they varied in length between 1 m and 3 m and in thickness from 5 mm to 19 mm and 20 mm, resp. [21]. Sree et al. conducted 2D experiments with viscoelastic sheets motivated by wave-ice interaction of broken ice [22]. They used oil-doped Polydimethylsiloxane (PDMS, target density 940 kg/m^3) and PP as material for their models. The model length was 1 and 2.45 m in an 8 m long, 0.3 m wide wave flume. The tests were carried out in a 2D setup with specimens covering the full width of wave flume. Ultrasonic height probes were used to measure the structural elevation along the centerline of the model. Sree et al. concluded that under the stiffer PP model, wave lengthening occurred compared to the open water wave, while for the longer waves in the experiment, wave shortening was observed under the viscoelastic PDMS models. Both effects could be related to a viscoelastic model presented by Wang and Shen [23]. With common hydroelastic theory, only wave lengthening is expected for conventional VLFS as reported e.g. by Ertekin and Kim [24].

One of the most governing parameters in hydroelastic response of floating structures is the characteristic length λ_c defined by Suzuki and Yoshida [25], which expresses the ratio of structural bending stiffness EI to hydrostatic stiffness of the support k_c according to Eq. (1) [26].

$$\lambda_c = 2\pi \left(\frac{EI}{k_c}\right)^{\frac{1}{4}} \tag{1}$$

where the hydrostatic stiffness of the support is given by Eq. (2) from the density ρ of the water, gravitational acceleration g, and the width B of the beam.

$$k_c = \rho g B \tag{2}$$

Using the material properties reported for the aforementioned experiments at model scale, the characteristic length was calculated. Based on these results, the relative wavelength λ/λ_c was set out against the relative structural length L/λ in Fig. 1 with wavelength λ and structural length L.

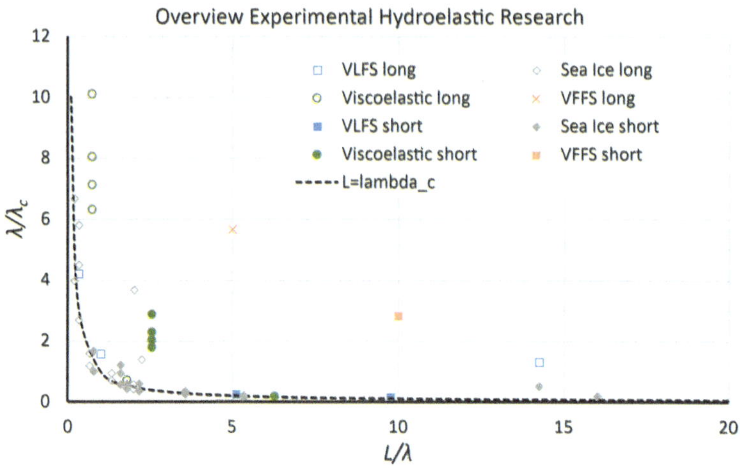

Fig. 1 Overview of experimental hydroelastic research. Open symbols are results with the longest wavelength reported in the references, closed symbols pertain to the shortest wavelength

From the overview in Fig. 1, it becomes apparent that most experiments were conducted at conditions where the characteristic length λ_c was close to the structural length L as indicated by the dashed line representing $L/\lambda_c = 1$. For sea ice, this allocation can be interpreted that ice floes break up at lengths close to their characteristic length. For VLFS, the structural stiffness resulted from the intended application as floating airport and matched well with the characteristics of natural ice. The clustered outliers indicated by the open and closed circles are the viscoelastic models investigated by Sree et al. [22]. The individual open and closed circle located along the dashed line are the reference models of that study made from PP material.

The orange x and square in Fig. 1 mark the model tests of VFFS reported in this study. From this overview it becomes obvious that VFFS have much shorter characteristic length compared to the wavelength than both sea ice and conventional VLFS. Therefore, exploratory model tests with VFFS were carried out to obtain a first impression of their hydroelastic response in waves. For this purpose, we employed a 5 mm thick sheet of neoprene foam in long-crested regular waves and measured the vertical deflection using 3D Digital Image Correlation (DIC).

2 Model Experiments

2.1 Test Setup

The experiments were carried out in the No. 2 towing tank of the Ship Hydromechanics Lab of Delft University of Technology. The tank is 80 m long and

2.75 m wide. Water depth for the experiments was 1.00 m and water temperature was 17 °C. The tank is equipped with a single-paddle wave maker with adjustable hinge point at one end and a wave-damping beach at the other. We used the wave maker in regular flap-type mode with hinge point at the bottom of the tank.

The model structure was a closed-pore neoprene foam rubber sheet of $L = 4.95$ m long and $B = 1.02$ m wide with a thickness of $h = 5$ mm. The density of the neoprene material was 116 kg/m^3. The plate bending stiffness D of the sheet was determined from a static deflection test with a localized load to $D = \frac{Eh^3}{12(1-\nu^2)} = 6.9 \cdot 10^{-3}$Nm. With an assumed Poisson ratio for soft materials of $\nu = 0.4$, the Young's modulus became $E = 560$ kPa and the characteristic length of this model was $\lambda_c = 0.17$ m.

In transverse direction, the model was moored in the center of the tank by four mooring lines connected to the sidewalls of the tank. The front mooring lines towards the wave maker were Dyneema lines with very high stiffness and diameter of 0.2 mm. The aft mooring lines towards the beach were soft mooring lines made from an elastic sewing thread with <1 mm diameter. The aft mooring lines were pre-stretched with an elongation of $\epsilon = \Delta L / L_0 = 15 - 20\%$ resulting in a pretension of 0.1 N and a mooring line stiffness at this elongation of $dF(\epsilon)/d\epsilon = 0.475$N. The mooring lines were attached to the corners of the neoprene sheet by small pad eyes made from a loop of adhesive tape to avoid the load concentration of a hole punched through the model. To further limit the influence of the mooring lines on the structural elevation in waves, the mooring lines were 5 m long and connected to the tank walls to either side of the model at the height of the undisturbed waterline. The longitudinal positions of the mooring connectors on the tank wall are given in Table 1. The setup with stiff front mooring lines and elastic aft lines was chosen to keep the model in the field of view of the cameras without applying excessive pretension on the model.

With the available DIC cameras and the geometric constraints of the laboratory building, it was not possible to bring the whole model structure into the field of view of the cameras. Due to the elaborate camera setup and time-consuming alignment and calibration procedure of the DIC system, we decided to leave the cameras in place and shift the model in the tank to observe the deflections of the full model via repeated test runs. The basic setup is shown in Fig. 2 with the model in its most forward position (position 1), in which the cameras saw the aft section of

Table 1 Longitudinal position (in meters) of wave probes and mooring points with respect to the wave maker flap in its upright position

Position		1	2	3
Wave probe 1	*dwp1* (m)	25.84	25.84	25.84
Wave probe 2	*dwp2* (m)	29.72	29.72	29.72
Wave probe 3	*dwp3* (m)	38.35	38.35	38.35
Front mooring line	*df* (m)	20.85	22.55	24.25
Aft mooring line	*da* (m)	35.94	37.64	39.34
Front model edge	*def* (m)	25.77	27.47	29.17
Aft model edge	*dea* (m)	30.72	32.42	34.12

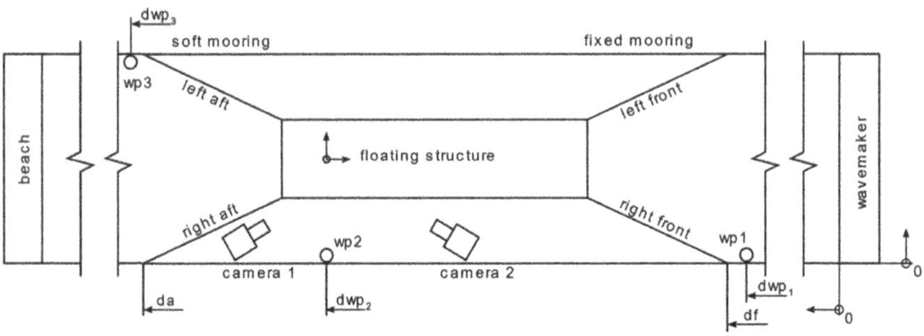

Fig. 2 Plan view of the test setup

the model. In position 2, the mid-section of the model was seen by the cameras, and in position 3, the front section of the model was in the field of view. The model was shifted by relocating the attachment points of the mooring lines on the tank walls in longitudinal direction.

To measure the wave elevation around the model, three resistance-based wave probes WP1-WP3 were placed along the tank wall. WP1 was located between the wavemaker and the model. WP2 was placed next to the model at the field of view of the DIC cameras. WP3 was positioned behind the model towards the beach. The longitudinal positions of wave probes (*dwp1-3*) and mooring points (*df*, *da*) with respect to the wave maker flap in its upright position are summarized in Table 1. A picture of the test setup in the towing tank is shown in Fig. 3.

2.1.1 Wave Probe Calibration

The wave probes were calibrated every morning before the first test run by submerging the probes over a range of 180 mm in steps of 20 mm into the water with 30 s immersion time per step and averaging the results per step over 15 s. The sensitivity was then determined by linear regression of the resulting data points. The correlation coefficient was $R^2 \geq 0.9999$ for all calibrations. The largest offset of a calibration point from the resulting regression line was 0.7% with respect to the calibration range of 180 mm. After calibration, the height of the wave probes was adjusted such that the undisturbed water surface was in the middle of the calibrated range.

2.2 Optical Surface Deformation Measurements

The shape of the floating structure and its deformation under the influence of the incoming waves is determined via Digital Image Correlation (DIC). This is a fully non-intrusive, optical full-field technique, where images of the object under

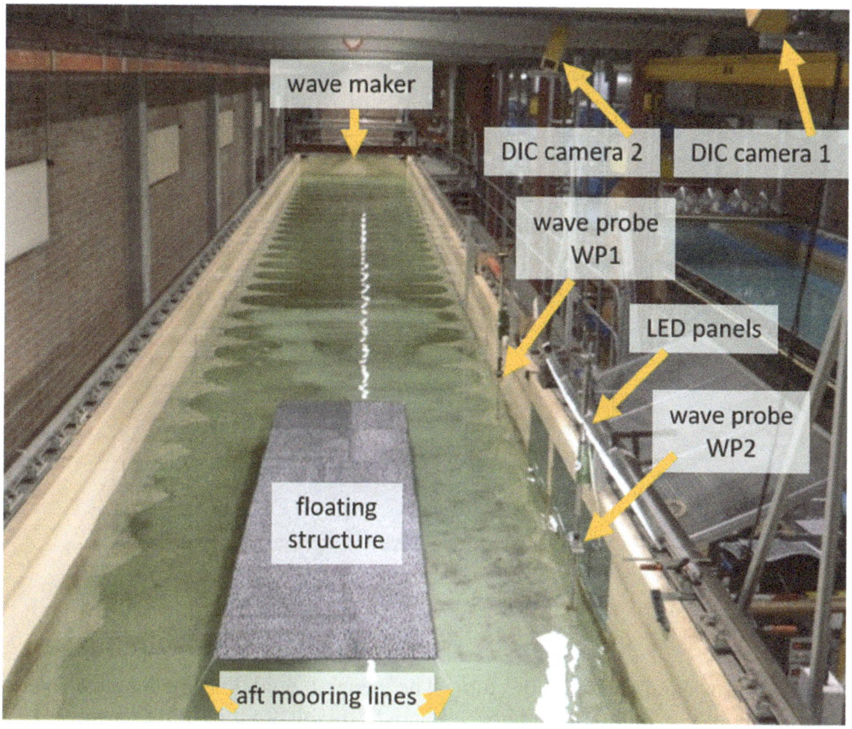

Fig. 3 Test setup in the towing tank

investigation are recorded, and a point-to-point mapping from an undeformed image to the deformed image allows the measurement of the surface displacement [27]. For a successful correlation of successive images, the object surface needs to have a distinct pattern that produces varying intensities of the reflected light. The recorded images are split into subsets to determine the shift of each subset with respect to the reference images. While a single camera setup allows for the measurement of planar deformations, the usage of two or more cameras allows for a reconstruction of the complete three-dimensional deformation vector of the structure by making use of the stereovision principle. The latter was used for the present study, where the initial position of the structure was found from the triangulation of the images of two cameras, and subsequently the deformation was obtained via cross-correlation.

Figure 4 shows the optical setup used for the analysis in this study. The cameras were mounted to the side of the towing tank with a distance of 6 m with respect to each other at a height of 2.5 m above the water surface.

The cameras were LaVision imager MX models with a resolution of 4 megapixels. Equipped with 28 mm lenses and a Scheimpflug adapter, the cameras' field of view covered a model section of 1.8 m. In the current setup, the image

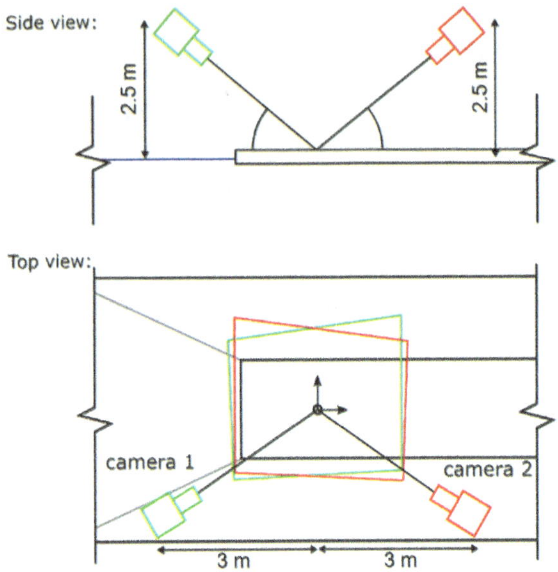

Fig. 4 Optical setup of the DIC system

pixel size was equivalent to 0.7 mm. The general guideline is that the size of speckles should be in the range between 3 and 7 pixels. In this study the size of the speckles applied to the model structure was 9 mm. This yielded an average size of 11–13 pixels per speckle. The model structure with the applied speckle pattern is depicted in Fig. 5. To minimize modification of the mechanical parameters of the neoprene sheet, the speckles were applied with flexible rubber spray paint. For the application of the random pattern, a template of 600 mm × 900 mm was used. To enhance the contrast of the speckle pattern, the measurement area was illuminated with two 45-Watt LED panels.

Fig. 5 Model structure with speckle pattern in the towing tank. Wave propagating from right to left

2.2.1 DIC Calibration

The accuracy of the reconstructed surface elevation depends on the exact determination of the intrinsic and extrinsic parameters of the optical system. These were obtained with a pinhole camera model and a dot-pattern calibration target, which was placed into the field of view prior to the experiments. The size of the target was 1000 mm × 1000 mm with black circular dots on white background in a square grid of 75 mm × 75 mm center to center. The target was carefully aligned with the centerline of the towing tank and the water surface. To enable a comparison of the surface elevation with the data from the wave probes, the center of the calibration target, which defined the model coordinate system, was placed at the longitudinal position of wave probe WP2. Based on the calibration, the images obtained by the DIC cameras were transformed from the image plane back to the real-world coordinate system. In these de-warped images, the model appeared as a rectangle again. Figure 6 shows a de-warped image of the model with its speckle pattern in the tank obtained by camera 1.

2.3 *Test Conditions*

In this study, we investigated the model structure in regular waves with a wavelength λ in the range of $L/\lambda = 5 \ldots 20$ and wave steepness in the range of $H/\lambda = 0.02 \ldots 0.05$ based on wave height $H = 2A$, A being the wave amplitude. In this paper we concentrated on the test runs with wavelength $L/\lambda = 5$ and $L/\lambda = 10$. The wave conditions are summarized in Table 2.

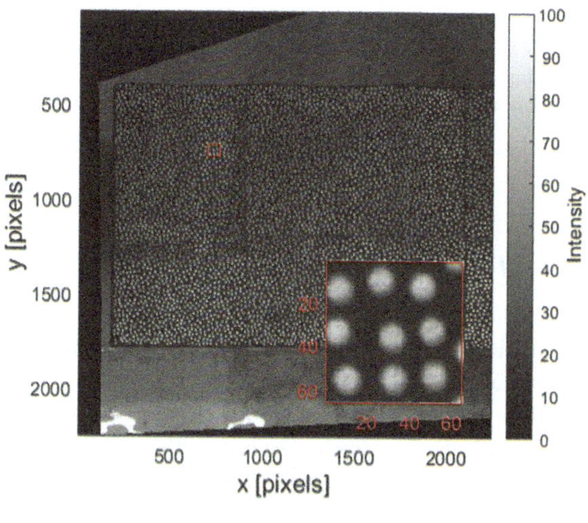

Fig. 6 De-warped image of camera 1 showing the aft section of the model with speckle pattern. Enlarged box illustrates the subset size for DIC measurement and the size of speckles

Table 2 Summary of nominal test conditions

Wave condition	Period	Wavelength	Rel. structure length	Rel. wavelength	Wave amplitude	Wave steepness
	T (s)	λ (m)	L/λ (–)	λ/λ_c (–)	A (mm)	$2A/\lambda$ (–)
W05	0.563	0.495	10.00	2.83	10	0.040
W10	0.796	0.990	5.00	5.67	20	0.040

2.4 *Data Acquisition*

2.4.1 Wave Probes and Trigger

Each wave probe was connected to its own amplifier, which provided a ±10 V analogue output signal. For synchronization, a TTL trigger signal with low level of 0 V and high level of 3.5 V was generated, which was linked to the steering signal of the wave maker. This trigger signal was connected to the DIC image acquisition system and was recorded along with the wave probe data.

The wave probe and trigger signals were fed through an analogue low-pass filter with a cut-off frequency of 100 Hz. The filtered signals were then recorded by a data acquisition PC with a National Instruments A/D converter card PCI-6033E at a sampling rate of 1000 Hz and with 16-bit resolution.

2.4.2 DIC Acquisition

The DIC images were recorded by a dedicated image acquisition computer, which controlled the DIC cameras by the DaVis 10 software package from LaVision via the timing unit also provided by LaVision. For sufficient temporal resolution, images were recorded at an acquisition rate of 125 frames per second. This translated to a number of images per wave period of 70.4 for wave condition W05 and 99.5 for wave condition W10. The rising flank of the aforementioned trigger signal started image acquisition.

2.5 *Test Procedure*

After the model was in position, air bubbles under the model and water accumulations on the model were removed to ensure that all measured deformations were due to the wave and not due to preloading by accidental bubbles under or water on the model. Before each test run with waves and DIC measurements, zero measurements were taken for the wave probes.

The DIC measurement runs were only started when the amplitude of any residual waves in the tank was less than 0.5 mm. Once this condition was met, wave generation and data acquisition for the wave probes and trigger signal were started. The wave generation began with a start-up ramp of 5 s, extended to the next integer multiple of the wave period. After the start-up ramp, monochromatic waves were generated for a duration of 150 wave periods, which was followed by a ramp-down over the same duration as the start-up ramp.

The start trigger for the DIC acquisition was given after the first 50 waves had theoretically passed the wave probe WP2. With the trigger delay, we allowed the wave to develop fully and any start-up effects on the structural deflection to have settled. DIC images were acquired over a duration of 10 s resulting in 1250 images covering 17.8 wave periods for wave condition W05 and 12.6 wave periods for wave condition W10. Figure 7 shows the time series of wave probe WP2 and the trigger starting the DIC acquisition on its rising flank. Furthermore, the timeframe of DIC acquisition is marked with the two red vertical bars.

Data acquisition for wave probes was stopped after 300 s when the last wave had passed the model in all conditions. After each test run, a waiting time was observed until the water in the tank was settled enough to start the next measurement.

Due to the limited camera field of view, the model was shifted in steps to measure the deflections over the entire model. The measurements began with the model in the foremost position 1, in which the DIC cameras were focused on the aft end of the model. After that, we shifted the model in 2 steps of each 1.7 m aft to

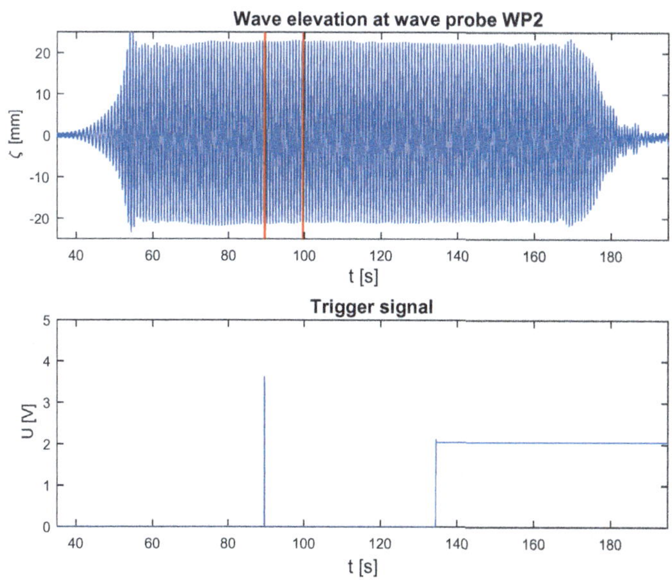

Fig. 7 Time series of wave probe WP2 and DIC trigger showing the DIC acquisition window as the two red vertical bars. Wave condition W10, $L/\lambda = 5$

Table 3 Allocation of test conditions to test runs

Wave condition	Test runs with model in position			
	1	2	3	No model
W05	R48*, R58, R73	R75	R83	R87
W10	R28, R69	R77	R85	R90

*Only DIC data available

positions 2 and 3, where the cameras looked at the mid and front section of the model, respectively.

For reference, one test run per condition was performed without the model in the tank. The allocation of test runs to wave conditions is summarized in Table 3.

2.5.1 Repeatability of Wave Conditions

With the approach to build up the deformation information over the full model from repeated test runs with a shifted model, the test conditions needed to match closely for the individual runs to obtain reasonable results. This repeatability is demonstrated in Figs. 8 and 9 for wave conditions W05 and W10, respectively, by plotting the section of the test runs, which was used for this study, in the same plot. In both figures, the top graph shows the wave elevation ζ at wave probe WP1 in front of the model normalized by the actual wave amplitude per test run, while the bottom graph shows the normalized wave elevation at wave probe WP2 at the longitudinal center of the camera field of view. In all four plots of these figures, the waves show good repeatability. Per plot, the five lines were in almost perfect agreement with the same period and amplitudes, which were within 1 mm of each other.

2.6 Data Processing

For the wave probes, the processing was limited to removing the initial offset in the water level, trimming the signals to the DIC acquisition window, and removing a small time shift between the signals. The resulting signals are shown in Figs. 8 and 9 normalized by the actual wave amplitude per run, which are given in Fig. 11.

The recorded DIC images were processed with the Strain Master package from LaVision. For this processing, the subset size was chosen to 63 pixels with a step size of 21 pixels. With one pixel corresponding to 0.7 mm, this resulted in a final spatial resolution of 15 mm or 33 data points along the shorter wavelength investigated in this paper. Over the model in the camera field of view, a total of 65×120 data points was obtained.

The surface elevation reconstruction was carried out in two steps. First the initial location of the surface was obtained by a triangulation of the reference image taken before the test run. Subsequently, the displacement per subset area was obtained by

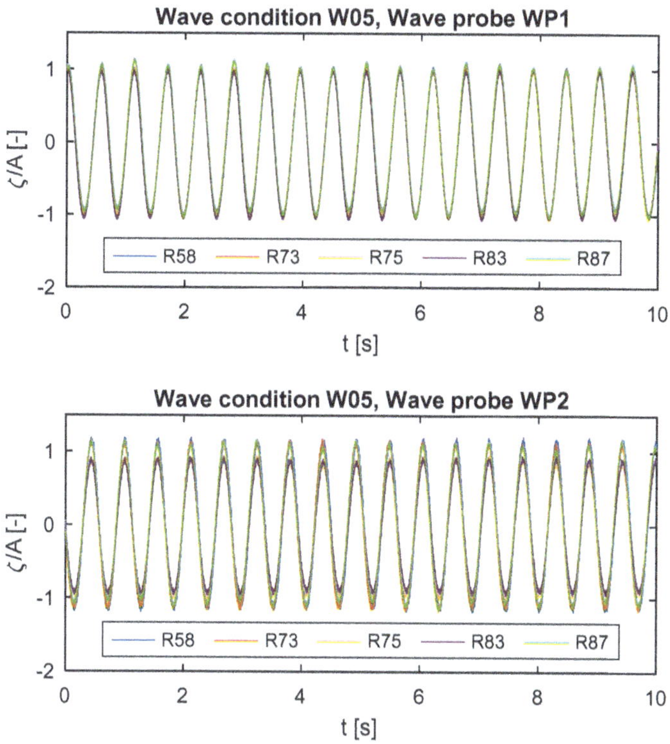

Fig. 8 Repeatability of generated waves. Wave condition W05, $L/\lambda = 10$

cross correlation of the images. Based on this displacement per subset, the elevation of the model surface was reconstructed. Figure 10 shows a 3D representation of the deformed model with the speckle pattern overlaid with the color-coded surface elevation.

For further assessment, the surface elevation fields obtained from Strain Master calculations were exported to Matlab. As a first step, the elevation was plotted as surface plots over the observed area of the model per model position. Figure 11 shows a snapshot per test run with wave condition W05 in the left column and wave condition W10 in the right one. From top to bottom, the plots show the aft, mid, and front section of the model, respectively. In all graphs, the structure elevation was normalized by the actual wave amplitude *A* of the incoming wave.

Along the edges of the model, the displacement calculations were somewhat spotty, depending on the distance of the first speckles from the edge. These frayed edges can be seen in Fig. 10. For further presentation of the model surface elevation, the longitudinal edges were trimmed down to the area with continuous data available. At the front and aft edge, the blank spots remained in the data sets and can be seen e.g. on the right edge in Fig. 11 for runs R83 and R85.

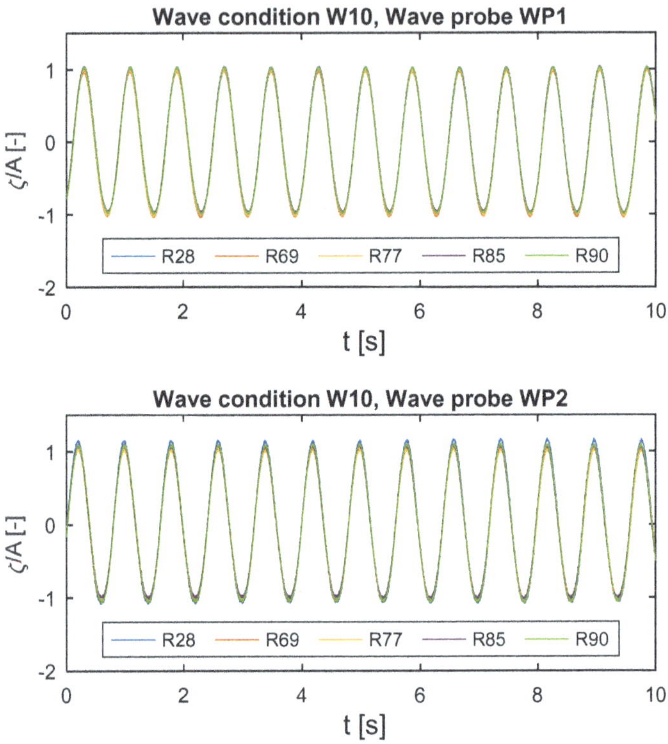

Fig. 9 Repeatability of generated waves. Wave condition W10, $L/\lambda = 5$

Fig. 10 Superposition of DIC results for model surface elevation and deformed model in waves. Wave propagation from right to left. Wave condition W05, $L/\lambda = 10$

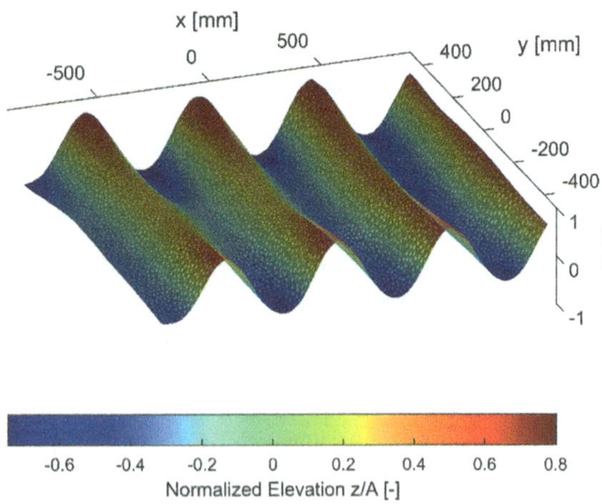

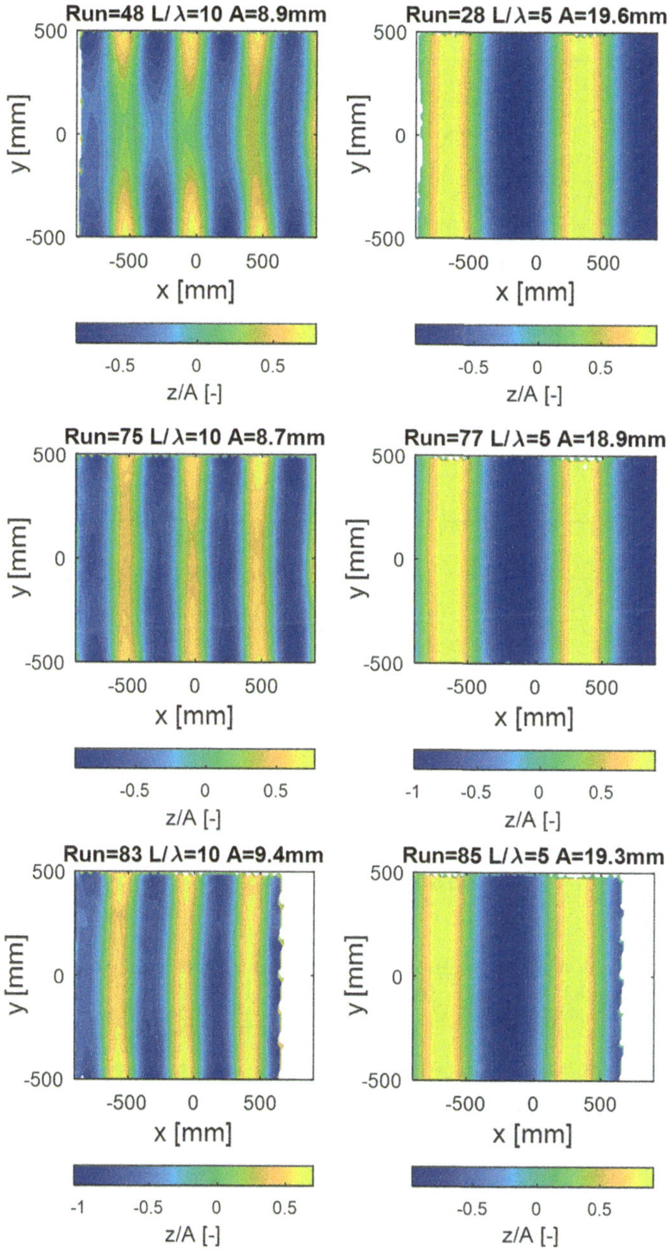

Fig. 11 DIC results of model surface elevation. From top to bottom, snapshots of individual DIC runs looking at the aft, mid, and front section of the model. Left column: wave condition W05, $L/\lambda = 10$, right column: wave condition W10, $L/\lambda = 5$. Wave propagation from right to left

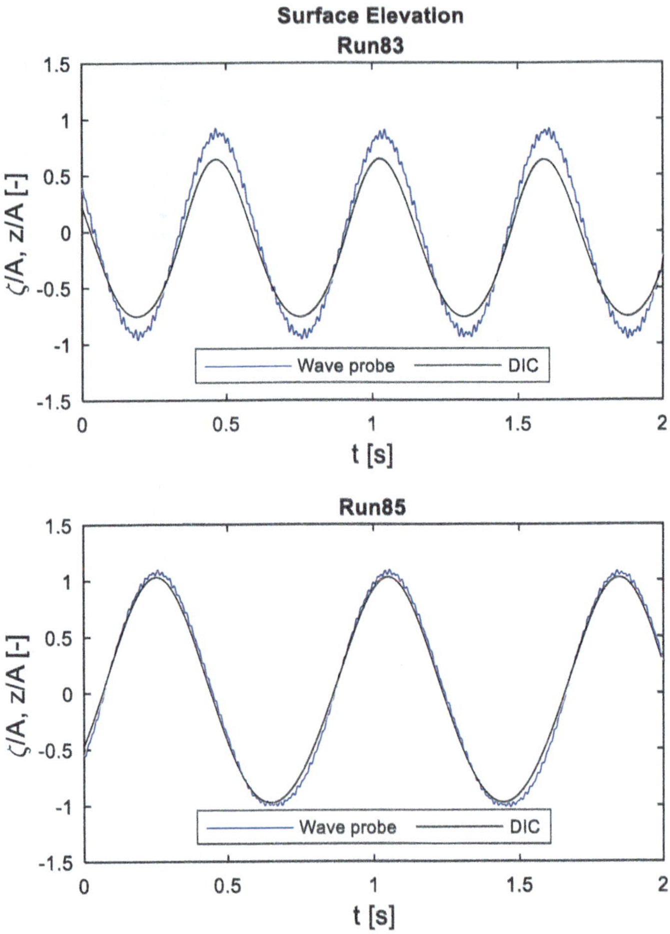

Fig. 12 Comparison of structure elevation and wave elevation at longitudinal position of wave probe 2. Top: run R83, W05, $L/\lambda = 10$; bottom: run R85, W10, $L/\lambda = 5$

To compare the structure elevation with the wave elevation around the model, a time trace of the structure elevation at the centerline of the model at the longitudinal position of wave probe WP2 was generated from the DIC results. This time trace was then plotted together with the time trace of the wave elevation obtained from wave probe WP2. The resulting plot is shown in Fig. 12.

For the investigation of the surface elevation over the entire length of the model, the three elevation fields per model of Fig. 11 were stitched together in Matlab to one elevation field per model. Given that the camera field of view was fixed with respect to the tank as well as the wave probe WP2 and the model was moved aft through the field of view of the cameras, there was a phase shift between the model and the repeated wave trains. During the DIC acquisition window, the waves were in a

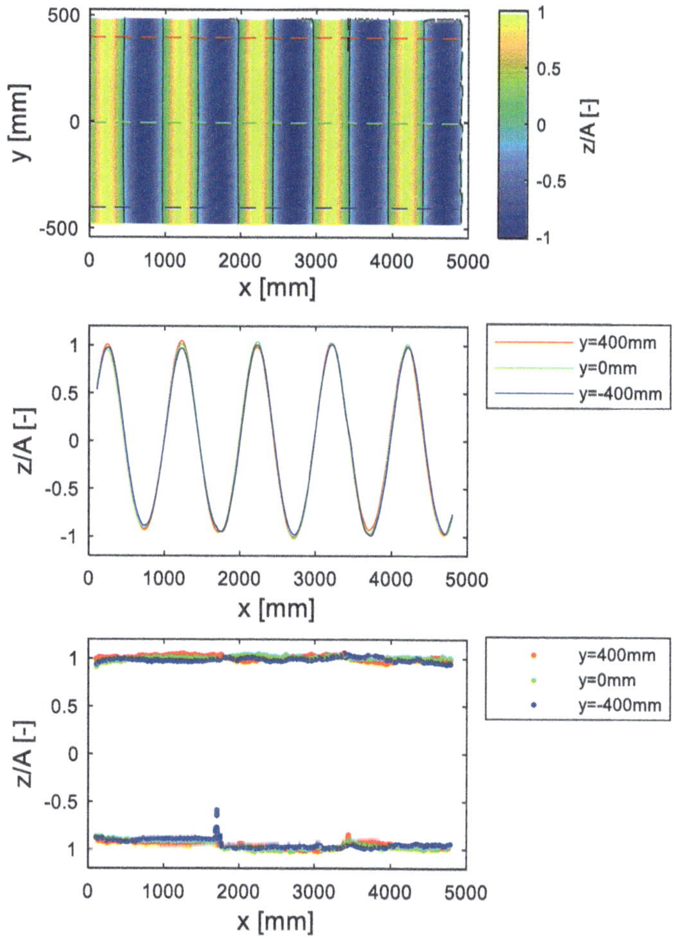

Fig. 13 DIC results for wave condition W10, $L/\lambda = 5$. From top to bottom: surface elevation, elevation profiles, and envelope curves of elevation profiles. Wave propagation from right to left. Surface elevation normalized by actual wave amplitude A

steady-state condition as can be seen from the constant amplitudes in Figs. 8 and 9 as well as Fig. 7. Therefore, the displacement fields were only phase-shifted to match the phase of the neighboring displacement field and not exactly the particular wave of the repeated wave train. To highlight the wave shape of the structure elevation, the $z/A = 0$ contour was marked in the model surface elevation plots. Snapshots of the resulting images are shown in the top graphs of Figs. 13 and 14.

To look into possible variations over the width of the model, three elevation profiles along a longitudinal line down the length of the model were generated. The profiles were taken at the centerline as well as 400 mm to either side of it. The location of the elevation profiles in the model are marked with the dashed lines in

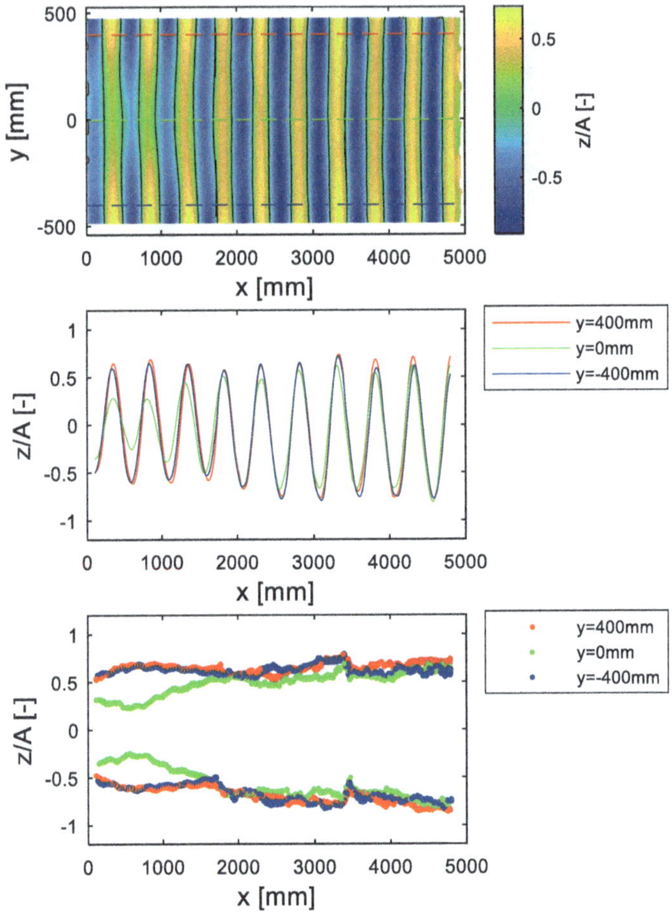

Fig. 14 DIC results for wave condition W05, $L/\lambda = 10$. From top to bottom: surface elevation, elevation profiles, and envelope curves of elevation profiles. Wave propagation from right to left. Surface elevation normalized by actual wave amplitude *A*

the surface elevation graphs. Snapshots of the surface elevation profiles are shown in the middle graphs of Figs. 13 and 14.

The elevation profiles were also used to assess the repeatability of the structure elevation in repeated wave conditions. For this purpose, the elevation profiles of the aft section of the model were plotted together in one plot for two repeated test runs of both wave condition W05 and W10 in Fig. 15.

For the evaluation of the spatiotemporal evolution of the surface elevation over several wave periods and the entire length of the model, the envelope curves of the elevation peaks and troughs along the profile lines were drawn. These envelope curves were generated by applying a Matlab peak finding algorithm to the elevation

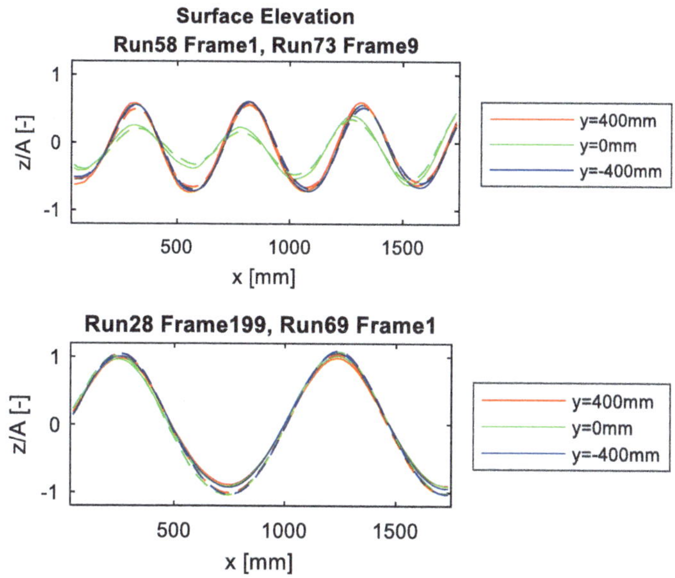

Fig. 15 Repeatability of structure elevation profiles of the aft section of the model for each two test runs with repeated wave condition. Top: runs R58 (continuous lines) and R73 (dashed lines), W05, L/λ = 10; bottom: runs R28 (continuous lines) and R69 (dashed lines), W10, L/λ = 5. Wave propagation from right to left

profiles per time step. The accumulated data of all time steps is shown in the bottom graph of Figs. 13 and 14.

3 Results

The results are presented in four steps. First, the elevation of the structure and the waves is evaluated in one location for both wave conditions W10 and W05. Then the surface elevation over the entire model is assessed in consecutive order for wave condition W10 and W05. Eventually, the repeatability of the structural response in addressed.

As can be seen from the right column of Fig. 11, the surface elevation of the model from the DIC measurements for wave condition W10 showed an amplitude closely matching the wave amplitude. This is also visible in the bottom graph of Fig. 12 showing the results of test run R85 with wave condition W10, while the top graph was taken from test run R83 with wave condition W05. For both wave conditions, both wave elevation (blue line) and surface elevation (black line) of the model showed a harmonic oscillation with the same frequency. The elevation amplitude in both cases was smaller than the measured wave amplitude. This effect was more pronounced with the smaller and shorter waves in the top graph of

Fig. 12, and was barely visible for the longer waves of run R85 in the bottom graph of this figure. This was a clear indication of a hydroelastic interaction between the model structure and the shorter wave. However, for this particular model, the details of this interaction are not known yet and require further investigation.

3.1 Wave Condition W10

The overview of the results from the DIC measurements with wave condition W10 with $L/\lambda = 5$, $\lambda/\lambda_c = 5.67$ and nominal wave amplitude $A = 20$ mm is given in Fig. 13. In the top graph of this figure, the surface elevation of the model is shown over the entire measured surface of the model with yellow and blue colors indicating wave crests and troughs, respectively. In all three plots of this figure, the origin of the x-axis coincides with the aft edge of the model. The waves propagated into the negative x-direction from right to left. From this graph, it is clearly visible that there were five wavelengths covering the length of the model. The $z/A = 0$ elevation contours are straight vertical lines in this graph, indicating that the waves propagated under the model as the 2D long-crested waves as they arrived at the model.

This assessment is supported by the elevation profiles over the length of the model depicted in the middle graph, with their location indicated in dashed lines in the top graph. The color of the profile line corresponds to the color of the dashed line, with a green line indicating the profile along the centerline and red and blue lines representing the profiles at $y = 400$ mm and $y = -400$ mm, respectively. In this wave condition, all three elevation profiles match perfectly with normalized amplitudes of 1.0 and exactly the same wavelength. There are only minor deviations at the crests and troughs. These are especially visible at the crest around $x = 1250$ mm and the trough around $x = 3700$ mm. Looking at the five troughs in this graph, a slight upward trend can be noted following the waves from right to left. This indicates that the wave trough got slightly shallower over the length of the model.

The bottom graph of Fig. 13 confirms that there was little change of the waves over the length of the model. The envelope curves of crests and troughs in the model surface elevation are depicted with the same colors as their corresponding elevation profiles. There are hardly any differences between the three lines over the width of the model. The few outliers in the curve for $y = -400$ mm at $x = 1700$ mm are an artifact that can be attributed to the stitching of the results of different test runs. At closer inspection, the rising trend of the troughs in wave propagation direction is confirmed, though the difference from first to last wave trough is only 5–10% of the actual wave amplitude. Close to the aft edge of the model, both elevation crests and troughs show a reduced magnitude. However, also this is only a minor effect.

Based on these results, the structure appeared to have little impact on the waves passing underneath it for the wave condition W10 with the longer waves considered in this study at relative structural length of $L/\lambda = 5$ and relative wavelength of $\lambda/\lambda_c = 5.67$.

3.2 Wave Condition W05

In Fig. 14, the overview of the DIC measurement results is shown for the wave condition W05 with $L/\lambda = 10$, $\lambda/\lambda_c = 2.83$ and nominal wave amplitude $A = 10$ mm. The figure follows the same structure and color-coding as Fig. 13.

In the top graph of Fig. 14, the 10 wavelengths over the length of the model are clearly visible. Following the wave propagation along the model from right to left, increasing 3D effects on the wave shape become apparent. While the crests were highest and the troughs were deepest at the sides of the model, the elevation amplitude reduced towards the centerline. Furthermore, the $z/A = 0$ elevation contours were not straight vertical line anymore but were bent towards the center of the model. For the first 4 wavelengths, i.e. for $x > 3000$ mm, the wave front showed a convex shape with the contour line at the center being more advanced than towards the sides of the model. Then there was a zone of 2000 mm $< x <$ 3000 mm, where the wave contours were almost straight lines. Towards the aft end of the model, the 3D effects became more pronounced with more clearly visible convex shape of the wave contours at 600 mm $< x <$ 2000 mm. The very last wave in the model even shows a concave shape with the crest at the center lagging behind the crest at the sides. These increased 3D effects went along with a reduced wave amplitude, which was more pronounced around the centerline compared to the sides of the model.

The elevation profiles and the envelope curves confirm the presences of 3D effects in the surface elevation. Over roughly the front half of the model, there was only a small phase shift between the elevation profile at the centerline and the two profiles at the sides. The elevation amplitude was only slightly smaller at the centerline than at the sides. In the aft part, for 600 mm $< x <$ 1800 mm, the green curve of the surface elevation at the centerline was clearly shifted towards the left in this graph, thus indicating an increased wavelength at the center of the model compared to the sides. At the same time, the elevation amplitude at the centerline is only half that at the sides of the model. Over the whole length of the model, there is good lateral symmetry in the surface elevation of the model, with the elevation profiles at $y = \pm 400$ mm matching well with each other.

The envelope curves underline the good lateral symmetry of the model surface elevation and only show small differences between the sides and the center over the front 2/3 of the model. Over this part of the model, the elevation crests and troughs reduce almost linearly with the distance of the waves travelled under the model. Over the aft 1/3 of the model length, the sides and center showed opposing trends. While the crest height and trough depth continued to reduce at a higher rate in the center for 500 mm < x < 1500 mm, the crest height and trough depth increased at the sides. Thus, the differences between center and sides increased. Over the aft 500 mm, i.e. the last wavelength under the model, both trends reversed again and the difference in crest and trough height reduced towards the aft edge of the model. Around x = 500 mm, the difference was largest, such that the elevation amplitude at the center was only about half that at the sides. This confirmed that the profile of the snapshot in the upper two graphs were representative of the general evolution of the model surface elevation.

For this wave condition W05 with relative structural length $L/\lambda = 10$ and relative wavelength of $\lambda/\lambda_c = 2.83$, a clear impact of the structure on the waves underneath it is visible from the DIC measurements.

3.3 Repeatability of Structure Elevation

The repeatability of the wave conditions was established in Figs. 8 and 9. In Fig. 15, the surface elevation profiles were plotted for two repeated test runs with wave condition W05 (top) and wave condition W10 (bottom) over the aft section of the model. As in Fig. 14, also in the top graph of Fig. 15, there was a clear difference for condition W05 between the profile along the centerline shown with the green line and the elevation profiles at the side of the model, indicated by the red and blue lines. However, at the same position in the model, there was only a small difference between the results of test run R58, shown by the continuous lines, and test run R73, represented by the dashed lines. In the bottom graph of Fig. 15, the results for wave condition W10 from test run R28, shown in continuous lines, and test run R69, plotted with dashed lines, agree even better. Furthermore, as expected from Fig. 13, there was hardly any difference in the results for wave condition W10 over the width of the model. This indicates that repeatable test conditions indeed led to repeatable test results.

4 Discussion

Compared to the results with wave condition W10, the shorter waves of wave condition W05 clearly experience a stronger hydroelastic interaction with the structure. The increased wavelength and reduced elevation amplitude at the center towards the aft edge of the model suggest that wave spreading occurred for the shorter waves.

While the wave steepness was the same for both conditions, the curvature of the waves and thus also of the neoprene sheet was larger for the shorter waves. Furthermore, there were twice as many wavelengths over the length of the model for the shorter waves. Therefore, there was more time for each wave to interact with the structure. Adding to this effect is the shorter relative wavelength compared to the characteristic length, which means that the shorter waves led to relatively stronger bending reaction of the structure.

The build-up of 3D effects was also shown by Ertekin and Kim [24] in their study. However, at the aft edge, they report a larger elevation amplitude in the center than at the sides, which is the opposite of what we found in our experiment. In this comparison it needs to be noted that the structure investigated in by Ertekin and Kim had a characteristic length of $\lambda_c = 400$ m for a structure length and width of $L = 5000$ m and $B = 1000$ m, respectively. Therefore, the ratio of structure width to characteristic length was $B/\lambda_c = 2.5$ for their model as compared to $B/\lambda_c = 6$ for the model employed in this paper. Thus, the model presented here is more flexible in transverse direction.

Since the focus of this paper is on the application of the DIC technique for the measurement of surface elevation of flexible floating structures, the results are not discussed in more detail against available theoretical results.

5 Conclusions

In this study, we applied DIC technique to measure the deformation of a floating flexible neoprene foam rubber sheet of 5 mm thickness and 4.95 m length in regular waves in a 3D setup. We demonstrated that wave conditions and results were repeatable. The measurement results showed that the model could follow long-crested waves with an amplitude of 20 mm and wavelength of 0.99 m ($L/\lambda = 5$) without noticeable effects on the waves.

For shorter waves with the same wave steepness, there was considerable hydroelastic interaction between the wave and the model, and 3D effects were detected in the surface elevation of the model. These 3D effects partially matched the effects reported in literature. However, there is disagreement with the literature results regarding the distribution of the elevation amplitude over the width of the model close to the aft edge. This and other effects of wave scattering require further investigation in future studies.

In the presented study, the DIC technique was found to be a promising tool to measure the surface elevation of large flexible floating structures in a towing tank. However, especially with deformations much larger than the thickness of the model, careful consideration needs to be given to the accuracy of this measurement technique. The development and application of appropriate procedures to ensure accurate measurement of large out-of-plane deformations will be the next steps to make DIC a versatile measurement technique for very flexible floating structures.

Acknowledgements The authors express their gratitude to LaVision GmbH for providing the DIC license to the DaVis software.

References

1. Karmakar D, Bhattacharjee J, Sahoo T (2011) Contemporary approaches in the hydroelastic analysis of floating and submerged structures. Mar Technol Eng 1:461–478
2. Squire VA (2007) Of ocean waves and sea-ice revisited. Cold Reg Sci Technol 49:110–133
3. Squire VA (2011) Past, present and impendent hydroelastic challenges in the polar and subpolar seas. Philos Trans R Soc A Math Phys Eng Sci 369:2813–2831
4. Chen X, Wu Y, Cui W, Jensen JJ (2006) Review of hydroelasticity theories for global response of marine structures. Ocean Eng 33:439–457
5. Lamas-Pardo M, Iglesias G, Carral L (2015) A review of very large floating structures (VLFS) for coastal and offshore uses. Ocean Eng 109:677–690
6. Squire VA (2008) Synergies between VLFS hydroelasticity and sea-ice research. Proc Int Offshore Polar Eng Conf 8:1–13
7. Suzuki H (2005) Overview of megafloat: concept, design criteria, analysis, and design. Mar Struct 18:111–132
8. Trapani K, Redón Santafé M (2015) A review of floating photovoltaic installations: 2007–2013. Prog Photovoltaics Res Appl 23:524–532
9. Trapani K, Millar DL, Smith HCM (2013) Novel offshore application of photovoltaics in comparison to conventional marine renewable energy technologies. Renew Energy 50:879–888
10. Trapani K, Millar DL (2014) The thin film flexible floating PV (T3F-PV) array: the concept and development of the prototype. Renew Energy 71:43–50
11. Jamalludin MAS et al (2019) Potential of floating solar technology in Malaysia. Int J Power Electron Drive Syst 10:1638–1644
12. Yago K, Endo H (1996) On the hydoroelastic response of box-shaped floating structure with shallow draft. J Soc Nav Archit Japan 1996:341–352
13. Yago K, Ohmatsu S, Endo H (1997) On the Hydroelastic response of box-shaped floating structure with shallow draft. J Soc Nav Archit Japan 1997:307–317
14. Kagemoto H, Fujino M, Murai M (1998) Theoretical and experimental predictions of the hydroelastic response of a very large floating structure in waves. Appl Ocean Res 20:135–144
15. Ohta M, Ikegami K, Yamaguchi Y (1998) Experimental study on elastic behavior of a huge floating structure in waves (in Japanese). Trans West-Japan Soc Nav Archit 99–108
16. Ohmatsu S (2008) Model experiments for VLFS. In: Wang CM, Watanabe E, Utsunomiya T (eds) Very large floating structures, pp 141–164. Taylor & Francis
17. Squire VA (1984) A theoretical, laboratory, and field study of ice-coupled waves. J Geophys Res 89:8069–8079
18. Meylan MH (1993) The behaviour of sea ice in ocean waves. University of Otago
19. Sakai S, Hanai K (2002) Empirical formula of dispersion relation of waves in sea ice. Ice Environ Proc 16th IAHR Int Symp Ice 327–335
20. Montiel F et al (2013) Hydroelastic response of floating elastic discs to regular waves. Part 1. wave basin experiments. J Fluid Mech 723:604–628
21. Bennetts LG et al (2015) An idealised experimental model of ocean surface wave transmission by an ice floe. Ocean Model 96:85–92
22. Sree DKK, Law AWK, Shen HH (2017) An experimental study on the interactions between surface waves and floating viscoelastic covers. Wave Motion 70:195–208
23. Wang R, Shen HH (2010) Gravity waves propagating into an ice-covered ocean: a viscoelastic model. J Geophys Res 115:C06024
24. Ertekin RC, Kim JW (1999) Hydroelastic response of a floating mat-type structure in oblique, shallow-water waves. J Sh Res 43:241–254
25. Suzuki H, Yoshida K (1996) Design flow and strategy for safety of very large floating structure. Proc Int Workshop Very Large Floating Struct VLFS 96:21–27

26. Suzuki H et al (2006) Very large floating structures. In: Proc ISSC2006: volume 2: structures, safety and reliability; petroleum technology symposium vol 2, pp 391–442. ASMEDC
27. Sutton MA, McNeill SR, Helm JD, Chao YJ (2000) Advances in two-dimensional and three-dimensional computer vision. In: Rastogi PK (ed) Photomechanics, topics in applied physics, pp 323–372. Springer

F-STES: Floating Seasonal Thermal Energy Storage and Thermal Potential of Lakes

Eduard Voelker, Vasiliy Glazov, and Wolfgang Schmidt

Abstract The impact of climate change is already clearly noticeable. The growth of world population and their demands of living space have been increasing rapidly. In this case, there is an increasing need for thermal energy, even in industrialized areas where population declines. The development of renewable energy in Germany and all over the world is growing steadily at different levels and provides a clear insight into ecological energy production. The generation of renewable energy varies by season and thus causes to the difference between the supply and demand of energy in different seasons. Seasonal energy storage is one of the most important components of future energy supply. The aim of research is to describe the thermal potential of lakes. Using the novel floating seasonal thermal energy storage (F-STES) with efficient smart temperature control can be stored high energy density for a long time. Wind and solar excess from the summer months carry the greatest capacity in the winter months. The thermal energy potential of the F-STES is sufficient to provide the entire floating settlement. Lusatian Lakeland in Germany where has different riverbank structures and water quality is particularly a suitable cxperiment area to use the lake's thermal potential. This technology will be applied in a concept for the floating settlement structures on the lake in Großraeschen, Germany.

Keywords Seasonal thermal energy storage · Renewable energy · F-STES · Floating settlement

E. Voelker (✉) · W. Schmidt
Department of Building Physics and Building Technology, Brandenburg Technical University, Cottbus, Germany
e-mail: eduard.voelker@b-tu.de

W. Schmidt
e-mail: wolfgang.schmidt@b-tu.de

V. Glazov
Department of National Research University, MPEI, Moscow, Russia
e-mail: mamurik@bk.ru

Ł. Piątek et al. (eds.), *WCFS2020*, Lecture Notes in Civil Engineering 158,
https://doi.org/10.1007/978-981-16-2256-4_22

1 Introduction

Lakes have been used thermally worldwide for a long time with the heat pumps [1]. In most cases, these are open circulation systems without additional storage. However, there is a lack of conceptual solutions and global assessment of the potential of water bodies with regard to the storage of excess energy. A practicable guideline for the implementation of possible variants is not yet available. The further development of seasonal thermal energy storage is innovatively advanced by means of conceptual and technological development as well as basic research on the thermal use of seawater. The Lusatian Lakeland with its different bank structures and water qualities is particularly well suited as an experiment and testing field for:

- Creation of artificial thermal lakes as a low-temperature source for heat pumps for heating and cooling urban quarters and industrial areas
- Utilization of the potential for heating and cooling from the main lakes by means of underwater heat exchangers
- Development and testing of floating seasonal thermal energy storage.

2 Thermal Potential of Lakes on the Lusatian Lakeland

The region of Lusatia has a lake landscape with an area of 14,000 ha [2]. This will gradually create Europe's largest artificial water landscape with cultural and economic potential. The place is ideal for the construction of floating buildings and floating settlements. The floating architecture is a part of the integral concept to develop the Lusatian Lakeland [3]. Numerous tourists have been enjoying the floating residential buildings on the Geierswald and Partwitz Lakes. Interest in the floating buildings is growing steadily both in the region and in the world. It is imaginable and advantageous to supply the buildings on the water excellently with the energy from the lake. This stored energy of the water can also supply the surrounding villages where are close to the lakes. Enormous resources such as natural water reservoirs are available for this purpose. The map section of the former open-cast-mining landscape shows the extent of the water bodies in Lusatia (Fig. 1).

It is well-known that water has the highest specific heat capacity compared to naturally occurring substances. Even with the temperature difference of only 1 K, it becomes 157,125 MWh of thermal energy stored such as in a lake (Großräschener See) with a volume of 135 million m^3 of water. With this novel technology, this potential can be expanded and used effectively. At the Brandenburg University of Technology and the Institute for Floating Structures, innovative technologies in particular including underwater heat exchanger and floating seasonal thermal energy storage have been developed that enable the long-term use of the thermal potential of the lakes.

Fig. 1 Lusatian Lakeland. *Source:* lmbv.de/index.php/lausitzer-seenland

3 Underwater Heat Exchangers

With effective underwater heat exchangers, all buildings located nearby water bodies can use the energy of water for heating or cooling. The high flexibility of the underwater heat exchangers enables universal application possibilities. One example is the integration of a heat exchanger into the surface of a pontoon. This variant is particularly suitable to keep ice-free on the surface of pontoons and to cool a floating house. Further compact model of the heat exchanger allows using the thermal potential of water for heating and cooling floating buildings. In connection with the F-STES, the underwater heat exchangers form a unit for loading the storage with thermal energy (Figs. 2 and 3).

3.1 Spiral Heat Exchanger

One of simple and practical solutions to generate energy from the water is the Spiral heat exchangers from Babben Company. It is a simple design made of turns that are of any length assembled into a spiral. The power of the heat exchanger we can define by using numerical calculations for varying parameters such as the material, mass flow and diameter of the pipelines (Table 1). The results that we get from this analysis are checked and verified with a Thermal-Response-Test.

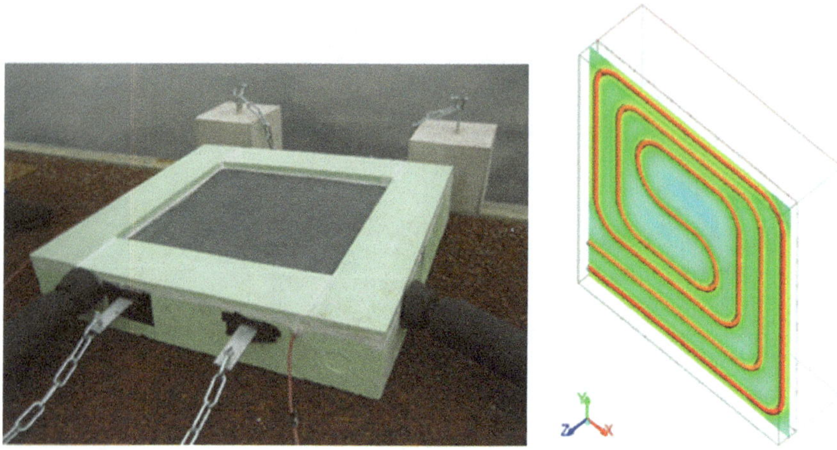

Fig. 2 Heat exchanger on the surface of a pontoon test samples with a simulation model

Fig. 3 Prototype of spiral underwater heat exchanger

4 F-STES Heat Storage

Seasonal energy storage is one of the most important components of future energy supply. From the current point of view, the usual seasonal thermal energy storage is not practical and is associated with high effort. Large heat storage in heating networks can only be found occasionally. Usually, these are thermal heat storage units made from steel or concrete colossus. An alternative is the usage of a seasonal F-STES (floating seasonal thermal heat storage) is suggested. This intends to supply regional energy providers and local industrial with heat from the surrounding natural and artificial lakes. The development of such an F-STES started several years ago at the Brandenburg University of Technology. The first calculations have

Table 1 Variant comparison

Variant	A	B	C
Spiral length [m]	3	1.5	3
Internal pipe diameter [mm]	26	26	51
Mass flow [kg/s]	0.181	0.09	3.1
Thermal conductivity [W/(m K)]	0.33	1.5	20
Return flow temperature [K]	281.89	286.33	286.25
Heat flux [W/m²]	836	1765	15,832
Power [W]	2827	3070	104,997
Temperature distribution of the brine (water)	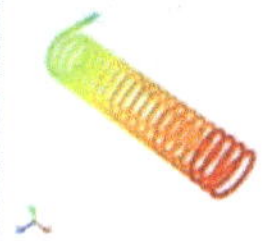	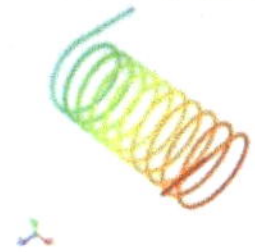	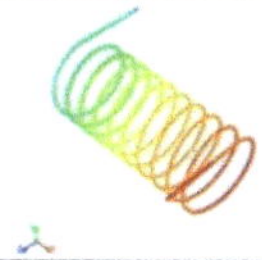

been investigating the effect of direct solar radiation on loads and the temperature stratification of a floating storage tank, which has a transparent cover at the top.

4.1 *Solar Gains of an F-STES Within One Year with Transparent Cover*

ANSYS FLUENT enabled detailed numerical simulations to investigate the temperature curve in a floating storage tank. According to important boundary conditions such as radiation absorption and resulting free convection, we get better accuracy in the results. The plausibility of the results was checked constantly by analytical ways on simplified models.

$$\frac{\partial}{\partial \tau} T(x,\tau) = a\left(\frac{\partial^2}{\partial x^2} T(x,\tau)\right) + \frac{kqe^{-kx}}{c} \quad (1)$$

Partial differential equation of heat transport with the processes of heat conduction and heat radiation.

With the aid of simulation, the weaknesses of the construction were determined in order to develop further improvements (Figs. 4 and 5; Table 2).

The course of the year shows a significant warming of water in the storage without additional 278 energy sources.

Additional modifications of F-STES will enable the surplus of renewable wind and solar energy to be stored during the productive months. With an adaptable temperature level, we increase the efficiency of the storage. In the midsummer months, when there is barely any heat requirement for heating and the energy surplus is at its peak, the temperature in the storage tank will be adjusted to the demand for domestic hot water. In the winter, the storage tank serves as an energy

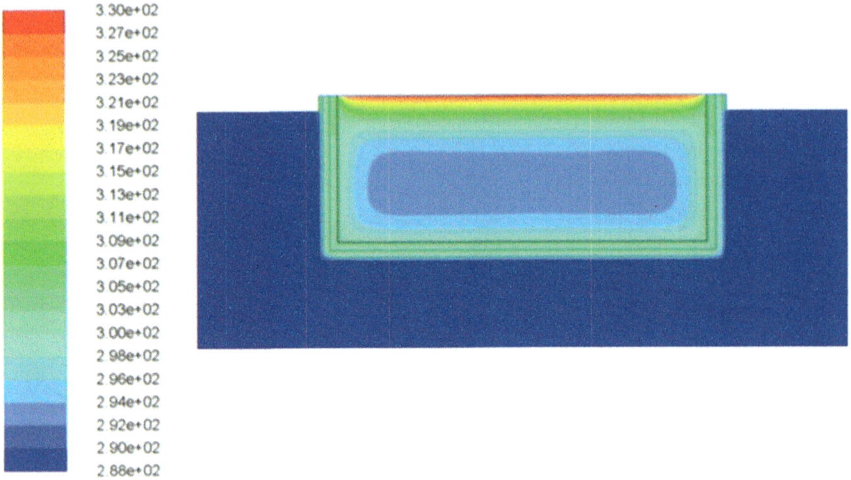

Fig. 4 Results of numerical simulation of floating seasonal thermal energy storage (F-STES). Contours of static temperature (K) (Time = 4.32003+05)

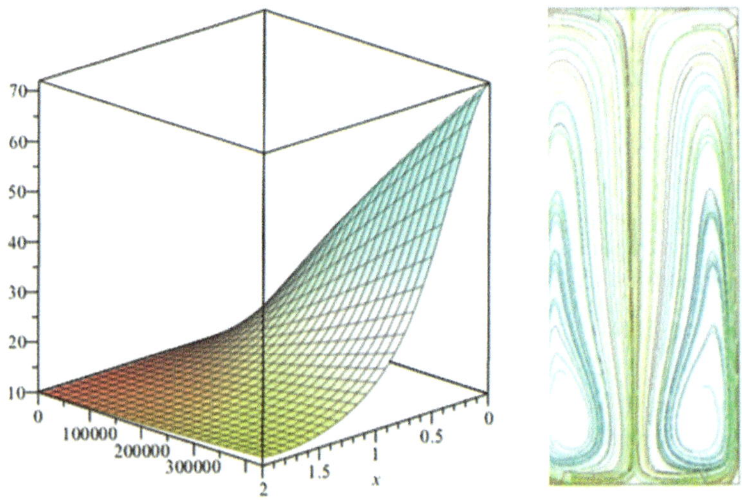

Fig. 5 Results of numerical simulation of floating seasonal thermal energy storage (F-STES). The results show a slight circulation of water, which creates a temperature stratification in the floating storage tank

source for the use of heat pumps. With approx. 10–20 °C on the primary side of the heat pump, high COP values are achieved and the available storage quantity is distributed for the entire heating season. This mode of operation allows an efficient shifting of excess energy from summer months to the winter months with the

Table 2 Energy balance sheet of the thermal energy storage

Month	Outdoor-temp.	q with g = 0.6	Gains	Losses in Wh Q = h * ΔT * A * t		Water-temp.
	Kelvin	W/m²	Wh	Glas	Wall	Kelvin
January	275	15	50,240	−47,112	−27,599	280
February	274	28	92,349	−36,954	−21,648	283
March	277	55	183,381	−40,313	−23,616	296
April	284	90	296,471	−69,184	−40,529	315
May	286	127	419,175	−169,255	−99,153	330
June	289	122	402,154	−243,859	−142,858	332
July	289	131	433,530	−253,257	−148,363	335
August	289	109	361,626	−271,498	−159,049	328
September	287	71	235,103	−241,982	−141,758	313
October	283	44	146,869	−179,934	−105,409	299
November	278	18	60,084	−124,074	−72,685	285
December	277	11	37,927	−45,012	−26,369	281

g = solar heat gain coefficient, SHGC windows

reduction of losses. The hot water is loaded into the floating storage tank primarily by means of a solar field that is installed on the cover of floating storage tank. As a further source of energy, the wind turbines and their surplus energy will be available for hot water. A sufficiently large F-STES with a dimension 100 × 100 m would be able to supply heat for over 145 households (10 MWh/per household) during the year in Germany, even at low-temperature differences of dT 20 K. This requires an area of approximately 3296 m^2 of solar collectors to load the storage only with solar heat. When the surplus wind energy is loaded, 1/3 of the 1655 full-load hours of a 3 MW wind turbine [4] provides enough energy to supply completely the storage.

5 Concept for the Floating Settlement Structures on the Lake in Großraeschen

Water is one of the basic prerequisites for settlement construction from the beginning. In low lying coastal areas where water gradually displaced the landmass, isolated buildings and constructions were initially built on the water. In many countries all over the world. From today's point of view, there are several reasons for water such as climate change and economic reasons to build floating buildings and floating settlement. Particularly global warming and rising sea levels not only directly affect on cities in coastal regions but also leads to more frequent floods. This reduces the potential settlement area, which without it will become increasingly scarce. In addition, there is an important reason from an economic point of

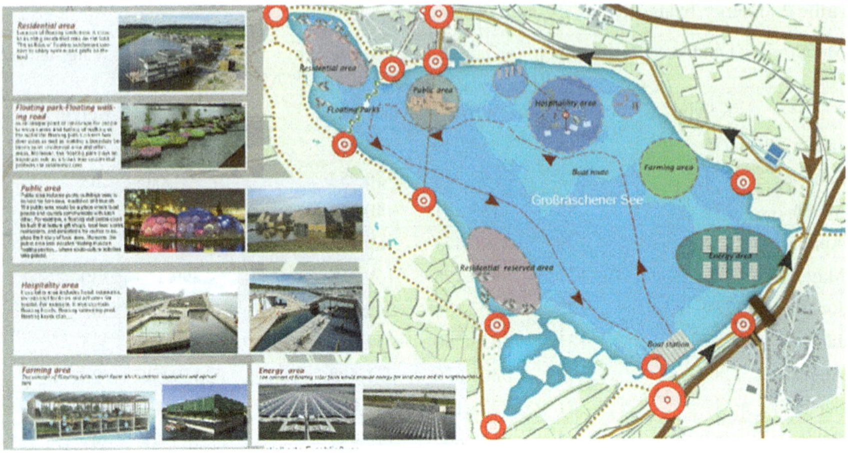

Fig. 6 Concept of a floating settlement with energy generation on the water in Grossraeschen

view to promote construction on water. Researchers from BTU proposed a concept for a floating settlement, located on the lake in Großraeschen, is used to demonstrate the possibilities of urban infrastructure on the water. The concept includes a residential area, public area with parks, cultural centers, tourism area and industrial area. An important component of this settlement is the autarkic energy supply through solar farms and thermal utilization of the lake with seasonal thermal energy storage (Fig. 6).

6 Conclusion

6.1 Creating Water Landscape as Re-cultivate Post-mining Areas

Creating a water landscape as a re-cultivation of post-mining areas brings a new opportunity to create sustainable and ecological settlements that would be able to use water as a renewable energy source. The thermal potential of lakes could extend far beyond floating settlements. By combining the F-STES with renewable energy sources, large urban districts could be supplied. Surplus income from the wind and solar energy is stored seasonally. It balances the temporally fluctuating yields from locally available energy sources and the energy demand over time. Therefore, the goal is getting closer of an effective and economically justifiable CO_2-free energy supply for buildings. The results of the research provide an advantage for future technological development in the thermal use of water surface. In the context of economic growth and climate change mitigation, F-STES would be novel technology and innovative export products.

References

1. Henriette Kammer (2017) Thermische Seewassernutzung in Deutschland, Bestandsanalyse, Potential und Hemmnisse seewasserbetriebener Wärmepumpen, Springer Vieweg, pp 55–65. ISBN 978-3-658-20900-1
2. LMBV, Lausitzer Seenland Lausitzer und Mitteldeutsche Bergbau-Verwaltungsgesellschaft mbH (2019) Flyer on website https://www.lausitzerseenland.de/de/die-seen/artikel-grossraeschener-see.html. Accessed 6 Sept 2020
3. IBA 2010, Concept Internationale Bauausstellung (IBA) Fürst-Pückler-Land 2000–2010. http://www.iba-see2010.de/en/verstehen/konzept.html. Accessed 26 June 2020
4. Fraunhofer Institute for Energy Economics and Energy System Technology, Windenergie Report Deutschland (2017), p 7

Low-Cost Utility Scale Offshore Energy Storage

Rohit Fenn, Remy Dygert, and Mike McDermott

Abstract In order to curb anthropogenic climate change, deep decarbonization of the electric grid is essential. Although the economical front of technologies like wind and solar power has improved, the fundamental intermittency of these sources of energy remains a challenge. Cheap, reliable and scalable storage solutions are urgently required. Most energy storage technologies are either expensive (Lithium Ion) or geographically constrained (Pumped Hydro). Low cost bulk energy storage could be a vital catalyst in decarbonizing our current grid infrastructure and would increase the competitiveness of offshore renewables considerably. This paper explores the feasibility of a large scale offshore floating Osmotic Energy Storage (OES) system. OES stores electrical energy by desalinating a clean, mixed solution to create a chemical potential between NaCl brine and freshwater in a closed loop system. It recovers this energy in a controlled membrane based mixing process called Pressure Retarded Osmosis (PRO). An offshore digitally operated Reverse Osmosis-Pressure Retarded Osmosis based OES system could be scaled up to upwards of 6 MWh to function as storage for coastal utilities or offshore communities. The core value proposition is in its economical, modular and environmentally benign design that could increase access to large scale energy storage integration to the grid globally.

Keywords Energy · Storage · Offshore · Economical · Desalination

R. Fenn (✉)
Sterling College, Craftsbury, VT, USA
e-mail: rohitfenn@gmail.com

Community College of Vermont, Burlington, VT, USA

R. Dygert
SUNY Albany, Albany, NY, USA
e-mail: redygert@gmail.com

M. McDermott
Rensselaer Polytechnic Institute, Troy, USA
e-mail: defenestrated23@gmail.com

Ł. Piątek et al. (eds.), *WCFS2020*, Lecture Notes in Civil Engineering 158,
https://doi.org/10.1007/978-981-16-2256-4_23

1 The Importance of Energy Storage to Decarbonization

To slow down rapid anthropogenic climate change the deep decarbonization of electricity production is an imperative challenge for society to overcome. The pace of progress in renewable energy development is improving but a compelling scalable remedy for the fundamental intermittency of renewables still remains largely absent [1]. Intermittency of production can be observed in both diurnal and seasonal cycles. Apart from a need for stable baseload power, peak production and consumption are offset in the infamous ‘duck curve’ [2]. The concept of electrical energy storage is simply to store energy when it is abundant and recover it when it is scarce There is consensus that for energy storage to scale, it must be much cheaper, almost an order of magnitude cheaper than prices today. Analysis has determined that energy storage would have to cost roughly US $20 per kilowatt-hour (kWh) for the grid to be 100% powered by a wind-solar mix [3]. Pumped hydro systems currently dominate energy storage globally and can often be economical. However, the fact that it is geographically constrained to hilly terrain makes them limited in their ability to scale [4].

This paper presents a novel configuration for an economical, scalable and environmentally benign off-shore osmotic energy storage system (OES). The technology stores electrical energy by creating a chemical potential through desalinating a solution into concentrated brine and freshwater. Energy is then recovered from the system in a controlled mixing of the two produced solutions by a membrane based process, which in turn runs a turbine. All processes performed in a closed loop. It essentially utilizes a concentration gradient to generate a potential and recover useful work.

Given that a mature industry is established around the desalination of saltwater and that the materials used in the system are low cost, it could be economical. The size and capacity of the storage can be arbitrarily increased by adding membranes, pumps and using larger reservoirs for the solutions, making it scalable. The use of saltwater as the solution in which the chemical potential is created while floating offshore makes it environmentally benign. As the prospects of offshore wind and solar gain momentum, a cheap energy storage system could further increase their competitiveness [5, 6].

1.1 Osmotic Energy Storage: Parallels to a Flow Battery

A flow battery typically is a configuration where a chemical potential is created and then recovered by two chemical components dissolved in liquids contained within the system and separated by a membrane. However, most flow batteries are different because their solutions are electrolytes and have an active electrochemical role. In comparison, this system is largely mechanical. A mixed NaCl solution is desalinated using modified Reverse Osmosis and then the energy is recovered using a

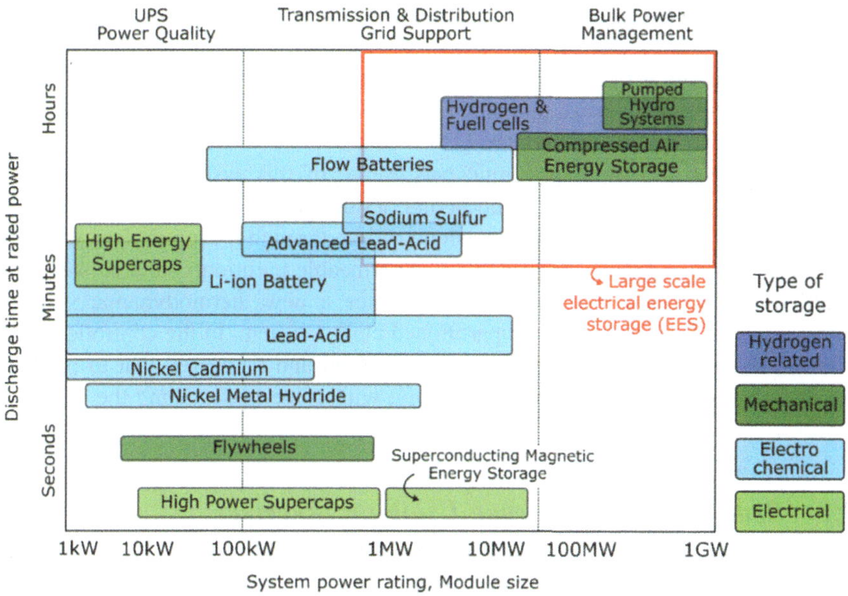

Fig. 1 Characteristics of different energy storage technologies adapted from Taylor et al. and Akhil et al. The red square highlights the required discharge time and power rating of a large-scale electrical energy storage system [7–9]. Image adapted from van Egmond (2018)

mixing process called Pressure Retarded Osmosis. In the recovery two solutions with different osmotic pressures are pumped on opposite sides of a semi-permeable membrane. Due to the difference in osmotic pressure, water flows from the solution of lower salinity (the feed solution) to the solution of higher salinity (the draw solution). If volume is kept constant, this effect can be harnessed as increased pressure which in turn spins a hydro-turbine.

Similar to a flow battery, the capacity and discharge rates of an OES battery is determined by the volume of the reservoir and the active surface area of the membranes involved. Therefore, they scale similarly (Fig. 1).

1.1.1 Storing Energy in a Salinity Concentration Gradient

Natural and anthropogenic salinity gradients have been identified as significant sources of sustainable energy that are currently untapped. The International Energy Agency estimates there are 2000 TW of osmotic power in the mixing of rivers at the oceans [10]. The Gibbs free energy released during mixing can be harnessed by technologies such as reverse electrodialysis [11, 12], capacitive mixing [13], and pressure-retarded osmosis (PRO) [14].

1.1.2 Mixing Theory

Gibb's free energy is the thermodynamic potential that can be used to calculate the maximum of reversible work that may be performed by a thermodynamic system at a constant temperature and pressure [15].

In many ways, a thermodynamic perspective of this system yields a clearer picture. Starting with a mixed solution and desalinating it to brine and freshwater artificially reduces the entropy of the whole system through the input of energy. The thermodynamic operation of removal of impermeable partition(s) and between the solutions and introducing a membrane, creates a new thermodynamic state of internal equilibrium in the new unpartitioned closed system. In the controlled PRO mixing, the net entropy of the system is increased and harnessed that to produce useful work in the form of electrical energy, with some losses along the way.

1.2 Osmotic Energy Storage Using the Reverse Osmosis-Pressure Retarded Osmosis (RO-PRO) Combination

Storing electrical energy in the form of a chemical potential through the differences in solute concentration, means honing in on an efficient way to create and recover this potential. Despite having a few options like Electro dialysis (ED), Reverse ED and thermal distillation to increase salinity, Reverse Osmosis and PRO seem to emerge as clear winners for efficiency of electrical energy conversion to a chemical potential [16].

Reverse Osmosis is a mature technology and has been commercialized for use in desalination all over the world. It is advanced enough that our current many RO desalination systems utilize only 25% higher energy than the thermodynamic minimum to separate salt from water. In 1970, they consumed 90% the minimum threshold for a single stage process [17]. The commercial success of the industry translates to optimized membranes, robust supply chains, global expertise in repair and maintenance and ultimately lower costs due to economies of scale (Fig. 2).

Pressure Retarded Osmosis was pioneered a half century ago by Sidney Loeb, one of the co-inventors of RO but many key developments and innovations that make it viable didn't unfold untill recently [14, 19–23]. Besides thermodynamic analysis, these developments mostly have to do with the performance of membranes.

Membrane materials and consideration include an active membrane layer, a support layer and spacers on either side. Commercially available FO membranes today can achieve practically usable power densities. However, carefully selected support layers and spacer geometry will be key to translating these lab bench scale tests to successfully deploy conventional spiral wound membranes that can pack much more area. Some important considerations are increasing water flux, reducing

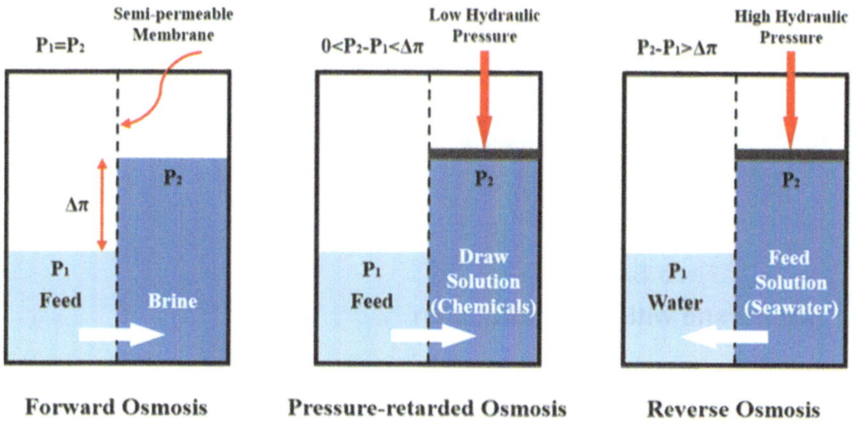

Fig. 2 A graphical representation of the key differences between the three osmotic processes [18]

salt flux and Internal Concentration Polarization. Membrane deformation can be reduced with the right spacer geometry that provides it structural integrity at high pressures [24] (Fig. 3).

The first public experimental installation of PRO technology was at the StarKraft plant in Norway in 2009 [26]. It was designed to generate 10 KwH in an open system that employed a naturally low and high salinity stream respectively to generate power. Although practical power densities from the membranes used at the time in an open loop system were not achieved, many learnings came out of that experiment.

Some reasons for low power densities achieved at the early Starkraft installation had to do with low performance membranes, the salinity difference between streams and the pressures utilized in the process.

It is evident that multiple roadblocks to commercialization still remain. Beyond membrane performance, the optimal configuration of a PRO system is crucial to its success. Attempting open systems for power generation might pose more challenges around fouling and performance than a closed loop system that can cycle cleaner feed and draw solutions.

1.2.1 PRO Membrane Qualification Test

Numerous lab bench scales have confirmed the higher membrane performance with high draw salinity and high pressure. There is increasing consensus in the research community that High-Pressure PRO Holds Promise as an Economical Means of Salinity Gradient Power Generation [27]. After qualifying the membrane, support layer and spacer combination, the fabrication of the PRO spirally wound membranes is essential to packing the most active membrane surface area. Achieving comparable power density and performance in this configuration when compared to

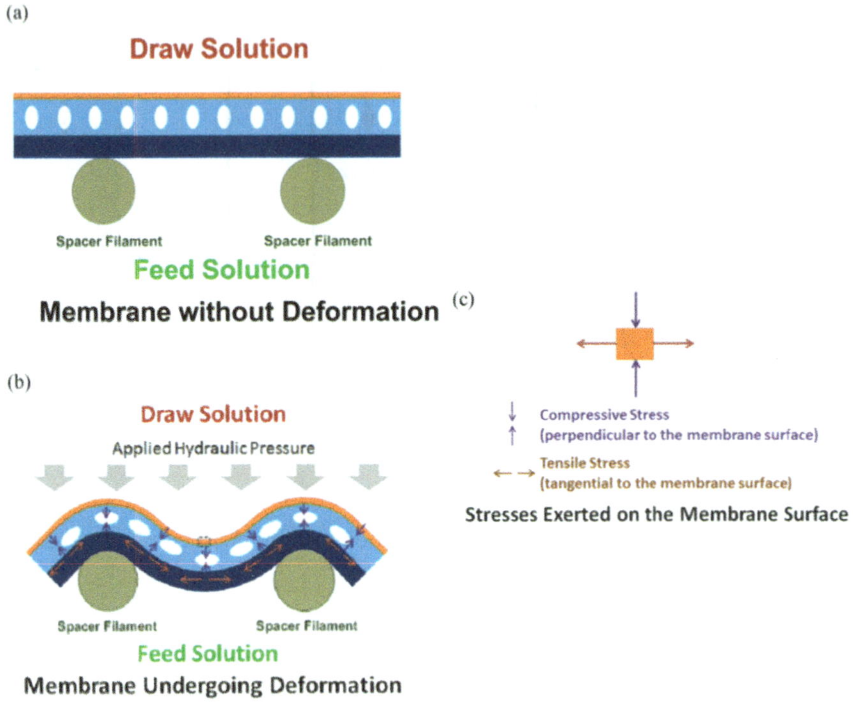

Fig. 3 Schematic illustration of the coupled effects of hydraulic compaction and tensile stretching on the membrane deformation in the PRO operation. **a** Membrane is in the original state without deformation, **b** membrane is undergoing deformation in the PRO testing, and **c** magnified image of dotted area in (**b**) to illustrate the coupled compressive stress and tensile stress exerted on the active layer surface when membrane is undergoing deformation. The coupled effects of compaction and stretching could influence the overall membrane separation and structural parameters in the PRO process [24, 25]. Adapted with permission from She et al. (2016). Copyright (2016) B.V Elsevier

lab testing might pose a challenge. At the StarKraft plant, 5 W/m^3 was identified as a target power density for economical power generation, bench scale tests have access 10 times that [26] (Fig. 4; Table 1).

1.3 Multi-stage Processes: Hybrid Systems

Theoretical modelling has made evident that to increase energy density and roundtrip efficiency, multi stage systems can be helpful [31]. This means having membrane modules connected in parallel allows for a larger processing volume but having successively smaller sets of membrane modules connected in series can enhance the systems capability to store and recover energy more efficiently. Designing in the ability to digitally switch membrane modules from parallel flow to

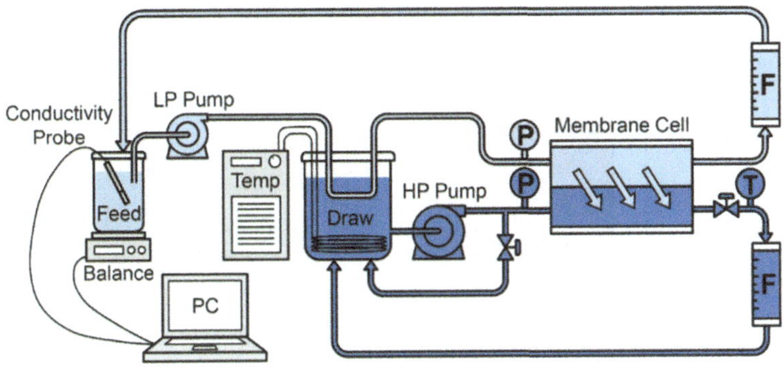

Fig. 4 A typical lab-bench experimental assembly to test membrane performance [27]. Adapted with permission from Straub et al. [28]. Copyright (2014) American Chemical Society

Table 1 Studies that have investigated membrane performance at high pressure and salinity with existing Forward Osmosis membranes [27, 29]

Study	Membrane	Brine conc.	Pressure	Power density
Madsen et al. [30]	HTI Spiral wound membrane	3 M (180 g/L) NaCl draw solution	40 bar	30 W/m^2
Madsen et al. [30]	FTS CTA (cellulose acetate)	3 M (180 g/L) NaCl draw solution	60 bar	31 W/m^2
Straub et al. [28]	HTI TFC (thin-film composite)	3 M (180 g/L) NaCl draw solution	48.3 bar	59.7 W/m^2

Some tests employ different test cell designs, spacer geometries and support layers

series flow can enable a variable battery power rating and discharge rates. Figure 5 depicts the useful work input and recovered from the system in both single and multi stage configurations.

1.4 *Energy Density and Roundtrip Efficiency: Tradeoff*

It is worth noting that theoretical models suggest an energy density of an OES system with multi-stage processes can be similar to that of a 500 m high pumped hydro plant. Albeit, with a lower round trip efficiency. While PH can achieve about 80–90% energy recovery, this technology can achieve close to 60% round trip efficiency [31]. However, this lower round trip efficiency can be compensated for as renewably generated energy becomes much cheaper as they scale. This energy is currently curtailed when it is overproduced. Utilities operating offshore generation can still find value by integrating energy storage. The competitiveness of offshore wind and solar can both be considerably enhanced with the ability to load balance over the grid [28, 30].

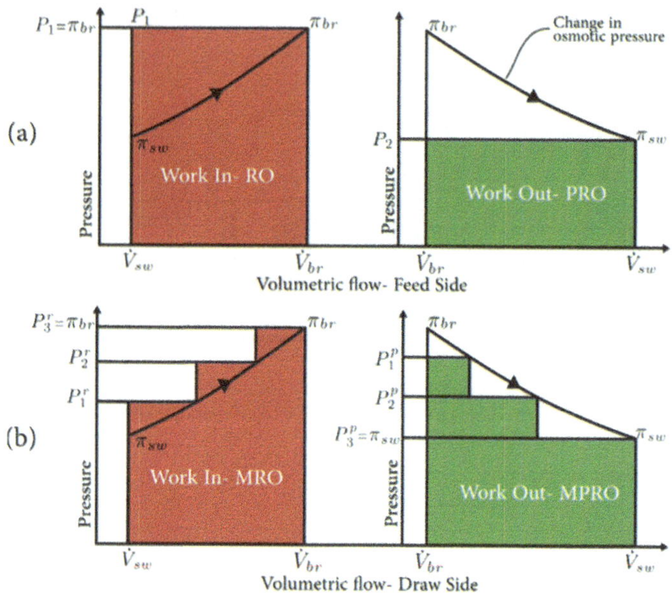

Fig. 5 Multi stage systems can yield more efficient RO and PRO processes [31]. Adapted with permission from Bharadwaj and Struchtrup [31]. Copyright (2018) Royal Society of Chemistry

1.5 Closed Loop, High Pressure and Hyper Saline Systems

A synthetic NaCl solution in a closed loop system introduces the ability to optimize salinity to access maximum power density and membrane performance. In turn, reducing the capital cost further as it reduces the total number of membrane modules required.

A closed loop system also enables significantly reduced membrane fouling as a cleaner solution is run and fluxed through the membranes, increasing their longevity.

2 Offshore Battery: Scaling Up an OES System Capacity

An OES system can be scaled up with the integration of more pumps, plumbing, membranes and solution reservoirs. Theoretical modelling of multi stage systems suggest that a storage plant capable of storing 6MWhs of electricity would have to store 5000 m^3 (roughly two olympic swimming pools) of total solution. A multi stage plant of this scale can have an energy density of 1.2 kWh m^3 [31]. This is on the lower scale of energy densities with existing battery technologies having energy storage densities ranging from 50 to 500 kWh m^3 to fuel cells that have an energy density of 500–3000 kWh m^3 [4]. However, with a system out in the open ocean, there is certainly a reduced need to optimize for energy density in order for it to scale.

A larger system with a capacity of 1 GWh of electrical energy storage would require storing at least 106 m^3 (1000 million liters) of liquid [31]. This scale of storage is conceivable with floating bags in the ocean but would be impractical to achieve on land. However, it is worth noting that the GWh scale would incorporate additional losses to the system. Some optimum scale of components and desired energy storage need that is most economical can be determined and the system can be made modularized to integrate multiple units (Figs. 6 and 7).

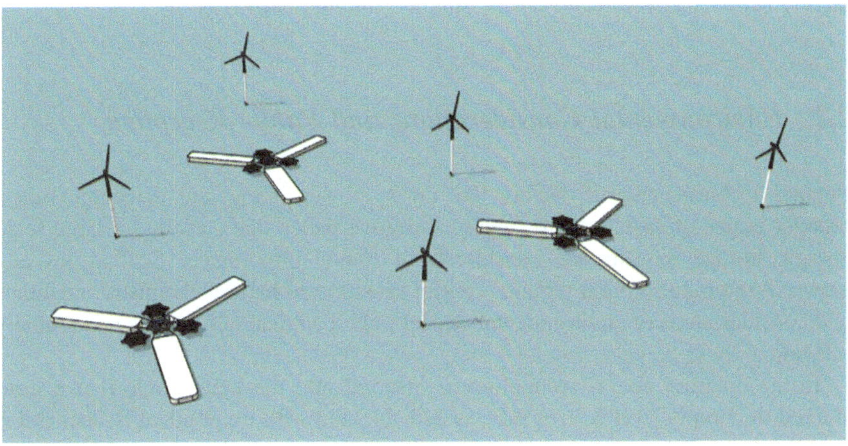

Fig. 6 Rendering of a mixed offshore wind and solar farm integrated with OES energy storage

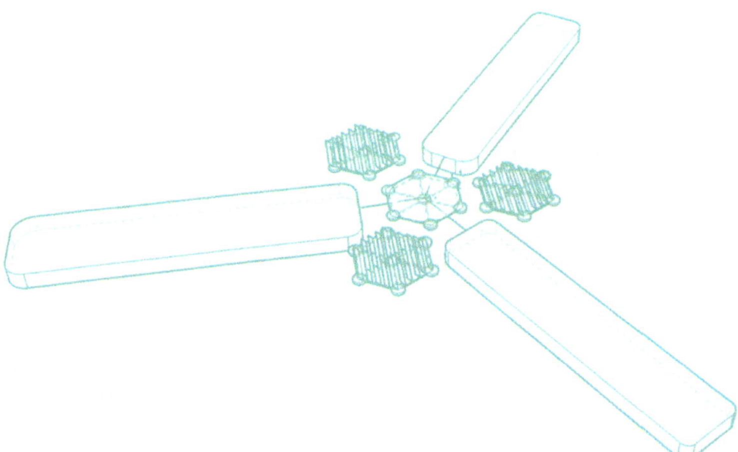

Fig. 7 Large scale OES system with a large floating bag for the mixed solution and two smaller bags for brine and freshwater. Plumbed into a central floating platform with the membrane modules, pumps and turbine

2.1 Floating Tensegrity Platform and Bags as Storage

Cost considerations for the floating platform are important to consider for the system at large. Using a barge will likely increase capital costs. 'Dweep' is a low cost, modular floating tensegrity based platform with efficient stress distribution. Developed and tested by Sea6 Energy in Bangalore, India. This platform could also enable offshore solar farms.

Floating bags for offshore water storage have already been tested and developed both in the United States [28] and Europe [30] for the transport of freshwater in the ocean using tugboats. For the OES application, the volumes are comparable but it is largely stationary and hence can be simpler, durable bags (Fig. 8).

2.2 Environmental Considerations and Public Reception

Besides techno-economic factors like costs, power rating and discharge times, broader issues of sustainability, safety and governance that surround a large scale energy storage system are considered of major importance. Scaling up any grid-scale storage solution requires careful selection of battery chemistry, materials and considerations of environmental impact to be commercially viable and globally relevant.

In the diagram above, technology is inserted into the inner circle if it is considered to comply (if it is very safe, sustainable and with no political issues) and is inserted into the outer circle if it complies with caveats. Note that the distance between dots has no meaning. The intersection of the inner circles represents the golden area with any storage technology that complies with all three aspects and is

Fig. 8 Floating solar powered desalination plant on Sea6's Dweep tensegrity based platforms [25]

considered desirable from a societal point of view [7]. Although currently, no technology is located in this 'sweet spot', OEs may be a worthwhile technology to consider in this domain.

Floating OES systems can have a lower ecological impact by not displacing land. The system's components are environmentally benign in comparison with some other storage systems that may include rare elements and metals. Its primary solution chemistry being NaCL makes it relatively less toxic to extract, use and discard once the battery has completed its life cycle.

More importantly, an OES system can have crucial parts with a lower lifecycle like RO and PRO membranes replaced to increase the overall longevity of a storage system considerably.

2.3 Systems Integration and Conclusion

An offshore OES system may be autonomously computer controlled with sensors that enable analytics, predictive maintenance and remote operation, all of which will enhance such a storage system's value proposition.

Advanced smart battery grid integration has benefited utilities that can buy power when it's cheap and abundant to sell the energy when it's scarce and more expensive.

A low cost- environmentally benign energy storage solution that scales well and uses predominantly available technology is a viable candidate for economical energy storage in the twenty-first century. Careful selection of the membrane stack for the PRO process, fabrication of large floating bags and the viability of a low cost floating platform will all be integral challenges to overcome for this storage solution to be a success.

References

1. Arbabzadeh M, Sioshansi R, Johnson JX, Keoleian GA (2019) The role of energy storage in deep decarbonization of electricity production. Nat Commun 10(1)
2. Denholm P, O'connell M, Brinkman G, Jorgenson J (2015) Overgeneration from solar energy in California: a field guide to the duck chart [online]. Available at: https://www.nrel.gov/docs/fy16osti/65023.pdf
3. Ziegler MS, Mueller JM, Pereira GD, Song J, Ferrara M, Chiang Y-M, Trancik JE (2019) Storage requirements and costs of shaping renewable energy toward grid decarbonization. Joule 3(9):2134–2153
4. Chen H, Cong TN, Yang W, Tan C, Li Y, Ding Y (2009) Progress in electrical energy storage system: a critical review. Prog Nat Sci 19(3):291–312
5. Sioshansi R (2011) Increasing the value of wind with energy storage. Energy J 32(2)
6. Bermudez V (2017) Electricity storage supporting PV competitiveness in a reliable and sustainable electric network. J Renew Sustain Energy 9(1):

7. van Egmond WJ (2018) Concentration gradient flow batteries : salinity gradient energy systems as environmentally benign large scale electricity storage [online]. Available at: https://doi.org/10.18174/444261. Accessed 16 Sep 2020
8. Taylor PG, Bolton R, Stone D, Upham P (2013) Developing pathways for energy storage in the UK using a coevolutionary framework. Energy Policy 63:230–243
9. Akhil A, Huff G, Currier A, Kaun B, Rastler D, Chen S, Cotter A, Bradshaw D, Gauntlett W (2015) SANDIA REPORT DOE/EPRI electricity storage handbook in collaboration with NRECA [online]. Available at: https://prod-ng.sandia.gov/techlib-noauth/access-control.cgi/2015/151002.pdf. Accessed 16 July 2020
10. Bard J (2007) Implementing agreement on ocean energy systems (IEA-OES), Annual report. International Energy Agency
11. Post JW, Hamelers HVM, Buisman CJN (2008) Energy recovery from controlled mixing salt and fresh water with a reverse electrodialysis system. Environ Sci Technol 42(15):5785–5790
12. Długołęcki P, Gambier A, Nijmeijer K, Wessling M (2009) Practical potential of reverse electrodialysis as process for sustainable energy generation. Environ Sci Technol 43 (17):6888–6894
13. Brogioli D (2009) Extracting renewable energy from a salinity difference using a capacitor. Phys Rev Lett 103(5)
14. Achilli A, Childress AE (2010) Pressure retarded osmosis: from the vision of Sidney Loeb to the first prototype installation—review. Desalination 261(3):205–211
15. Perrot P (2008) A to Z of thermodynamics. New York Oxford University Press, Oxford
16. Yip NY, Elimelech M (2014) Comparison of energy efficiency and power density in pressure retarded osmosis and reverse electrodialysis. Environ Sci Technol 48(18):11002–11012
17. Elimelech M, Phillip WA (2011) The future of seawater desalination: energy, technology, and the environment. Science 333(6043):712–717
18. Forward Osmosis (FO) (2017) From https://shiqiangzou.com/fo/. Accessed 08 July 2020
19. PATTLE RE (1954) Production of electric power by mixing fresh and salt water in the hydroelectric pile. Nature 174(4431):660–660
20. Yip NY, Brogioli D, Hamelers HVM, Nijmeijer K (2016) Salinity gradients for sustainable energy: primer, progress, and prospects. Environ Sci Technol 50(22):12072–12094
21. Straub AP, Deshmukh A, Elimelech M (2016) Pressure-retarded osmosis for power generation from salinity gradients: is it viable? Energy Environ Sci 9(1):31–48
22. Reimund KK, McCutcheon JR, Wilson AD (2015) Thermodynamic analysis of energy density in pressure retarded osmosis: The impact of solution volumes and costs. J Membr Sci 487:240–248
23. Banchik LD, Sharqawy MH, Lienhard JH (2014) Limits of power production due to finite membrane area in pressure retarded osmosis. J Membr Sci 468:81–89
24. She Q, Wei J, Ma N, Sim V, Fane AG, Wang R, Tang CY (2016) Fabrication and characterization of fabric-reinforced pressure retarded osmosis membranes for osmotic power harvesting. J Membr Sci 504:75–88
25. OFFSHORE INFRASTRUCTURE (n.d.) From https://www.sea6energy.com/offshore-infrastructure. Accessed 01 July 2020
26. Thorsen T, Holt T (2009) The potential for power production from salinity gradients by pressure retarded osmosis. J Membr Sci 335(1–2):103–110
27. Straub AP, Yip NY, Elimelech M (2013) Raising the bar: increased hydraulic pressure allows unprecedented high power densities in pressure-retarded osmosis. Environ Sci Technol Lett 1 (1):55–59
28. www.waterbag.com (n.d.) Spragg bags [online]. Available at: http://www.waterbag.com/. Accessed 1 Jul 2020

29. Madsen HT, Nissen SS, Muff J, Søgaard EG (2017) Pressure retarded osmosis from hypersaline solutions: investigating commercial FO membranes at high pressures. Desalination 420:183–190
30. The big bag theory: A cheap and innovative way to transport fresh water (2016) https://ec.europa.eu/research/infocentre/article_en.cfm. Accessed 01 July 2020
31. Bharadwaj D, Struchtrup H (2017) Large scale energy storage using multistage osmotic processes: approaching high efficiency and energy density. Sustain Energy Fuels 1(3):599–614

Assessing the Influence of Floating Constructions on Water Quality and Ecology

Rui L. P. de Lima, Floris C. Boogaard, and Vladislav Sazonov

Abstract Large floating projects have the potential to overcome the challenge of land scarcity in urban areas and offer opportunities for energy and food production, or even for creating sustainable living environments. However, they influence the physical, chemical, biological and ecological characteristics of water bodies. The interaction of the floating platforms affect multiple complex aquatic processes, and the potential (negative/positive) effects are not yet fully understood. Managing entities currently struggle with lack of data and knowledge that can support adequate legislation to regulate future projects. In the Netherlands the development of small scale floating projects is already present for some years (e.g. floating houses, restaurants, houseboats), and more recently several large scale floating photovoltaic plants (FPV) have been realized. Several floating constructions in the Netherlands were considered as case-studies for a data-collection campaign. To obtain data and images from underneath floating buildings, underwater drones were equipped with cameras and sensors. The drones were used in multiple locations to scan for differences in concentrations of basic water quality parameters (e.g. dissolved oxygen, electrical conductivity, algae, light intensity) from underneath/near the floating structures, which were then compared with data from locations far from the influence of the buildings. Continuous data was also collected over several days

R. L. P. de Lima (✉) · F. C. Boogaard · V. Sazonov
Indymo, Molengraaffsingel 12, 2629 JD Delft, The Netherlands
e-mail: rui.plima@gmail.com

R. L. P. de Lima
MARE, Coimbra, Portugal

F. C. Boogaard
Hanze University of Applied Sciences, Groningen, The Netherlands
e-mail: f.c.boogaard@pl.hanze.nl

Deltares, Delft, The Netherlands

V. Sazonov
Witteveen en Bos, Rotterdam, The Netherlands
e-mail: vlad.hro@gmail.com

Ł. Piątek et al. (eds.), *WCFS2020*, Lecture Notes in Civil Engineering 158,
https://doi.org/10.1007/978-981-16-2256-4_24

using multi-parameter water quality sensors permanently installed under floating structures. Results show some differences in concentrations of water quality parameters between open water and shaded areas were detected, and some interesting relations between parameters and local characteristics were identified. Recommendations are given, in order to minimise the undesired impacts of floating platforms. Considering the complexity of the interactions between water quality parameters and the influence of the surrounding environment it is recommended to continue and to improve the monitoring campaign (e.g. include new parameters).

Keywords Water quality · Ecology · Environmental impacts · Floating structures · Monitoring tools · Aquatic drones

1 Introduction

Floating developments are promising climate change adaption solutions, as they offer flood proof constructions and opportunities for creating sustainable living environments or for energy and food production. Floating structures influence the physical, chemical, biological and ecological characteristics of water bodies. The interaction of the floating platforms affect multiple complex aquatic processes, and the potential (negative/positive) effects are not yet fully understood. This lack of knowledge about their impact on the water quality and ecology often hinders their implementation, as water boards and municipalities often encounter difficulties and challenges for their regulation and licensing.

Floating houses block the incident short wave solar radiation, depending on their size and the sun's position [1]. This shade effect impedes the growth of phytoplankton and macrophytes below the platform [2] and hence photosynthesis is reduced. As a floating house is a barrier for wind and waves, the re-aeration of the water body is weakened on the lee side. Between two houses a tunnel effect may occur at higher wind velocities causing better mixing of the water column [3]. The surfaces of platforms get colonized by sessile organisms [4] which use oxygen for respiration and therefore may deplete the dissolved oxygen content of the surrounding water. Excreted nutrients get dispersed increasing the nutrient concentration in the water as well as at the water bottom. Dead mussels also fall down and get decomposed, which may increases the oxygen demand at the water-sediment interface [1, 5]. Kitazawa et al. [5] reports that no decrease in current velocity nor variations in temperature and salinity were observed; however, the concentration of dissolved oxygen was slightly lower in the deeper column below the platform, but did not reach hypoxic or anoxic levels, not even in summer. Foka [3] detected a reduction of dissolved oxygen by 1 mg/l between two floating houses, compared to open water. These differences occurred only in the upper layers (<1 m depth) and mainly around noon, whereas in the morning and evening at both sites similar values were recorded. For water temperature the difference was 0.5 K, temperature variations in depth were very small. Hartwich [1] and de Lima et al. [6–8] found that oxygen content decreased with greater depth, stronger than in open water. Also organic enrichment, higher nitrogen and organic carbon content was determined, in comparison to open water.

Despite these studies, managing entities currently struggle with lack of data and knowledge that can support adequate legislation to regulate future projects, and therefore further monitoring and studies are necessary. In the Netherlands the development of small scale floating projects is already present for some years (e.g. floating houses, restaurants, houseboats), and more recently several large-scale floating photovoltaic plants (FPV) have been realized.

2 Methodology

To obtain data and images from underneath floating buildings, underwater drones were equipped with cameras and sensors (Fig. 2; [8]). The mobile drones were used as platforms to position the sensors underneath the floating objects. Sensors were able to monitor basic water quality parameters (e.g. dissolved oxygen, electrical conductivity, nutrients and algae/chlorophyll-a) from underneath/near the floating structures, which were then compared with data from locations far from the influence of the buildings (Fig. 1). As some of the sensors take time to adjust to local conditions, the drones were kept in each position for several minutes.

Several floating constructions in the Netherlands were considered as case-studies for a data-collection campaign. Table 1 provides an overview of these locations, including some characteristics of the structures and of the water body. This methodology was repeated in multiple locations around the Netherlands (Table 1; Fig. 3), during spring/summer period.

3 Results

The measurement campaigns generated data that can be represented as indicated in Fig. 4. This figure compares the data from open water (two graphs on the left) with the data from underneath the structure (two graphs on the right), in a single location

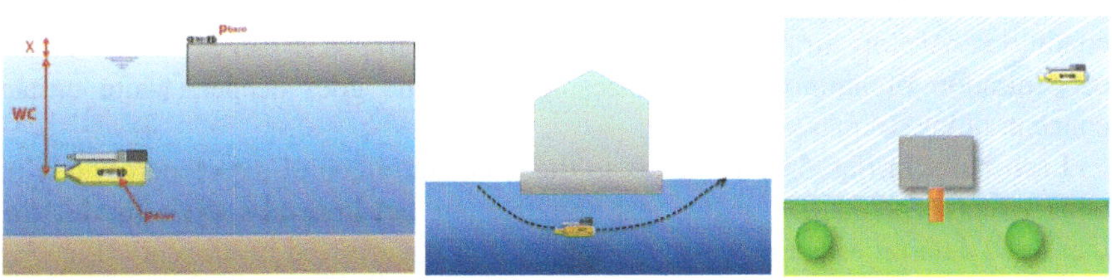

Fig. 1 Illustration of the position of underwater drones when collecting data **a** near (*left*), **b** under (*center*), and **c** far away (>8 m) from floating structures (*right*)

Fig. 2 Impression of underwater drones and sensors used in this research

Table 1 Information regarding the locations with floating structures in The Netherlands where measurements were collected

Name	ID	Location	Type	Year	Size (m^2)	Water system
Drijvende Kas	Naald	Naaldwijk	Greenhouse	2005	900	Pond/storage
Warande	Lely	Lelystad	Houses	2012	800	Canal (dredged/widened)
Harnaschpolder	Harn	Delft	Houses	2013	540	Pond/polder
Expo Sealife Almere	Alm	Almere	Housing complex	2010	500	Harbour
Havenpaviljoen Schiedam	Sch	Schiedam	Support Pavilion	2009	64	Canal (urban)
Sea Palace, Chinese	SPAms	Amsterdam	Restaurant	1984	900	Harbour (urban)
Zwameneiland	Meer	Groningen, Meerstad	Houses	2013	390	Lake
Oolderhuuske Marina's	Roer	Roermond	Houses	1998	5760	Lake
Maasvillas	OeL	Ohé en Laak	Houses	2010	618	Lake (connected to river)
Gouden Wok	GW	Rotterdam	Restaurant	n/a	1700	Harbour (urban)

(Floating Pavilion, Rotterdam). It can be observed that dissolved oxygen is lower under the structure than it is in open water. However, the difference is small, and measured values are above the (healthy) minimum of 5 mg/L of dissolved oxygen (never lower than 7.5 mg/L). The variation of temperature seems to be mainly affected by water depth, considering that the same pattern occurs in open water conditions. As for nitrate and ammonium, on the graphs it can be seen that the concentration of ammonium increases as the drone goes deeper, while nitrate concentrations decrease.

Each point in Fig. 5 corresponds to the averaged value of dissolved oxygen in open water (y-axis) and under/near the floating structures (x-axis), for all the measured data (at different water depths). A linear regression of this data (Fig. 5a) places the fitting line slightly above the 1:1 line indicating that the differences in the dissolved oxygen values (lower under/near the house) is small. In Fig. 5a, the

Fig. 3 Impression of some of the floating structures visited (restaurants, floating houses, pavilions and greenhouses)

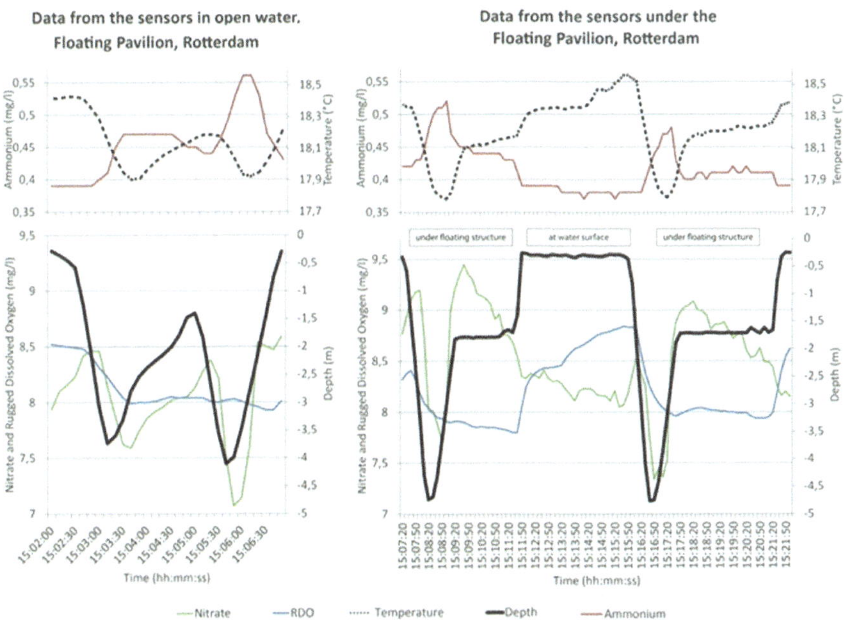

Fig. 4 Plotting of various parameters in open water and under the Floating Pavilion

different colours correspond to different locations, whereas in Fig. 5b the colour gradient relates to the depth interval.

Figure 6 shows a compilation of the measured differences data, organized in a circular graph with sections corresponding to the water depth of the measurements. This reveals that the bigger differences were detected in lower water depths (closer to the bottom surface of the floating structures), whereas for depths higher than

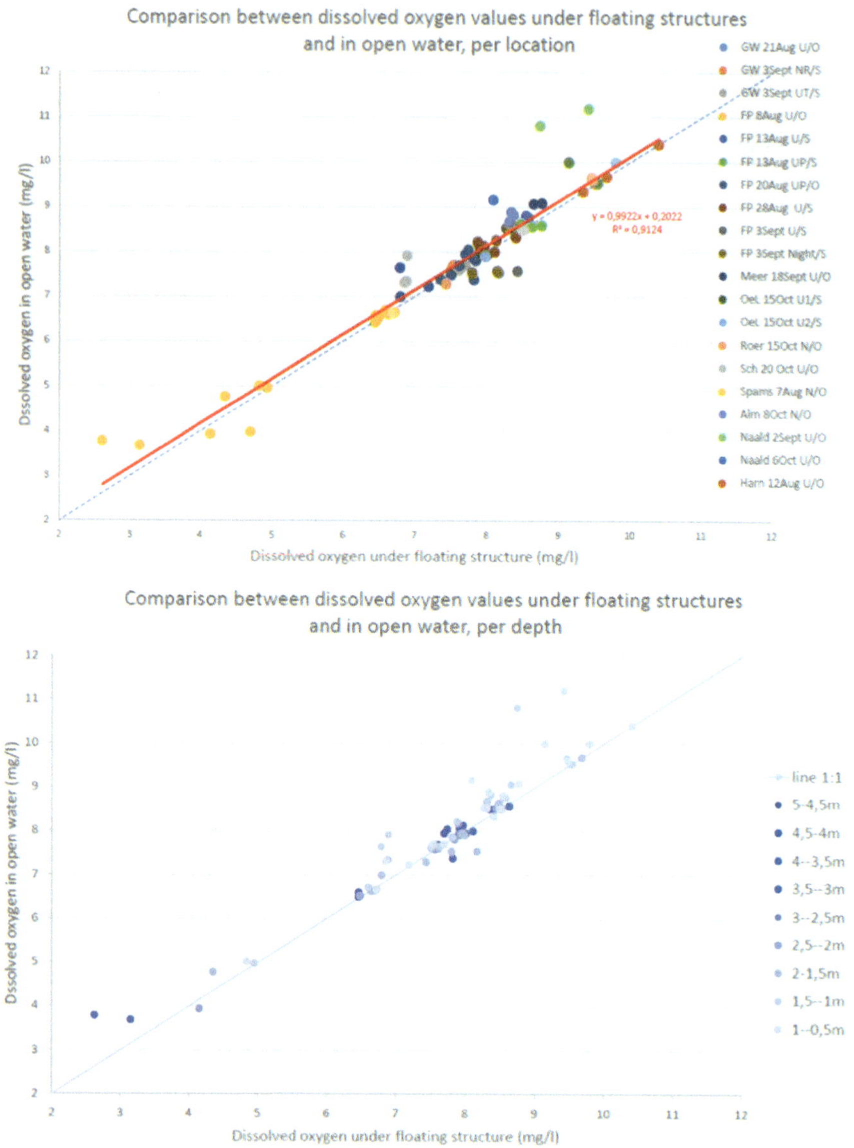

Fig. 5 Comparison between dissolved oxygen values under floating structures and in open water: **a** per location (*top*), and **b** per depth (*bottom*)

1.5–2 m there isn't a noticeable difference in dissolved oxygen, at most locations. However, there is also more data available from lower depths, because when the underwater drone was collecting data under the floating constructions, it was usually positioned right under it (drone is positively buoyant). It was noticeable that locations with greater water depth under floating structures show lower differences

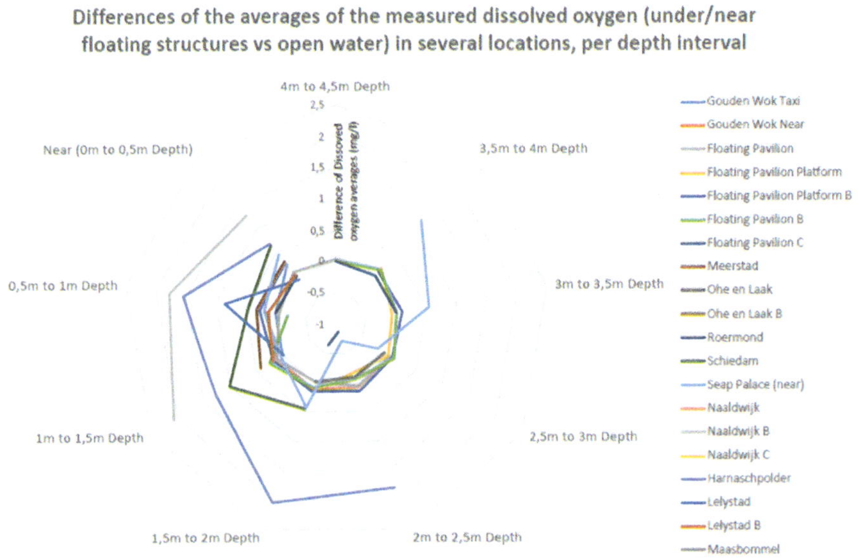

Fig. 6 Differences in dissolved oxygen concentration between open water and underneath structures in several locations per depth interval

in the amount of dissolved oxygen when compared to more shallow locations. It is likely that the depth of water below floating structures, as well as its position (near the shore, or in deeper parts of the water body) influence water mixing and water flow/currents underneath the structure, and therefore lower the amount of time for renewal of water under the houses, hence having an effect on the amount of dissolved oxygen.

With regard to the size of the floating structure, it was not possible to establish strong correlations as represented on Fig. 7. Different platform did not result in higher differences in concentrations, which were detected in both large and smaller platforms. The same was observed for lower differences in concentrations.

Another factor that may contribute to changes in the availability of dissolved oxygen is the vegetation/benthic/bivalves ecosystem that is present under/nearby floating structures. The vegetation can produce dissolved oxygen due to photosynthesis, therefore influencing water quality under floating buildings. The underwater drones, equipped with cameras, allowed to capture underwater footage of the aquatic ecosystems in the vicinity of the floating structures (Fig. 8). Although some of the visited locations had high water turbidity, in some locations lively ecosystems were visible, with bivalves hanging from the structures, as well fish (of different sizes, inclusively present underneath the floating structure) and aquatic pants (mostly around the structures).

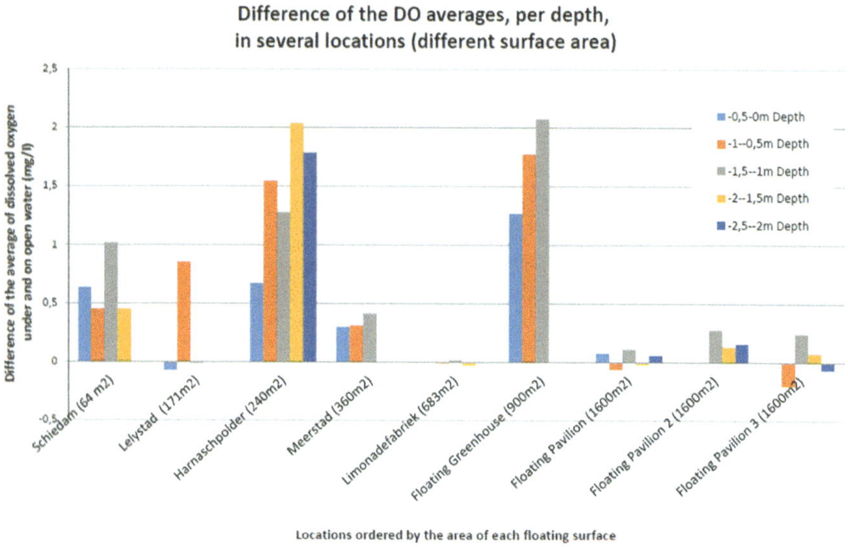

Fig. 7 Differences in dissolved oxygen concentration in different locations with multiple platform sizes, per depth interval

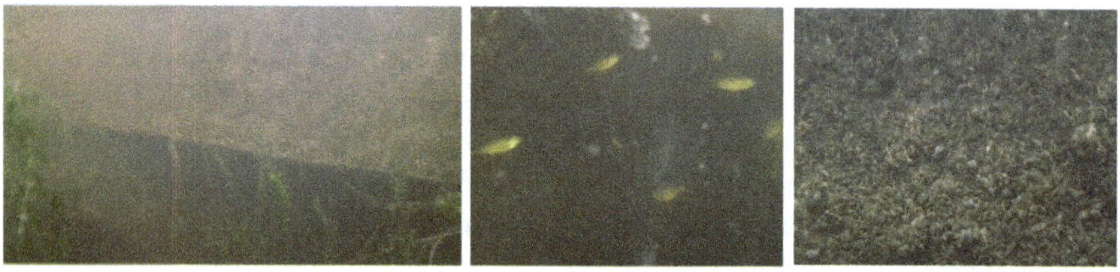

Fig. 8 Example of underwater images collected by underwater drones: **a** macrophytes/vegetation (*left*), **b** fish under platform (*center*), **c** mussels attached to wall (*right*)

4 Conclusion

Results of this study indicate that there are detectable variations in concentrations of water quality parameters between open water and under/near floating structures, but they were consistently low. For instance, in most of the cases, dissolved oxygen did not vary more than 1–2 mg/l between each position, and the minimum values detected are above the required for a healthy habitat. Regarding the nitrate and ammonium measurements (not available in all locations), the measured concentrations were within the expected total nitrogen concentrations and the differences were also small (also lower than 1–2 mg/l).

The collected datasets in each location corresponded to a specific moment in time (few hours or measurements), and therefore it did not take into consideration the daily/seasonal variability of water quality. Additionally, no physical/hydrodynamic characteristics of the water bodies were analysed. Continuous water quality data collection (for several days or months), and additional knowledge about the characteristics water body where the floating structure is built (e.g. currents, flow velocities) are important for follow-up studies. Considering the complexity of the interactions between water quality parameters and the influence of the surrounding environment it is recommended to continue and to improve the monitoring campaign (e.g. include new parameters).

During this research, it became clear that the characteristics of water body is a decisive factor for the extent of impact of water quality structures on water quality, as currents and mixing capacity highly influence the renewal rate of water under the structures. This aspect varied considerably from location to location, as the floating structures were located in different water bodies such as ponds, streams, or lakes. For this reason, it was not possible to establish clear relationships between water quality and characteristics such as the area of the structure, coverage of the water body, or the available space below the house.

Besides physicochemical water quality measurements, the underwater footage allowed to observe and to demonstrate the presence of fish under and nearby the floating structures, which is also an indicator for the health of the water system. A substantial amount of fish and organisms were found attached to this kind of structure, creating a new habitat where otherwise there wouldn't be much bio-diversity.

The research was limited to the available small scale floating structures that currently exist in the Netherlands. Despite the detected variations in water quality parameters, these structures do not seem to have a significant negative impact in the water quality, and may even be regarded as opportunities for building with nature, considering all the new habitat that was unveiled with the underwater images. However, this might be different in the future if the scale and number of projects increase. In order to ensure that bigger scale projects continue not to have an adverse impact in the environment, further research is necessary to infer about recommendations and best practices for the development of larger scale floating urbanization. Aspects such as determining the acceptable platform density ranges, acceptable coverage ratio of the water body, how to minimize the blockage of sunlight (e.g. best positioning of constructions), evaluation of the best materials to use in these structures (ecological/chemical point of view), or how to improve water movement/circulation (prevent water to remain for long periods in the same place) should be taking into future plans and design for the floating development. By integrating floating wetlands or other designs/solutions that enhance the development of underwater habitats (Fig. 9), floating projects could potentially improve water quality and biodiversity, be an opportunity for the implementation of green solutions, and contribute to enhance the connection of the cities with the nature, and in particular, the water.

The collected data, information and videos from the several studied locations around The Netherlands are available in an online tool (www.climatescan.org).

Fig. 9 Examples of floating green solutions (floating wetlands and gardens) that can potentially be combined with floating constructions for minimising undesired impacts of floating platforms and stimulating ecosystems and biodiversity

References

1. Hartwich H (2016) The impact of floating platforms on the benthic community structure in Dutch freshwater ecosystems. Universiteit van Amsterdam
2. Burdick DM, Short FT (1999) The effects of boat docks on eelgrass beds in coastal waters of Massachusetts. Environ Manag 23(2):231–240
3. Foka E (2014) Water quality impact of floating houses: a study of the effect on dissolved oxygen levels. TU Delft
4. Cole VJ, Glasby TM, Holloway MG (2005) Extending the generality of ecological models to artificial floating habitats. Mar Environ Res 60:195–210
5. Kitazawa D, Tabeta S, Fujino M, Kato T (2010) Assessment of environmental variations caused by a very large floating structure in a semi-closed bay. Springer Science and Business Media B. 165:461–474
6. de Lima RLP, Boogaard FC, de Graaf RE (2015) Innovative dynamic water quality and ecology monitoring to assess about floating urbanization environmental impacts and opportunities. International waterweek, vol 3
7. de Lima R, Sazonov V, Bogaard FC, de Graaf R, Dionisio M (2015) Monitoring the impacts of floating structures on the water quality and ecology using an underwater drone, Den Haag, IAHR world conference
8. de Lima RLP, Boogaard FC, de Graaf-van Dinther RE (2020) Innovative water quality and ecology monitoring using underwater unmanned vehicles: field applications, challenges and feedback from water managers. Water 12:1196

Policies and Guidelines

Eco-engineering for Climate Change—Floating to the Future

Tomer Hadary, Jorge Gutiérrez Martínez, Ido Sella, and Shimrit Perkol-Finkel

Abstract Oceans make 71% of our planet's surface and coastal areas are home to over 50% of the worlds' population, resulting in coastal hardening. This replacement of natural habitats with urban and industrial waterfronts, cannot provide ecosystem services similar to those offered by undisturbed coastlines. As a result, coastal infrastructures are considered as sacrificed zones with no environmental value. Ecological engineering is an evolving discipline with the aim of building more resilient and safer coastal and marine structures for people and nature, while maximizing ecosystems, social, and economical benefits. As humans are starting to exhaust the land resources in urban waterfronts, eyes are set to the open oceans. Floating offshore structures will add significant amounts of hard surfaces to the marine environment that will inevitably be colonized by marine life. Due to a combination of structural and hydrodynamic considerations, the communities that develop on floating structures are mainly filter feeding organisms, and are typically dominated by invasive and nuisance species. This is where ECO-engineering must come to play, as planners, designers, and developers must take concrete action towards reducing the ecological footprint of future floating structures. Novel floating urban marine environments must be carefully crafted to generate productive multifunctional structures that are teeming with life. This paper addresses current knowledge gaps that must be addressed with respect to the unique nature of marine life developing on floating structure.

T. Hadary (✉)
SeArc Ecological Marine Consulting, 4, Yehoshua' Bin Nun Street, Tel Aviv, Israel
e-mail: tomerh@searc-consulting.com

J. G. Martínez
ECOncrete, Ronda de Sant Pere 52, Barcelona, Spain
e-mail: jorge@econcretetech.com

I. Sella · S. Perkol-Finkel
ECOncrete, 8 Tversky Street, Tel Aviv, Israel
e-mail: ido@econcretetech.com

S. Perkol-Finkel
e-mail: shimrit@econcretetech.com

Ł. Piątek et al. (eds.), *WCFS2020*, Lecture Notes in Civil Engineering 158,
https://doi.org/10.1007/978-981-16-2256-4_25

Keywords Ecological engineering · Bioprotection · Floating infrastructure · Ecological footprint · Bio-enhancing concrete

1 Introduction

Ocean makes 71% of our planet's surface. Coastlines, represent about 15% of our land areas, and yet are home to over 50% of worlds' population [1–3]. This is one of the key reasons for increased anthropogenic stress on our shorelines and on precious coastal habitats. As humans are starting to exhaust the space and resources in urban waterfronts, eyes are set to the oceans for more space. Space for utilities and industries including offshore ports and airports, power plants, wind farms, bridges, floating energy devices, aquaculture facilities, and ambitious plans for offshore floating cities as well as massive floating installations to make and/or protect these structures and others, like floating breakwaters [4–8].

While the core engineering challenges associated with such complex, often, mega-infrastructure projects are relatively advanced (offshore equipment, vessels, technology and techniques of construction coming from the offshore mining and Oil and Gas industries), there are still many technological gaps and challenges. For example, installation of fewer larger turbines in a given project size, and standardization in the installation process to get cost efficiency [9]; equipment efficiency for construction and installation of offshore bigger wind turbines, innovative solutions that need to be proved (e.g. by means of numerical and physical model tests) and environmental impacts [10].

With respect to floating offshore projects, the technological barriers are even greater. Monopile and gravity-based foundations are usually effective in water depths of 0–25 m, tripod-jacket in 20–50 m and floating wind in 50 m + depths [11], and demonstration projects are needed for testing different floating concepts with the objective to reduce cost or upscale previous demonstrators [12]. Currently, monopiles remain the most installed foundation, with 4258 units (81%) up to date. This includes all foundations installed with and without grid connection. The jackets share (8.9%), Gravity base (5.7%), tripod (2.4%), and tripile (1.5%) follow the cumulative share [12]. Nonetheless, in light of the expected proliferation of floating offshore facilities: the average rated capacity of turbines installed in 2019 was 7.8, 1 MW larger than in 2018. The average size of wind farms in construction almost doubled in one decade (621 MW). The average distance to shore (59 km) and water depth (33 m) continue to increase even though most wind farms are bottom-fixed. Between other aspects, there is a depletion of nearshore locations and the turbine size has significantly increased in floating projects, reaching same capacities as bottom-fixed wind farms. Furthermore, areas suitable for the deployment of offshore wind turbines based on bottom-fixed foundations are not only constrained by water depth but also from numerous other factors: visibility from shore, fisheries, shipping lanes and routes, oil and gas extraction infrastructure [13], and consideration shall be taken to environmental impact at the sea bottom and marine habitats. A big advantage of floating offshore solutions is that floating substructures can fit in areas with prime wind resource, largely independent of soil conditions or water depth. Greater wind resource further out at sea leads to higher capacity factors and less wasted time with no wind generation. This also means less issues with balancing power which in turn translates to reduced costs for consumers. In this regard, the US Bureau of Ocean Energy Management modelled the development of the Levelised Cost of Electricity (LCOE) for floating offshore wind structures and came to the conclusion that floating breaks even with bottom fixed offshore wind around 2027 [13]. These considerations are highly important especially as over the next three years there will be a significant increase in the yields from floating offshore farms, with the installation of floating offshore projects in the UK, France, Norway and Portugal [11].

In summary, there is a growing demand for novel offshore technologies and systems that allow for an efficient use of space. This is where multi-use offshore platforms and advanced marine spatial planning programs (MSP) must come to play, with offshore wind energy and aquaculture/fisheries being the predominant combination in MSP scenarios as for example in the Belgian coast of the North Sea [14].

1.1 The Nature of Floating Habitats

There is a major knowledge gap regarding marine communities associated with floating habitats and their species assemblages. Several researches have shown that the prevailing life forms on artificial floating structures are filter feeders like tunicates, sponges, tube worms, and hydrozoans, that enjoy the constant water flow delivering food and oxygen [15, 16] and despite the ongoing proliferation of artificial floating habitats worldwide [16 and reference therein], little is known of the effect of the motion capabilities, lack of zonation and shading on species assemblages developing on and around floating structures. It is important to note, that there are no natural floating habitats that can be compared to these artificial floating ecosystems, besides perhaps kelp forests, although these are not solid surfaces which effect their hydrodynamic regime, hence not really comparable [17], meaning that the floating artificial structures create novel habitats for both benthic and pelagic organisms [17, 18].

Buoys [19], pontoons [20, 21], docks [22], Fish Aggregation Devices (FADs) [23] and offshore infrastructure [24] are all artificial substrates that can have positive or negative impacts on both sessile and mobile species. Floating pontoons for example, add shaded areas that serve as shelter for fish and mobile invertebrates, and the underside of the pontoon is known as a preferable habitat for sessile communities [25]. On the other hand, a series of studies on piers in the Hudson River estuary, demonstrated that overwater structures can negatively impact fish and lead to reduced abundance of pelagic fish [26]; change the composition of invertebrate species that make fish prey, with reduced abundances of larger species [27]. Moreover, cage experiments suggested that shading of habitat by piers caused a reduction in growth opportunities by limiting prey detection and consumption by demersal fish [27, 28].

New floating offshore structures will add significant amounts of hard surfaces to the marine environment that will inevitably be colonized by marine life. The communities that grow on floating structures are very unique due to several aspects mainly associated with hydrodynamics with the combination of loads from wind, waves and currents around floating structures, the pressure exerted by the flow onto the floating body and the motion of the body. For instance, offshore structures are generally subject to high Reynolds number flows [29]. These high Reynolds number flow conditions ($Re > 10^6$) are in the range of a turbulent flow [29]. In addition, the composition of these artificial structures is very different to any natural habitat (be it fixed or floating) with the vast majority of structures being made of concrete, steel, or (more historically) wood [30].

2 The Role of Ecological Engineering in Floating Infrastructure

As engineers, developers, policy makers, and visionaries are looking into a future with a new dimension of offshore floating structures, principles of Ecological Engineering must be an integral part of the planning, design, and implementation of these floating structures, if we are to navigate into a more sustainable future.

Ecological engineering of shoreline schemes is an evolving discipline [31] with the aim of building more inclusive, resilient and safe coastal and marine structures for people and nature that maximize benefits for ecosystems, society and economies. Over the past decade, the scientific community has studied, experimented, and published, numerous research projects dealing with the topic of ecological enhancement of coastal infrastructure, greening (or rather "Bluing") the gray, and multi-function structures in the marine space [32]. Nonetheless, most of these are not commercial, and the need to scale up ecological engineering projects is on the rise [33, 34].

2.1 *Floating Challenges*

In order to scale ecological engineering and apply such considerations into floating Concrete Based Coastal and Marine Infrastructure (CB-CMI), several knowledge and technology gaps will need to be bridged. Most of these are related to our ability to develop robust structures in rough marine environments, especially with climate change impacts like increased storm intensity and frequency.

Structural performance

Currently, most of the floating structures are made of steel and/or concrete [30]. Structural soundness of these structures is key to assure a good performance and a long lifespan. The integration of bioenhancing materials can improve the structural performance of these structures. For instance, the growth of marine organisms that attach to concrete can improve the compressive strength [35] and the tensile strength [36] and reduce the Chloride penetration [35, 37].

Longevity

To be applicable and competitive in the market, floating structures must have the same lifespan as the non-floating structures. Therefore, materials that comply with the standards and that are suitable to marine environment are to be used. Marine organisms that produce Calcium Carbonate ($CaCO_3$) has proved to improve the protection of the concrete by creating a biofilm that protects the concrete from abrasion and from Chloride penetration in a process so called bioprotection [37]. In addition, the presence of mollusks on the surface of a steel structure can inhibit corrosion and their removal is not always beneficial [38]. There are evidence that biofouling can serve as a protective layer against corrosion with higher efficiency than any paint, thus should not be removed [38, 39]. Therefore, one of the protective measures to the structures can be the marine growth.

Dynamic movement

The motion of the floating structure is a challenge to be addressed during the design phase. Wind, waves and currents should be carefully studied. Long wave periods should be analyzed since might affect the performance of the floating structure. The mooring system must keep the floating structure in place while allowing some displacement (e.g. tidal regime). It should allow dampening effect, minimizing peak loads on the system [40]. Furthermore, fatigue analysis of the mooring system should be addressed as structure will be subject to constant movement throughout their design life. In addition, the mooring system should avoid contact to the seabed in order to prevent erosion and minimize negative ecological impact [40].

Drag and load

The addition of marine growth to the structure implies the gain of extra weight and can increase the drag forces around the structure. These aspects should be considered when designing the structure [38]. Nevertheless, marine growth might reduce the drag forces. Some experimental studies of flow around piles compared clean/smooth piles versus piles with marine growth attachment, especially oysters. As a result, it was discovered that the oysters actually caused a decrease in resistance, as they disrupted the flow of current around the pile and instead produced turbulent eddies [41].

Maintenance

Bioprotection of the structure can also reduce the maintenance costs thanks to increased lifespan. Although such marine intense growth might disrupt visual surveys of the infrastructures' state, inspection can be achieved by scraping off sections of the growth at random (typically, no more than 10% of the surface), which will re-grow with time [42]. Reducing the magnitude and frequency of structural maintenance can be translated into improved ecological stability (reduced anthropogenic intervention), as well as a higher ROI (reduced maintenance costs).

2.2 *Nature Inclusive Designs*

When diving into nature inclusive design of floating CB-CMI, we must consider three key pillars that strongly influence the marine life that will adhere to the structure, and subsequently, influence local biodiversity and productivity.

Material composition

Portland cement-based concrete, which is widely used for marine construction, is known as a poor substrate in terms of biological recruitment, due to high surface alkalinity and leaching of compounds that are toxic to marine life [43]. Thus, the ability of floating CB-CMI to provide ecosystem services similar to those offered by natural habitats is severely compromised. An environmentally sensitive alternative that can and should be applied in floating CB-CMI is Bio-enhancing concrete admixtures tailored to double the biodiversity and species richness compared to standard Portland cement based concrete elements [35, 42, 44, 45].

Texture

The traditional smooth surface CB-CMI does not provide suitable conditions for the development of diverse biological assemblages [46]. As a result, these are often dominated by nuisance and invasive species [47], and do not function as surrogates to natural reef environments. Eco engineered surfaces that have high surface rugosity, helps marine larvae adhere and attach to the concrete surface, by creating micro turbulence, thus breaking the laminar flow across the element [48, 49] and facilitating settlement processes.

Macro-design

Non-enhanced CB-CMI are highly homogeneous and offer limited shelter, leading to low diversity assemblages [32, 43, 50, 51]. Ecologically engineered CB-CMI can include different designed habitats, and shelter, that can be tailored to specific species, and even to accommodate specific needs of life history traits of desired organisms (e.g., larval stages, juveniles, and adults might have different habitat needs). For example, ecologically modified Antifer units that were deployed on a breakwater in the Mediterranean, presented more than double the diversity and species richness and lower ratio of invasive species, compared to adjacent standard Antifer units, both for invertebrate and fish species [35].

The capacity to add biological niches through design features is most obvious with respect to addition of water retaining features to CB-CMI that are traditionally designed to drain water. For example, Perkol-Finkel and Sella, 2015 [45] have demonstrated increased biodiversity in ecologically enhanced Tide-Pool Armor integrated into a constructed riprap beach in NYC, supporting variable community including algae, mobile and sessile invertebrates and fish, compared to adjacent rocky riprap that had extremely low live cover and virtually no life. Water retaining elements that were installed on a UK seawall ('Vertipools') has supported 24 taxa after 5 years compared to only 8 on the vertical seawall [52].

2.3 *Additional Considerations*

Apart from the above-mentioned design consideration, there are additional design aspects that are associated with the unique physical and environmental conditions of floating substrates, that must be addressed.

No tidal effects/Fixed depth

Floating habitats offer ample of surface area at the waterline, however these areas, as opposed to fixed structures and obviously natural fringing and rocky intertidal reefs, do not experience tide. This influences zonation patterns of marine flora and fauna, and might cause changes to classic behavioral patterns of benthic intertidal communities like limpets or barnacles, that are normally driven by phototactic and geotactic patterns [51, 53, 54]. In practice, organisms growing on floating structures experience the same depth permanently. While biofouling communities on ship hulls and pontoons have been experiencing such conditions since ancient times, to the best of our knowledge this aspect has received no scientific attention, and calls for future exploration and experimentation, especially if we are to expand the presence of floating structure in the coming years.

Overhangs and Shaded Habitats

Floating structures like docks, platforms, aquaculture facilities and in the future floating housing/utilities solutions will offer vast amounts of shaded environments and overhangs, that are subject to different levels of motion. Such conditions can hardly be found in natural habitats. To anticipate the biological assemblages that will be associated with such habitats we can look into biofouling communities on ship hulls and pontoons, nonetheless the scale of floating CB-CMI will be much more massive. These aspects must be carefully considered, and addressed through ecological engineering, in order to avoid strong dominance of invasive species that typically thrive in such conditions [47]. By integrating principle of ecological engineering to the design and construction of future floating structures we can not

Fig. 1 Floating dock with sub-sea protrusions designed for fish, Eilat, Red Sea

only offset negative environmental effects, but actually provide valuable ecosystem services.

By adding elements like moon pools into docks and piers, shading effects can be diminished, and through different design solutions overhangs can include 3D complexity offering shelter for juvenile fish (Fig. 1). Light penetrating features, for example have been included in the ecological design of a recent seawall replacement in Seattle (WA, USA). These features alongside with designed seawall panels, aimed to improve habitat conditions for several species of juvenile Pacific salmon, considered endangered in the area [55].

As many of these floating structures are public, and are found at the interface between humans and the ocean, ecological engineering can also provide education

Fig. 2 Rendering of ecologically enhanced seawall elements and an underwater observatory

and outreach opportunities. For example, ecologically enhanced floating structures that will be teeming with life, can integrate moon pools and observation docks for the public, to increase engagement, and educate the users of the structure to preserve valuable urban marine nature (Fig. 2).

Connectivity challenges: stepping stones versus island theory

The addition of these hard surface habitats, either near or offshore, offer habitat connectivity, promoting spatial distribution of marine communities [24, 56, 57]. However, artificial structures in general, differ significantly from natural habitats with respect to material, complexity, surface area, age, orientation, movement and disturbance regimes [58–61] and the ecological communities they support are heavily dominated by invasive and nuisance species when compared to adjacent natural habitats [17, 21, 35, 44, 62, 63]. Additionally, by affecting light availability, flow, wave energy or leaching chemicals floating structures may modify the communities of adjacent habitats [50, 64].

Anchoring systems—Ecological Engineering from Top to Bottom

Floating structures, although in motion, are fixed to the seabed, and thus require extensive mooring systems. The impact of mooring is characterized by the destruction of habitats by the anchor and connection system [65]. The system (typically chains, cables, ropes) should be designed for minimal contact with the seabed in order to prevent erosion and destruction of valuable habitats (in soft bottom—vegetation, clams, infauna and in hard bottom—coral reefs, sessile communities). Nevertheless, if the chain system is provided by floaters loses its catenaria effect [40]. Similarly, any mooring/anchoring used should minimize the footprint on the seabed, and avoid releasing harmful pollutants into the ecosystem [40]. Concrete gravity based anchors or sinkers have a greater eco-engineering potential than other mooring solutions since they can be designed to provide habitat, shelter, and even nursing grounds for target species like lobsters, oysters, octopus and alike and generally be designed to offer greater ecosystem services for fish (both resident and transient species). Researchers like Wilson and Elliott [66] and Lacroix and Pioch [67] have reviewed this topic, and presented theoretical schemes that can be implemented in offshore wind projects, actual field studies are severely lacking, let alone full-scale implementations of such enhancement measure. To unlock this potential, and reduce the ecological footprint of future offshore projects, such enhancement measure should turn into reality and use bio-enhancing concrete for gravity-based moorings, designed to match the ecological needs of local species.

3 ECO-engineering for Climate Change. Concluding Remarks

In light of current trajectories of human population growth, and unfortunately as humanity keeps depleting our natural resources on land, floating CB-CMI might be increasingly dominant. It is our duty to ensure that we develop our oceans in a sustainable manner. Through ecological engineering, floating CB-CMI can serve multiple ecological, environmental, and operational goals. With the right materials and design, these structures have the capacity to induce rich and diverse communities, that increase the local biodiversity and productivity. Filter feeding communities like oyster, sponges, and tunicates, can positively impact water quality [68].

Bio-enhancing concrete elements can also induce the growth of ecosystem engineers [69] that have profound impacts on the way communities develop and, ultimately, on biodiversity. Many of these ecosystem engineers have an additional environmental advantage, with respect to climate change mitigation and adaptation. Species like oysters, tube worms, corals and alike, that secrete $CaCO_3$ skeletons onto the substrate are serving multiple benefits. Apart from adding structural complexity and heterogeneity, this "biological crust" serves as an active carbon sink, as carbon is assimilated into skeletons of these organisms in a process called biocalcification [70]. When applied at large-scales, this natural process can provide a substantial mitigation tool. Finally, with time, as technology gaps are bridges, we can strive for a future that combines disciplines and industries on floating CB-CMI to generate highly efficient marine spaces with a blend of renewable "blue energy" resources (wind, waves, tides and sun), multi trophic farming opportunities taking advantage of the entire water column, and space for living.

References

1. Vitousek PM, Mooney HA, Lubchenco J, Melillo JM (1997) Human domination of Earth's ecosystems. Science 277(5325):494–499
2. Neumann B, Vafeidis AT, Zimmermann J, Nicholls RJ (2015) Future coastal population growth and exposure to sea-level rise and coastal flooding—a global assessment. PLoS ONE 10(3):
3. He Q, Silliman BR (2019) Climate change, human impacts, and coastal ecosystems in the anthropocene. Curr Biol 29(19):R1021–R1035
4. Japan Port Consultants ltd. (2018) Performance category: port access roads, offshore airports. http://www.jportc.co.jp/en/genre/port_access_roads. Accessed 23 Jul 2020

5. Doe J (2019) Upper Zakum offshore oil field development, Abu Dhabi. https://www.offshore-technology.com/projects/upper-zakum-offshore-uae/. Accessed 20 Jul 2020
6. Equinor (2020) Hywind Tampen: the world's first renewable power for offshore oil and gas. https://www.equinor.com/en/what-we-do/hywind-tampen.html. Accessed 23 Jul 2020
7. Profloating (2020) Developer and producer of floating solar mounting systems. https://www.profloating.nl/. Accessed 15 Jul 2020
8. van Hoeken A (2019) A world's first: offshore floating solar farm installed at the Dutch North Sea. In: Oceans of energy. https://oceansofenergy.blue/2019/12/11/a-worlds-first-offshore-floating-solar-farm-installed-at-the-dutch-north-sea/. Accessed 17 Jul 2020
9. MacLeay A, den Hollander R (2018) Construction challenges in future offshore wind developments. Paper presented in offshore technology conference, 30 April–3 May, Houston, Texas, USA
10. Li R, Wang X (2011) Status and challenges for offshore wind energy. Paper presented in international conference on materials for renewable energy & environment, 20–22 May 2011, Shanghai, China
11. DNV (2014) Design of offshore wind turbine structures, standard number OS-J101, p 238
12. Walsh C (2019) Offshore wind in Europe—key trends and statistics 2018. Wind Europe, Brussels
13. Friends of Floating Offshore Wind (2018) Offshore energy. The future's floating
14. Degraer S, Brabant R, Rumes B (2011) Offshore wind farms in the Belgian part of the North Sea. Royal Belgian Institute of Natural Sciences Management. Unit of the North Sea Mathematical Models, Marine Ecosystem Management Section
15. Perkol-Finkel S, Zilman G, Sella I, Miloh T, Benayahu Y (2008) Floating and fixed artificial habitats: Spatial and temporal patterns of benthic communities in a coral reef environment. Estuar Coast Shelf Sci 77(3):491–500
16. Connell SD (2000) Floating pontoons create novel habitats for subtidal epibiota. J Exp Mar Biol Ecol 247(2):183–194
17. Perkol-Finkel S, Zilman G, Sella I, Miloh T, Benayahu Y (2006) Floating and fixed artificial habitats: effects of substratum motion on benthic communities in a coral reef environment. Mar Ecol Prog Ser 317:9–20
18. Holloway MG, Connell SD (2002) Why do floating structures create novel habitats for subtidal epibiota? Mar Ecol Prog Ser 235:43–52
19. Giulio Relini MR, Montanari M (2013) An offshore buoy as a small artificial island and a fish-aggregating device (FAD) in the Mediterranean. Island, Ocean and Deep-Sea Biology: Proceedings of the 34th European marine biology symposium, held in Ponta Delgada (Azores), Portugal, 13–17 September 1999. Springer Science & Business Media
20. Connell SD, Glasby TM (1999) Do urban structures influence local abundance and diversity of subtidal epibiota? A case study from Sydney Harbour, Australia. Mar Environ Res 47 (4):373–387
21. Glasby TM, Connell SD, Holloway MG, Hewitt CL (2007) Nonindigenous biota on artificial structures: could habitat creation facilitate biological invasions? Mar Biol 151(3):887–895
22. Caine EA (1987) Potential effect of floating dock communities on a South Carolina estuary. J Exp Mar Biol Ecol 108:83–91
23. Dempster T, Taquet M (2004) Fish aggregation device (FAD) research: gaps in current knowledge and future directions for ecological studies. Rev Fish Biol Fish 14(1):21–42
24. Macreadie PI, Fowler AM, Booth DJ (2011) Rigs to reefs: will the deep-sea benefit from artificial habitat? Front Ecol Environ 9(8):455–461
25. Toh KB, Ng CSL, Wu B, Toh TC, Cheo PR, Tun K, Chou LM (2017) Spatial variability of epibiotic assemblages on marina pontoons in Singapore. Urban Ecosyst 20(1):183–197
26. Able KW, Grothues TM, Kemp IM (2013) Fine-scale distribution of pelagic fishes relative to a large urban pier. Mar Ecol Prog Ser 476:185–198
27. Duffy-Anderson JT, Able KW (2001) An assessment of the feeding success of young-of-the-year Winter Flounder (*Pseudopleuronectes americanus*) near a municipal pier in the Hudson River estuary, USA. Estuaries 24:430–440

28. Duffy-Anderson JT, Able KW (1999) Effects of municipal piers on the growth of juvenile fishes in the Hudson River estuary: a study across a pier edge. Mar Biol 133(3):409–418
29. Ong MC (2017) CFD applications in offshore engineering. In: EPJ web of conferences. EDP Sciences
30. The Editors of Encyclopedia Britannica (2018) Encyclopedia Britannica, Naval architecture. https://www.britannica.com/technology/naval-architecture. Accessed 20 Jul 2020
31. Mitsch WJ (2012) What is ecological engineering? Ecol Eng 45:5–12
32. Firth LB, Thompson RC, Bohn K, Abbiati M, Airoldi L, Bouma TJ, Bozzeda F, Ceccherelli VU, Colangelo MA, Evans A, Ferrario F, Hanley ME, Hinz H, Hoggart SPG, Jackson JE, Moore P, Morgan EH, Perkol-Finkel S, Skov MW, Strain EM, van Belzen J, Hawkins SJ (2014) Between a rock and a hard place: environmental and engineering considerations when designing coastal defense structures. Coast Eng 87:122–135
33. Pioch S, Relini G, Souche JC, Stive MJF, De Monbrison D, Nassif S, Simard F, Allemand D, Saussol P, Spieler R (2018) Enhancing eco-engineering of coastal infrastructure with eco-design: moving from mitigation to integration. Ecol Eng 120:574–584
34. Morris R, Loke L, Lau E, Strain E, Airoldi L, Alexander K, Bishop M, Coleman R, Cordell J, Dong Y, Firth L, Hawkins S, Health T, Kokora M, Lee S, Miller J, Perkol-Finkel S, Rella A, Leung K (2019) Design options, implementation issues and evaluating success of ecologically-engineered shorelines
35. Sella I, Perkol-Finkel S (2015) Blue is the new green–Ecological enhancement of concrete based coastal and marine infrastructure. Ecol Eng 84:260–272
36. Risinger JD (2012) Biologically dominated engineered coastal breakwaters. California State University
37. Coombes DW, Naylor LA, Viles HA, Thompson RC (2013) Bioprotection and disturbance: Seaweed, microclimatic stability and conditions for mechanical weathering in the intertidal zone. Geomorphology 202:4–14
38. British Standards Institution (2000) BS 6349-1: Maritime structures—part 1: code of practice for general criteria. BSI London
39. Palanichamy S, Subramanian G, Eashwar M (2012) Corrosion behavior and biofouling characteristics of structural steel in the coastal waters of the Gulf of Mannar (Bay of Bengal), India. Biofouling 28:441–451
40. Wilhemsson R, Sella I (2020) Webinar anchoring sustainability. Ecological mooring solutions. Seaflex—ECOncrete
41. Santos K, Angellucci N, Fitzgerald AM, Miller J, Rella A (2015) The effects of invertebrate colonization on water flow around Pier Pilings: implications on homeland security. New Jersey University; Stevens Institute of Technology
42. Perkol-Finkel S, Sella I (2014) Ecologically active concrete for coastal and marine infrastructure: innovative matrices and designs. In: From Sea to Shore—meeting the challenges of the Sea. ICE Publishing, Edinburgh, UK
43. Lukens RR, Selberg C (2004) Guidelines for marine artificial reef materials. In: A joint publication of the Gulf and Atlantic States Marine fisheries commissions
44. Perkol-Finkel S, Hadary T, Rella A, Shirazi R, Sella I (2017) Seascape architecture—incorporating ecological considerations in design of coastal and marine infrastructure. Ecol Eng
45. Perkol-Finkel S, Sella I (2015) Harnessing urban coastal infrastructure for ecological enhancement. In: Proceedings of the institution of civil engineers-maritime engineering. Thomas Telford Ltd.
46. Firth LB, White FJ, Schofield M, Hanley ME, Burrows MT, Thompson RC, Skov MW, Evans AJ, Moore PJ, Hawkins SJ (2016) Facing the future: the importance of substratum features for ecological engineering of artificial habitats in the rocky intertidal. Mar Freshw Res 67(1):131–143
47. Mineur F, Cook EJ, Minchin D, Bohn K, MacLeod A, Maggs CA (2012) Changing coasts: marine aliens and artificial structures. Oceanogr Mar Biol 50:189–234
48. Chase AL, Dijkstra JA, Harris LG (2016) The influence of substrate material on ascidian larval settlement. Mar Pollut Bull 106(1–2):35–42

49. Zilman G, Novak J, Benayahu Y (2008) How do larvae attach to a solid in a laminar flow? Mar Biol 154(1):1–26
50. Dugan JE, Airoldi A, Chapman MG, Walker S, Schlacher T (2011) Estuarine and coastal structures: environmental effects. A focus on shore and nearshore structures. In: Elliott M, Dugan JE (eds) Treatise on estuarine and coastal science. Elsevier, New York, pp 17–41
51. Guy-Haim T, Rilov G, Achituv Y (2015) Different settlement strategies explain intertidal zonation of barnacles in the Eastern Mediterranean. J Exp Mar Biol Ecol 463:125–134
52. Hall AE, Herbert RJH, Britton JR, Boyd IM, George NC (2019) Shelving the Coast with vertipools: retrofitting artificial rock pools on coastal structures as mitigation for coastal squeeze. Frontiers in marine science, vol 6(456)
53. Foster B (1971) Desiccation as a factor in the intertidal zonation of barnacles. Mar Biol 8 (1):12–29
54. Schonbeck M, Norton TA (1978) Factors controlling the upper limits of fucoid algae on the shore. J Exp Mar Biol Ecol 31(3):303–313
55. Sawyer AC, Toft JD, Cordell JR (2020) Seawall as salmon habitat: Eco-engineering improves the distribution and foraging of juvenile Pacific salmon. Ecol Eng 151:
56. Coolen JW, Boon AR, Crooijmans R, Van Pelt H, Kleissen F, Gerla D, Beermann J, Birchenough SNR, Becking LE, Luttikhuizen PC (2020) Marine stepping-stones: connectivity of Mytilus edulis populations between offshore energy installations. Mol Ecol 29 (4):686–703
57. Bishop MJ, Mayer-Pinto M, Airoldi L, Firth L, Morris RL, Loke LHL, Hawkins SJ, Naylor LA, Coleman RA, Chee SY (2017) Effects of ocean sprawl on ecological connectivity: impacts and solutions. J Exp Mar Biol Ecol 492:7–30
58. Airoldi L, Bulleri F (2011) Anthropogenic disturbance can determine the magnitude of opportunistic species responses on marine urban infrastructures. PLoS ONE 6(8): https://doi.org/10.1371/journal.pone.0022985
59. Bulleri F, Chapman MG (2010) The introduction of coastal infrastructure as a driver of change in marine environments. J Appl Ecol 47(1):26–35
60. Chapman MG, Underwood AJ (2011) Evaluation of ecological engineering of "armored" shorelines to improve their value as habitat. J Exp Mar Biol Ecol 400(1–2):302–313
61. Airoldi L, Bulleri F (2011) Anthropogenic disturbance can determine the magnitude of opportunistic species responses on marine urban infrastructures. PLoS ONE 6(8):
62. Dafforn KA, Johnston EL, Glasby TM (2009) Shallow moving structures promote marine invader dominance. Biofouling 25(3):277–287
63. Firth L, Hawkins SJ (2011) Introductory Comments-Global change in marine ecosystems: patterns, processes and interactions with regional and local scale impacts. Elsevier, Netherlands
64. Dafforn KA, Glasby TM, Airoldi A, Rivero NK, Mayer-Pinto M, Johnston EL (2015) Marine urbanization: an ecological framework for designing multifunctional artificial structures. Front Ecol Environ. https://doi.org/10.1890/140050
65. Jackson EL, Griffiths C, Durkin O (2013) A guide to assessing and managing anthropogenic impact on marine angiosperm habitat-part 1: literature review
66. Wilson JC, Elliott M (2009) The habitat-creation potential of offshore wind farms. Wind Energy 12(2):203–212
67. Lacroix D, Pioch S (2011) The multi-use in wind farm projects: More conflicts or a win-win opportunity? Aquat Living Resour 24
68. Paalvast P, van Wesenbeeck BK, van der Velde G, de Vries MB (2012) Pole and pontoon hulas: An effective way of ecological engineering to increase productivity and biodiversity in the hard-substrate environment of the port of Rotterdam. Ecol Eng 44:199–209
69. Coleman FC, Williams SL (2002) Overexploiting marine ecosystem engineers: potential consequences for biodiversity. Trends Ecol Evol 17(1):40–44
70. Hily C, Grall J, Chauvaud L, Lejart M, Clavier J (2013) CO_2 generation by calcified invertebrates along rocky shores of Brittany, France. Mar Freshw Res 64(2):91–101

Floating Cities and Equitable Grafting onto Marine Ecosystems

Ioana Corina Giurgiu

Abstract Based on current predictions of sea level rise and other climatic changes, in the near future, large urban coastal areas will no longer provide viable urban environments for human habitation. In response, floating city design proposals of various scales, degrees of connectivity and mobility are emerging as potential adaptations to predicted climate changes which partially or completely replace existing coastal urban typology. The prospect of utilizing large floating structures for urban habitation and energy production seems highly appealing from an urban development perspective as it allows for a certain degree of continuity between the existing coastal typology and its envisioned floating counterpart. However, transplanting the large scale, static and perpetually connected land-based city model within an aquatic environment may prove challenging in terms of maintaining an equitable relationship between urban growth and marine ecosystem maintenance. By analyzing the floating city model as a graft onto marine ecosystems, the paper explores site selection, scale and mobility, highlighting opportunities for future floating city proposals to enhance and contribute to the host marine environments they inhabit.

Keywords Floating city · Ecosystem graft · Environmental benefits · Design principles · Climate refugia

1 Introduction

In April 2019, the first UN High-Level Round Table on Sustainable Floating Cities was held in New York. The UN Deputy Secretary-General Amina Mohammed remarked that cities "are often our testing ground for new ideas and solutions" and acknowledged that we currently "live in a time when we cannot continue building cities the way New York or Nairobi were built" [1]. In this context, sustainable

I. C. Giurgiu (✉)
Sea Cities Lab, Cities Research Institute, Griffith University, Gold Coast, QLD, Australia
e-mail: ioana.giurgiu@griffithuni.edu.au

Ł. Piątek et al. (eds.), *WCFS2020*, Lecture Notes in Civil Engineering 158,
https://doi.org/10.1007/978-981-16-2256-4_26

floating city proposals were introduced as "part of our new arsenal of tools" [1] that help us respond to changes in climatic conditions but also to social and economic issues which emerge as consequences of urban growth trends.

Although various floating solutions exist [2, 3], modern floating cities which, in terms of location and size, fall under the category of very large floating structures (VLFS) have yet to be fully implemented and tested. Compared to land-based urban design, the design of floating infrastructure as urban habitation is less established in terms of urban and environmental management and other regulatory frameworks.

From a design perspective, sustainable floating cities, therefore afford an opportunity for re-thinking existing strategies and implementing knowledge gained via land-based testing. This is especially relevant in terms of the relationship between modern city infrastructure and the natural environment.

Land-based designs have, thus far, been slow to shift towards a more sustainable and equitable relationship. For example, while cities have been relying on natural ecosystem services since the agricultural revolution, the term "ecosystem services" was only coined in 1981 [4] and recognized within policy frameworks much later. The same can be observed in case of the of vegetation bridges which allow for continuity in terms of animal migration paths over highway infrastructure. Initially, transport infrastructure design did not account for ecosystem continuity, leading to the development of separate infrastructure to support wildlife migration rather than achieving an integrated design solution which serves both needs.

When expanding design criteria to include ecosystem requirements, it can be argued that the resulting relationship between man-made and natural systems is similar to that of graft and host organism.

In the fields of medicine and biology a graft is defined as a "a piece of healthy skin or bone cut from one part of a person's body and used to repair another damaged part, or a piece cut from one living plant and attached to another plant so that it grows there" (Oxford Dictionary). As a verb, grafting signifies "to join or add something new" (Oxford Dictionary) .

In looking at how cities perform as grafts in relation to the host ecosystems they inhabit, traditional land-based designs often create inequitable relationships where, although striving to minimize detrimental impacts, cities often do not actively and positively contribute to establishing growth at the ecosystem level. However, if the graft function of healing deriving from the medical field would be considered as a main design driver, this relationship could become more balanced and potentially enhance both natural and anthropic environments (Fig. 1).

In this context, floating solutions create opportunities for "equitable grafting", where the design of human infrastructure and habitat generates hybrid environments that could help repair and enhance both the natural and anthropic environments.

In order to achieve this, several directions that diverge somewhat from typical land-based cities should be explored. The basic parameters of location, size and mobility are further discussed, highlighting opportunities afforded by floating city designs to improve the relationship between new floating development and the natural marine environments they will become a part of.

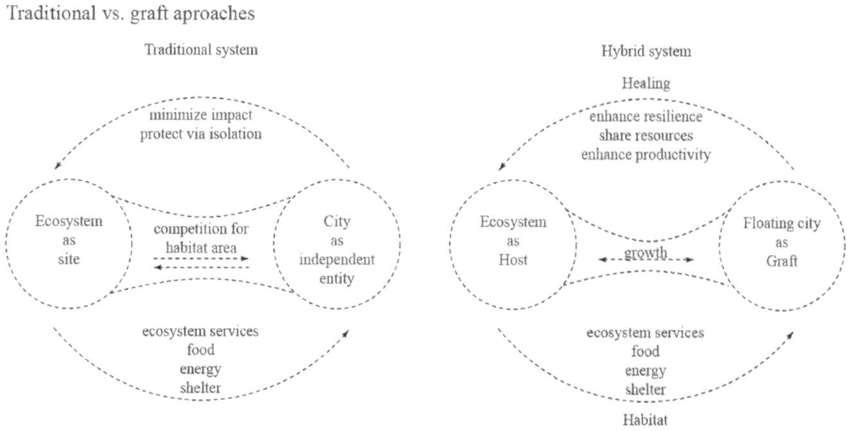

Fig. 1 Conceptual diagram showing differences of approach and resulting relationships between natural and built environments in traditional land-based cities and floating cities performing as grafts

On the basis of an interdisciplinary, qualitative literature review regarding the relationship between cities and ecosystems, this paper presents a conceptual approach to identifying design directions and considerations for future floating solutions in order to enhance this relationship.

The concept of grafting, via its associated functions of healing and creating hybrid configurations, was used to focus an initial analysis, identifying possible points of overlap between urban and environmental agendas.

The following sections, regarding location, scale and mobility, are used as conceptual experiments, exploring how the grafting approach could be applied in order to achieve positive environmental impacts through the design and engineering of floating cities.

The analysis highlights ways in which this approach would diverge from current conventional design considerations, provides examples of opportunities for floating cities to enhance ecosystem growth and speculates on how the identified design considerations could be applied within future floating cities.

Lastly, the speculative design considerations identified via location, scale and mobility examples are compiled as a set of criteria and opportunities which support the use of the graft conceptual approach as a potentially viable future research direction.

2 Location

One of the first parameters to consider in the design of future floating city structures is their location and the strategic thinking behind site selection. Based on their location and the specific technological and logistic challenges encountered, VLFS can be split into two categories: Offshore and Nearshore structures [3].

Site selection also reflects the underlying function of the structure. For example, floating runways such as the Mega-float in Tokyo Bay or floating energy farms (e.g. floating solar arrays or wind farms) are traditionally considered for nearshore areas. This location provides shorter connections (e.g. to decrease transmission losses for energy delivery) which increase mobility and shorten transport routes between existing infrastructures (e.g. between coastal city and floating airport runway). On the other hand, structures such as oil-rigs or Mobile Offshore Bases (MOB) can be sited offshore, in locations of military tactical importance (for MOB) or near resources (for oil extraction).

The UN Habitat press release regarding the 2019 High-Level Round Table on Sustainable Floating Cities mentions that discussions revolved around floating solutions which "would be situated close to cities and could be used to house those fleeing from rising sea levels and other threats from natural or climate-related disasters in their home areas and overcome housing shortages" [5].

Negative climate-related impacts are, however, not limited to human development. Given our relative reliance on various ecosystem services and the fact that human activity has a strong impact on adjacent ecosystems, mitigation strategies relating to the preservation and enhancement of ecosystem services could play an important role in the successful establishment of future floating communities.

For example, the 2019 IPCC report [6] relating to marine ecosystems and dependent communities, shows that loss of wetland habitat, a type of ecosystem providing important services for both human and natural activity, is "primarily caused by non-climatic drivers" which derive from human disturbance.

In 2013, research findings [7] showed that the effect of wetland loss not only impacts local ecosystems but has far-reaching effects via their connectivity to migratory species routes. The study shows that, depending on the location of the wetland loss, a predicted 30% loss in wetland ecosystems could lead to up to 72% population reduction in migratory bird species. This in turn would have significant impacts on the local distribution of species in each location and ultimately on the communities depending on local fisheries and other ecosystem services.

From a strategic point of view, given the need for connection to existing physical and economical infrastructure, floating communities will likely be located near existing human activity hubs. There is, therefore, a degree of overlap between locations which hold potential for human development and locations in which natural ecosystems are most affected by human activity. This presents an opportunity for the design of new floating structures to enhance and protect natural systems.

Specifically, in nearshore applications, floating solutions could incorporate floating wetland areas which could help to preserve vital nesting and feeding grounds for local and migratory species. While sea level rise (SLR) is predicted to cause species redistribution and landward relocation of wetland areas [6], especially in areas close to urban development, this survival strategy is often limited due to competition between expansion of urban and wetland areas. The inclusion of floating wetlands within nearshore zones could, thus, provide additional foraging and nesting grounds for migratory species and allow for the sea-ward expansion of

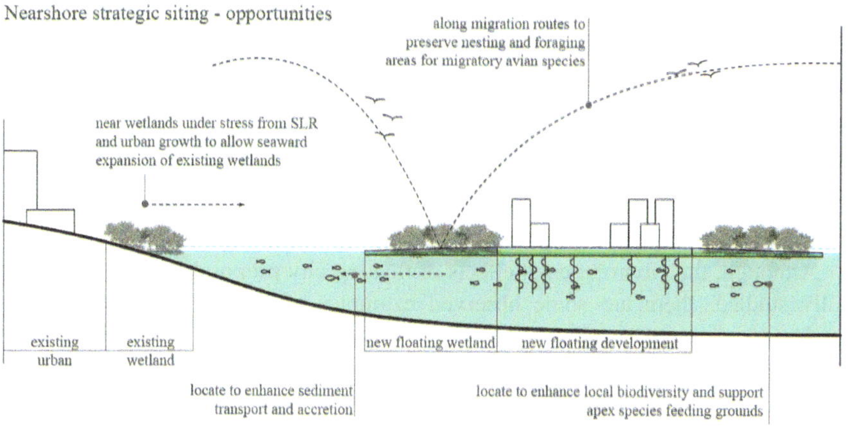

Fig. 2 Conceptual diagram showing opportunities for positive environmental impacts via strategic site selection

existing wetlands by engineering designs to enhance sedimentation and allow for the natural processes of wetland land formation to occur (Fig. 2).

The notion of integrating the "healing" function of grafts as a design parameter which influences site selection creates opportunities to support and enhance ecosystem functioning and implicitly, the ecosystem services that can be relied upon. The strategy for site selection, could therefore revolve around identifying locations where human and environmental needs, resultant from climate change and urban growth, intersect.

3 Scale

A second basic aspect which can have a significant impact on how future floating cities will perform is scale. In studying and predicting the environmental impacts of global warming, effects are explored across multiple scales ranging from global to local and micro scale impacts. It is proposed that, in order to adequately respond to the challenges deriving from changing climatic conditions, floating city designs should consider effects and opportunities to enhance natural system resilience across all scales.

Partly due to the engineering challenges as well as the specific construction methods developed for floating solutions, VLFS and implicitly some floating city solutions are designed as modular systems of relatively small scale when compared to their land-based counterparts.

A recent example of a modular design was the Oceanix proposal by architect Bjarke Ingles, which was promoted at the aforementioned UN Roundtable discussions. The proposal consists of several 4.5-acre hexagonal platforms, each forming a village and combining to form full scale floating cities [8]. In terms of scale, the platforms can be categorized as VLFS which, by definition, range in sizes between 1 and 10 km in length [3].

At this localized scale, some studies on ecosystem adaptation and survival in the context of predicted climate change highlight the conservation and management of microclimate areas [9] and localized climate refugia [10] as potentially productive approaches towards increasing the resilience and chance of survival for marine ecosystems.

Although, the environmental effects of floating city proposals have not yet been fully studied, there are some observed examples of man-made climate refugia having positive impacts on specific species survival [11] as well as positive impacts of small floating structures on water quality and species richness [12] and positive impacts of offshore windfarm structures [13].

While the positive impacts mentioned above are mostly unintended consequences, they highlight a potential opportunity to design future floating cities as man-made climate refugia which contribute to and enhance their surrounding environments (Fig. 3).

In identifying past and present climate refugia that occur naturally, some of the known traits of such climate havens, present a good indication of design criteria which could be applied to create man-made refugia. Climate refugia may be identified via their "higher species richness than the surrounding landscape", "relatively high number of endemic species", high diversity of resources, topography and localized climatic conditions [14].

In simpler terms, man-made climate refugia could potentially be achieved by locally engineering the resulting habitat and environmental parameters (lighting levels, oxygenation, surface water temperature, etc.) to increase species richness, support endemic species, diversify topography, resources and habitat type.

At a local scale, design approaches for enhancing resource diversity and endemic species richness could include diversifying aquaculture, agriculture and biofuel crops based on endemic vegetation and allowing for shared resource allocation between human and ecosystem demands in order to maintain and support local food webs.

One possible implication is that design sites might include areas larger than those necessary for meeting human needs, allocating zones which can support productive ecosystem engineer species such as whales (help cycle nutrients and water [15]) and reef building bivalves (help prevent coastal erosion and filter nutrients [16]).

At the micro-scale, building material choices that allow for oxygenation (for example by providing diffuse natural light to allow photosynthesis) and create micro-habitats via a rich topography towards the water environment [17] could help to encourage the development of floating cities as biodiversity hotspots.

Adding to the idea of interdependence and locating future floating cities at the intersection of human and natural needs, further opportunities for enhancing human and ecosystem growth emerge from introducing the graft function of healing and enhancement across various scales. At the micro and local scale, floating designs

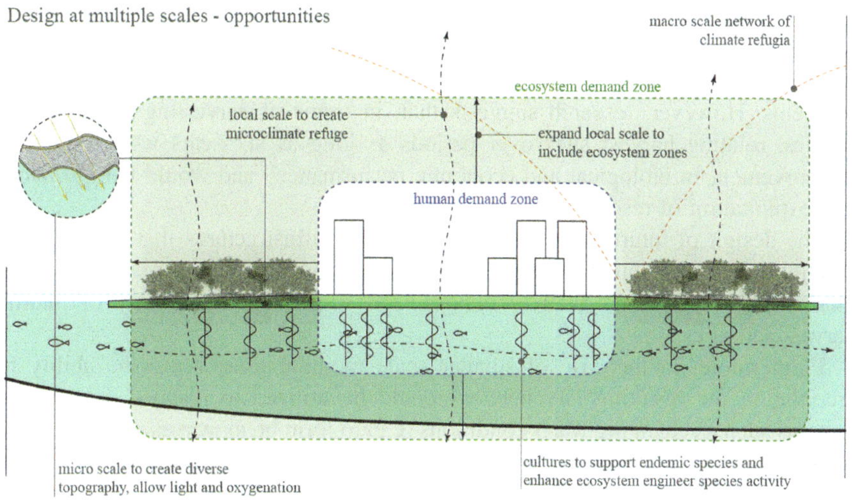

Fig. 3 Conceptual diagram showing opportunities to enhance ecosystem productivity and create climate refugia by designing across multiple scales (from micro to macro)

could help enhance natural productivity and diversity, ultimately providing refugia from the effects of the changing climate.

At the global (macro) scale, by allowing mobility via strategic site selection along migration routes and creating a network of floating refugia, future floating communities could aid to increase the overall resilience of multiple species, ecosystems and ecosystem services available.

Mobility is, therefore a third basic parameter which can drive the sustainable development of floating solutions. The local aspects of mobility and opportunities deriving from the intrinsic qualities of floating structures are further discussed.

4 Mobility

At both local and global scales, mobility can have an impact at multiple levels: basic transport infrastructure required, degree to which floating communities are able to connect and rely on existing land-based infrastructure and economy, ability of the community to relocate to more favorable sites. For example, the Oceanix proposal is scaled in order to facilitate minimization of transport infrastructure and encourage pedestrian access while, via its modular design, it allows the community to easily relocate and expand [8].

In terms of the relationship between floating communities and their surrounding ecosystems, however, there are further opportunities which could enhance the availability of resources by taking advantage of the ease of mobility afforded by floating solutions (Fig. 4).

One potentially productive avenue emerges from the field of marine fishery management. While in land-based scenarios, the use of crop rotations is an established management strategy, it is a less common practice for marine environments. However, research suggests that, in terms of harvesting some marine species, rotating harvest sites over periods as long as six years would generate "improvement in biological and economic performance" and would help to avoid overexploitation of resources [18].

The design of future floating cities as mobile infrastructures that change their location over time, could therefore minimize the impact of human exploitation on marine environments as well as enhance the economic productivity of marine harvest activities.

Furthermore, in the case of smaller scale modular assemblies, the ability to reconfigure the resulting city network could be utilized to respond to seasonal patterns such as changing water conditions (e.g. to form break waves when needed). Reconfiguration could also provide an optimal solution to temporarily accommodate urban and natural spatial requirements (e.g. seasonal nursery space for food and energy crops or temporary public space for human activity).

In terms of the relationship between city and ecosystem, the main advantage of the enhanced mobility of modular floating solutions is, therefore, the ability to optimize the use of space and resources by reconfiguring and relocating the structures in seasonal or longer cycles that correlate with natural growth patterns.

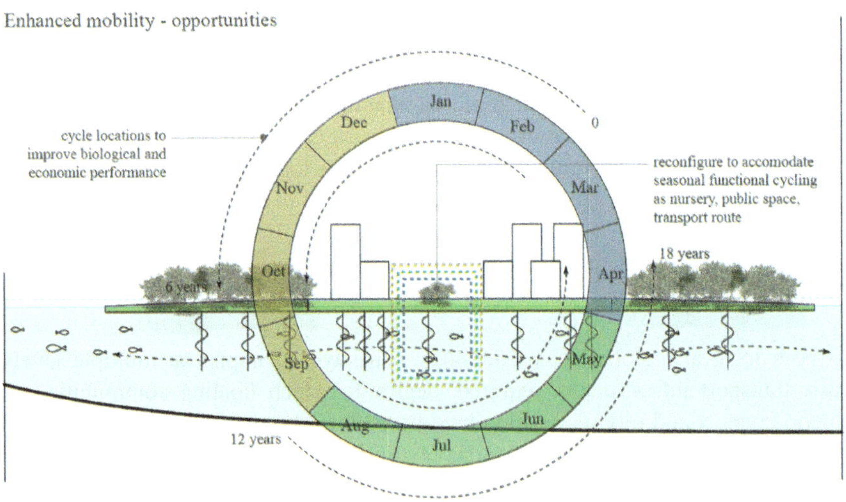

Fig. 4 Conceptual diagram showing opportunities for optimized resource distribution and spatial use by taking advantage of the enhanced mobility of floating structures

5 Conclusions

Coming back to the notion of grafting, when looking at floating city designs through the lens of cities performing as grafts which heal, add to and create hybrid symbiotic relationships to the natural host environments, several opportunities and associated design criteria have been identified.

Figure 5 summarizes potential design criteria in relation to the basic parameters of location, scale and mobility. Strategic site selection, design across multiple scales and enhanced mobility (via reconfiguration and cyclic relocation) could thus create opportunities to support wetland ecosystems located near urban areas, create marine biodiversity hotspots and microclimate refugia, optimize resource and space allocation as well as sustain global resilience by maintaining migration corridors.

Given the reliance of human communities on ecosystem services for habitat, energy, food and leisure activities, it is important to expand design criteria across disciplines and identify common points between urban growth and environmental agendas as well as develop strategies for future cities to become a productive partner in the relationship between built and natural environment.

In this context, the notion of future floating cities performing as grafts and positively contributing to the environmental balance to achieve equitable relationships seems to generate opportunities for sustainable growth and should be further explored.

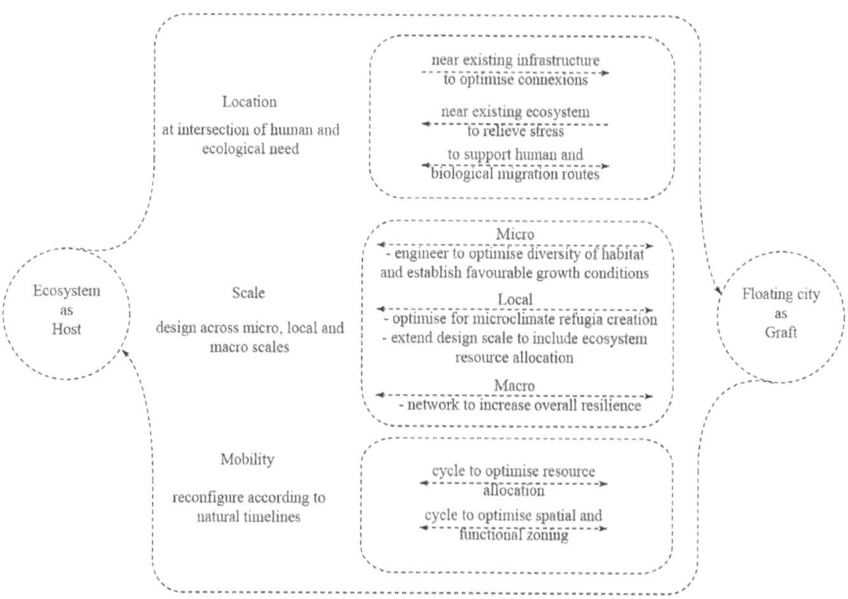

Fig. 5 Summary of design criteria identified that could help to establish equitable relationships between floating cities as grafts and ecosystems as host organisms

References

1. United Nations (2019) Press Release DSG/SM/1269-ENV/DEV/1936-HAB/248, Available via https://www.un.org/press/en/2019/dsgsm1269.doc.htm. Accessed 03 Nov 2020
2. Wang CM, Wang BT (eds) (2015) Large floating structures: technological advances, Springer Singapore Ocean Engineering & Oceanography, vol 3. Springer, Singapore
3. Lamas-Pardo M, Iglesias G, Carral L (2015) A review of very large floating structures (VLFS) for coastal and offshore uses. J Ocean Eng 109:677–690. https://doi.org/10.1016/j.oceaneng.2015.09.012
4. Ehrlich P, Ehrlich A (1981) Extinction: the causes and consequences of the disappearance of species. Random House, New York
5. UN Habitat (2019) Press release Available at https://unhabitat.org/roundtable-on-floating-cities-at-unhq-calls-for-innovation-to-benefit-all/. Accessed 03 Nov 2020
6. Bindoff NL, Cheung WWL, Kairo JG et al (2019) Changing ocean, marine ecosystems, and dependent communities. In: Pörtner HO, Roberts DC, Masson-Delmotte V et al (eds) IPCC special report on the ocean and cryosphere in a changing climate (in press)
7. Iwamura T, Possingham HP, Chadès I et al (2013) Migratory connectivity magnifies the consequences of habitat loss from sea-level rise for shorebird populations. J Proc R Soc B 280 (1761). https://doi.org/10.1098/rspb.2013.0325
8. Schwab K (2019) Floating cities once seemed like sci-fi. Now the UN is getting on board. Available online via https://www.fastcompany.com/90329294/floating-cities-once-seemed-like-sci-fi-now-the-un-is-getting-on-board. Accessed 03 Nov 2020
9. Woodson CB, Micheli F, Boch C et al (2019) Harnessing marine microclimates for climate change adaptation and marine conservation. J Conserv Lett 12(2). https://doi.org/10.1111/conl.12609
10. Morelli TL, Barrows CW, Ramirez AR et al (2020) Climate-change refugia: biodiversity in the slow lane. J Front Ecol Environ 18(5):228–234. https://doi.org/10.1002/fee.2189
11. Morelli TL, Smith AB, Kastely CR et al (2012) Anthropogenic refugia ameliorate the severe climate-related decline of a montane mammal along its trailing edge. J Proc R Soc B 279 (1745). https://doi.org/10.1098/rspb.2012.1301
12. Lima R, Boogaard F, De Graaf R, Dionisio Pires M, Sazonov V (2015) Monitoring the impacts of water quality and ecology using an underwater drone. In: E-proceedings of the 36th IAHR world congress, The Hague, the Netherlands, 28 Jun–3 July 2015
13. Draget E (2014) WWF-Report, environmental impacts of offshore wind power production in the North Sea. A literature overview, WWF-World Wide Fund for nature (formerly world wildlife fund), Oslo, Norway. Available via https://www.wwf.no/assets/attachments/84-wwf_a4_report___havvindrapport.pdf. Accessed 03 Nov 2020
14. Keppel G, Van Niel K, Wardell-Johnson G et al (2011) Refugia: identifying and understanding safe havens for biodiversity under climate change. J Global Ecol Biogeogr 21(4):393–404. https://doi.org/10.1111/j.1466-8238.2011.00686.x
15. Roman J, Estes J, Morissette L et al (2014) Whales as marine ecosystem engineers. J Front Ecol Environ 12(7):377–385. https://doi.org/10.1890/130220
16. Ysebaert T, Walles B, Haner J, Hancock B (2019) Habitat modification and coastal protection by ecosystem-engineering reef-building bivalves. In: Smaal A, Ferreira J, Grant J et al (eds) Goods and services of marine bivalves. Springer International Publishing, Cham
17. Smith LJ (2019) Envisioning and designing the floating future. Available via https://www.smithsonianmag.com/innovation/designing-floating-buildings-with-eye-to-marine-species-living-underneath-180973285/. Accessed 03 Nov 2020
18. Plagányi EE, Skewes T, Murphy N, Pascual R, Fischer M (2015) Crop rotations in the sea: increasing returns and reducing risk of collapse in sea cucumber fisheries. Proc Natl Acad Sci 112(21):6760–6765. https://doi.org/10.1073/pnas.1406689112

Legal Framework for Sustainable Floating City Development: A Case Study of the Netherlands

Fen-Yu Lin, Otto Spijkers, and Pernille van der Plank

Abstract Sustainable floating city development has recently gained increasing popularity as a serious solution to climate change threats and land scarcity faced in urban areas. While many design and engineering aspects have been widely studied and tested, social acceptance and legal issues have been relatively underemphasized. The legal aspects of floating city development are multifaceted, contextual and rather complicated. It involves different scales and levels of legislation and branches of law. This paper aims to identify the current legal framework at different levels, as well as the knowledge gaps that still need to be filled in order to make living (i.e. human settlement) at sea possible and regulated. Taking the Netherlands as a host nation example for floating city development, the research investigates into the status-quo and future challenges regarding international law (United Nations Law of the Sea Convention [LOSC]), national laws and property law. The results shed light on the complex interrelations between different scales and levels of laws that need to be taken into account for expanding cities on water. Recommendations on future research and regulatory actions needed to overcome the challenges and facilitate the realization of sustainable floating city development are provided.

F.-Y. Lin (✉)
Blue21 B.V., Molengraaffsingel 12, 2629 JD Delft, The Netherlands
e-mail: vicky@blue21.nl

O. Spijkers
China Institute of Boundary and Ocean Studies (CIBOS), Research Institute of Environmental Law (RIEL), and International Water Law Academy (IWLA), Wuhan University, Hubei Province, 299 Bayi Road, Wuchang District, 43007 Wuhan, P.R. China
e-mail: ottospijkers@whu.edu.cn

P. van der Plank
Molengraaff Institute for Private Law, Utrecht University School of Law, Utrecht, The Netherlands
e-mail: p.j.vanderPlank@uu.nl

P. van der Plank
Utrecht Centre for Water, Oceans and Sustainability Law, Utrecht, The Netherlands

Ł. Piątek et al. (eds.), *WCFS2020*, Lecture Notes in Civil Engineering 158,
https://doi.org/10.1007/978-981-16-2256-4_27

Keywords Legal framework · Floating city development · United Nations Law of the Sea Convention · National law · Property law

1 Introduction

Land cultivation is often followed by human settlement near areas where natural resources are rich, and eventually leads to rural–urban migration. Cities grow due to the rapid increase in world population and urbanization. Globally it has been estimated that by 2050, 66% of the world's population will live in urban areas [1]; in Europe, this number rises to even 80% [2]. Extraordinary pressure on finding more land for people to live, produce food, energy and other ecosystem services has thus been created. Traditionally, cities in countries with limited space such as Singapore and the Netherlands have resorted to land reclamation, a process of creating new land from the sea by filling the area with large amount of rock and soil to raise the elevation, or by draining submerged wetlands [3]. Lately some even take a giant step forward and investigate into the feasibility of moving to Mars. Instead of colonizing Mars, how about taking a look at somewhere closer to us, the resource of which 70% of our planet consists: water? Why not introduce floating cities, creating new space on top of large-scale platforms that float on water, which has (by far) hardly been used for urbanization?

Particularly land reclamation has become a questionable proposition due to environmental concerns over the decremental impacts of sand mining, increasing scarcity of sand resources and land subsidence being a significant problem [4, 5]. In comparison, floating platforms provide many advantages which are absent in land reclamation, including cost efficiency in large deep water, environmentally friendliness in marine ecosystem, ease and rapidness of construction, and adaptability to water level changes [6, 7]. By 2050 the global land scarcity is estimated to be between 13 and 36 million km^2, marine floating city development could potentially be a more sustainable alternative solution to address land scarcity issues [8]. While numerous technical studies have been conducted and proofs have shown that floating city development is technically feasible [9–11], legal issues, financial implications and social acceptance have generally been underinvestigated. To facilitate and catalyze the development of floating cities, the status-quo of these topics in relation to floating development must be scrutinized in depth.

This study aims to identify the current legal framework of floating city development at the international, regional and national level, as well as the knowledge gaps that still need to be filled in order to make living (i.e. human settlement) at sea possible and properly regulated. The research takes the Netherlands as a host nation example for large-scale floating city development. The country is renowned for its water engineering and management. It has successfully created many new lands, or polders,[1] with water pumped out or drained by opening sluices at low tide. The country carried out major land reclamations since the 70s, marking the start of the modern era of land reclamation [3]. This study intends to shed light on how the Netherlands could take a leading role in innovating and taking land creation to the next level by "creating land on water".

1.1 The Past, Present and Future

Historically, humans have settled at sea or large inland water bodies for different reasons. Hundreds of years ago, in southeast Fujian province of Luoyuan Bay, China, fishermen formed a floating village by the sea where their livelihood, fish farming, took place on a daily basis. In Lake Titicaca in South America, floating islands were created by the Uros, the indigenous people of Peru and Bolivia, after their escape from fierce assaults by the ruling tribes (Fig. 1). Since 1960s, the concept of creating habitable conditions for humans at sea was successfully replicated in the offshore oil and gas industry in the form of oil-rigs or naval sea forts [12]. During the same period, Kenzo Tange's Plan for Tokyo envisaged man-made islands on Tokyo Bay for the first time, followed by Buckminster Fuller's Triton City. In Europe, Hall Moggridge's Sea City with sheltered floating marinas was designed for Dogger Bank in the North Sea [13]. Visionaries have never ceased to come up with plans to expand cities on water since.

In recent years floating has gradually gained ground in urban environment as an innovative and climate-adaptive building solution in face of increasing floods or rising sea level. At building level, numerous references could be found around the world. In 2017, the world's largest floating villa was constructed in Finland (Fig. 2); at district level, the most sustainable floating district in Europe, consisting of 46 floating homes, began construction in the same year in Amsterdam, called

Fig. 1 Floating village in Fujian Province, China (left) and floating islands between Peru and Bolivia (right). *Source* Dailymail and Natgeotraveller, retrieved in May, 2019

[1]low-lying land reclaimed from the sea or a river and protected by dikes.

Fig. 2 The world's largest (privately commissioned) floating villa constructed in Finland (left), and the most sustainable floating district, Schoonschip, in Amsterdam, the Netherlands (right). *Source* ADMARES and Gemeente Amsterdam, retrieved in July 2020

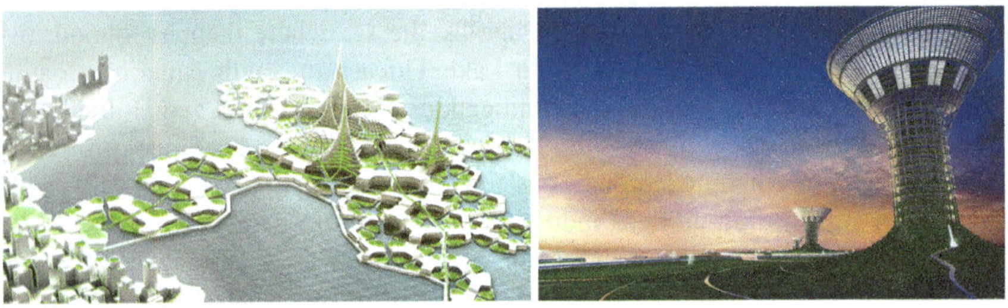

Fig. 3 Blue Revolution of Blue21 (left) and Green Float of Shimizu Corporation (right). *Source* Blue21 and Shimizu Corporation, retrieved in July 2020

"Schoonschip" [Clean ship]; at city level, for the first-time ever, UN-Habitat convened a roundtable discussion at the UN Headquarters in New York in 2019, where architects, engineers, designers, academics and entrepreneurs gathered to discuss how floating cities could be a viable solution to urban challenges such as climate change and lack of affordable housing [14].

There have been several visions illustrating what mega floating cities could look like, in internal waters, in the territorial sea, in the Exclusive Economic Zone (EEZ) or even on the high seas. These include the Blue Revolution from Blue21, Green Float from Shimizu Corporation (Fig. 3) or the Floating city within the Future World Vision of the American Society of Civil Engineers. On the one hand, floating city development has gained unprecedented momentum with many realizing its potential and feasibility; on the other hand, many questions still need to be answered for such an innovative way of urban development, particularly the legal aspects of floating cities.

1.2 Research Background and Questions

Floating cities consist of superstructures and substructures. A superstructure refers to the part of the structure that is constructed above the "ground level" (i.e. the

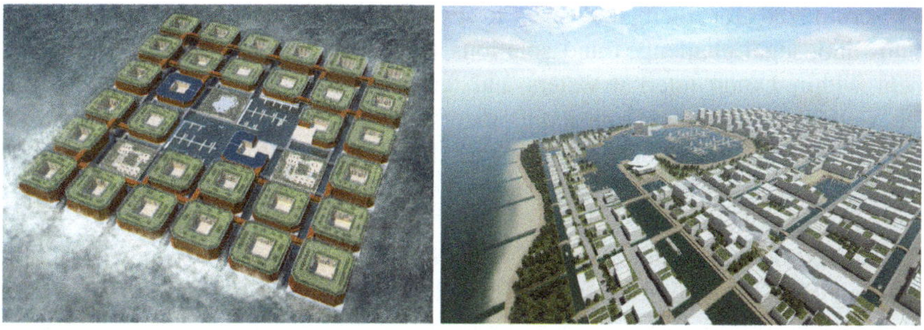

Fig. 4 Visualizations of large-scale floating city development: 2,000 people (left) and 50,000 people (right). *Source* Waterstudio.Blue & Blue21, 2019

buildings); whereas, a substructure refers to the "foundation" (i.e. floaters), or the part of the structure that is built below the "ground level". The superstructures and substructures could both be constructed elsewhere, towed to the installation site and be assembled. The substructures would then be connected and moored to the seabed, which would limit the movement of the structure and ensure stability and safety.

While the superstructure of a floating city resembles buildings on land or accommodation units on offshore platforms, it remains unclear what the substructure is. Different names have been used to address this substructure, including (artificial) islands, installations, platforms, structures, perhaps even vessels or ships. However, which of these terms or categories do floating cities belong to? What are the legal consequences? Are these substructures considered movable or immovable properties? What legal consequences do these different labels have? And does it matter, again from a legal point of view, whether they are to be connected to the coast, to be situated in internal waters, the territorial sea, the EEZ, or even in the high seas?

These questions arose in Living@Sea work package within the 3-year EU Horizon2020 funded research project, Space@Sea (2017–2020). Living@Sea aimed to conceptualize a large-scale floating city development for nearshore and offshore community (Fig. 4). During the research, the work package came across critical legal challenges that needed further investigation but were outside of the research scope of Space@Sea, as well as the expertise of the consortium. Thus, external experts with international law and private law backgrounds have been invited to probe further into the legal issues of floating city development.

This study aims to answer the following questions:

- How are floating cities currently defined legally?
- Which laws and regulations are relevant to investigate into for the governance of floating cities at different levels and legal systems?
- What are the legal consequences of the locations of floating cities?
- Why is it important to find the right legal status for floating cities?

In Sect. 2, the governance of floating cities in the territorial sea or EEZ is discussed from the perspective of the international law of the sea. Different labels are scrutinized for their relevance to floating cities; Section 3 describes the status-quo, legal challenges and solutions to enable large-scale floating development within Dutch internal waters and territorial waters from the property law perspective. In Sect. 4, permit requirements and spatial planning (urban and maritime) from the Dutch context are reviewed and presented. Section 5 summarizes the topics discussed in this paper, and provides some recommendations for future research and policy.

It should be noted that different terms are used throughout this study, including "floating cities", "floating development", "floating houses", "floating structures" or "floating platforms". None of these terms exists in international law, but they are nonetheless used in this paper, as they are used frequently by most professionals working in this field. "Floating platform" is further discussed in Sect. 3. The approach of the research is qualitative, focusing on literature review and using the Netherlands as a case study. Scientific literature that is peer-reviewed and grey literature from governmental sectors and research institutes are both main sources of references.

2 International Law

This section looks at the relevance of the international law of the sea for the governance of floating cities. The term "floating city" does not exist in the international law of the sea. Depending on the precise characteristics and purpose of a floating city, we need to identify the proper label to attach to it. We can label it an "artificial island", "installation" or "structure", "permanent harbor work", "ship" or "vessel". Floating cities are often characterized as "platforms", but the latter do not constitute a separate category in the law of the sea. Platforms may fall into different categories, depending on whether they are fixed to the seabed or floating (more on this below).[2] From the perspective of international law, this choice of label is not without consequences: the label we attach to the floating city might determine what the legal rights and obligations are, primarily of the coastal State.[3]

Another crucial question is the location of the floating city: the international law of the sea is zonal, i.e. different rights and obligations govern different maritime zones. In what follows, the focus is on floating cities situated either in the territorial sea or in the Exclusive Economic Zone (EEZ), because it is most likely that they will be placed there. Floating cities on the high seas are beyond the scope of this analysis.

Floating cities may also be situated in the internal waters of the Netherlands. Those are the waters situated "on the landward side of the baseline of the

[2]In Article 1 of the United Nations Law of the Sea Convention (LOSC), concluded in Montego Bay, on 10 December 1982, entry into force 16 November 1994. The Netherlands signed the LOSC on 10 December 1982 and ratified it on 28 June 1996.

[3]Note that the international law-based label does not always determine its legal qualification under Dutch property law.

territorial sea".[4] Internal waters are subjected to the same sovereignty a coastal State has over its land territory. Internal waters are not regulated by the law of the sea *stricto sensu*—i.e. the regime created under the Law of the Sea Convention (LOSC). They thus fall outside the scope of this section. Internal waters may include both saltwater areas as well as freshwater areas, such as rivers and lakes. The use of transboundary watercourses, including rivers and lakes shared with other states, is regulated by international water law, and is also beyond the scope of this research.

Generally speaking, a coastal State has sovereignty over all maritime features situated inside its territorial sea, including artificial islands, installations, and structures. In that zone, the legal reality is thus quite clear and straightforward. The same can be said of maritime features situated in the internal waters. Over maritime features other than naturally formed islands situated outside its territorial sea, e.g. on its continental shelf or in the EEZ, the coastal State has only sovereign rights. In what follows below, the legal consequences of this distinction are mentioned, where relevant.

2.1 Finding the Right Label

What label must we attach to a "floating city"? We can choose between "island", "artificial island", "installation" or "structure", "permanent harbor works", and "ship" or "vessel'. After a brief introduction of all these labels, an explanation is provided of the reasons why these terms must be distinguished for the purpose of the present research.[5] Of note is that the LOSC itself does not provide a definition of any of the labels listed. Some are defined in other treaties, but that is of only limited help. After all, unless we find evidence suggesting otherwise, the meaning of a term in one treaty may not correspond with the meaning of the same term in another treaty. For example, Article 2(4) of the International Convention for the Prevention of Pollution from Ships, concluded in London in 1973, defines a "ship" as "a vessel of any type whatsoever operating in the marine environment and includes hydrofoil boats, air-cushion vehicles, submersibles, floating craft and fixed or floating platforms". As we shall see below, defining fixed platforms as ships is problematic, because the LOSC labels such fixed platforms as "installation", "structures", or possibly even as "artificial island". To complicate matters further, some conventions introduce entirely new terms, which are not used at all in the LOSC. For example, in Article 1(3) of the Protocol for the Suppression on Unlawful Acts against the Safety of Fixed Platforms located on the Continental

[4] Article 8 LOSC.

[5] See also Mohammad Ali Zohourian, 'The Real Nature of Artificial Islands, Installation and Structures from Perspective of Law of the Sea', in the *Asia–Pacific Journal of Law, Politics and Administration*, Vol. 2, No. 1 (2018), pp. 13–26.

Shelf, signed at Rome in 1988, a "fixed platform" is defined as "an artificial island, installation or structure permanently attached to the seabed for the purpose of exploration or exploitation of resources or for other economic purposes". The LOSC does not use this term at all. On the other hand, the LOSC does use the terms contained in the just-cited definition of a fixed platform, i.e. artificial island, installation or structure.

2.1.1 Island

Can a floating city be defined as an island? According to Article 121 LOSC, an island is a "naturally formed area of land, surrounded by water, which is above water at high tide".[6] Since islands must be naturally formed—and not be man-made—we can quickly conclude that this is not the appropriate label for our floating cities.

2.1.2 Artificial Island

Can a floating city be defined as an artificial island?[7] An "artificial island" is constructed by human beings; it is not naturally formed. This distinction does not relate to the materials of which the island is made, but to the process of its becoming. In other words, islands consisting of natural materials—such as sand, gravel, and stone—but manufactured by human beings are not naturally formed, and thus belong to the category of "artificial islands". Artificial islands are areas of land, surrounded by water, above water at both high and low tide (that basically means they are always above water), and made by human beings. A city constructed on a platform cannot be said to constitute an artificial island, primarily because it is not made of land-like materials, and is thus not an "area of land" (see also "installation", discussed immediately below).

2.1.3 Installation or Structure

Can a floating city be labelled as an installation or a structure? The term "installation"—but not the term "structure"—was already used in Article 5 of the Convention on the Continental Shelf, concluded in Geneva in 1958. Interestingly, at the time of drafting of this Convention, the Dutch delegation noted that the term

[6]See also Myron H. Nordquist, 'Textual Interpretation of Article 121 in the UN Convention on the Law of the Sea', in Holger Hestermeyer and Rudiger Wolfrum (editors), *Coexistence, Cooperation and Solidarity: Liber Amicorum Rudiger Wolfrum*, Brill Nijhoff, 2012.

[7]See also Alex Oude Elferink, 'Artificial Islands, Installations and Structures', in the *Max Planck Encyclopedia of Public International Law*, September 2013; and Alex Oude Elferink and Alfred Soons, 'Recht van de Zee', in Nathalie Horbach, René Lefeber & Olivier Ribbelink (editos), *Handboek Internationaal Recht*, 2007, pp. 748–750.

"installation" was normally used to refer only to "fixed structures", and floating structures would not be considered "installations". This is worth noting, because Article 60 LOSC—the central provision on installations and structures situated in the EEZ, more on this below—is very similarly phrased as Article 5 Convention on the Continental Shelf, with the difference that it refers to both installations and structures, without really distinguishing between the two.

There is no definition of the terms "installation" and "structure" in the LOSC. Some definitions have been proposed during the drafting process of the LOSC, and those we find in the *travaux préparatoires*. For example, the United States proposed to define installations as "all offshore facilities, installations, or devices other than those which are mobile in their normal mode of operation at sea".[8] The United States, like the Netherlands in the 1950s, clearly wanted to exclude floating platforms from the category of installations. Belgium agreed, and proposed to regard floating installations as ships instead.[9]

Of course, these documents of the *travaux préparatoires* are all very interesting; but one should be careful to draw any conclusions from them. After all, a treaty must be interpreted in good faith, in accordance with the ordinary meaning to be given to the terms of the treaty.[10] In short, remarks made by the Netherlands, Belgium and US delegations at the time the treaty was being drafted, are not decisive in determining the meaning of the terms "installation" and "structure" in the LOSC today.

A distinction between "artificial islands" on the one hand, and "installations" and "structures" on the other, which was proposed by Fred Soons in 1974, is still often quoted in literature. According to Soons, artificial islands are "constructions which have been created by the dumping of natural substances like sand, rocks and gravel" on the seabed; and installations are "constructions resting upon the seafloor by means of piles or tubes driven into the bottom" or "concrete structures".[11] The reference to piles or tubes was not meant to exclude other methods of attaching such constructions to the seafloor. The key message to take from Soons' distinction, is that installations and structures are not artificial "land areas", i.e. they do not consist of natural substances dumped on the seafloor. It must be noted, however, that this view is not universally accepted. Soons' approach focuses on the materials of which the thing is made; and not on the purpose it is meant to serve. What both installations/structures and artificial islands have in common, is that they are not naturally formed—like islands proper—but made by human beings.

---

[8]United States of America, in the 'Selected Documents from the Meetings Held from July 20 to August 24, 1973 (Artificial Islands, Land-Locked States, Settlement of Disputes, Territorial Sea, Continental Shelf, Straits, Fisheries, Economic Zones, Archipelagos)', in International Legal Materials, vol. 12 (1973), p. 1236.

[9]Belgian note, published in *idem*, pp. 1210–1213.

[10]*Cf.* Articles 31 and 32 Vienna Convention on the Law of Treaties, concluded in Vienna on 23 May 1969, entry into force on 27 January 1980.

[11]Fred Soons, *Artificial Islands and Installations in International Law*, Law of the Sea Institute, 1974, p. 3.

As said, the LOSC does not establish a special regime for "platforms", but the term is used in the treaty occasionally. Most importantly, in Article 1 LOSC mention is made repeatedly of "platforms or other man-made structures at sea", which suggests that platforms are best seen as a sub-category of structures.

The LOSC specifically refers to a number of subcategories of installations, such as "lighthouses",[12] "port installations",[13] and "scientific research installations"[14]; and it distinguishes "installations" from "equipment".[15] This accords with the way in which these terms are generally used: an installation is a place with equipment that is put there for a particular purpose.

All the above considered, we can conclude that the category of installations and structures includes fixed platforms. It might also include floating platforms which have been (temporarily) anchored into the seafloor with mooring lines, although there is some room for different opinions on this (cf. with the definition of a "ship" or "vessel", discussed below).

Some installations or structures, situated in the EEZ, may have their own harbor, and this might very well be the case for our floating cities.[16] In that case, one may wonder whether such a harbor is legally part of the installation. Paragraph 5 of Article 60 LOSC, which regulates the use of installations and structures in the Exclusive Economic Zone, states that the safety zone of installations shall be "measured from each point of their outer edge".[17] It could be argued that this outer edge refers to the low-water line along the installation.[18] This would mean that the waters of the port of a floating city, situated on a fixed platform in the EEZ, are part of its safety zone. Consequently, the coastal State would not have full jurisdiction over these waters, but could exercise only the limited rights that the coastal State has in safety zones around the city, where it has been established. These limited rights are enumerated in Article 60 LOSC. However, there are good reasons to believe that the port of a floating city is part of that city, and that the waters of the

---

[12] See Articles 7 and 47 LOSC.

[13] See Article 129 LOSC.

[14] See Articles 249, and 258–262 LOSC.

[15] See Article 249(1)(g), and 258–262 LOSC.

[16] And some may have their own airport. On this see, Henri Wassenbergh, 'The Status and Use of an Airport on an Artificial Island', in *Air & Space Law*, Vol. XXIV, Number 4/5 (1999), pp. 177–180.

[17] See also Sebastian tho Pesch, 'Coastal State Jurisdiction around Installations: Safety Zones in the Law of the Sea', in *International Journal of Marine and Coastal Law*, vol. 30 (2015), pp. 512–532; Pauline van der Meer Mohr, 'Measures to Prevent Collisions with Offshore Installations on the Dutch Continental Shelf', in the *Leiden Journal of International Law*, volume 1 (1988), pp. 222–230. For more info specifically about safety (zones) of offshore oil rigs, see *e.g.*, Stuart Kaye, 'International Measures to Protect Oil Platforms, Pipelines, and Submarine Cables from Attack', in the *Tulane Maritime Law Journal*, vol. 31(2007), pp. 377–423; Hossein Esmaeili, 'The Protection of Offshore Oil Rigs in International Law', published in two parts in the *Australian Mining and Petroleum Law Journal*, in vol. 18 (1999), pp. 241–252, and vol. 19 (2000), pp. 35–43, respectively.

[18] *Cf.* Article 5 LOSC.

port are thus also part of the city itself, and not part of the city's safety zone. What are these good reasons? Firstly, it can be assumed that the port of a floating city is normally constructed together with the city itself. Secondly, considering that ordinary ports—i.e. ports situated on a State's land territory or on a naturally formed island—are considered an integral part of the coastal State's territory, we can assume, applying analogous reasoning, that ports of a floating city are an integral part of that city. Thirdly, Article 11 LOSC says that permanent harbor works "are regarded as forming part of the coast".[19] Again, there is no reason to treat the harbor works of a floating city any different.

2.1.4 Permanent Harbor Works

Some platforms in the territorial sea are located very close to the port, which makes it difficult to determine whether they still belong to the port, or whether they are stand-alone installations/structures. Article 11 LOSC suggests that "the outermost permanent harbor works" still "form an integral part of the harbor system". However, "[o]ff-shore installations and artificial islands shall not be considered as permanent harbor works". There is no definition of "permanent harbor works" in the LOSC. Already in 1989, the UN Office for Ocean Affairs and the Law of the Sea defined them as "permanent man-made structures built along the coast which form an integral part of the harbor system such as jetties, moles, quays or other port facilities, coastal terminals, wharves, breakwaters, sea walls, etc.".[20] This means that offshore loading and unloading areas, meant to service ships that are too large to enter the port, are not to be considered harbor works, and therefore they must fall within the category of either "artificial islands" or "installations/structures".

Note that there is no definition of "off-shore", and thus it can be assumed to simply mean "away from or at a distance from the coast". There is no reason to suppose that the distance must be considerable, or that it is an implicit reference to a particular maritime zone.

Since floating cities have an entirely different purpose compared with ports, it seems unlikely that they can be qualified as "permanent harbor works", no matter how closely they are situated to an existing port, and no matter whether they form an integral part of the harbor system or not.

The difference between artificial islands and installations/structures, on the one hand, and permanent harbor works, on the other, is important because the latter, which form an integral part of the harbor system and are regarded as forming part of the coast, can cause a seaward shift of the coastal State's baseline. The same effect can be achieved by artificially enlarging a coastal State's land territory, as was done

---

[19] See also Articles 25(2), 50 and 218–220 LOSC.

[20] UN Office for Ocean Affairs and the Law of the Sea, *Baselines: An Examination of the Relevant Provisions of the United Nations Convention on the Law of the Sea*, Appendix I (Glossary of Technical Terms), 1989, p. 56.

with the construction of the Maasvlakte in the Netherlands. Some floating cities, situated in a State's territorial sea, may be regarded as such, i.e. as land reclamation, if their connection with the coast is sufficiently dense. The core question, in both cases, is how to distinguish between an artificial island or installation/structure, and an artificial extension of the port or natural coast. Only the latter leads to a shift in the coastal State's low-water line. Off-shore artificial islands, installations and structures are not considered to be part of the coast or a port system and expressly do not have such an effect.[21] When a floating city is artificially connected to the mainland, for example by means of a bridge, tunnel, or land road (dirt road), then the question arises whether it has thereby become an integral part of that mainland. The LOSC does not provide much guidance here. This lack of clarity is unfortunate, because this may turn out to be a crucial question for floating cities situated close to a State's land territory, and in some way connected with it.

2.1.5 Ships

The most important characteristic of a "ship" or "vessel" is its purpose: it must be used primarily to navigate from one place to another. In other words, it must be designed and intended for transportation on water, and it must actually be used for that purpose. Transportation can be described as the conveyance of things or persons from one place to another.[22] Floating oil platforms and similar structures do not have this purpose, and are therefore not considered to be a ship or vessel. This is equally true of our floating cities.

The difference between installations/structures on the one hand, and ships and vessels on the other, is important, because ships do not fall under the regime of Article 60 LOSC, which regulates the use of installations and structures situated in the EEZ. As explained above, objects made by humankind, which move (independently) in the marine environment, and which serve to navigate from one place to another, fall under the freedom of shipping or navigation. However, as soon as such an object is moored, submerged, or anchored for purposes other than what falls within "the normal activities of ships" as provided for in Article 58 LOSC, it no longer falls under Article 58, but under Article 60, and it can be regarded as an installation or structure. This distinction between vessels/ships and installations/structures is thus crucial for floating cities.

---

[21] 11 LOSC.

[22] *Cf.* Ryan C. Schmidtke, 'Artificial Islands of the Future: The Seasteading Movement and the International Legal Regimes Governing Seasteads in EEZs and on the High Seas', in the *Asian-Pacific Law & Policy Journal*, vol. 21, no. 1, Fall 2019, p. 1–28, at p. 14.

2.1.6 Floating Cities?

In the above section on "installations" and "structures", we already noted that installations primarily include constructions resting upon or fixed to the seafloor. This suggests a firm attachment; they are fixed. This raises the question whether a floating city, anchored (temporarily) into the seafloor only with a couple of mooring lines, can qualify as an installation. And if not, whether it is then a 'vessel'. But from the above section on "ships" or "vessels", we must conclude that floating cities cannot be categorized as such, because they do not navigate from one place to the other. So, what are they?

It has been suggested in literature that none of the labels listed above can be properly attached to floating cities. In other words, it is argued that they constitute an entirely new category, unregulated in the LOSC. Floating cities consist of a substructure (the foundation, the floater), on top of which one or more superstructures are built, i.e. a single or multiple buildings (see also Sect. 1.2). They could be qualified as "barge", which is a term not used in the LOSC, and the supposed consequence of this would arguably be that their use is as yet unregulated.[23] Since the LOSC was meant to regulate all uses of the sea, this is a conclusion we should not accept too easily.

Another term one finds in the literature is "floating asset" or "offshore asset", which refers to structures like a drifting abode, process plant, recreational facility, or luxury yacht.[24] Again, this is a term not used in the LOSC. The term "asset" is used frequently in that Convention, but primarily in the economic sense, not as an alternative classification to the ones discussed above—i.e. "island", "artificial island", "installation" or "structure", "permanent harbor works", and "ship" or "vessel". The term "floating asset" could thus be used when reference is made to structures at sea in the economic context, but that does not resolve our classification problem.

So, what would be the alternative? It has also been suggested that cities located on floating platforms should be qualified as "vessels" when they are being moved (towed) to their location, and that they become "artificial islands" when moored.[25] However, considering the distinction made above between "artificial islands" and "installations/structures", it appears that floating cities, when moored, are best

[23]See *e.g.*, Ryan C. Schmidtke, 'Artificial Islands of the Future: The Seasteading Movement and the International Legal Regimes Governing Seasteads in EEZs and on the High Seas', in the *Asian-Pacific Law & Policy Journal*, vol. 21, no. 1, Fall 2019, p. 1–28, at pp. 14–15.

[24]See e.g., 'Floating asset; Ambrosia III offers a life of luxury on the ocean for Euro 40 million', in the South China Morning Post, November 27, 2007.

[25]See *e.g.*, Ryan H. Fateh, 'Is Seasteading the High Seas a Legal Possibility: Filling the Gaps in International Sovereignty Law and the Law of Seas', in the *Vanderbilt Journal of Transnational Law*, vol. 46, no. 3, May 2013, pp. 899–932, at p. 909; Max K. Morris and John W. Kindt, 'The Law of the Sea: Domestic and International Considerations Arising from the Classification of Floating Nuclear Power Plants and Their Breakwaters as Artificial Islands', in the Virginia Journal of International Law, vol. 19, no. 2, Winter 1979, pp. 299–320.

qualified as "installations" or "structures".[26] After all, they are not made with natural and local materials such as rocks, sand, and soil. If floating cities, when moored, become non-movable constructions mounted on piers resting on the seabed, then they are probably best labelled as "installations".[27] The term 'non-movable' does not mean that the installation has to remain completely static. If the mooring lines allow the floating city a very limited degree of motion, for example to move with the changing tides, as a means to ensure the safety and comfort of the people living in the floating city, this is not considered movement. It also does not mean to suggest that the city loses its capacity to move; it only means that it is attached to the seafloor in such a way that it is, until it has been detached again, non-movable.

3 Property Law Aspects of Floating Development Within Dutch Territorial Waters

The previous section discussed which label should be attached to a floating city located in the different maritime zones of the Netherlands, based on the international law of the sea and presented the legal consequences of that according to international law. This section provides an overview of the legal issues that would arise if a floating city were to be located within Dutch territorial or internal waters.

Why using the Netherlands as a case study? The Netherlands and water are inextricably linked: almost one third of the country lies below sea level and the mills, polders and dikes are part of the national heritage. Large parts of the Netherlands were created by land reclamation: a large part of present-day Amsterdam used to be water, the Beemster (an area above Amsterdam) was reclaimed around 1600 to serve as agricultural land for fast-growing Amsterdam and at the beginning of the nineteenth century a whole new province, Flevoland, was created by reclaiming the Zuiderzee (now: IJsselmeer). Maasvlakte mentioned in the previous section is another example. And if you look at Schiphol Airport now, you can hardly imagine that this was once the largest lake in the Netherlands: the Haarlemmermeer.

---

[26]So-called "Seasteads", *i.e.* permanent dwellings at sea, have been defined as installations; but those are located on fixed platforms. See *e.g.*, Megan Binder, 'Taking to the Sea: The Modern Seasteading Movement in the Context of Other Historical Intentional Communities', in the *Indiana Journal of Global Legal Studies*, Vol. 23, No. 2 (Summer 2016), pp. 765–794, at p. 790. On seasteads more generally, see e.g. Surabhi Ranganathan, 'Seasteads, land-grabs and international law', in the *Leiden Journal of International Law*, vol 32 (2019), pp. 205–214.

[27]Not everybody agrees with this distinction. See e.g., James Grimmelmann, 'Sealand, Havenco, and the Rule of Law', in the *University of Illinois Law Review*, vol. 2012, no. 2, 2012, p. 405–484. Somewhat confusingly, he refers to Sealand, situated on a gigantic platform, built during World War II for antiaircraft defense purposes, resting on the seabed of the North Sea with the help of a pair of concrete legs, as artificial island.

Large parts of the Netherlands were therefore created by land reclamation. The image as water pioneers is also reflected in the fact that the Netherlands (together with Belgium) has the largest dredging fleet in the world and that they participated in large land reclamation projects, such as The Palm in Dubai. For this reason, it is perhaps not surprising that the Netherlands is also one of the frontrunners in the field of floating urbanization; presumably the land reclamation of the twenty-first century. One of the global leading companies on floating urban projects is based in the Netherlands: Blue21.[28] In recent years, they have conducted research into all kinds of different facets of floating urbanization: not only into the question of whether it is technically possible to build floating platforms on which several buildings—or even high-rise buildings[29]—, but also, for example, into the influence of large floating platforms on water quality.[30] Although the issues mentioned above are all vital, they will not be further discussed in this analysis as we will focus on legal issues surrounding floating construction.

The central question in the section is: "Is it possible under current Dutch property law to legally design floating platforms holding several buildings?" It is evident that floating development would be hampered if the answer to this question would be negative.

To answer this main question, we will discuss: 1) whether a floating platform is movable or immovable property under Dutch law and why this qualification is relevant for the legal design of floating development. In Sect. 3.3 we will discuss whether it is possible to use existing limited real rights to legally design a floating platform: is it possible to divide ownership of different buildings on one floating platform? Can a floating platform be burdened by existing limited real rights? At the end of this section, an overview of the current legal challenges concerning floating development is presented and this section will be concluded with a recommendation.

3.1 The Location of Floating Platforms

As mentioned above, the analysis in this section assumes that the floating platforms are located within territorial or internal waters. The North Sea is legally divided into several maritime zones, each of which has its own legal regime.

On the basis of LOSC, the Dutch state has various sovereign rights and jurisdiction with respect to the EEZ. However, these rights are not so far-reaching that the seabed of the EEZ is property of the Dutch state. The Dutch Civil Code

[28]www.blue21.nl

[29]K.K.M. Ko, Realising a floating city, A feasibility study of the construction of a floating city, knowledge base Blue21.

[30]Research shows that floating houses can have a positive impact on ecosystems: Boogaard, F.C.; de Graaf, R.E.; Foka, E.; Rutten, M.; de Lima, R.L.P.; Giessen, N. The effect of floating houses on water quality. In Proceedings of the Amsterdam International Water Week, Amsterdam, The Netherlands, 2–6 November 2015; p. 7.

(DCC) provides that the seabed in the EEZ is not an object of property. As a result, rules concerning accession and the distinction between movable and immovable property for example do not apply fully.[31]

Because the focus of this section is on Dutch property law, it is assumed that the floating platform is located in inland waters, such as lakes, canals or rivers, or at sea, as long as it is in territorial waters (i.e. up to 12 nautical miles from the coast).[32]

Although the ultimate goal of floating developments is the creation of floating cities, it is expected that such a development goes in stages: first a few houses on one platform, then a block of houses, then a somewhat larger platform with perhaps an entire floating district, with the aim of creating entire floating cities. Because the size of the platforms will (probably) change over the years, the following will not always refer to floating cities, but to "floating development".

As we will see below, the size of the platform is irrelevant to the legal qualification under current Dutch property law. Therefore, all that is discussed below with regard to the legal status of floating platforms applies equally to a small platform with only a single dwelling, as it does to platforms holding an entire neighborhood or city.

3.2 Floating Houses: Movable or Immovable Property

Living on water is not unknown in the Netherlands: the houseboats in Amsterdam's canals are an integral part of the streetscape. Characteristic of floating houses in the Netherlands is that it is actually always a single dwelling and not, for example, a floating semi-detached house, floating terraced houses or a floating apartment complex. The reason for this has to do with the Dutch Supreme Court qualifying floating houses as movable property, a judgment that has (perhaps unexpectedly) major implications for the legal design of floating development, which will be explained in more detail below. But first the case that led to this judgment will be discussed.

In 2010, the Dutch Supreme Court was asked to rule on whether a houseboat is movable or immovable property. It concerned a houseboat located in a canal in the city of Almere. The houseboat consisted of a concrete foundation with a wooden structure. The houseboat was attached to a bollard by means of two metal brackets, which were anchored in the ground. It was located in a residential area, between two bridges, which were so low that the houseboat could not pass underneath. The houseboat also had no engine or other propulsion equipment.

[31]For a detailed discussion see: A.R.P.M. Davits & M.M.G.B. van Drunen, 'Goederenrechtelijke aspecten van offshore windparken', WPNR 2017/7135, F.J Vonck & R. Bos, 'Eigendom van offshore windparken', WPNR 2018/7212, L.W.J. Hoppenbrouwers en K.A. de Groot, 'Reactie op "Eigendom van offshore windparken" van mr. F.J. Vonck en R. Bos LL.B. in WPNR 2018/7212', WPNR 2019/7224 en R.A.B. Cobussen, 'Waardepapieren en windparken', WPNR 2019/7223.

[32]Article 2 LOSC.

It was assumed[33] that the Supreme Court would qualify houseboats as immovable property based on the Portacabin Decision of 1997.[34] In the Portacabin decision the Dutch Supreme Court ruled that a building or construction is immovable for the purposes of Article 3:3 DCC if according to its character and construction the object is destined to remain permanently in situ. Whether it is technically feasible for the building or construction to be moved is of no importance in answering the question of whether it is permanently attached to the land in the view of the Court. There was no doubt that the houseboat at issue was destined to remain permanently in situ. However, the Dutch Supreme Court ruled that floating structures fall under the statutory definition of a ship within the meaning of Article 8:1 DCC and for that reason is qualified as movable property. Article 8:1 DCC provides the definition of a ship: "In this Code 'vessels' are all things, other than aircraft, which, according to their construction, are destined to float and which float or have done so."[35]The wording of Article 8:1 DCC is so broad that it actually includes every floating object, including the abovementioned floating villas, which at the first glance have little in common with a ship, apart from the fact that it floats.[36]

As abovementioned, until the Supreme Court ruled its judgment in the Woonark Decision, there was discussion on the question of whether floating homes should be classified as movable or immovable property. In 2002, the Dutch Supreme Court had ruled another judgment on the question whether floating jetties located in a marina in The Hague should be regarded as movable or immovable property. This concerned various connected jetties on concrete floats, which were connected by braces to mooring posts anchored in the ground, or quay. At that time, the Dutch Supreme Court ruled that the floating jetties should be regarded as immovable property. Based on this, it is therefore not surprising that around the time that IJburg arose in the Amsterdam IJmeer lake, the "Tijdschrift Bouwrecht" (the Journal on Construction Law) stated that "it is plausible that these houses will be qualified as immovable property by the court".[37]

The Woonark Decision put an end to this legal uncertainty, however. All case law that has appeared since then has confirmed that everything that falls under the definition of a ship (on the purpose of Article 8:1 DCC) is regarded as movable property.[38]

[33]See e.g. A.R.G. van Dijk-Barkmeijer e.a., 'Waterwoningen in IJburg: tussen wal en schip? Enkele privaatrechtelijke, fiscale en ruimtelijk bestuursrechtelijke aspecten van wonen op het water', BR 2007, afl. 2, p. 111.

[34]Dutch Supreme Court 31 October 1997, ECLI:NL:HR:1997:ZC2478 '*Portacabin Decision*'.

[35]Translation derived from: Warendorf e.a., Warendorf Legislation/ Article 1 CC Bk 8.

[36]See Sect. 2.1.5, where the meaning of a ship, or vessel, is discussed under LOSC.

[37]A.R.G. van Dijk-Barkmeijer e.a., 'Waterwoningen in IJburg: tussen wal en schip? Enkele privaatrechtelijke, fiscale en ruimtelijk bestuursrechtelijke aspecten van wonen op het water', BR 2007, afl. 2, p. 111.

[38]The Dutch Supreme Court confirmed its judgment once more on March 9, 2012, ECLI:NL: HR:2012:BV8198 '*Marina Decision*'.

3.3 The Impact of the Woonark Decision on Floating Development

The Woonark Decision, qualifying a houseboat as movable property, is an important complicating factor for floating development on a larger scale (more than one floating house per platform). Article 5:3 DCC lays down the principle of unity; one of the basic principles of Dutch property law.[39] On the basis of this principle, it is not possible to be the owner of a component part of a movable property. Exceptions to the rule that the owner of an item is also the owner of its component parts are provided for in our Civil Code are only for immovable property. Under Dutch law, any floating building is considered to be movable property.[40] As a result, if one intends to build several dwellings on one floating platform, this will be regarded as one (large) piece of movable property. Ownership of this cannot be divided. For this reason, it is not possible to transfer one of the dwellings to a third party. Nor is it possible to establish a right of mortgage or a right of pledge on one of dwellings on behalf of a bank or other financier. Only the entire platform can be encumbered with a right of pledge or mortgage. As a result, a floating platform with several houses on it is not or hardly financed. It goes without saying that this seriously hampers the floating development.

The qualification in movable or immovable property is not only important for the possibility of encumbering it with a security right; the qualification as movable excludes the application of other limited real rights. For example, a floating home cannot be encumbered with a right of superficies (in order to divide ownership vertically), it cannot be encumbered with a leasehold and on movable property a right of easement (e.g., a right of way) cannot be established. Finally, it is assumed that movable property cannot be divided into apartment rights.[41]

This means-in short-that all legal instruments that a real estate lawyer normally uses to design a real estate project do not apply to floating urbanization. In our opinion, this is a highly undesirable consequence.

---

[39]It states: "To the extent that the law does not provide otherwise, the owner of a thing is owner of all its component parts." Translation derived from: Warendorf e.a., Warendorf Legislation/ Article 3 CC Bk 5.

[40]Article 8:1 DCC provides that ships this Code 'vessels' are all things, other than aircraft, which, according to their construction, are destined to float and which float or have done so. The Supreme Court ruled that all things that fall under this definition are movable property. When a ship has floated once and is then moored it legally stays a ship and thus movable.

[41]In this contribution I have argued that according to current law it is possible to divide a houseboat into apartment rights: P.J. van der Plank, 'Rechtsvragenrubriek: Kan een drijvende woning in appartementen worden gesplitst?', Weekblad voor Privaatrecht en Notarieel Recht, 2017/7150.

3.4 A Necessary Legislative Amendment

As discussed above, current Dutch law qualifying all floating objects movable property is a major obstacle to the development of floating construction on a large(r) scale. The question is how this could be solved? One of the possible solutions is currently being worked on: a legislative amendment, which allows for floating platforms to be entered in the land registry, by which it then qualifies as immovable property. This would solve some of the problems described above: the floating platform could be encumbered with both a mortgage right and the other limited real rights mentioned. Furthermore, it allows floating platforms to be divided into apartment rights. The necessary legislative amendment is in the making and will hopefully come into force in the coming years.

However, this legislative amendment does not solve all problems: floating platforms with, for example, five houses on it will become immovable property but will not be treated in the same way as land. Land can be parcelled out, land can form public space and land is used as a starting point for almost all land registration systems. Preferably floating platforms referred to in this study would one day be put on an equal legal footing with land. For example, by adding the platforms as land to the land registration. In that case a floating platform would form "land", and all references in the Dutch Civil Code to "land" would also apply on floating structures. In order to achieve this, however, the entire method of land registration will have to be redefined. After all, not all floating constructions need to be legally equated to land; the current legal system is working fine with regard to single houseboats. The legal challenges are particularly acute once we develop floating buildings on a larger scale (e.g., one platform containing several buildings).

When equating floating platforms to land, several questions arise: what criteria must be met in order to be considered as a floating plot? Can one actually have land (i.e. the floating building plot) on top of other land (i.e. the water plot)? Does the floating plot have to remain in place, what if it is moved? All highly relevant questions that require more research.

4 Permit Requirements and Spatial Planning of Floating Development in the Netherlands

In addition to the legal status of floating cities as seen from the perspective of international law and private law, it is also important to know how floating development fits within current policies, laws and regulations in the Netherlands, particularly regarding building requirements and spatial planning. In this section, permit requirements for floating structures have been described, followed by the status-quo of land use zoning plans and maritime spatial planning regarding floating development. The term "floating structure" is used here as it is commonly used in several governmental documents in the Netherlands. This section aims to provide an overview of relevant regulations and the process of how some strategies and plans came to being.

To introduce floating development, it is compulsory to apply for permits for the floating structures and to make sure that such type of development fits into the local spatial planning. Land use zoning plans and maritime spatial planning are both made in order to promote the efficient, safe and sustainable use of water or land areas. Zoning plans are powerful spatial planning tools that include detailed rules on how a certain plot of land can be used, what type of buildings can be established and where; whereas, maritime spatial planning is a means of fostering sustainable use of the seas while simultaneously allowing for private sector initiatives. They are both relevant, depending on the location of the floating development. Sections 4.1 and 4.2 discuss permit requirements and zoning plan for floating development in internal waters; whereas, Sect. 4.3 explains permit requirements and maritime spatial planning for floating development in the territorial sea and Dutch EEZ.

4.1 Permit Requirements

According to the Informatieblad Drijvende Bouwwerken [Information sheet of floating structure], an environmental permit is required for building a floating structure or placing it at a specific location in the Netherlands, with the intention to use (or let it be used) for a long time. This permit is granted by the municipality under the condition that the floating structure fits into the zoning plan and meets the following regulations: the building regulations in Woningwet [Housing Act], de Wabo [General Provisions on Environmental Law], het Bouwbesluit 2012 [Building Decree 2012], de gemeentelijke bouwverordening [the Municipal Building Act], het gemeentelijk bestemmingsplan [the Municipal Zoning Plan] and de gemeentelijke welstandsnota [the Municipal External Appearance of Buildings Policy] [15].

In general, municipalities will include spatial rules for floating structures (if necessary) in their zoning plans and management regulations. The same environmental permit may be re-used if the floating structures have to be moved temporarily due to essential maintenance or dredging work on the waterway but returned to the original place. However, if a floating structure is permanently placed elsewhere in the water, an environmental permit for building is required again. Depending on the requirements set by the municipality, the province or the water board, a new berth permit may also be required for placing the floating structure at the new location, for example with a view to the efficient use of berth capacity, public order and smooth and safe passage. This is separate from the environmental permit for building a structure. This is apparent, for example, from the judgments of the Administrative Jurisdiction Division of the Council of State.

While there are still some uncertainties and the process of permit application for floating structures might not always seem clear, the Netherlands has endeavored to integrate many legislation and regulations on construction, the environment, water, spatial planning and nature, into one Environmental Act which bundles and modernizes the laws for the living environment [16]. This is believed to be the largest legislative change since 1848 and will enter into force from 1 January, 2022 [17]. By then, an Environmental Desk will also be accessible for contractors, entrepreneurs, governments and local residents to inquire about what is allowed in the living environment and to apply for a permit. It is speculated that such integration could reduce bureaucracy and facilitate the process of permit application for floating structures as well.

In terms of building regulations that floating structures need to comply with, the Dutch government has created several documents over the years to provide supplementary information needed regarding particularly the safety of floating structures (see Deliverable 7.2 of Space@Sea) [18]. In 2017, Wet verduidelijking voorschriften woonboten [Act on the Clarification of Regulations for Houseboats] was even created with the purpose of amending the Housing Act and General Provisions on Environmental Law in order to make it clear on which rules apply to floating structures [15].

With regard to the technical requirements for a floating structure that will be built after the entry into force of the Act, it will have to comply with the requirements set out in the Building Decree 2012. A floating structure that meets these technical requirements may be moved to another location without having to be renovated. The building may be placed elsewhere in the existing technical state, if the requirements are met in areas such as prosperity and spatial planning. For more detailed overview on the development of building regulations of floating structures in the Netherlands, an overview can be found in the report that is publicly accessible from Living@Sea within Space@Sea, Deliverable 7.2 A catalogue of technical requirements and best practices for the design [18], and will not be discussed in this section.

4.2 *Land-Use Zoning Plans for Urban Development*

The competences for spatial planning lie on a national-, sub-national or local level of a coastal State. The EU itself has no general competence assigned within this field [19]. Spatial planning decisions are made at the national, regional, and local levels in the Netherlands. The national government, provinces and municipalities make a structural vision together, describing the spatial developments they expect for infrastructure and space, as well as how these developments will be directed or implemented. The municipalities for instance, further develop the vision into regional land-use zoning plans (Fig. 5). Such plans set down where construction may take place, what may be built, the size of the structure and what it may be used for. The fixed components of a land-use plan include the rules and regulations for the area concerned and an illustration (planning map) that indicates and explains the various zones. When the interests of both national and provincial governments are at stake, they could come up with an integration plan.

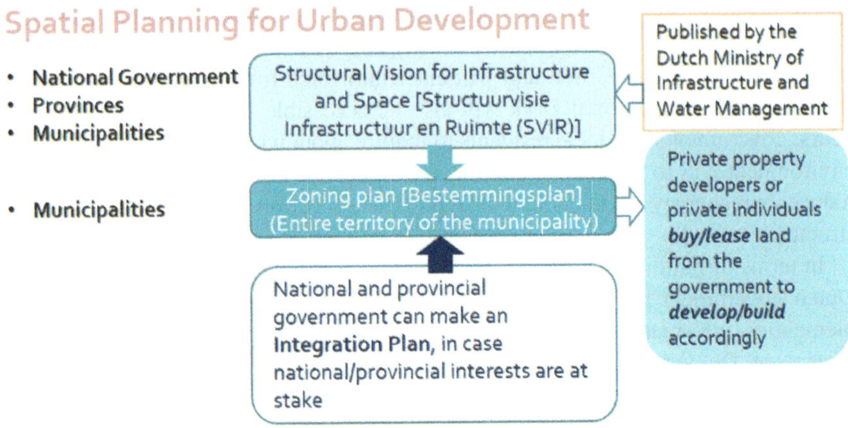

Fig. 5 The process of spatial planning for urban development come to being in the Netherlands

Land-use zoning plans allow for desired changes. In the face of increasing interests in floating development, the Municipality of Amsterdam, for instance, drafted a Bestemmingsplan Drijvende Bouwwerken [Zoning Plan Floating Structures], indicating a technical legal amendment to the Houseboat Clarification Regulations, which came into effect on 1 January, 2018 [20]. In the document, the Municipality of Amsterdam has assigned 59 prevailing zoning areas for building on water within the city of Amsterdam. In 2019, the Municipality of Amsterdam announced Vaststelling paraplubestemmingsplan Drijvende Bouwwerken [The Adoption of the Umbrella Zoning Plan Floating Structures], expressing that an umbrella zoning plan with an updated framework has been introduced to assess applications for environmental permits for the building activities with regard to floating structures. This was necessary in order to optimize the evaluation process [21]. It can be observed that there have been ongoing efforts from the local government to take into account development on water in its zoning plans.

4.3 *Maritime Spatial Planning*

When floating development will be situated in the territorial sea and the Dutch EEZ, it then becomes necessary to refer to maritime spatial planning. The competition for the use of maritime space has been ever-increasing and require nations to manage their waters more coherently. In 2014, the EU Directive 2014/89/EU on Maritime Spatial Planning was given to the coastal Member States of the European Union by the European Parliament and Council of the European Union. According to Lisbon Treaty Article 288, a directive shall be binding, as to the result to be achieved, upon each Member State to which it is addressed, but shall leave to the national authorities the choice of form and methods [22].

In response to the directive, the Netherlands updated the National Water Plan in 2015. The National Water Plan was firstly introduced in 2010 as a strategic framework based on the Dutch Spatial Planning Act, the Marine Strategy Framework Directive and the Water Framework Directive. It replaced certain policy sections of the National Spatial Strategy and included the spatial plan for the North Sea. In 2014, North Sea 2050 Spatial Agenda had been published [23]. In 2015, the Netherlands created the Policy Document on the North Sea 2016—2021, summarizing the long-term vision (2050) of the Netherlands and incorporated a maritime spatial plan [24]. In 2020, Het Akkoord voor de Noordzee [the North Sea Agreement] has also been drafted, indicating agreements between central government and stakeholders until 2030 with a view to the development of wind energy in the long term [25]. In short, the process of the development can be seen in Fig. 6.

One may be curious about the types of floating development have been included in these documents, and whether floating for living purpose was one of them. In the North Sea 2050 Spatial Agenda, floating constructions at sea for harvesting tidal and wave energy were included in the wind energy areas as a long-term energy solution since it is believed that combining energy generation technologies will offer financial, logistical and spatial opportunities. What is also interesting to note is that a group of primary school students were asked by the Ministry of Infrastructure and Water Management to think about the future of the North Sea. One of the ideas that the students came up with was to introduce "floating hotels" to the North Sea, showing that our next generation seemed to consider living on water a possible activity in the North Sea in the future.

Fig. 6 Maritime spatial planning for activities on water, giving the example of the Netherlands

In the Policy Document on the North Sea 2016–2021, amongst all the policy choices laid down and detailed, various interests for marine activities were addressed, including shipping, defene, fishing, aquaculture and mariculture, underwater cultural heritage, tourism and recreation, etc. In terms of floating, only "floating trans-shipment" was included as a potential use in the shipping sector. Nothing related to living on floating platforms was ever mentioned.

It has been indicated that industrial freedom and market forces prevailed during discussions on marine spatial planning in the Netherlands for years [26]. With the new knowledge gained and in response to the urgent needs to create more space in a more sustainable manner, it might be high time that floating cities development be taken into account in the next round of revision of Maritime Spatial Plan. In the Policy Document, an assessment framework for activities in the North Sea has also been developed and outlined for central government to use for ascertaining whether activities at sea are permitted. The assessment framework is a policy regulation and obliges the competent authority to act in accordance with this framework when issuing permits [24]. It would be highly possible that floating cities would be evaluated under this framework if proposed to be included in the North Sea.

There are different approaches to address the needs of large-scale sustainable floating city development. For the European Union, according to Lisbon Treaty Article 188, to exercise the Union's competences, the institutions shall adopt regulations, directives, decisions, recommendations and opinions to the Member States. For floating city development to be brought into the regional agenda of spatial planning, different interests groups must work together, express their interests and demonstrate the needs and urgency to regard floating city development as a serious option for future urbanization and as a better alternative to land reclamation. Such interests should be conveyed to the Council of the European Union and the European Parliament, who would then evaluate and make decisions upon. Depending on the sense of urgency and level of interests, in case the EU finds it necessary, it might address floating city development to its Member States in a certain format (e.g., regulation, directive, decision, recommendation or opinion) and have "living" or "urbanization" activities considered in maritime waters. It should, however, be noted that while promoting floating development is needed at all levels, not at least from international organizations like the EU, first and foremost attention should be paid to defining and circumscribing the concept/term of floating cities more clearly as discussed in Sect. 2.

5 Conclusions and Recommendations

Floating urban development is gaining ever-increasing recognition particularly for its climate adaptivity, flexibility (movability), feasibility and environmentally friendliness. Floating development at a building scale or neighborhood/community scale for residential purpose has been experimented all over the world. However, it has yet to be realized at a city scale, which has to do with several legal challenges.

While most countries still attempt to find out how to deal with regulations of floating structures, the Netherlands has become one of the frontrunners in floating urbanization. The country continues to clarify rules and regulations of floating structures over the years, making it an interesting country for a case study regarding the legal framework for floating city development. This study endeavored to investigate into different scales and levels of legislation and branches of law and identifies the knowledge gaps that still need to be filled, in order to facilitate the making of large(r)-scale sustainable floating cities.

5.1 A New Category to Be Defined in International Law?

The LOSC was analyzed, to gain an understanding of the governance of floating cities from the perspective of the international law of the sea. Different labels have been tentatively applied to floating cities situated in different maritime zones, such as internal waters, the territorial sea or the EEZ,[42] where floating cities are most likely to be situated when expanding from existing coastal cities. These labels included artificial island, installation or structure, permanent harbor works, ship or vessel. It has been concluded that currently none of the abovementioned labels can be comfortably attached to floating cities, although installation or structure come close.

There is the possibility that floating cities may have different status when in different state. For instance, they can be qualified as vessels when being moved (towed), and qualified as installations when being fixed (moored) to the seabed. Floating cities can also be regarded as barge as they float; however, such term is not used in the LOSC, and thus this would mean that they are as yet unregulated.

Another highly relevant question that still needs to be answered in future research is when floating cities will be situated close to the coastline and artificially connected to land territory in various ways (e.g., by bridge, tunnel, or road), will they become an integral part of the mainland and artificially enlarge a coastal State's land territory? The answer can unfortunately not be found in the LOSC. More research is necessary to shed light on such complicated issues and the possibilities of experimenting innovative governance.

5.2 Floating Platforms: Revolutionize the Perception of "Land"

The legal status of floating platforms under Dutch law has also been investigated, with a focus on whether a floating platform is qualified to be a movable or

[42]Floating cities situated on high seas are out of the scope of this analysis.

immovable property. The consequences of such status has to do with the possibility of ownership division, security rights and existing limited real rights. When the Woonnark Decision came to being, all floating objects were determined to fall under the definition of a ship (on the purpose of Article 8:1 DCC) and would be qualified as movable property. This turned out to be a serious obstacle for larger floating development (more than one house per floating platform), making it impossible to divide ownership between floating platforms and buildings on top, or divide apartment rights, let alone applying for mortgage from a bank/financier.

A starting point to solve the abovementioned problems is a legislative amendment, which would allow floating platforms to be entered in the land registry and thus be qualified as immovable property. Nevertheless, only when floating platforms can be perceived as "land" (as parcels in the public registry), can floating development take a giant leap forward in urbanization on water.

For this to happen, it would require the Dutch land registration to be changed entirely. As there will be land (i.e., platform) above land (i.e., seabed), a way to survey water areas and to be able to register these platforms will have to be set up, e.g., register "water parcels" or introduce 3D land registry. How the cadastral registration will be transformed when floating platforms become equal to land still needs to be studied and tested in real life. Additionally, rules on the kind of people that can build or introduce floating platforms, or consequences after the service life of the floating platforms still all need to be investigated further.

5.3 Floating Cities on Global Urban Development Agendas

Like many other urban development projects, applying for a permit and fitting into the spatial plan are the very first steps of introducing floating development. For floating cities that will be situated in internal waters in the Netherlands, an environmental permit is required and under the condition that the floating structure meets the designated building regulations and policies. These developments generally need to fit into the land-use zoning plan of the municipality, which are developed based on the structural vision on infrastructure and space by the national government, the provinces and the municipalities. Evidence has shown that floating developments have been increasingly taken into account in zoning plans such as by the Municipality of Amsterdam.

Whereas when floating cities will be situated in territorial sea or EEZ, central government would need to evaluate the proposed activities using the assessment framework provided in the Policy Document on the North Sea 2016–2021, and give permits accordingly. Although different types of floating development have been taken into account in the Policy Document, North Sea 2050 Spatial Agenda and The North Sea Agreement, focusing on the energy and/or shipping sector, floating urbanization for living and working purposes is relatively new and has not yet been taken into account in the maritime spatial planning.

Findings from this study suggest that stakeholders join forces and demonstrate the needs and urgency of regarding floating city development as a serious and better alternative for expanding existing coastal cities in comparison to conventional land reclamation. Such climate adaptive spatial strategy contributes to solving 21 century challenges such as land scarcity, climate change, urbanization and overpopulation. As more floating city prototypes are being tested for the time being, it is paramount to put floating city development onto global urban development agendas and pour efforts into researching and developing a more robust legal framework that could provide guidance and facilitate sustainable floating city development.

References

1. United Nations (2015) World population prospects: the 2015 revision, key findings and advance tables
2. Eurostat (2016) Urban Europe—Statistics on cities, towns and suburbs. Luxembourg
3. Stauber JL, Chariton A, Apte S (2016) Chapter 10—Global change. In: Blasco J, Chapman PM, Campana O, Hampel M (eds) Marine ecotoxicology. Academic Press, pp 273–313
4. Peduzzi P (2014) Sand, rarer than one thinks. Environ Dev 11:208. https://doi.org/10.1016/j.envdev.2014.04.001
5. Hackney C, Bendixen M, Best J, Iversen L (2019) Time is running out for sand. Nature 571:29–31. https://doi.org/10.1038/d41586-019-02042-4
6. Zhang H, Xu D, Xia S et al (2020) Dynamics of super-scale modularized floating airport. In: Wang CM, Lim SH, Tay ZY (eds) WCFS2019. Springer, Singapore, pp 113–134
7. Otto W, Waals O, Bunnik T, Ceneray C (2020) Wave induced motions of a floating Mega Island. WCFS2019. Springer, Singapore, pp 173–189
8. Roeffen B, Dal Bo Zanon B, Czapiewska KM, de Graaf RE (2013) Reducing global land scarcity with floating urban development and food production. In: International water week, Rotterdam
9. Ko KKM (2015) Realising a floating city. TU Delft Repositories. TU Delft
10. Czapiewska KM, Roeffen B, Dal Bo Zanon B, de Graaf RE (2013) Seasteading implementation plan—Final report. DeltaSync
11. Wang G, Goldfeld Y, Drimer N (2019) Expanding coastal cities—Proof of feasibility for modular floating structures (MFS). J Clean Prod 222:520–538. https://doi.org/10.1016/j.jclepro.2019.03.007
12. Ng C, Jiang R (2020) Classification principles for very large floating structures. WCFS2019. Springer, Singapore, pp 235–251
13. Anderson D (2017) Imaginary cities: a tour of dream cities, Nightmare Cities, and everywhere in between. University of Chicago Press
14. United Nations (2019) Sustainable floating cities can offer solutions to climate change threats facing urban areas, deputy secretary-general tells first high-level meeting. Meetings Coverage and Press Releases. https://www.un.org/press/en/2019/dsgsm1269.doc.htm. Accessed 8 Jul 2020
15. Ministerie van Binnenlandse Zaken en Koninkrijksrelaties (2017) Informatieblad- Drijvende bouwwerken
16. Waterstaat M van I en (2013) Nieuwe omgevingswet maakt omgevingsrecht eenvoudiger—Omgevingswet—Rijksoverheid.nl. https://www.rijksoverheid.nl/onderwerpen/omgevingswet/vernieuwing-omgevingsrecht. Accessed 8 Jul 2020

17. Koninkrijksrelaties M van BZ en (2020) Nieuwe datum inwerkingtreding Omgevingswet: 1 januari 2022—Nieuwsbericht—Rijksoverheid.nl. https://www.rijksoverheid.nl/actueel/nieuws/2020/05/20/nieuwe-datum-inwerkingtreding-omgevingswet-1-januari-2022. Accessed 22 Jun 2020
18. Lin F-Y, Czapiewska, Karina, Iorga, Gheorghe, et al D7.2 A catalogue of technical requirements and best practices for the design
19. Böhme K, Neugebauer W, Gaugitsch R et al (2018) Spatial planning and governance within EU policies and legislation and their relevance to the New Urban Agenda
20. Gemeente Amsterdam (2018) Bestemmingsplan Drijvende Bouwwerken
21. Gemeente Amsterdam (2019) Vaststelling paraplubestemmingsplan Drijvende Bouwwerken, gemeente Amsterdam
22. Croner-i (2020) Article 288. Croner-i Tax and Accounting. https://library.croneri.co.uk/cch_uk/btl/euro-it-treaty-tfeu-art-288. Accessed 23 Jun 2020
23. Ministry of Infrastructure and Environment and Ministry, of Economic Affairs (2014) North Sea 2050 Spatial Agenda
24. Waterstaat M van I en (2015) Policy document on the North Sea 2016–2021 (printversie)—Policy note—Government.nl. https://www.government.nl/documents/policy-notes/2015/12/15/policy-document-on-the-north-sea-2016-2021-printversie. Accessed 8 Jul 2020
25. Waterstaat M van I en (2020) Het Akkoord voor de Noordzee—Rapport—Rijksoverheid.nl. https://www.rijksoverheid.nl/documenten/rapporten/2020/06/19/bijlage-ofl-rapport-het-akkoord-voor-de-noordzee. Accessed 15 Jul 2020
26. De Vrees L (2019) Adaptive marine spatial planning in the Netherlands sector of the North Sea. Mar Policy 103418. https://doi.org/10.1016/j.marpol.2019.01.007

Architectural Design Guideline for Sustainable Floating Houses and Floating Settlements in Vietnam

Thi Thu Trang Nguyen

Abstract Vietnam has a long historical development of floating settlements with a major number of floating villages, where residence and workplace are combined and people have been living in floating homes for centuries. However, these floating villages exist out of dilapidated or simple floating construction. The existence of floating villages has been facing with many drawbacks of environment, socio-culture and economy. Thus, new methods and innovation tools need to be studied regarding to the sustainable concept of floating houses and floating settlements. Based on the traditional experience and vernacular architecture of floating houses in Vietnam, as well as based on the lessons and innovations of floating architecture development in other countries, the goal of this research is to create architectural design guideline for sustainable floating houses and floating settlements that ensure a safe, stable and permanent living on the water for inhabitants in Vietnam. The guideline intends to guide the applicant in the principles of planning and floating building designs according to considerations of social, economic and environmental sustainability.

Keywords Floating houses · Floating settlements · Floating villages · Sustainability

1 Introduction

In terms of geography, Vietnam has an abundant amount of inland waters and wetlands, in which locations for fishing are almost ubiquitous. Most farmer also use a variety of aquatic resource for family consumption, livestock feed and sale, and, just like fishers, many farming families living around inland water bodies and in coastal areas make a living by capture fishing [1]. The fishermen tribes built

T. T. T. Nguyen (✉)
Brandenburg University of Technology (BTU) Cottbus-Senftenberg, Postfach 101344, 03013 Cottbus, Germany
e-mail: Thutrangnguyen101@gmail.com

Ł. Piątek et al. (eds.), *WCFS2020*, Lecture Notes in Civil Engineering 158,
https://doi.org/10.1007/978-981-16-2256-4_28

complete water dwellings at the edges, and also a few at the middle of the lake, the river or the sea. The river system and the sea do not only provide water for rice cultivation, the travel and trade of water transportation and fishing grounds for inhabitants, but also have become the traditional living environment for a large share of the population over many generations. The first floating village in Vietnam were established for inland fisheries, mainly along the lower and middle reaches of rivers in the north, and around coastal lagoons in the north-central part of the country [1]. While exert figures were not given, it was observed that early in the twentieth century, Vietnam had many 'floating villages' which included groups of fishers or boatmen [2]. Other sources indicate that the number of floating villages was not high. At the beginning of the nineteenth century, there were 70 floating units in 12 old towns in the provinces of the Red River Delta and along the coast, from Quang Ninh Province to Ha Tinh Province. In the 1930s, there were about 90 floating hamlets or villages located in rivers, and about 21 ones along the coast in the region from the Vietnamese/Chinese border to the Tien Yen area. In contrast, in the Central Region of Thanh Hoa to Binh Thuan provinces, where the rivers are shorter in length and the land is mountainous, freshwater bodies cannot sustain fisheries big enough to support 'floating villages'. Therefore, such communities are concentrated in estuaries and lagoons" [1]. In the South of Vietnam, especially in Mekong delta, floating settlements has been existing along with the culture of river and the development of human settlement in Mekong Delta since eighteenth century.

Due to the environmental and cultural characteristics, water dwellings are emerged and developed through the centuries and become a common type of accommodation in Vietnam. Through long-term response and adaptation to water environment and changes with the rise and fall of the tides, the water-based settlements have been built in the water or on the water, including the four types of houses: houses with half on stilts and half on the ground; stilt houses; boat houses; and floating houses [3]. All these kinds of houses or some kinds of them link together to form a "floating village".

Since 1859, the French had started developing road transportation, therefore the canals were gradually filled up, and the image of the urban river began changing to a new form. However, rivers and canals still played an important role. In 1956, a law banning foreigners from rice trade and shipping that led to the rapid degradation of waterway system. Water bodies were no longer the center as well as the urban areas of the cities as before. Rivers and canals have become the backyard of urban civilization which was worthless and neglected, eventually they have become the shelter for the poor and the homeless people. Slums along the canals seem like the last step of the degradation of waterway system [4]. In conclusion, due to the rapid growth of urban civilization and road transportation as well as due to the result of the reorganization of rural management, irrigation development, the destruction of riverine resources, the number of "floating village" in Vietnam has decreased. The configuration and the structure of water-based settlements have been changed. The interaction between the settlement and water has been reduced, as the results more and more water dwellers have moved on to the land.

Nevertheless, since the early 1960s, small-scale aquaculture has been developing for domestic use in Vietnam [5]. The social and economic changes have opened a new trend for living on the water with the development of fishing farm. The development of aquaculture as the new way of livelihood for water dwellers opened a new prospect for floating villages. Nowadays, fishing farms integrated with floating houses has been existing in most provinces and cities in Vietnam. Most floating settlements/floating villages are located concentrating in Mekong River Delta and Red river Delta. However, larger aquaculture areas produced greater production volumes but also generated more pollution affecting the surrounding environment [4]. In addition, uncontrolled rapid growth of fishing farms in floating villages as well as the existence of dilapidated water dwellings without sanitation system, planning and organization has caused to water pollution and affected negatively to environment, landscape and life quality standard of inhabitants. The government and experts wish to research a concept of floating communities towards sustainable development, including sustainability in urban planning, local culture and the environment. The sustainable concept of floating housing units being buoyant on a floating community which ensures a stable and permanent living for inhabitants and adapting to climate change and rising sea levels would be a long-term solution for preservation and promotion aquatic lifestyle of Vietnamese inhabitants. Therefore, the goal of this research is to create architectural design guideline for sustainable floating houses and floating settlements that provides necessary, comfortable and convenient living conditions, as well as, to protect surrounding environment and secure sustainable livelihood for water dwellers. The future floating settlements would ensure a safe, stable and permanent living for inhabitants adapting to climate change and rising sea level. Livability, sustainability and resilience are three intertwined elements that together will define the quality of life of current and future residents [6]. The development of sustainable floating settlements requires respect for environment keeping the symbiotic relationship between man and nature, and the achievement of local economic objectives, as well as preserve and promote local culture and vernacular architecture. This needs to optimizes the advantages and limits the weaknesses of local architecture by the use of new construction techniques and materials (Fig. 1).

2 Method

The research employs both quantitative and qualitative data collection and analysis methods. The main methods employed were: Site visits, Review related literatures; Case study research/comparison; Expert interviews/water dweller interviews and questionnaire; Technical feasibility model study; Participatory action research; Synthesis.

A literature survey was done to have an overview of theoretical concepts of sustainable floating buildings and floating communities. Based on this overview, sustainable floating buildings and sustainable floating communities were defined as frameworks. These framework was subsequently applied and adjusted for floating

Fig. 1 A floating village in an Giang province Vietnam [7]

houses and floating settlements in Vietnam. Action research in case studies was done to develop ideas and redefine the sustainable framework based on analyzing history development and typical characteristics of floating houses and floating settlements in Vietnam. The situation in Vietnam was in particular interesting with regard to climate, economy and socio-culture adaptation of floating houses and floating settlements. Thus, the general sustainable framework needs to be redefined to fit with environmental, economic and socio-cultural conditions in Vietnam.

The sustainable framework is further tested through national case studies. According to Robert K. Yin, the case study research method is defined as "an empirical inquiry that investigates a contemporary phenomenon within its real-life context; when the boundaries between phenomenon and context are not clearly evident; and in which multiple sources of evidence are used" [8]. Therefore, a case study methodology was chosen for applications of floating house in Vietnam. Three case studies in three different regions of Vietnam were done to evaluate the feasible designing solutions of the sustainable framework for floating settlements in practice. Based on creating the framework and the architectural strategies for sustainable floating settlements, the author proposes general architectural design guideline for sustainable floating houses in Vietnam.

3 Results

The main question in this research to be answered was defined as "How to create sustainable floating settlements in Vietnam". To answer the question, the research creates a framework and architectural guidelines for the sustainable development of floating settlements in Vietnam.

3.1 Sustainable Framework for Floating Settlements in Vietnam

The most prominent issue of floating settlement development in Vietnam is the lack of governance. The lack of specific regulations and policies for operation and management of floating houses is one of the main causes that leads to other weaknesses in environmental, social and economic issues. Therefore, the framework for the sustainable development of floating settlements in Vietnam will be established in accordance with global practice, covering three aspects of economy, society and environment, however with governance as an overarching principle (Fig. 2).

In addition, it need to be emphasized that one of the most serious problems have been affecting to the development of existing floating settlements in Vietnam is

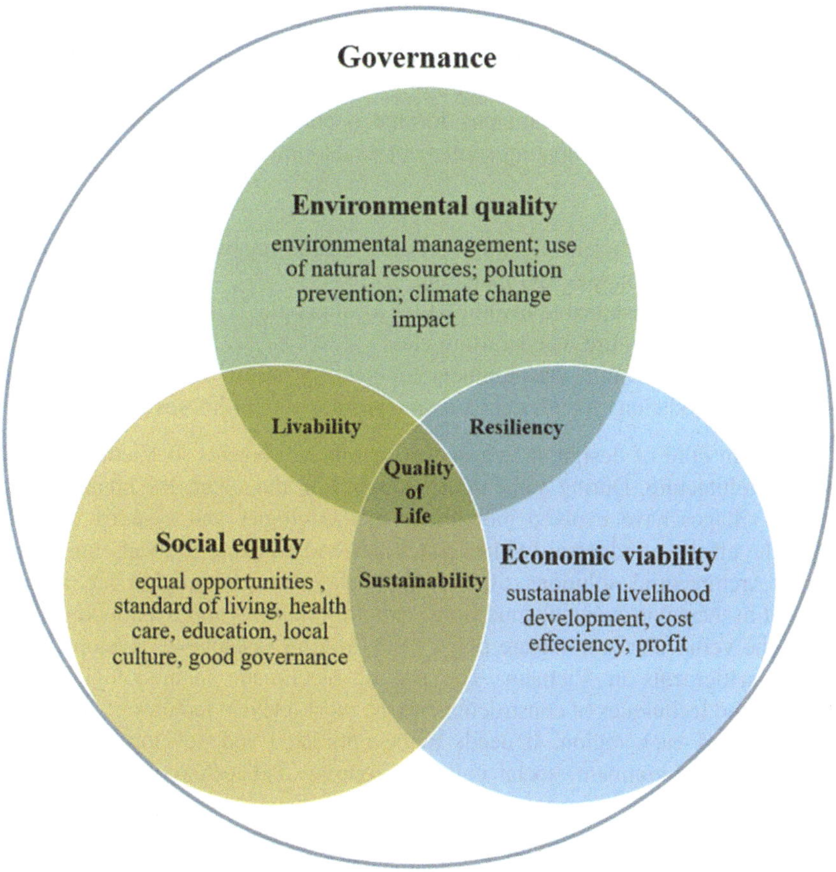

Fig. 2 Framework for sustainable floating settlements in Vietnam. *Source* By author

environmental pollution raised by the rapid development of aquaculture and economic condition of local people. Therefore, environmental performance assessments engaging with the development of local livelihood can play a key role in forming and planning of sustainable floating settlements.

To secure and enhance the quality of life for water dwellers, the main considerations of social, economic and environmental sustainability as below:

- Environment
  - Protection of water environment and aquatic ecosystem: Promoting waste management systems and sanitation system to reduce the air and water pollution; Promoting durability of materials and accessibility
  - Promotion and management of natural resource use
  - Promotion of renewable energy resource use
  - Improving climate change adaptation of floating settlements
- Socio-culture
  - Improving social facilities, health care service and education for water dwellers (providing playground, kindergartens and schools for children)
  - Preserve and promote local culture
  - Establishing good governance for the operation and management of the floating house and floating settlement development
- Economy
  - Design cost-efficient floating buildings using innovative and affordable floating technologies and materials
  - Developing sustainable livelihood for local people: ecotourism, aquaculture, floating agriculture, handcrafting etc.
  - Improving potential development for floating houses such as new kind of accommodations for residence and tourism; resilient houses etc.

The main core of design strategies for floating settlements in Vietnam is vernacular architecture. During some decades living on the water, the inhabitants of floating villages have evolved bioclimatic methodologies and systems for mitigating the effects of adverse weather conditions and adapting to local climate and nature. Architectural solutions of floating houses for each region are directly reflected in the environmental conditions, cultural traditions and livelihood of local areas. The vernacular architecture plays a vital role in sustainable development of floating settlements in Vietnam. The use of bioclimatic methodologies, local materials and techniques of construction are the most relevant features of vernacular architecture of each region. It needs to be optimized and developed following responses of environment, social culture, economy and technology approach to sustainability.

The main design strategies for sustainable floating settlements in Vietnam should contain five factors: (1) Adaptation: including adaptation with local conditions (environment, socio-culture, economy, architecture), and adaptation with climate

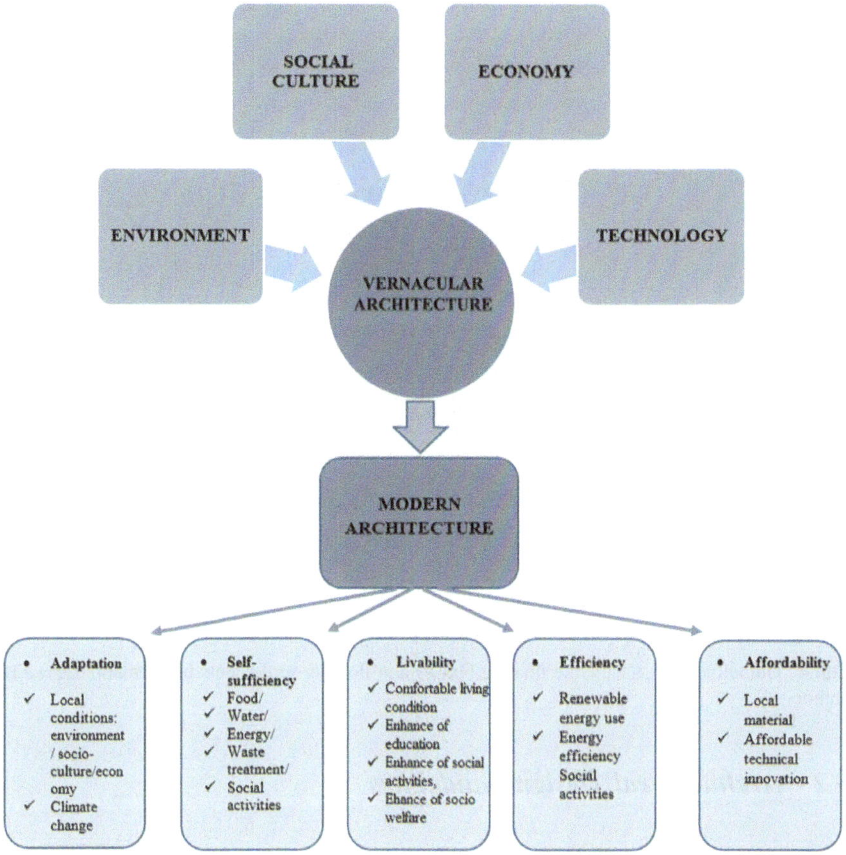

Fig. 3 The architectural design strategies of floating settlements in Vietnam. *Source* By author

change and social change; (2) Self-sufficiency: including abilities to secure sufficient food, sufficient water, sufficient energy and sufficient waste treatment; (3) Livability: including the provision of comfortable living conditions, enhance of education, social activities and socio welfare that contribute to community, and protection of human health; (4) Efficiency: including effective protection of environment, prudent use of nature resources/renewable energies, and increasing profitability and productivity of floating buildings; (5) Affordability: including affordable construction, operation, and maintenance of floating houses (Fig. 3).

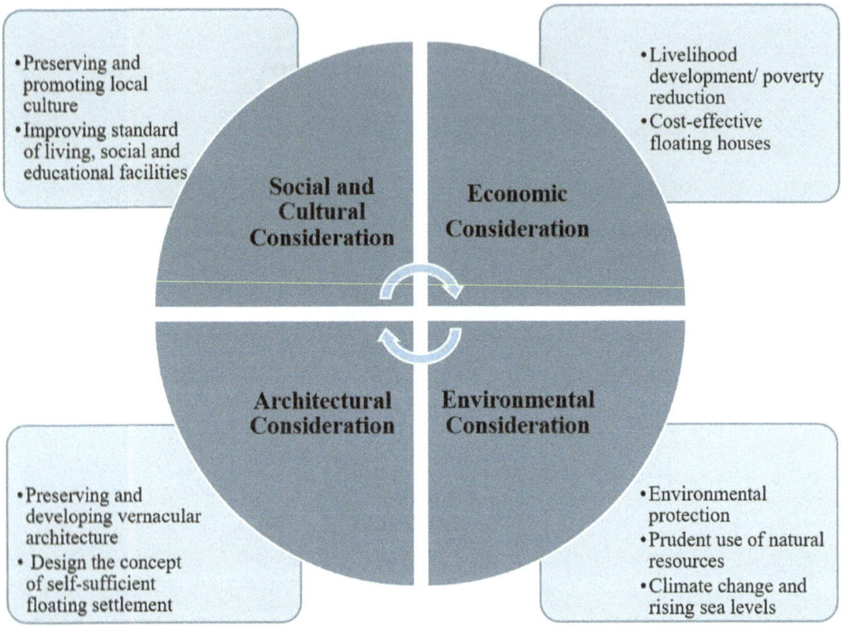

Fig. 4 Guidelines for sustainable floating houses and floating settlements in Vietnam. *Source* By author

3.2 Architectural Design Guidelines

Based on creating the framework and the architectural strategies for sustainable floating settlements, the author proposes general architectural design guideline for sustainable floating houses in Vietnam. The guideline intends to guide the applicant in the principles of planning and floating building designs according to considerations of socio-culture, economy, environment and vernacular architecture (Fig. 4).

3.2.1 Social and Cultural Considerations

A sustainable concept for floating settlements needs to consider about the following:

- Preserving and promoting local culture
  Local authorities and local people need to preserve and improve traditional occupation and the traditions of aquatic lifestyle and develop indigenous culture of floating settlement as a tourist attraction which could be one of the essential factors of ecotourism development.

- Improving social facilities and social activities
  Developing the concept of floating settlement which provides all adequate facilities for a sustainable community such as educational and recreational facilities, health care service, school, maket, place of worship and community areas.
- Establishing official management regulations
  Creating the official management regulations of living on the water, as well as, increasing education, awareness and responsibility toward to environmental protection and sustainable livelihood development for water dwellers.

3.2.2 Economic Considerations

- Livelihood development and poverty reduction
  Based on the report from Arlene Christy D. Lusterio [9] and developed by author, livelihood development in floating settlements can occur at three levels: (1) large-scale infrastructure development for floating agriculture, aquaculture production and ecotourism in the support of nation's economic objectives (2) small-scale support for water dwellers directly benefits the individual households that livelihood space is allocated with the residential plots, for example homestay, fishing farms, handcraft or business services is allocated in individual floating family house (3) household member skills training support directly benefits the individual household. According to the specific social, economic and environmental characteristics of local areas, local authorities and local people would chose suitable levels and methods to develop their livelihood and their economic potentials.
- Cost-effective floating houses
  The following are strategies to create the affordable floating houses:
  - Creating self-sufficient floating houses, using renewable energies from natural resources and using local environmental materials in order to reduce energy consumption and greenhouse gases as well as to improve waste and waste water treatment system of floating houses.
  - Smart design: Innovative designs and master planning can help to use space efficiently and effectively that save on cost and space as well as creating more community interaction, increasing the benefits of creating mixed-use developments with community spaces and commercial areas in floating houses.
  - Self-construction: Self-construction requires simple and assemble prefabricated structures which can be constructed by local labor.

3.2.3 Environmental Considerations

- Environmental protection
  The environmental protection in floating communities is to control water damage and reduce water pollution affecting to aquatic ecosystem, livelihood and human life and property. The innovative solutions of sanitation and solid waste management are considered.
- Prudent use of natural resources
  Living on the water being far away from grid on the mainland has difficulties in the installation of infrastructures and utilities, therefore innovative solutions using natural resources such as wind, solar, water, even biomass which improve energy efficiency and energy self-sufficiency should be considered.
- Climate change and rising sea levels
  The design of floating settlement needs to consider the climate change effects such as intensifies wind energy and fetch, surface-temperature, increasing tidal heights, flooding, wave strength, and salt-water intrusion, that are negative environmental factors affecting to the architecture of floating houses, as well as, the livelihood and the life of water dwellers. That requires sustainable concept designs within innovative architectural features and new construction techniques and materials which would reduce the damage of climate change and rising sea levels in Vietnam.

3.2.4 Architectural Considerations

Based on the vernacular architecture features combining with the use of new construction techniques and materials, the floating building has to be self-sufficient, has to reduce its carbon footprint and ecological impact in order to approach to the sustainability.

3.2.5 Site Selection

Based on the report from Arlene Christy D. Lusterio [9] and developed by author, site selection has to be primarily in accordance with local planning administered by the Provincial People's Committee. Other criteria include: (1) having good water quality for living and aquaculture development; (2) convenient accessibility to social infrastructures and the source of livelihood; (3) safety from natural hazards such as floods, erosion and typhoon, even preparing the plan of an effective response to an emergency such as rescue and evacuation procedures for water households.

- Accessibility (boat parking/docking/harbor structures): In floating settlements, access to water and to land is important for efficient movement of people and goods between settlements, especially during construction. Suitable docking infrastructure should be provided to accommodate water transport as well as to connect between floating settlements with facilities on the land. Boat parking areas need to be considered in the concept of planning.
- In case, the floating settlements which are located nearby the mainland can be connected to the mainland for transportation and import/export reasons by floating dock system. The floating dock system also links floating houses with each other. And in this case, the pipes of utility system including energy system, watering and sewage system can be installed under the floating dock and connect to existing utility system on the mainland.
- Floating breakwater/wave protection system or protective buffer of trees shield need to be built to protect the floating settlements from the serious effect of climate change such as tidal heights, flooding and strong wave.

3.2.6 Planning

According to Housing Construction Law [10], floating house construction planning must:

- In accordance with the objectives of the strategy and overall planning for socio-economic development; ensure national defense and security, and create motivation for sustainable socio-economic development; being appropriate with the development planning of related industries.
- Organize and arrange the water surface on the basis of rational exploitation and the reasonable use of marine resources, historical relics, cultural heritages and resources in accordance with natural, socio-economic conditions, as well as, historical, cultural, scientific and technological characteristics of each region.
- Protect environment, prevent natural disaster, response to climate change and mitigate adverse impacts on floating village communities.
- Preserve, embellish and promote the values of cultural heritage, beliefs, religions, vernacular architecture of floating villages.

Moreover, the planning concept will be created for both living and working that includes family floating houses within working place, educational and recreational facilities such as health care service, school, market and place of worship as well as offering public water-oriented recreation areas, community areas or the places to enjoy water leisure, natural view, landmark and social activities for water dwellers.

3.2.7 Building Design

The general requirement of floating house is to secure a safe, comfortable and convenient living for water dwellers (Table 1).

Table 1 Sustainable design guidelines for floating settlements. *Source* By author

Building design
Landscape/Organisation of space • The location of floating houses: less influenced by strong winds, have a slow water flow, have a low salinity and which are convenient for transportation • The organisation of architectural spaces should be flexible and open to increase ventilation and to reduce the level of humidity • The main block frequently faces south to welcome the cool prevailing wind. The southern direction also prevents the house from solar radiation from the east and west and a cold wind from the North in winter in the North and central coastal zones • Offering spaces for social support provision to enjoy water leisure, nature view, landmark and social activities • Developing floating garden concepts
Architectural features • Considering about local socio-culture and vernacular architecture • Considering design solutions adapting to climate change and rising sea level • Considering bioclimatic architecture, architectural elements should be designed for natural ventilation, minimum insulation standards and adapted to tropical climate (patio, porch, door, window, roof etc.) • The form of floating houses should ensure their stable position on the water • Safety equipment: The floating building must have appropriate life safety devices and firefighting equipment suitable for marine use
Material • Natural and local materials • Lightweight materials, durable materials, recycled materials with high corrosion protection
Structure • Load-bearing structure: prefabricated, modular structures and lightweight structures • Floating structures: safe, stable, high strength, high buoyance, high corrosion protection • Moring systems: flexible, high strength, high corrosion protection
Utilities
Energy • Using renewable energy (from water, the solar, wind, recycling waste, etc.) • Energy efficiency: minimise energy providing for heating and cooling system, air-condition, thermal isolation, sound insulation
Waste • Installing waste treatment system • Waste recycling for transferring to energy
Water • Using natural water resource • Treating and recycling wastewater for usage in daily life

4 Conclusion

In developed countries, the main reason of living on the water surface is due to social change, when the demand of a higher quality life is increasing. People want to live in a healthy and fresh environment close to nature and enjoy activities on the water. As the result, residents in developed countries choose floating houses as an attractive type of residence for living or relaxing purposes. By the contrast, in Vietnam, as well as, some other developing countries in South East Asia, the main reason for living on the water surface is due to resident's livelihood. The initial inhabitants of the floating villages were fishermen who earned their living from fishing. The floating villages are not only houses, but also places where they work and earn their living. In recent years, the floating villages have been revived and the number of villages has been increased due to the booming development of aquaculture. Subsistence is the main reason for living on the water, thus in order to create a stable, permanent and sustainable living, the prerequisite issue is to ensure sustainable livelihoods for water dwellers. Therefore, it cannot be denied that sustainable livelihood development combined with environmental protection must be the main core of strategic solutions towards to the sustainable development of floating settlements. The sustainable livelihood would increase the income of water dwellers, ensure food security, improve infrastructures and floating technologies.

In conclusion, in the context of the economy, socio-culture, environment, indigenous architecture and the policy mechanism of Vietnam, the author analyzed, evaluated and accumulated learnt lessons from the experience of floating house construction, as well as, valuable traditions and river culture in the process of forming and developing floating communities in Vietnam. Based on the inheritance and promotion of the available values of floating villages, as well as, based on the innovative principles of sustainable floating architecture, the research has created architectural guidelines for sustainable floating settlements in Vietnam. On hand results of the dissertation, the government can establish official regulations and laws to manage floating settlements, that not only preserve the traditional culture of living on the water in Vietnam, but also develop floating houses to become an alternative accommodation adapting current and future to rising sea levels.

References

1. Nguyen DT, Ruddle K (2010) The van chai system of social organization and fisheries community management. In: Ruddle K, Satria A (eds) Managing coastal and inland waters. Springer, pp 129–160 (Chapter 6)
2. Dao D (1938) Outline of the cultural history of Vietnam. Bon Phuong Publishing House (in Vietnamese), Hanoi
3. Hanh VTH, Duong V (2017) Morphology of water-based housing in Mekong delta, Vietnam. In: Floating settlements in Mekong delta proceedings of an international competition and a symposium series 2017. INIK e.V, ISBN 978-3-00-059920-0

4. Nguyen K (2017) River valley civilizations and Saigon urban rivers. https://ktds.vn/n52-sai-gon-xuoi-nguoc/n2834-van-minh-song-nuoc-va-do-thi-song-nuoc-sai-gon.html. Accessed 25 Mar 2018 (in Vietnamese)
5. Steinbronn S (2009) A case study: fish production in the integrated farming system of the Black Thai in Yen Chau district (Son La Province) in Mountainous North-Western Vietnam-current state and potential. Ph.D Dissertation. Faculty Agricultural Sciences at University of Hohenheim
6. Jamnický P (2006) The strategic management of a city development. Slovak J Civil Eng 1:30–33
7. Trang NTT (2020) Promoting sustainability and resilience in Vietnam's floating communities: the assessment of innovative housing units and materials for adaptation to climate change. Ph. D Dissertation. Faculty Architecture, Civil Engineering and Urban Planning, Brandenburg University of Technology. https://doi.org/10.26127/BTUOpen-5455
8. Yin RK (1984) Case study research: design and methods. Sage Publications, Newbury Park
9. Lusterio ACD (2018) Living with water: formal models of settlements in Vietnam Mekong delta. https://nanopdf.com/download/sustainable-urban-coastal-human-settlements-in-southern-vietnam_pdf
10. Ministry of Construction Vietnam (2014) Law No. 65/2014 / QH13, Housing Law. https://vanbanphapluat.co/law-no-65-2014-qh13-on-housing

Design of Floating Terminals as Integrated Project for Multi-machine Systems

Jovana Jovanova, Wouter van den Bos, and Dingena Schott

Abstract Design of floating terminals requires integrated approach as it requires multi-machine systems. Master students in mechanical engineering from Multi-machine engineering track at TU Delft were assigned design of floating terminals as part of their Integration Project course. Each of seven student groups designed a specific piece of port equipment that was later integrated in the floating terminal design. This required different design approaches: a detailed one for the equipment design (structure and functionality), and conceptual one for the floating terminal (overall layout and operational strategy). This encouraged the students to develop skills needed in real working environment, managing the design process and decision making within their own group and discussing setup, basic designs and dimensions together with the other groups. Owning their design throughout the entire process was in particularly important to the students, as they wanted other groups to use their equipment design. For the terminal design they needed to make a case for the feasibility of the floating terminal, including logistics simulations and cost. This paper shows the benefits of integrated design project course, the methods used for its implementation, as well as addressing current challenges of online group design work and supervision. Being part of European Horizon 2020 project motivated the students even more to contribute to an overall bigger objective.

Keywords Integrated project · Design · Floating terminal · Port equipment

J. Jovanova (✉) · W. van den Bos · D. Schott
Mechanical, Maritime and Materials Engineering, Department of Maritime and Transport Technology, TU Delft, Mekelweg 2, 2628 CD Delft, Netherlands
e-mail: j.jovanova@tudelft.nl

W. van den Bos
e-mail: W.vandenBos@tudelft.nl

D. Schott
e-mail: D.L.Schott@tudelft.nl

Ł. Piątek et al. (eds.), *WCFS2020*, Lecture Notes in Civil Engineering 158,
https://doi.org/10.1007/978-981-16-2256-4_29

1 Introduction

Floating platforms for different and multiple purposes have gained major interest in the last years because on land challenges occur from lack of space, lack of resources or natural and manmade disasters [1]. Designing floating platforms is an engineering challenge in development, as required competences are found in different engineering disciplines [2, 3]. However, future engineers need to become aware of this trend for design of floating platforms and to be ready to design machines ready to operate in floating conditions, taking in consideration design requirements based on the application. Large floating islands for living [4], aquaculture [5] or renewable energy production [6, 7] will need to be equipped with floating terminals to enable access and exchange of goods. Different designs of multipurpose floating platforms [8] and mega floating cranes [9, 10] have shown potential in the design of full size floating terminals.

To prepare the students for the future design engineering challenges, project-based learning has become part of curricula in all highly ranked engineering programs. Active learning is at the core of any design course as it requires participation and contribution from the students to deliver their results. Both Problem-Based and Project-Based Learning enable educators to prepare their students for their future professional life as opposed to simply being able to pass exams [11].

The curriculum for Master studies (MSc) in Mechanical Engineering (ME) at TU Delft is based on a solid scientific foundation, deep engineering knowledge and agile engineering design skills. Courses, projects and other modalities are designed to be mutually stimulating. For example, knowledge from courses is applied in projects and, conversely, in their design projects students experience the need for and utility of basic knowledge and engineering methodology. The MSc ME program focusses on three connected didactic goals: (1) To give students an understanding of all mechanical engineering disciplines, with a firm root in theory and a wide focus on applications; (2) To train students to handle the entire process of innovative design, manufacturing, and operation; (3) To coach students to perform research on mechanical engineering topics at an academic level. The essence of the Multi-Machine Engineering (MME) track is to develop, design, build and operate maritime and transport equipment and systems. The mechanical analysis of transport equipment and the interaction between transported material and equipment are fundamental topics for MME. The MME track addresses challenges related to efficiency, sustainability, and safety of complex processes with an integrated perspective that combines core (mechanical systems) design with real-time operation and distributed machine-machine interactions. Specifically graduates in MME track are able to analyze and explain the characteristics and mechanical behavior of material during transport and storage, analyze and model different types of transport equipment and transport facilities, analyze and model the logistics of complex transport systems and networks.

Design of a floating container terminal was a challenge accepted in the master design course Integrated Project Multi-Machine Systems at the MME track. In this course the students were assigned different European port locations with variable local conditions and connections with land transportation (Antwerp, Genoa, La Spezia, Thessaloniki, Constanta and Hamburg), and one group designed a disaster relief floating container terminal for the Mediterranean Sea. This integrated project design course resulted in 7 floating terminal designs in the framework of the Horizon 2020 Space@Sea project (spaceatsea-project.eu). The methodology of the course, the terminal designs and the students' feedback are presented in this paper.

2 Methodology: Integrated Project Multi-machine Systems

The course Integration Project Multi-Machine Systems is obligatory in the MSc ME for the MME track. The course is project-based and runs over the entire spring semester in the first year of the master program, as one of the final courses the students take before going into their second year focusing on their personal assignments (literature, research and graduation). The course brings together most of the skills students develop throughout their studies, from theory to methodology and engineering practice. The students' groups work together to develop designs of complex systems (in this year example that is floating terminals). Students are encouraged to use acquired knowledge (theory, simulation skills, calculations, etc.) from their previous courses and apply it to solve design problems. They need to make multiple decisions in the design process and justify it based on their previous knowledge and experience, which increases their critical thinking.

The course is organized by two lecturers that support the students in their design process. The course Integration Project Multi-Machine Systems simulates an engineering working environment and encourages the students to apply their already accumulated knowledge. The students are highly motivated as the design process resembles real-live engineering work and they take ownership of their designs. The student groups were given a task to design a piece of equipment and then all groups had to design a terminal with multiple equipment from other groups. This mimics real engineering environment: teamwork, understanding requirements and deliver results, communicate, iterate on design improvements, responsibility, etc. There were 44 students following the course Integrated Project Multi-machine Systems in spring semester 2020. They formed 7 groups each consisting of 6–7 people.

The learning objectives of the course are: Apply design methods for multi-machines; Use standards for equipment design; Study and recommend system integration including market availability and custom designs; and Design a project for a multi-machine system. The students applied design methods they have already learned in different courses to develop a piece of equipment. Then they exchanged their designs, went through round of improvements based on feedback, and at last they used their own and other teams' equipment designs for their floating terminal solutions. At the end of the course they showed highly integrated functional terminal designs and were able to communicate the benefits of floating terminal at their specific location.

The lecturers organized topic-specific lectures and provided additional literature once they noticed lack of knowledge. For example this semester, they invited a professor from the Ship Hydromechanics and Structures section to give a lecture on wave motions to help the students understand how to model floating terminal dynamics. In feedback sessions the lecturers supported the students in their design assignments with expertise, advice, calculations and suggestions for improvements.

After the first few weeks of the course, measures to go completely online were implemented because of the Covid-19 pandemic. The lecturers organized immediately everything online and the frequency of meeting with the students twice a week remained throughout the entire semester. Each group had sufficient time to get support in the design process. The students were directed where to look for design solution online, challenged to defend their decisions and given critical feedback on the feasibility on their designs.

At the end they prepared detailed drawings, reports, final presentation, banners and videos that showed high level of enthusiasm for their designs. The assessment of their designs was split into 4 equally weighed parts: terminal design, detailed design, report and presentation, see Fig. 1. The presentation was assessed by a jury

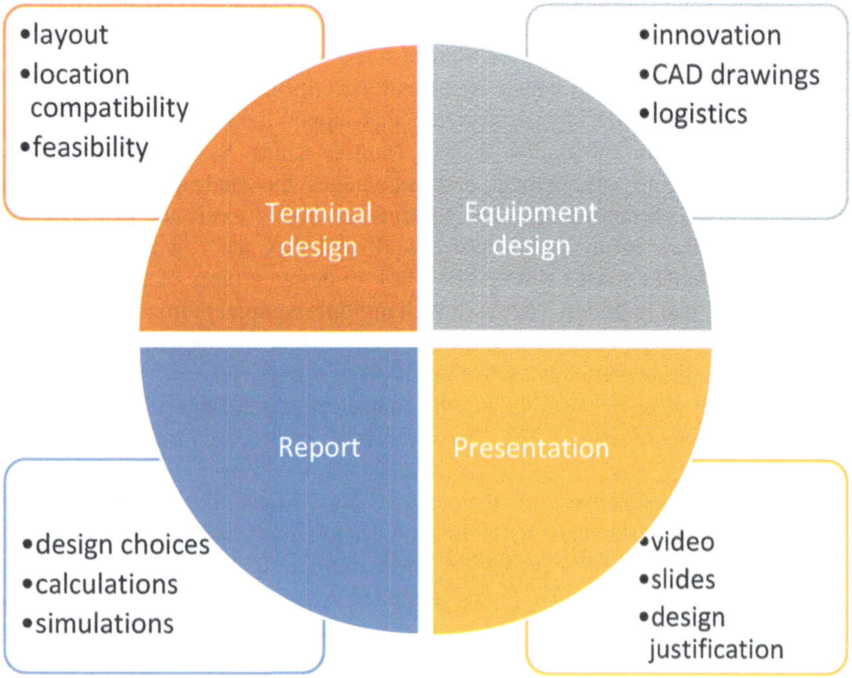

Fig. 1 Student assessments of the floating terminal designs

consisting of the 2 course lecturers, an external professor who is the TU Delft project coordinator for Space@Sea and the project leader of the Space@Sea project. Even though it was challenging to do it online, the course finished with successful presentations and at the final exam all students passed the course with high grades.

3 Port Equipment Designs

In the very first week of the semester the course structure and expectations were delivered to the students and the 7 detailed equipment design assignments, previously prepared by the lecturer, were presented. Each group picked their specific equipment assignment based on personal interest on first come-first served basis. The assignments included pontoons with crane designs, automated guided vehicles (AGVs) and rail bound electric container carriers (RECCs) with bridges to enable motion between pontoons. The pontoons are made of building blocks of minimum 50 m × 50 m with 5 m space in between, which usually came to pontoons with 45 m width and 95 length. Each of the specific designs is briefly presented below.

3.1 Design of Floating Stacking Modules

Design 1 is a dedicated pontoon with a rail mounted gantry (RMG) crane for stacking containers. In a floating terminal it will serve as a storage for containers until they are transferred further to/from the floating island. The configuration consists of two RMG cranes covering the complete surface of the platform, Fig. 2 (*left*). The cranes differ in height so they can move over each other. Three transport lanes in the middle provide space for AGVs to move in the terminal. Pontoon specifications: stack capacity 1,500 TEU (5 × 10 × 15 = 750 40 ft containers for an example length of 145 m); large RMG height 10 m, boom length 45 m, crane width 16 m; small RMG height 7.0 m, boom length 40 m, crane width 9.5 m.

Design 2 is a floating overhead container crane, Fig. 2 (*right*). The shape of the pontoon resembles the hull of a ship. A triple rail system runs along the length of the pontoon. The system consists of three lanes above each other, the top rails are for direct loading and unloading of RECCs along the entire length of the pontoon, while the lower lanes are for longer distance transport. The lanes enter and exit through the sides of the pontoon. The containers are stacked five high and fourteen wide. Seven containers fit in the longitudinal direction: six of which are reserved for 40 ft containers and one for 20 ft containers. Overall, there is a capacity of 490 containers on each pontoon.

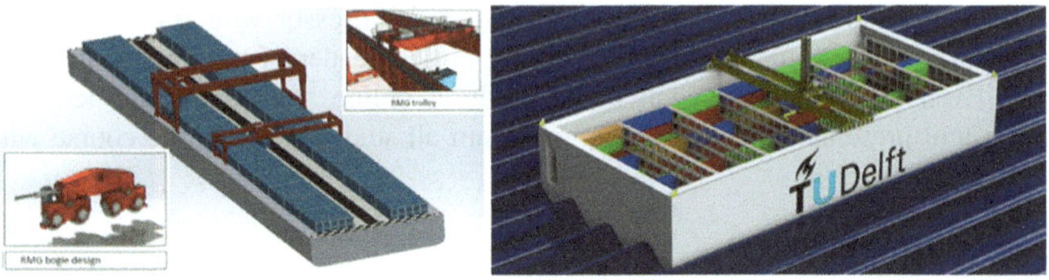

Fig. 2 Stacking module Stacking module with a rail mounted gantry (RMG) crane (*left*) with overhead container crane (*right*)

3.2 *Design of Horizontal Transportation Systems: AGVs and Bridge, Rail-Based Conveyor System and ECCs*

Design 3 was dedicated to AGV's that drive in the pontoons, either in a separate driving pontoon or in a pontoon with port equipment and they can drive onto other pontoons via a bridge. The AGV is designed to carry all standard container sizes and should also be able to drive on the road. Therefore there are a total of 8 axis per side, totaling a number of 32 wheels with standard truck tires and rims, Fig. 3 (*left*). To remain in contact with the road surface at all times, the AGV has active suspension on every axle with coupled hydraulic cylinders to keep the wheel loads constant. For individual steering angles every axle rotates around a king pin which is controlled by hydraulic cylinders. These cylinders are powered by a hydraulic system, located at a central location in the AGV. For the use of the AGV in combination with the ramp, these suspension pistons need to be able to displace a maximum of 200 mm.

The concept of the bridge is inspired by a stern ramp. The bridge has 2 stable positions: down (working position) or up (during storm). Because the terminal

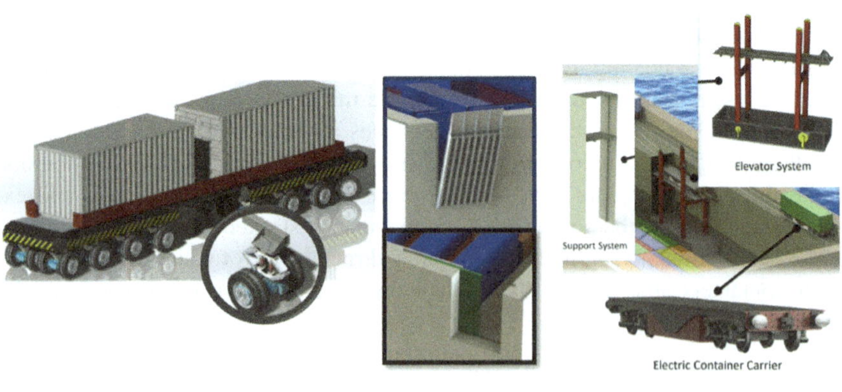

Fig. 3 AGV design (*left*), bridge design (*center*) and rail conveyor system with ECC (*right*)

being located at open sea, the bridges and pontoons will be sensitive to the formation of algae causing problems with slipperiness. Anti-slip coating was selected as the best option to provide high friction with an even surface. An important measure is the cleaning of the drive lanes of the pontoons and bridges once every three months, all formed algae should be removed with a "fleet-cleaner" AGV. The top deck is 5 by 7 m and has a thickness of 14 mm, Fig. 3 (*center*). The height of the bridge is 500 mm. The bridge is lifted using a hoisting mechanism that is placed in the side of the pontoon. The design of the connection point makes it possible to lift the bridge 90°. On the opposite side of the bridge, the inlet of the pontoon will be closed off by a rolling gate. The gate is designed to withstand the slamming pressure the waves of a rough sea.

Design 4 is a rail-based transport system that serves the crane pontoons and the stacking modules. The design includes the rail bound electric container carriers (RECCs) to serve the pontoons and an elevator to provide switch tracks for the carriers, Fig. 3 (*right*). The RECCs exists of a standard train container wagon adapted to fit a 48 V battery pack and 4 electric motors. The Lithium-ion battery pack has a capacity of 924 kWh, weighs 9.4 tons and can run the RECC for 8 h.

3.3 Floating Crane Designs: Rotate Crane, Double Sided Carrier Crane and Feeder Crane

Design 5 was a crane that rotates the containers, Fig. 4 (*left*). For some floating terminal concepts this concept can be beneficial as it offers a unique feature that the containers can be picked up and rotated for maximum use of space in stacking.

Design 6: The Double sided carrier crane, shown in Fig. 4 (*center*) is a variant of the carrier crane, Fig. 5, developed by the Marine and Transport Technology section of the TU Delft as a new concept to solve the problem of increase of cycle times per container moves due to the increase in travel distance with ever increasing vessel sizes [12]. By splitting the (un)loading cycle into 3 separate cycles (vertical transport at ship side, horizontal transport and vertical transport at quay side, the overall cycle time is drastically reduced. For unloading a vessel, the trolleys above the ship lift the containers on carriers which transport the containers along the main

Fig. 4 Floating cranes design: rotate crane design (*left*), double-sided carrier crane (*center*) and feeder crane (*right*)

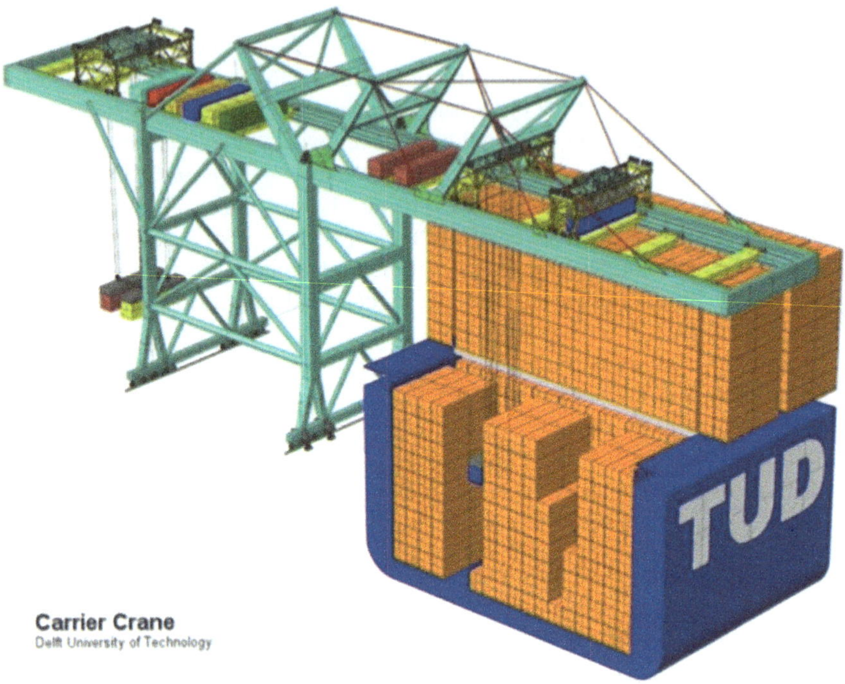

Fig. 5 Carrier crane

beam on the boom and bridge. The carriers are separately moving along 2 trail tracks on the beam, the top rail is for the loaded carriers, while empty carriers are moving on the lower track. The trolley on the quay side picks up the containers from the carriers and lowers them on the quay for further transport. The overall cycle time of the lifting and lowering at the quay and above the vessel is around 1/3 of the cycle time of a traditional full (un)load cycle. With a buffer of queuing carriers small mismatches in cycle times between the sea and land processes are leveled. The total productivity of the carrier crane is inverse of the cycle time and therefor 3 times higher than a traditional Ship to Shore crane (STS-crane).

The double-sided carrier crane shown in Fig. 4 (*center*) is equipped with 32 carriers to allow horizontal container transportation across the boom. Each carrier has its own drive system and serves both as a horizontal support and allows for a buffer function. The ship trolleys (1) place the containers onto horizontal carriers (2), which will transport the containers across the boom (3) until they are positioned above the pontoon (4). Here, a total of four trolleys will load the containers from the horizontal carrier onto the AGVs (5). The AGVs will transport the containers to their respective stacking area. The total technical unloading capacity of the carrier crane is up to 110 containers (or moves) per hour for each moored ship. The only uncontrolled motion is vertical lifting compared to a normal crane which need to combine the vertical motion with a movement to or from shore. The pendulum motion of the container or sway makes a traditional STS crane very vulnerable for wave motions of the pontoon.

To serve smaller ports, design 7 is a feeder crane with a variable level lifting platform, shown in Fig. 4 (*right*), which can handle ships with a capacity up to 5,000 TEU, 300,000 TEU per year and 8,880 TEU storage. Because of the reduced distance to the vessel the ship to shore motion can be controlled better which results in a 20% reduction of cycle time. The variable height platform is connected to the trolley with a scissor system. This allows the platform to remain stiff while being in lowered configurations during operations. Note that the scissor system is only there to provide stiffness by guiding the cables. It does not lift the spreader, this is all still done by the pulley systems for hoisting.

4 Floating Terminal Designs

Floating terminal locations were assigned to each of the students' groups taking in consideration different locations, throughput, boundary conditions and connections. Each of the layouts is discussed separately in this chapter.

4.1 Thessaloniki Port Extension

The port of Thessaloniki is the second biggest harbor of Greece. Its location gives the port a natural barrier for rough sea conditions. It is aiming to become a gateway to the Balkans and South Eastern Europe because it could serve over 20 million people in its direct hinterland and the capitals of 5 different countries within 600 km from the port. The port is connected to a double track railway to the national rail network of Greece. The bay of Thessaloniki, near the container terminal, is 9–12 m deep which has been a limitation for bigger ships to reach the port. Further away from the current port location the water depth reaches more than 20 m. At approximately 1.5 km the depth is 17 m which will allow ULCVs to dock there. The wind in Thessaloniki is 31% of the time dead calm coming from the North-Northwest. Only 0.1% of the wind is a strong breeze or more (meaning 6–8B). The waves that are encountered in the port are usually below 2 m, there is no current, the tide is less than 0.5 m and the significant wave height is approximately 0.5 m. The tide in the gulf is favorable, less than half a meter. This calm environment makes this location very suitable for a floating terminal. The proposed layout contains shown in Fig. 6 consists of: STS module (can handle ULCV, 450 m quay); Stack modules, ground slots (capacity of 5,000 TEU); Terminal transport (AGV or rail system that connects to the land). The specifications are: Quay length 800 m, Depth (without excavation) 16 m, 5 RMG stacking modules Total Capacity

Fig. 6 Thessaloniki port floating terminal

7,500 TEU, Average cycle time < 60 s, 1 Double-sided STS module Unloading capacity (one-side) 880 TEU/hr, 50 AGVs.

4.2 *Floating Terminal on the River Elbe, an Extension for Hamburg Port*

The waterway towards the port of Hamburg, the river the Elbe, is characterized by many tight curves. Recently, the government of Hamburg has invested in dredging and widening the way, thereby allowing for easier maneuverability for larger ships. However, it would be much easier if the ships of the largest size could navigate up to a point halfway, and let inland waterway transport do the rest of the route to Hamburg. This could be done by creating a new port location further downstream the Elbe at Brunsbüttel. Furthermore, since a bulk terminal already exists at the location, this means that certain connections with the hinterland are already present.

The layout of the terminal, Fig. 7, consists of two types of storage pontoons, and two pontoon types to unload the vessels. The pontoons will be connected via a rail system running through the terminal, via a vehicle bridge at the land side, and via overhanging cranes of some of the storage pontoons. Terminal specifications: 1,315 m quay; 2,510 TEU/hr; 7.3 M TEU/year; 13,408 TEU storage; fully

Fig. 7 Hamburg port floating terminal

automated; connected via truck and train. Because of the floating units the natural water flow in the river remains undisturbed and perturbations and swirls further downstream can be avoided.

4.3 Mediterranean Floating Terminal: A Mobile Port for Disaster Relief

The group started with their design requirements: the port should be Quick, Efficient, Smart and Safe, as after a disaster, relief should be available in the shortest time period. The disaster relief port should also be useful for temporary capacity increase (while waiting for port expansion for example) and use units which can be reused in another location afterwards. They set criteria: operating area is the Mediterranean Sea; port reaches a destination within 10 days; operates in different locations; modular design; has ship-to-shore pontoons, storage pontoons and driving pontoons, see Fig. 8.

The carrier crane has been chosen as Ship-to-Shore crane (STS crane), because of the possibility of direct transfer to another ship and the capability of serving all types of container vessels. The original design is made for a fixed quay but in the disaster relief port the carrier crane will be placed on a floating pontoon. For the storage the pontoon with the overhead crane is selected, because it has a compact structure that catches less wind during shipping and the possibility for the AGV's to drive through the pontoon. The transfer of the containers from the STS crane to the storage pontoon is done by AGV's. When the port is out of use as an emergency port, the modules can contribute to the port where it is stored (for example in a port in Greece). The shipment from the pontoons to another port is done with tugboats.

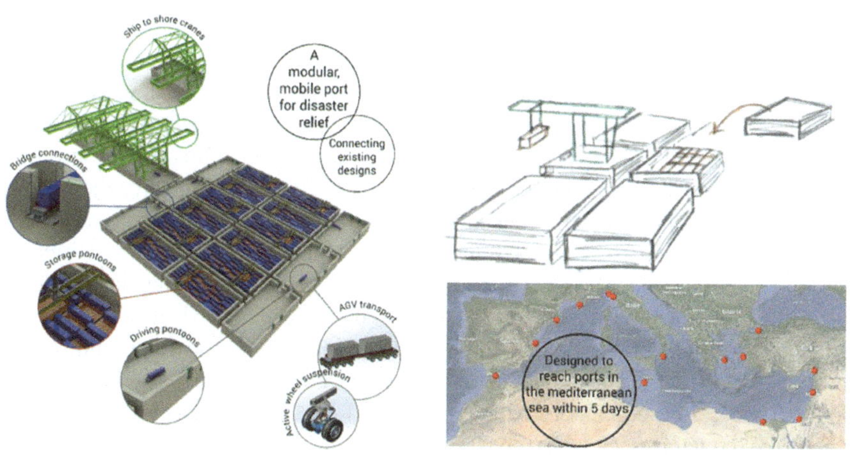

Fig. 8 Disaster relief port

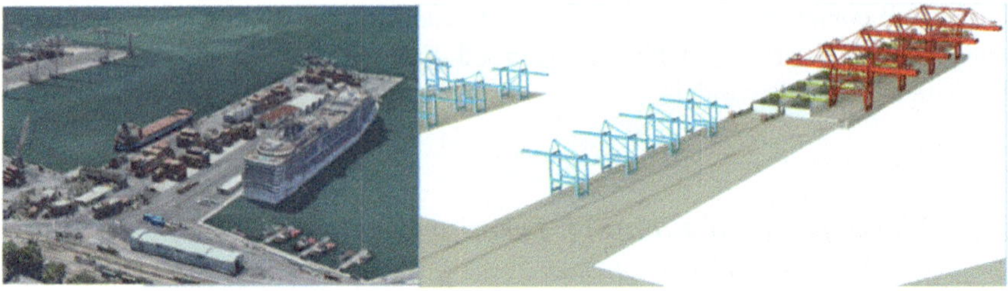

Fig. 9 La Spezia floating port extension

4.4 La Spezia Port Extension

The port in La Spezia, Italy is known for its connection to the rail network running from Italy across Europe. To meet the supply and demand of the container transport an expansion plan is created for the port of La Spezia. This plan includes not only changes to the existing harbor, but will include the use of a modular floating terminal. A floating terminal is more flexible than a standard expansion and the throughput and storage capacity can easily be expanded separately, Fig. 9. Terminal capacity: Stacking capacity 40,000 TEUs with 15 pontoons; handling capacity 2 million TEUs per year with 19 STS cranes (4 on floating pontoon) and 4 mobile cranes, 5 simultaneous vessel operations.

4.5 Port of Genoa Extension

The Genoa area is very crowded and free space is limited. A floating terminal gives the possibility to increase the port capacity without the need for expensive land reclamation in relative deep water. The position of the floating terminal in Genoa is selected close to the existing container terminal and rail service center. The floating extension, shown in Fig. 10, is feasible due to the elongation of the breakwater and the depth of the harbor is deep enough (20–40 m). The containers will be offloaded using double sided carrier cranes (yellow) which are able to offload two ships at the same time. The containers will be placed on top of RECCs which carry them along rail tracks from the quay through a 90° turn. After unloading, there are several locations the containers can be transported to. First location is the storage or stacking pontoon (green). In order to transport the containers to the stacking pontoon, the containers are picked up by the own developed Lift Carry Rotate (LCR) crane (light blue). Behind the LCR the containers are lowered and rotated onto another RECC. These RECCs transport the containers towards the stacking pontoon where the containers will (temporarily) be stacked. The second location where the containers can be transported to is the truck loading station. These containers can be transported directly from the quay or can be taken out of the

Fig. 10 Genoa floating port extension

storage pontoon. At the truck loading station several Rubber Tire Gantry Cranes (RTG's) are placed. These RTG's transport the containers from the RECCs to the trucks. The third location is the rail transport location. This location is located in the existing container terminal and is currently equipped with Rail Mounted Gantry Cranes (RMG's).

4.6 Floating Port Antwerp

The harbor of Antwerp is located in the center of the city, which means that the harbor cannot be expanded inland, even though the harbor has a yearly increase in cargo flow. Thus, a floating terminal design can be a good alternative for capacity extension. The design of such a terminal is based on many different aspects, such as the desired cargo flow that the terminal needs to handle, the environment in which the terminal is built, and the storage capacity needed. For the location of the port few criteria were analyzed: Antwerp is a Belgian city the port has to be placed in Belgian territorial waters; the port is outside of protected nature areas like the Vlakte van de Raan and the special protection zone in front of the harbor of Zeebrugge; stay clear of any existing North Sea wind farms; the water depth is taken into account, as the biggest container ships have a depth of about 15 m; major route from the North Hinder South route to the Westerschelde. The floating port, presented in Fig. 11, has handling capacity of 582,772 containers per year; storage capacity of 48,720 containers and the storage capacity is expendable as a respond to the expected annual growth of 5%. The floating terminal is completely focused on maritime transport and does not need a direct road or rail connection with the hinterland.

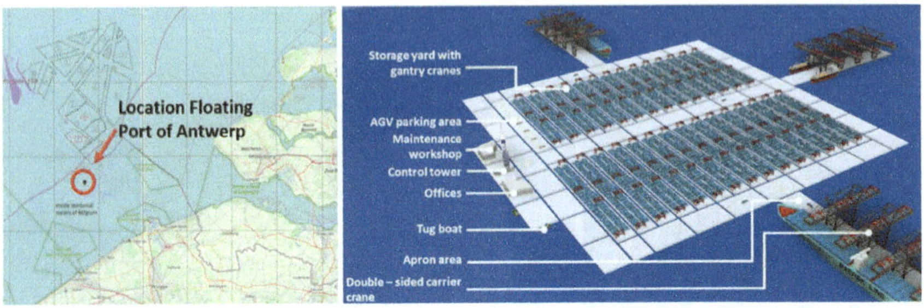

Fig. 11 Antwerp floating port

Fig. 12 Constanta floating port extension

4.7 *Port of Constanta, Romania*

The port of Constanta is located at the Black Sea and is the main port of Romania. It is connected to the hinterland via road, train and the Danube-Black Sea canal. Based on the development strategy the design of the proposed floating terminal, Fig. 12, will be able to process 300,000 TEU per year. Examination of existing terminal designs concludes that a storage capacity of 8,880 TEU is sufficient for this terminal.

The design consists of one large quay pontoon with three feeder cranes. Alongside are four storage pontoons installed and one service pontoon. A connection to land is established by a bridge connecting the quay pontoon to an existing dam. Via this dam the main container terminal is reached. To transport containers between the feeder cranes, storage pontoon and land, 12 AGVs are used. The quay length is 345 m, sufficient to process a feeder ship and a barge at the same time.

5 Students' Competences and Feedback

In this course the students were working on integrated projects where they designed machines and floating terminals. They were encouraged to critically think in the design process. They were able to look for solutions in literature and apply them in their designs. They had to communicate within their team and coordinate with other teams. In the final presentations they showed the benefits of their systems in current ongoing scenarios and put them into a commercial value.

The floating terminal design projects were inspired by an ongoing Horizon 2020 project Space@Sea that explores the viability of floating structures. TU Delft led the development of the Transport and Logistic hub with aspects ranging from selection of cargo, to design of terminals and coordination of multi-machines [13–15]. This motivated the students even more and was noticeable through their creativity and dedication to the course. The students' feedback stressed the importance of the course and the active learning design process they went through. They felt as they were part of a bigger challenge and felt as their contributions are valued beyond the scope of the course itself.

During the course durations the students spark discussions about trends in port design, environmental issues, energy efficiency, safety, sustainability and other societal challenges the world is facing right now. This makes them aware of the technical challenges and responsibilities they have once they practice engineering outside the student role. They developed critical thinking and reflection; carrying out research; designing; developing an academic approach; communication and collaboration in interdisciplinary and intercultural teams; taking into account the temporal and social context of technological solutions, which is expected from all Graduates of the TU Delft. This aligns perfectly with the TU Delft's vision: Making a contribution to solving global challenges by educating new generations of socially responsible engineers and by pushing the boundaries of the engineering science goes far beyond education.

6 Conclusions

In this work we have showed a successful active learning approach through integrated design course inspired by the Horizon 2020 Space@Sea project. The students raised their engineering design confidence in this course and learned how to combine the skills gained from other courses to design complex multi-machine systems, though project-based design and active learning. The first quarter of the course was first focused on single piece of equipment detailed design and the second quarter on the overall terminal layout. The students need both approaches when working in industry, a detailed design and more abstract conceptual design.

They showed at the end of the course that a graduate from MME track is able to model, calculate and simulate the interaction between equipment and containers, design unique port equipment for floating terminals, automate the transport equipment, design floating terminals taking in consideration location specific design constraints as well as the logistic systems. This course Integration Project Multi-Machine Systems brings them a step closer to a real engineering environment.

This course successfully changed into an online version due to the Covid-19 pandemic, which makes the course adaptable for any blended learning scenario.

The positive experience with online feedback showed potential that for active learning. The combination of online and in person activities in education is likely to stay in the future. Visiting port terminals and equipment manufacturers will remain as part of planned activities. European projects such as Space@Sea are a great inspiration for the project assignments and are beneficial for the students' motivation and commitment to the course.

Acknowledgements The authors would like to acknowledge all students from the course Integrated project Multi-machine systems in spring 2020, Master program in Mechanical engineering at TU Delft, track Multi-Machine Engineering for their innovative designs, commitment and dedication to their projects.

References

1. Flikkema M, Waals O (2019) Space@Sea the floating solution. Front Marine Sci 6:553
2. Wang G, Goldfeld Y, Drimer N (2019) Expanding coastal cities—proof of feasibility for modular floating structures (MFS). J Clean Prod 222:520–538
3. Yao T, Fujikubo M (2016) Buckling and ultimate strength of ship and ship-like floating structures. Butterworth-Heinemann, Oxford
4. Chih Lin Y, Tan HS (2019) Design and functions of floating architecture—a review. Mar Georesour Geotechnol 37(7):880–889
5. Olsen TO (2020) Fish farming in floating structures. In: WCFS2019. Springer, Singapore, pp 191–208
6. Ranjbaran P, Yousefi H, Gharehpetian GB, Astaraei FR (2019) A review on floating photovoltaic (FPV) power generation units. Renew Sustain Energy Rev 110:332–347
7. Díaz H, Soares CG (2020) Review of the current status, technology and future trends of offshore wind farms. Ocean Eng 209:
8. Ang KK, Dai J, Hellan O, Watn A, Si MBI (2020) Design and potential applications of floating structures in Singapore. In: WCFS2019. Springer, Singapore, pp 135–154
9. Ham SH, Roh MI, Lee H, Hong JW, Lee HR (2017) Development and validation of a simulation-based safety evaluation program for a mega floating crane. Adv Eng Softw 112:101–116
10. Lee HW, Roh MI, Ham SH, Ku NK (2018) Coupled analysis method of a mooring system and a floating crane based on flexible multibody dynamics considering contact with the seabed. Ocean Eng 163:555–569
11. Christie M, de Graaff E (2017) The philosophical and pedagogical underpinnings of active learning in engineering education. Eur J Eng Educ 42(1):5–16
12. Luttekes E, Rijsenbrij J (2002) Carrier crane. Patent NL1020675C2 / WO2001025131A1, TU Delft
13. Schott D, Dafnomilis I, Duinkerken M (2020) The Space@Sea transport & logistics hub: comparison to onshore and reclaimed land expansion. WCFS paving the waves (paper accepted and will be presented at WCFS 2020)
14. Souravlias D, Duinkerken MB, Morshuis S, Schott DL, Negenborn RR (2019) Stochastic floating quay crane scheduling on offshore platforms: a simheuristic approach. In: Proceedings of 21st international conference on harbor, maritime and multimodal logistics modeling and simulation, HMS
15. Dafnomilis I, Souravlias D, Ley J, Assbrock G, Duinkerken M, Negenborn RR, Schott DL (2020) Design framework for a modular, floating container terminal. Front Marine Sci (paper submitted)

Expectation of Floating Building in Java Indonesia, Case Study in Semarang City

Slamet Imam Wahyudi, Henny Pratiwi Adi, Jonathan Lekerkerk, Jorge Jansen, and Floris C. Boogaard

Abstract Many delta cities worldwide are dealing with the same kind of problems: rising of the sea level, land subsidence, scarcity of land and illegal housing. Multiple land use is one of these solutions that will help to reduce flooding and scarcity of land. An example of multiple land use is a floating community. This research used Semarang as location for the research into the social acceptance of floating houses. The data in this study were obtained through literature study and survey among inhabitants. The social acceptance of the inhabitants is determined with 35 respondents that have been done in the area of Kemijen, Semarang. In order to determine the social acceptance of floating houses, there are elements used, namely: knowledge of floating houses, perception of risk, urgency, implementation, chose for a floating house, requirements, positive and negative elements, self-sufficient system. According to the result of research, the social acceptance of the inhabitants is quite low, but there is potential because they see positive elements in a floating house. Low social acceptance is caused by the fact that the concept of floating houses is not well known in this community. With raising awareness on the

S. I. Wahyudi (✉) · H. P. Adi
Department of Civil Engineering, Sultan Agung Islamic University, Jl. Raya Kaligawe KM4, Semarang, Indonesia
e-mail: wahyudi@unissula.ac.id

H. P. Adi
e-mail: henni@unissula.ac.id

J. Lekerkerk · J. Jansen
Department of Water Management, Rotterdam University of Applied Sciences, G.J. de Jonghweg 4-6, Rotterdam, The Netherlands
e-mail: j.lekkerkerk@hr.nl

J. Jansen
e-mail: j.g.f.jansen@gmail.com

F. C. Boogaard
Hanze University of Applied Sciences, Zernikeplein 7, 9747 AS Groningen, The Netherlands
e-mail: f.c.boogaard@pl.hanze.nl

F. C. Boogaard
Deltares, Daltonlaan 600, 3584 BK Utrecht, The Netherlands

Ł. Piątek et al. (eds.), *WCFS2020*, Lecture Notes in Civil Engineering 158,
https://doi.org/10.1007/978-981-16-2256-4_30

challenges and informing the community on the possibilities on floating infrastructure will result in higher social acceptance.

Keywords Floating building · Expectation · Flooding · Polder area

1 Introduction

Dealing with water related issues has always been a challenge in many parts of the world [1, 2]. Especially in high density areas along rivers near the coast: The Delta Cities. With the rising of the sea level, climate change and land subsidence delta cities have to adapt to the changing conditions in order to keep the city safe from flooding. Only heighten the embankments is in most situations not enough. A combination of different solutions must be applied [3]. Climate change causes more heavy rainfall in a shorter period of time. Therefore, city planners should organize the city in a way that it can adapt to the extra water. This requires extra storage zones and demands more space and more flexible management [4].

Besides water related issues there is another problem where city planners have to deal with: limited space. In many cities there is an urgent need for urban development [5]. The number of city dwellers is expanding and this requires an increase of space for housing area.

The demand for extra water storage and space for housing are in conflict with each other. This conflict will increase in the future when the demands are growing through climate change and growth of the population [6]. City planners should therefore search for other possibilities; multiple use of space seems the solution for this problem. Floating houses are an example of multiple land use [7]. Combining different functions in a city can help to reduce the water problems and provides more space for living areas.

1.1 Case Semarang

The study case of this research is Semarang, a city with 1.7 million inhabitants in Central Java, Indonesia [8]. The delta city is one of the locations that is facing floods on a daily base. Through climate change rises the sea water level with 6 mm a year [3]. Nevertheless, the relative sea level rise is way higher. Through large-scale groundwater extraction and oxidation subsides the land in Semarang with average 9 cm per year [9]. The current water system (city rivers and sewerage system) are not dimensioned for this rapidly changing situation and resulting in flooding. Improper maintenance and a poor drainage system worsen this situation [10]. In order to deal with these problems, the government of Semarang and the Dutch government, started a cooperation in 2003 [9]. The purpose of this cooperation is to set up a water authority similar to the Dutch model and construct Banger polder system (Fig. 1) with embankments, a pumping station and dredging

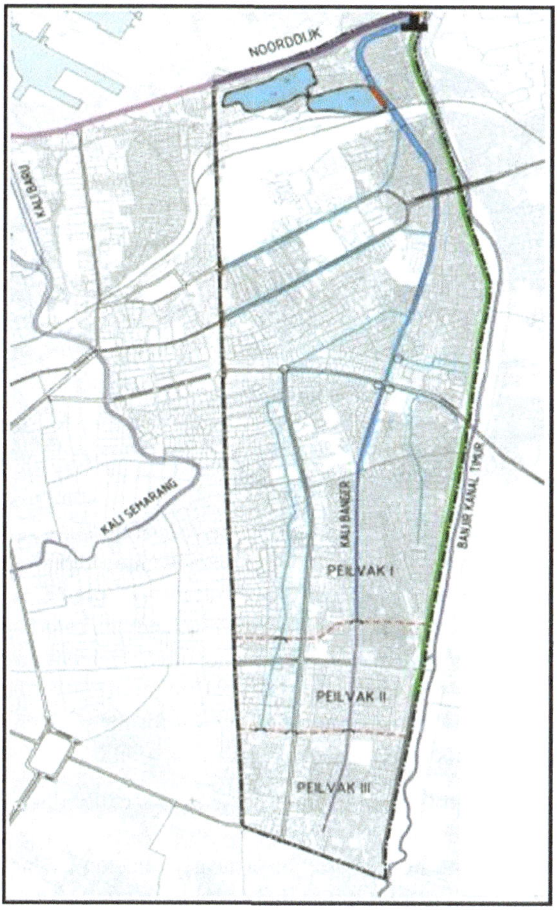

Fig. 1 The banger polder

schedule to reduce the water level of the river. This project should be finished in 2020 and will make Semarang less vulnerable for flooding.

1.2 *Problem Analysis*

The Banger Polder contains 2 retention basins in the Kemijen area (one of the sub districts in Semarang) which can be used to store water during heavy rainfall (see Fig. 2). The others retention basins are in Semarang Polder System, Tenggang Polder System and Sringin Polder System. Around these basins people build illegal houses. By making an electricity connection people are protected for replacement by the law that says that people who are connected to electricity aren't illegal anymore [11]. Through this people get in unsafe situations and the storage volume of the basin is being affected. Besides this, the current retention basins do not have the required capacity yet. The basins must be extended so that they can store more

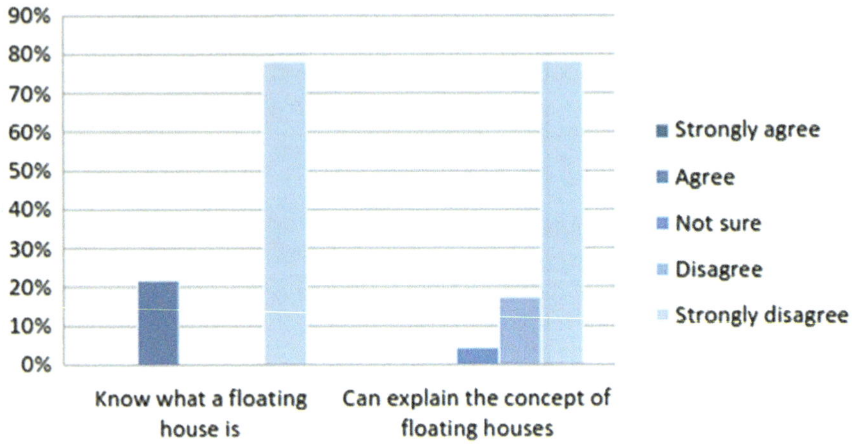

Fig. 2 Knowledge about the concept

water when needed and lower the risk of flooding. When the polder system and the extension of the basins is completed, the polder will deal with floodings with a returning period of every 8 years [11]. Without the extension of the basins, floods will occur every 2 years (which is already an improvement to the current situation). When the basins will be extended, the inhabitants who live around the basins should be replaced.

Floating houses can be a solution for the problems in Semarang. In this context floating houses could be an adaptive solution with several positive effects [12]:

- The application of multiple land use will reduce flooding problems and lack of space.
- Floating houses are not or nearly affected by land subsidence.
- Instead of replacing inhabitants for a possible expansion of the retention basins, inhabitants can continue living in the same area.
- Provide security of tenure and make inhabitants live on a legal basis.
- Floating houses are a positive addition to the pilot project the Banger Polder and can serve as a showcase for other vulnerable areas in Indonesia.

This research focus on the social elements that are required by technical improvements. Research into the social acceptance of a new technology of the target market is important for the development of the product. In this context the new technology is a floating house.

2 Research Method

In this research several methods are applied, such as literature study, field research and survey amongst inhabitants. These methods are based on scientific resources and implemented to the situation in Semarang. In order to determine the social acceptance of floating houses in Kemijen, Semarang, the elements that are necessary for the social acceptance for floating should be determined first. These elements are set up by using the literature resources and experience: (a) knowledge of floating houses, (b) perception of risk, (c) urgency, (d) Implementation, (e) chose for a floating house, (f) Requirements, (g) Positive and negative elements, (h) Self-sufficient system. These elements form the basis for the questionnaire to determine the social acceptance of the inhabitants of Kemijen, Semarang. Every element represents one theme, wherein every theme consists of several questions. With this questionnaire it is possible to determine the social acceptance for floating houses. The participants for the survey were selected according to the following criteria:

- They should be living in Kemijen, Semarang. Because Kemijen lies around the retention basin and is most vulnerable for flooding.
- They should be living in a lower laying house than sea level and street level. In Kemijen there are different kinds of houses. Some of them are higher than the streets, some are at the same level and some houses are even below the street level. The houses which are lower than the street level are most vulnerable for flooding. Therefore, these people are most interesting for this research. For this research the most vulnerable people are most interesting to do a survey because for them the utility is the highest.
- For this research it is impossible to interview the whole community. There-fore this research is making use of a sample. For a sample it is important that the participants represent the community. In order to do this the participants are selected on a variety of age and gender.

The results of the survey will be shown in graphics and tables with explanations. Methods derived from several literature resources will be used in order to make a clear overview.

3 Results and Discussion

The social acceptance of the inhabitants is determined with 35 surveys that have been done in the area of Kemijen, Semarang. This is only a small amount of the total population of Kemijen (13.000). But because this research is making use of a sample these surveys can represent the population and give a first impression of the acceptance of new technologies. Besides during the selection of the participants the variety of the inhabitants is taken into account. The surveys were hold in many different streets and areas in Kemijen. There is also a good balance between male and female; age; kind of house and income around IDR 1500 or 100 USD. The questionnaire consists of 2 parts: the first 4 themes based on the theory of Von Wartburg & Liew in [13] and a second part with additional information to determine the social acceptance. The results of the first four themes can be shown in diagrams that show the answers that have been given.

3.1 Knowledge About the Concept of Floating Houses

The first element is 'knowledge about the concept'. This is an important element because ignorance of a technology is mostly a negative factor for the social acceptance. Diagram 1 shows that almost 80% of the participants don't know what a floating house is. The 20% who said that they know what it means was wrong or couldn't explain what a floating house is exactly.

Every theme can be valued with a number. With these numbers it is possible to compare the social acceptance with other researches of social acceptance. If this research will be done in another area or in the same area in a few years, the numbers can qualify what the social acceptance is comparing other researches. The value of the number can vary from 1 to 5. Where in 5 is most positive for the social acceptance and 1 negative concerning the social acceptance.

The knowledge of the concept gets a 1.46 which is very low on the range of 1–5. For improving the social acceptance, this value should be higher by giving information or starting a pilot (Table 1).

After this question the concept of floating houses is explained. It is explained what a floating house is and how it can contribute to their lives. Besides the current situation is drawn wherein the inhabitants are being told about the flooding problems, the banger polder and the extension of the retention basins. Thereafter the next questions were asked.

3.2 Perception of Risk

Perception of risk is the second element. The perception of risk is important because it gives an indication of inhabitants will accept a floating house. According to Von Wartburg and Liew (1999) most people will probably accept the risk of doing something if the risk or not doing it is even greater. The risk of not doing anything is that people have flooding. A floating house will only be accepted if people see the risks of a floating house smaller than the risks of flooding that they have now. Analyzing the results shows that about half of the participants think a floating house is totally safe. But only 37% think that a floating house is more safe than living in a normal house. Less than 30% think they can live with their family safely on a floating platform. This question is a control question for question 1 because the form of the question is different, but the meaning is the same (Fig. 3).

Table 1 Score of knowledge about the concept

No	Statement	Score
1	Know what a floating house is	1.65
2	Can explain the concept of floating house	1.26
Average score		1.46

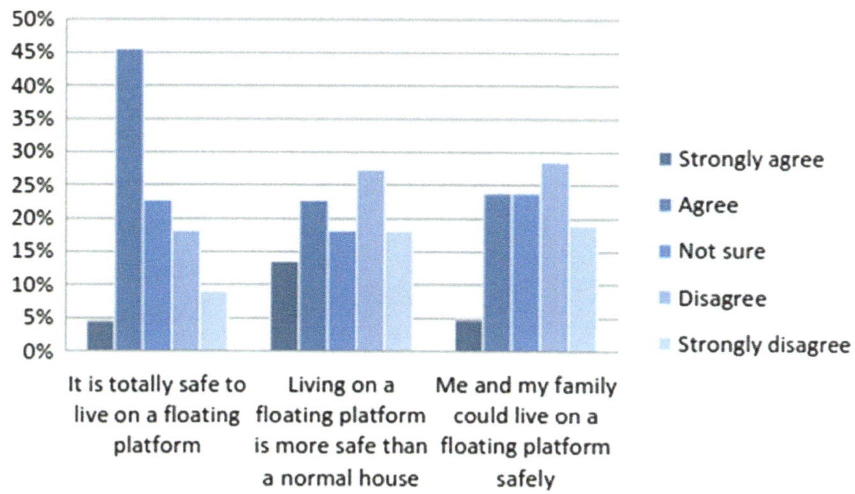

Fig. 3 Perception of risk

Table 2 Score of perception of risk

No	Statement	Score
1	It is totally safe to live on the floating platform	3.18
2	Living on a floating platform is more safe than a normal house	2.86
3	Me and family could live on a floating platform safely	2.67
Average score		2.90

Table 2 shows the values for the perception of risk. A high number indicates that people see low risks in the new concept. The perception is risk gets a score of 2.90. This is quite a low number and should be increased. The perception of risk is strongly linked with the knowledge about the concept. Because inhabitants don't know the concept, the anxiety for floating houses is also quite high. More information and a pilot can raise this number.

3.3 *Urgency*

According to the Dutch standards, flooding on a daily basis is far from accepted. If Dutch people would be living in a situation like in Kemijen where floods are very common, the perception of urgency would probably be very high. But the outcomes of the survey in Kemijen show that the urgency of Indonesians is not that high for an area where floods occur almost daily. 57% of the participants see the utility of floating houses and 58% thinks floating houses are necessary in their neighborhood.

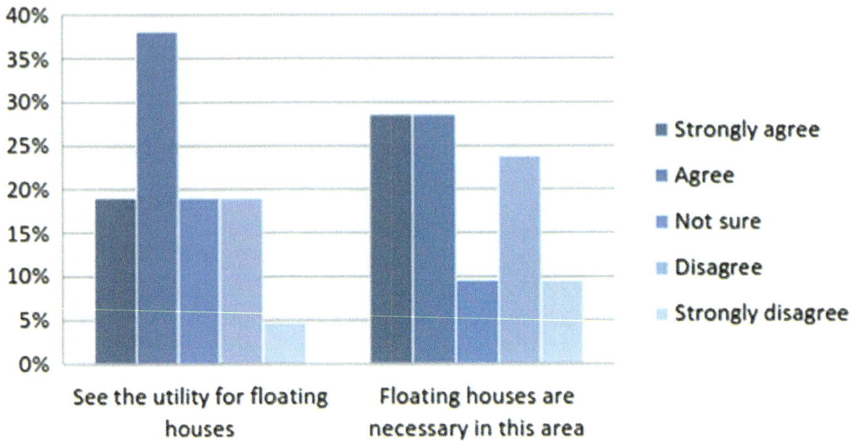

Fig. 4 Urgency

Table 3 Score of knowledge about the concept

No	Statement	Score
1	See the utility for floating house	3.48
2	Floating house are necessary in this area	3.43
Average score		3.46

The inhabitants who see the utility and who think it is necessary are mainly the ones who live in lower houses that are more vulnerable for flooding. The urgency is less by inhabitants with a higher house or in streets where there is less flooding (Fig. 4).

Urgency has scored 3.46. This is the highest score of this survey. This score is quite positive for the social acceptance. It shows that many people see that there is a need for a solution for the current situation (Table 3).

3.4 *Implementation*

When inhabitants are asked about the implementation, they are quite reluctant. About 1/3 of the participants think that floating houses can be implemented in their area. But if they are asked where they rather want to live, more than 80% give 'a normal house on the ground' as answer. 14% is willing to pay some extra money for a floating house. For a possible implementation of floating houses, this 14% should be the focus group (Fig. 5).

The most important score of the survey is the value for implementation. The score for this is a 2.27 which is low. Most inhabitants think the concept is good, but in they still rather live in a normal house (Table 4).

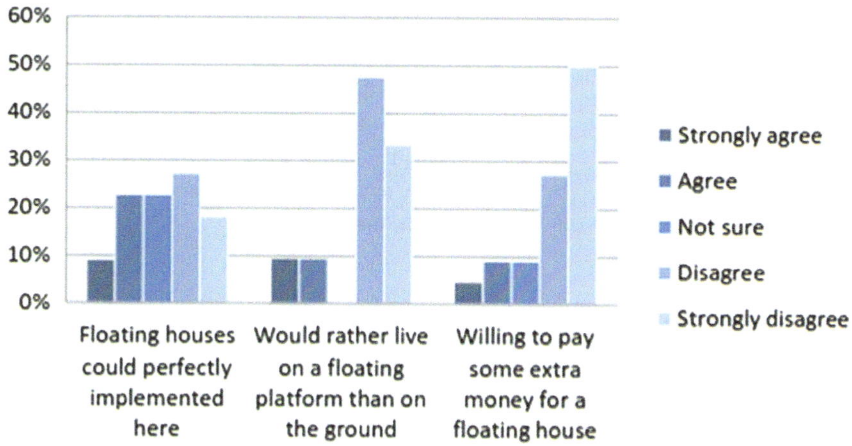

Fig. 5 Implementation

Table 4 Score of knowledge about the concept

No	Statement	Score
1	Floating houses could perfectly implemented	2.77
2	Would rather live on a floating platform than on the ground	2.14
3	Willing to pay some extra money for a floating house	1.91
Average score		2.27

3.5 *Choosing a Floating House*

Participants are asked in a different way (see theme 4) if they would choose a floating house instead of a normal house. Since this is an important question of the research, it is worthy to validate the answers given by the participants. The results are shown in Table 5.

These answers are in line with the answers given at theme four where about 80% would choose their current house above a floating house. The validation is checked and gives the same result. For the participants who give "not likely" or "totally not likely" as answer, these are their main reasons (Table 6).

Table 5 Choosing a floating house

No	How likely	Amount (%)
1	Very likely	4.3
2	Likely	4.3
3	Maybe	13.0
4	Not likely	52.2
5	Totally not likely	26.1

Table 6 Reason for no floating house

No	Main reason	Amount (%)
1	Don't need a floating house	25.0
2	Don't want a floating house	25.0
3	Satisfied with the current house	35.0
4	Can't afford a floating house	5.0
5	Don't want to pay for a floating house	5.0
6	Don't know the concept	5.0

The participants who were not likely to choose a floating house were asked what their main reason was. It turns out that most participants were satisfied with their current house and didn't need or want a floating house. These reasons where according to the inhabitants more important than financial reasons.

3.6 Requirements

In order to make a floating house more attractive it is important what the most concerning elements of a floating house are according to the participants. Table 7 shows what inhabitants concern most.

The main concerns of the inhabitants are price, safety, and material. Safety is especially for women and important issue. Many are concerned about their children who can fall into the water from the floating platform. Men are more thinking about the materials. Bamboo and wood are less accepted as building material for the house. Besides many are wondering if the materials are suitable for the salt in the water of the basins.

3.7 Positive and Negative Elements

It's important to review the positive and negative elements mentioned during this study. The positive elements can be highlighted and improvements can be made to the concept in response to the negative feedback on wooden floating houses.

Table 7 Main concern

No	Main concern	Amount (%)
1	Price	31.8
2	Maintenance	0.0
3	Safety	27.3
4	Material	36.4
5	Information	4.5

3.7.1 Positive Elements that Are Given by the Inhabitants

- They don't have to heighten their house anymore.
- They are safe from flooding.
- It is good that the house can fluctuate with the water level.

3.7.2 Negative Elements

- People don't want to live in houses made of wood and bamboo. The house should be made of concrete because the quality of wood and bamboo is not as good as a concrete house. Besides wood of good quality is more expensive than concrete.
- People want to build their own house so that they can build and adjust it how they want to.
- The salty water can affect the materials.
- It should be safe for children.
- Inhabitants are not familiar with floating houses, so they don't want to live in a floating house. For most inhabitants the house they are currently living in is fine for them.
- The price of a floating house is too much.

3.8 Self-sufficient System

Participants are asked what they think of the self-sufficient system. The response to this question *is not very high*. A floating is for most people already hard to imagine, the self-sufficient system is even harder to understand. Therefore, people didn't give much comment. The comments that have been given are listed below:

- It's a good system.
- I would spend some extra time and money in the system, but only if we don't get problems with it and the maintenance is low.
- People required that the toilet should be inside the house.

3.9 Results Recapitulation

Table 8 gives the outcomes of the social acceptance derived from the inhabitants of Kemijen.

Table 8 Results recapitulation of theme

Theme	Conclusion
Knowledge of floating houses	The knowledge about the concept is very low. Almost no one could explain what a floating house is
Perception of risk	People see relative much risk in the floating concept. This is negative for the social acceptance. This can be explained because people have never seen the concept
Urgency	About 60% of the participants see the urgency for floating houses
Implementation	Most inhabitants do not want to live in a floating house. Even less people want to invest some extra money in the concept
Choosing a floating house	Almost all participants chose a normal house above a floating house because they don't want or need a floating house and are satisfied with the house they are currently living in
Requirements	People consider price, material and safety as most important issues for a floating house
Positive and negative elements	Most people see the positive elements of a floating house: they are less vulnerable for flooding. But people also see negative points such as the used material, safety and the price
Self-sufficient system	Most inhabitants think it is a good system, but they say that they have insufficient knowledge about the system to judge about it

4 Conclusions and Recommendation

4.1 Conclusions

According to the results of research, the conclusions can be drawn as follows:

- In order to determine the social acceptance of floating houses in Semarang, the *elements* that are necessary for the social acceptance for floating should be determined first. These elements are setup by using the literature resources and experience: Knowledge of floating houses, Perception of risk, Urgency, Implementation, Choosing a floating house, Requirements, Positive and negative elements, Self-sufficient system.
- To determine the social acceptance, the *themes* that are described above are used. This has been brought into practice. The outcomes can be divided into the acceptance of the inhabitants and the perception of the other stakeholders and experts.
- It can be concluded that the social acceptance of the inhabitants is quite low, but there is potential because they see positive elements in a floating house. The main criteria why the social acceptance is low is because they don't know the concept. By improving this, the social acceptance will probably also become higher. In order to understand the circumstances, it is withal important to interview stakeholders and experts who can explain and clarify the context. Besides they can give their opinion about the project which gives a more nuanced view on the topic.

4.2 *Recommendation*

- Overall there are several recommendations concerning the improvement of the social acceptance of floating houses in Semarang.
- Start a pilot project so that people can see how the concept is working.
- Providing more information about floating houses and its advantages and disadvantages. Start a project of floating houses without a self-sufficient concept because the step is too far.

Acknowledgements This acknowledgment is addressed to all those who assisted with this research. Among them were Ministry of Public Works and Public Housing, Directorate of Research and Community Service, Ministry of Research and Technology, Higher Education, Erasmus + Grant, Rotterdam University of Applied Sciences and Hanze University of Applied Sciences Groningen, and Sultan Agung Islamic University, Semarang.

References

1. Boogaard FC, Groningen H, Muthanna TM (2019) International knowledge exchange on climate adaptation with the Climatescan platform. In: Proceedings of conference on 4th European climate change, Lisbon, July
2. Ki-moon B, Georgieva K, Gates B (2019) Adapt now: a global call for leadership on climate resilience. https://cdn.gca.org/assets/2019. Global Commission on Adaptation, 90p
3. Eccles R et al (2019) A review of the effects of climate change on riverine flooding in subtropical and tropical regions. J Water Clim Chang 10(4):687–707
4. Karim A (2020) Innovation and strategies to eliminate flood effects on rice farming. J Tanah Lingkung 13(April 2011):35–41
5. Wang CM, Watanabe E, Utsunomiya T (2008) Very large floating structures: applications, analysis and design. Taylor & Francis, Milton, pp 1–30
6. Asrasal A, Wahyudi SI, Adi HP, Heikoop R (2018) Analysis of floating house platform stability using polyvinyl chloride (PVC) pipe material. In: Proceedings of MATEC web conference, vol 195(02025), pp 1–8
7. Ishaque F, Ahamed MS, Hoque MN (2014) Design and estimation of low cost floating house. Int J Innov Appl Stud 7(1):49–57
8. Wahyudi SI, Adi HP, Schultz Bart (2017) Revitalizing and preparing drainage operation and maintenance to anticipate climate change in Semarang heritage city. J Environ Sci Eng B 6 (1):17–26
9. Adi HP, Wahyudi SI (2018) Tidal flood handling through community participation in drainage management system (A case study of the first water board in Indonesia). Int J Integr Eng 10:19–23
10. Boogaard FC, Groningen H, Restemeyer B, Groningen H, Venvik G (2019) Knowledge exchange on climate adaptation with nature-based solutions and best management practices for sustainable (ground) water management in resilient cities. In: 46th IAH congress, Malaga, Spain
11. Wahyudi SI, Rochim HPAA, Marot D (2014) Aspects of hydrology, tidal and water storage capacity for simulating dike model of channel and retention basin. Int J Civ Environ Eng 14 (05):6–10

12. Adi HP, Wahyudi SI, Sudarmono CS (2020) Comparison analysis of expanded polystyrene system (eps) and polyvinyl chloride (pvc) pipe as platform material of floating buildings in the coastal areas of Semarang. J Phys Conf Ser 1444:012047
13. Von Wartburg WP, Liew J (1999) Gene technology and social acceptance, vol 321. University Press of America, pp 1104–1109. ISBN 0-7618-1325-x

LaunchCenter for Amphibious Construction on a New Lake in a Transforming Region of Opencast Mines

Benjamin Casper

Abstract The LaunchCenter combines research, development, and construction with an artificial testing field for amphibious constructions at an artificial lake to be filled starting in 2031 with a final size of about 11 km^2 in Inden in the west of Germany. Amphibious means "can-float": Buildings or constructions that reside on ground when dry but can float up when the water level rises. Usually is useful when there is an inundation of more than 0.6 m for a longer period than one week. The *Rheinische Revier* area is to develop from an opencast lignite mining area to a region with three future lake locations. The region is facing enormous challenges due to the loss of jobs, the reduction of local tax revenues, the existing energy-intensive industries that are dependent on energy supply and the high pressure on the remaining land. Several thematic clusters have been identified, which should lead to economic diversification, sustainable regional development and transformation of infrastructure systems. The LaunchCenter is a concept for initiating and introducing a economic niche in this region with high ambitions regarding local and international vocational training, interdisciplinary research and high quality production of low and high tech products. The 30–40 years of filling up the lake will be used to train and learn how to amphibiate. Specific and outstanding amphibious, floating and hybrid construction concepts shall be realized in the rising water level partly as experiments and partly as permanent mobile or immobile facilities.

Keywords Opencast mine · Future lake · Artificial testing field · Research and development · Living with water

B. Casper (✉)
Zukunftsagentur Rheinisches Revier GmbH, Karl-Heinz-Beckurts-Str. 13, 52428 Jülich, Germany
e-mail: benjamin.casper@rheinisches-revier.de

Ł. Piątek et al. (eds.), *WCFS2020*, Lecture Notes in Civil Engineering 158,
https://doi.org/10.1007/978-981-16-2256-4_31

1 Governance and Formatting

Starting in 2020 until 2041 there will be a federal support program to let invest approx. 291 Mio. €/year from the country North Rhine-Westphalia and additionally approx. 506 Mio. €/year from the federal government in the *Rheinische Revier*. This funding is now just approved by law. The frame to guide investments and developments is the *WSP—Wirtschafts- und Strukturprogramm*—Economic and structural program [1], which is a document to be adapted in several steps until 2038+ to deal with unforeseeable challenges and changing settings. The 180-page program has been developed jointly by regional stakeholders using workshops and conferences to include the regional and local perspective. The Ministry of Economic Affairs, Innovation, Digitalization and Energy of the State of NorthRhine-Westphalia supports this process with resources and finance. To facilitate the complex process of transformation many institutions, besides the Ministry, are setting up staff units to support the structural transformation:

- District Government—Funding Management
- Regional Associations of the Rhineland—Governing Body between Ministry District Government and Municipalities
- Territorial Authorities—Local Authorities consisting of several municipalities
- Universities and Universities of Applied Sciences

Once the funding regulation is set up, there will be a constant quarterly release of calls for investive promotion. The *Zukunftsagentur* is the leading agency to program and coordinate the efforts of the region and beyond. They bundle the initiatives and form networks to progress ideas and processes. Therefore, they support the development of innovative ideas to diversify industries and generate jobs and build a sustainable region. The *Zukunftsagentur* is now organized with so-called *Revierknoten*—district nodes to include regional stakeholders in the process of setting up decentral expertise teams for distinct task fields (energy, industry, innovation & education, resources & agrobusiness, infrastructure & mobility and spatial development). These nodes are planned to be merged in the *Zukunftsagentur* in 2022.

The municipality of Inden has planning sovereignty over 90% of the area of the future Indesee. In the plans adopted for the Indesee, research facilities are planned in the lakeshore area. In addition, water-related businesses are also to be accommodated. The development prospects at the neighbouring Hambach open-cast mine (approx. 11 km away) are still completely open and may provide a further impetus for the establishment of the LaunchCenter (Fig. 1):

2 Aims

The development of the LaunchCenter on Lake Indesee has several goals and is embedded in a wider strategy, because each project must contribute to the WSP's objectives and should not mark just one point in the region. A broad perspective is

Fig. 1 View into the opencast mine of Inden with the village Schophoven in the background (photo by author)

needed to avoid a singular implementation without links to the areas around the lake, e.g. the two other opencast mining lakes Hambach and Garzweiler:

- Set up a network of international experts and companies to research, develop and produce (LaunchCenter) amphibious and floating constructions in a future lake environment.
- Induce a new industrial branch and thus diversify the economy in this region.
- Produce jobs for skilled labour, R&D and initiate a value-added chain.
- The clear ambition to experimental high-end amphibious and floating constructions as a "settlement" for living, working, aquafarming, energy production, etc. and thus adding value to the usage of the lake.
- Build a knowledge bridge to the Competence Center in the Lusatian Area (East Germany) for floating buildings [2].
- Export knowledge and products worldwide.

2.1 Definition of a LaunchCenter

The general idea of the so-called LaunchCenter [1] has been brought up in the regional conferences and can be explained with the following ideas:

- A LaunchCenter is a research, development and production facility in which products and processes can be transferred from R&D status to series production and approved on the market ("Launch Effect"). This ensures the loss-free transfer of R&D know—how to series production.

- Central management and bundling of R&D activities (key accounting with dedicated contacts), management of cross-industry innovations, joint representation of interests in cooperation with suppliers.
- Skill and training platform for employees from industry for a loss-free technology transfer from the LaunchCenter to upstream and downstream industrial companies and to international customers.
- Creation and consolidation of jobs and thus the prevention of emigration of well-trained skilled workers and experts from the *Rheinische Revier*.

2.2 IBTA—The Framework and Format for High Ambition and Next Practice

The IBTA as an advancement of the IBA-format is being developed in a process with regional stakeholders and also a group of external experts and advisors with the aim to produce a memorandum as a formatting of the IBTA for the *Rheinische Revier* until end of 2021 to be approved by the Government of North Rhine-Westphalia by mid 2022:

The "International Building and Technology Exhibition (IBTA) *Rheinisches Zukunftsrevier*" provides the content, quality and procedural framework for shaping structural change in the *Rheinische Revier* over the next three decades. The aim is to develop the *Rheinische Revier* into a sustainable, largely greenhouse gas-neutral, innovative industrial and economic region, an attractive residential location, working and living space as well as a multifunctional post-mining open-cast landscape by means of model concepts and projects which, as next practice approaches, point beyond their time, and by means of cooperation.

2.2.1 Leverage and Impulse Effect of IBTA

The IBTA-format as a framework and the associated ambitions and quality requirements formulate a perspective that triggers considerable private follow-up investments beyond public investments and directs further funds into the area. IBTA thus unfolds a leverage effect that creates value and is time-saving will have an effect well beyond the previously planned funding period for structural change in the region until 2038 [1].

3 Derivation

The idea and the aims are based on observation of planetary change, the discussions on the anthropocene as the age of humankind as a global force to alter the surface of the earth and trends in human settlements development in combination with the efforts to reach a more sustainable approach and adjust to climate change.

Amphibious and floating constructions have been a cultural technique since ages to live with changing water-levels [3, 4]. The knowledge is dispersed across the continents and deserves more attention due to rising sea-levels, ground subsidence in low-lying delta regions and seemingly random heavy rain events. There is a need both for low-tech pro-poor adjustment of existing houses and settlements in marginal settings in the Global South and livelihoods in constantly increasing flooding areas [5] and for high-tech to initiate a paradigmatic change towards a living with water.

The warming of the oceans leads currently already to higher amounts of water in the atmosphere. This leads to higher loads of precipitation in shorter periods of time and in a various spatial distribution [6]. With this reality we might have to deal with more water in areas that are not prepared yet.

Settlements are increasingly at risk due to higher pressure on spaces for development and thus have constructions in flood zones [4, 7]. This applies not only, but above all, to the global South. In addition to the need to implement planning strategies and improve living conditions, there will be more and more people who will have to live with water and will no longer be able to protect themselves from it, especially in coastal cities, where economic losses and impact for societies will multiply over the coming decades if efforts stay on a low level [8]. Buildings are nearly only built to be on dry land. Different options are useful to deal with changing water levels in settlement areas: Dry-proof, wet-proof, stilts and amphibious. The optimal choice of construction approach depends on the depth of water-level and the duration of the flooding. Whereas dry-proof and wet-proof techniques are useful for short periods or floods and building on stilts is useful for very long periods of high water, amphibious or can-float buildings offer an opportunity "to float when it floods" in seasonal or durations of 1–3 months [9, 10].

Until now there is no existing testing field for amphibious constructions worldwide. There is a need to investigate further on the technical issues of buoyancy, wind stability, wave resistance, floating debris management, material usage and questions of settlements on water.

The biggest impediments for amphibious constructions are insurance issues, governmental regulations and building codes.

4 Concept

The open pit mines offer possibilities to use the process of a slow filling with water to test constructions in an artificial but open lake environment repeatedly. Furthermore, the masterplan of the municipality of Inden shows that different usages of the shoreline will be developed according to the process of the lake-filling, but mainly touristic, sports and recreation-oriented. Therefore, the set-up of a new industry branch can bring jobs and a different usage to the water.

Several floating constructions are already agreed on to be implemented in the process of the filling and for the final stage of filling. These are pontoons to connect

Fig. 2 Masterplan of Lake Inden—no water [11]

interim uses of shoreline recreational activities as well as a memorial island to be installed in the lake to remember the place of the resettled village of "Pier". These ideas are officially accepted and approved but have not been approached further technically and in design (Figs. 2 and 3).

To install amphibious testing constructions (interim-uses) in the time of the filling, the mining company (RWE Power AG), the responsible municipality (mostly Ind en) and the mining supervisory authority need to be asked for permission. This is due to the fact that the Mining Law is of overriding importance and is applied on these areas. The Mining Law can make it possible to set up a special operating plan for the mining area and thus e.g. install PV-power on the water surface, which would be a real challenge in a natural lake.

Considering the typical section of the open pit with the many so-called "berms", the idea is to use this structure to have multiple possibilities for testing amphibious structures, its foundations and its environment (from aquafarming over houses to experimental installations and whole streets) without needing to wait for a real flooding. A useful impression is the map without water infill yet, where one can see the berms lines very well. This will be one core element of the LaunchCenter (Fig. 4).

Fig. 3 Masterplan of Lake Inden—final water level in 2050/60 [11]

Fig. 4 Principle dam and berms

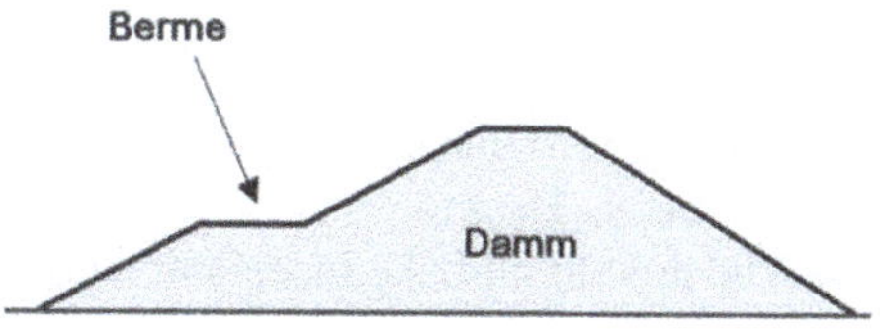

While testing of amphibious constructions in the filling time of the lake can help to improve design and constructions, there will be a need to also test the up- and down-move and the behaviour in simulated wave-conditions. This should be possible in a test-field next to the lake with a physical dock-like connection to the lake.

The detailing and equipment of the test basin and the test series must be planned with the stakeholders and in studies. The basis should be the preparation of 1:1 studies in the filling opencast mining lake. There, even very large-format

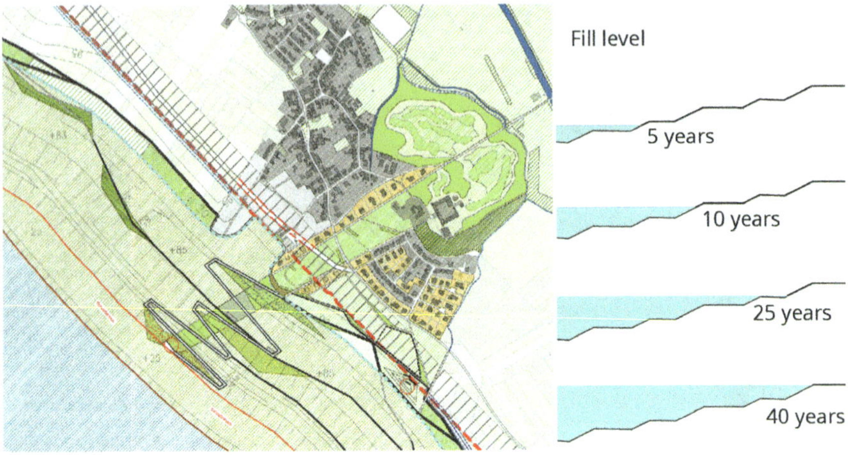

Fig. 5 Detail of the concept for village Schophoven and section with fill levels and berms after 5–40 years [11]

constructions could then be prototypically tested. It is possible to combine the future construction of a harbour area on the Indesee with the construction of the test basin.

The adjacent LaunchCenter combines research, development and construction facilities with a part being open to the public. A production unit should be set-up closely connected to the testing field and the LaunchCenter or included into this (Fig. 5).

Considering the timescale, the interim use in most parts will be possible already after 5 years. The filling will proceed very fast in the beginning due to slowing down of the groundwater-pumps that dried the open pit mine. Yet, the recent decision in the beginning of 2020 on phasing out lignite usage earlier affects the opencast mine Hambach, which is next to Inden and will stop extracting lignite in 2030 already. Therefore, the groundwater that would have been extracted further, can't be used to fill Lake Inden and will instead be used for Hambach itself. A level of 25 m infill will be reached after 5 years. Approx. 50 m depth will be reached after 10 years. The last infill will need a considerable amount of time (approx. 20–30 years) due to the restabilizing of the groundwater in the region around (Figs. 6 and 7).

In the following, the conceptual considerations are presented graphically and should be understood as principle sketches and not represent a localization (Figs. 8 and 9).

Fig. 6 Visualization of the intermediate filling state after 10 years [12]

Fig. 7 Visualization of the filled lake Inde after 30–40 years [12]

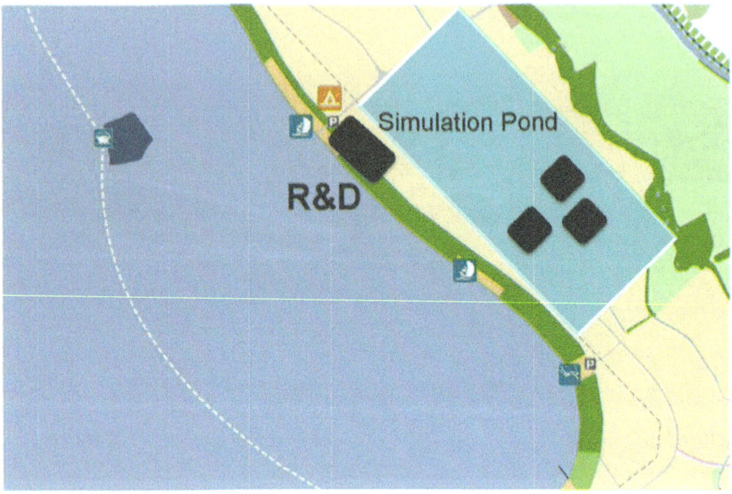

Fig. 8 Principle sketch LaunchCenter at the edge of the lake with test basin (photo by author)

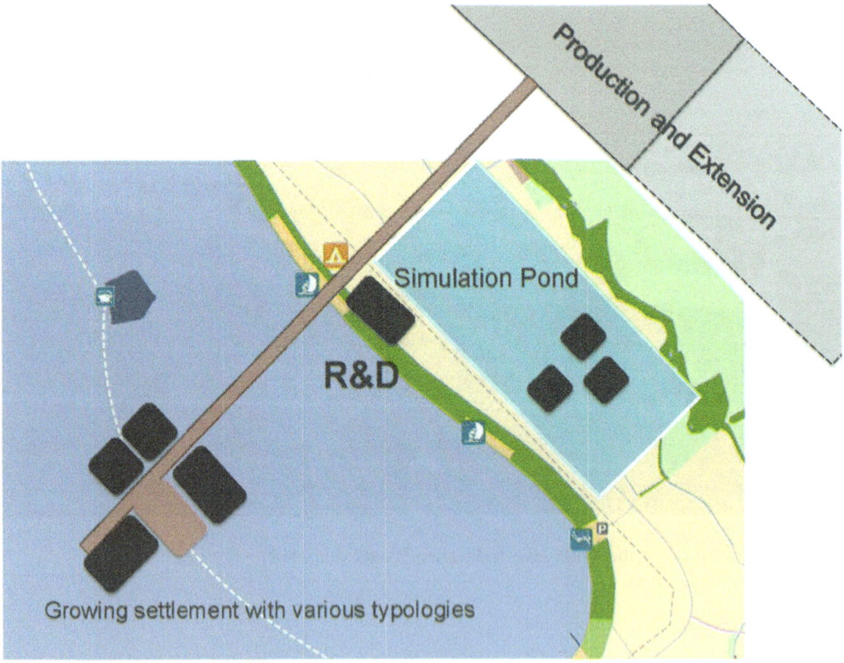

Fig. 9 Principle sketch of growing floating settlement and connection with production plants (photo by author)

5 Challenges and Conclusion

As mentioned above, the building law, mining law, regulatory and insurance issues are the predominant challenges in the construction of an amphibious structure or house. Furthermore, it needs a certain time and a good network to build up a new competence and thus a new economic path. Also, there is a significant need in research on the specifications of the testing field. Awareness needs to be carefully build up and intensified and an evaluation needs to follow up on the abilities and hidden competences of the existing production companies. Some might adjust to work together within a new field of amphibious construction, other construction partners dealing with further technical production processes need to be found then.

Considering the start of filling in with water in 2031 we count backward and realize the needs to start very soon to achieve some basic knowledge in this field:

- Which companies could contribute expertise and capital?
- Which institutions and actors should be involved in development?
- What barriers and guidelines must be considered with regard to insurance policies, mining law, government regulations and building codes?
- What is the aquafarming potential? How large and compatible can floating PV systems be? How great is the potential for living and working areas on the water?
- When will a feasibility study be required to connect the lakes by means of canals?
- How can a new branch of the construction industry be consolidated and established on and with the water?

The time frame for the following steps could be as follows, starting from 2020:

- Year 1–3 Research and network building, cooperation with the Hambach structural development company, since the coal mining operations in the Hambach opencast mine will be stopped almost simultaneously
- Year 4–8 Planning and construction of the competence center and the simulation/test facility and preparation of the berms for interim use as a simulation field with rising water levels
- Year 9–12 Testing and improvement of the prototypes and start of construction of the test fields in the discontinued opencast mine (Year 9: discontinuation of mining, Year 11: start of filling)
- Year 13–20 Production scaling (filling reaches usable level)

The LaunchCenter should achieve an international standing for excellence and thus develop a diversified use of the lakes, which will complement the tourist, sports and recreational uses.

References

1. MWIDE (2019) Wirtschafts- und Strukturprogramm 1.0 (in German). https://www.rheinisches-revier.de/media/wsp_1-0_web.pdf. Accessed 26 June 2020
2. IfSB—Institut für schwimmende Bauten btu Cottbus. https://www.b-tu.de/en/schwimmende-bauten/. Accessed 3 Jul 2020
3. Sumet J (1988) Naga. Cultural origins in Siam and the West Pacific. Oxford University Press, Singapore
4. Rijcken T (2005) Floating neighbourhoods as they were and will be; why dwellers would want to live on water. Paper presented at the conference doing, thinking, feeling home: the mental geography of residential environments, TU Delft, 14–15 Oct 2005
5. Elizabeth E et al (2018) An innovative strategy to increase the resilience of flood-vulnerable communities while reducing risk of population displacement and psychological trauma. Paper presented at the 8th international conference on building resilience—ICBR, University of Lisbon, 14–16 Nov 2018. https://www.buoyantfoundation.org/s/ICBR-174-DisplacementTrauma-20aug2019.pdf. Accessed 15 Jan 2020
6. IPCC (2014) IPCC assessment report 5. Working group III—final draft. http://mitigation2014.org/report/final-draft/. Accessed 24 Sept 2014
7. Barker R, Coutts R (2015) Aquatecture. Buildings and cities designed to live and work with water. RIBA Publishing, London
8. Hallegatte S, Green C, Nicholls R et al (2013) Future flood losses in major coastal cities. Nat Clim Chang 3(9):802–806. https://doi.org/10.1038/NCLIMATE1979
9. Nillesen A, Singelenberg J (2011) Amphibious housing in the Netherlands. Architecture and urbanism on the water. NAi, Rotterdam
10. Polpat N et al (2015) Amphibious architecture and design: a catalyst of opportunistic adaptation?—case study Bangkok. Paper presented at the conference urban planning and architecture design for sustainable development, UPADSD, University of Salento, Lecce, 14–16 Oct 2015
11. Entwicklungsgesellschaft indeland GmbH / RMP Stephan Lenzen
12. Entwicklungsgesellschaft indeland GmbH / HH Vision

The manufacturer's authorised representative in the EU is Springer
Nature Customer Service Centre GmbH, Europaplatz 3, 69115 Heidelberg,
Germany. If you have any concerns regarding our products, please
contact ProductSafety@springernature.com

Printed and bound by CPI Group (UK) Ltd, Croydon, CR0 4YY
30/09/2026
02225251-0016